RADIO WAVE PROPAGATION
AND ANTENNAS
An introduction

RADIO WAVE PROPAGATION AND ANTENNAS
An Introduction

JOHN GRIFFITHS

School of Electrical Engineering
University of Bath, UK

Prentice-Hall PHI **International**

Englewood Cliffs, New Jersey London Mexico New Delhi
Rio de Janeiro Singapore Sydney Tokyo Toronto

Library of Congress Cataloging in Publication Data

Griffiths, John, 1935–
  Radio wave propagation and antennas.
  Includes bibliographies and index.
  1. Antennas (Electronics)
  2. Radio wave propagation.
  I. Title.
  TK7871.6.G74   1987   621.3841'35   86–16961

  ISBN 0–13–752312–2

British Library Cataloguing in Publication Data

Griffiths, John, 1935–
  Radiowave propagation and antennas: an
  introduction.
  1. Radio wave propagation
  I. Title
  621.3841'1   TK6553

  ISBN 0–13–752312–2
  ISBN 0–13–752304–1

© **1987 Prentice-Hall International (UK) Ltd.**

PRENTICE-HALL INC., *Englewood Cliffs, New Jersey*
PRENTICE-HALL INTERNATIONAL (UK) LTD, *London*
PRENTICE-HALL OF AUSTRALIA PTY LTD, *Sydney*
PRENTICE-HALL CANADA INC., *Toronto*
PRENTICE-HALL HISPANOAMERICANA S.A., *Mexico*
PRENTICE-HALL OF INDIA PRIVATE LTD, *New Delhi*
PRENTICE-HALL OF JAPAN INC., *Tokyo*
PRENTICE-HALL OF SOUTHEAST ASIA PTE LTD, *Singapore*
EDITORA PRENTICE-HALL DO BRASIL LTDA, *Rio de Janeiro*

Printed and bound in Great Britain for
Prentice-Hall International (UK) Ltd,
66 Wood Lane End, Hemel Hempstead, Hertfordshire, HP2 4RG
at the University Press, Cambridge.

1 2 3 4 5   90 89 88 87 86

ISBN 0-13-752312-2
ISBN 0-13-752304-1 PBK

Contents

PREFACE   xi

LIST OF SYMBOLS   xiii

1 Antennas and Propagation   **1**
  1.1 Electromagnetic Waves in Space   1
  1.2 Antenna Definitions   6
  1.3 Propagation Equations   11
  1.4 Some Basic Antennas   13
      1.4.1 Isotrope and incremental linear radiator   13
      1.4.2 Short dipole and halfwave dipole   17
      1.4.3 Vertical monopole   19
      1.4.4 Small loop antenna   21
      1.4.5 Aperture-type antennas   23
      1.4.6 Lens antennas   25
  1.5 Radio Wave Propagation   27
      1.5.1 Modes of propagation   27
      1.5.2 General applications   29
      Problems   31
      Reference   32

2 Ground Wave Propagation   **33**
  2.1 Ground Parameters   33
  2.2 Propagation Equations   35
      2.2.1 Attenuation factor A 37
      2.2.2 Ground wave propagation charts   38
  2.3 Propagation Problems   41
      2.3.1 Land–sea boundary   41
      2.3.2 Transmitter coverage for broadcasting   43
      2.3.3 Sky-wave interference   48
  2.4 VLF Propagation   51
      2.4.1 Attenuation rates   51
      2.4.2 Electric field strength   52
  2.5 Low-frequency Antennas   55
      2.5.1 Transmitting antennas   55

2.5.2   Receiving antennas   58
Problems   60
References   61

3   Sky Wave Propagation   62
3.1   The Ionosphere   63
3.2   Primary and Secondary Parameters   66
        3.2.1   Attenuation   66
        3.2.2   Refractive index   67
        3.2.3   Conductivity and permittivity   68
        3.2.4   Earth's magnetic field   69
3.3   Electron Collision Frequency   70
3.4   D-layer Attenuation   72
3.5   The Propagation Path   74
        3.5.1   Snell's law of refraction   74
        3.5.2   Critical frequency and maximum usable frequency   76
        3.5.3   Curved ionosphere   80
        3.5.4   Maximum usable frequency factor and optimum working
                frequency   81
3.6   Propagation Equations   84
        3.6.1   Power at receiver   84
        3.6.2   Reception levels   87
        3.6.3   Noise levels   88
        3.6.4   Circuit reliability   89
3.7   High-frequency Antennas   93
Problems   98
References   99

4   Space Wave Propagation in the Troposphere   100
4.1   Line-of-sight Range   100
4.2   Reflection   101
        4.2.1   Flat earth reflection   102
        4.2.2   Reflection with variable distance   103
        4.2.3   Reflection with variable wavelength   104
        4.2.4   Reflection with variable heights and wavelengths   105
        4.2.5   Inverse distance equations   106
        4.2.6   Point of reflection on curved earth   108
        4.2.7   Divergence of reflected waves   110
        4.2.8   Fresnel's ellipsoid   111
4.3   The Troposphere   112
        4.3.1   Permittivity of the troposphere   113
        4.3.2   Refractivity of the troposphere   113
        4.3.3   Standard models of the troposphere   114
4.4   Curvature of Space Wave in the Troposphere   114
4.5   Modified Refractivity   119
4.6   Diffraction of Space Waves   122

4.6.1	Huygens' principle	123
4.6.2	Amplitude diagrams	125
4.6.3	Knife-edge diffraction	127
4.6.4	UHF propagation over rounded hills	129
4.7	Mobile Radio in an Urban Environment	130
4.8	Variation of Space Wave Signal Strength	133
4.8.1	Simple model of fast fading	134
4.9	Antennas for Space Wave Propagation	136
	Problems	139
	References	140

5 Antenna Arrays — **142**

5.1	The Halfwave Dipole	142
5.2	Dipole Impedance	144
5.3	Two-element Array	146
5.3.1	Array of two isotropes	146
5.3.2	Arrays of two halfwave dipoles	149
5.3.3	Antenna and reflector array	153
5.3.4	Parasitic antenna	157
5.4	Three-element Antenna Array	159
5.5	Linear Antenna Arrays	160
5.5.1	Linear arrays of isotropes	161
5.5.2	Array factor and normalized array factor	162
5.5.3	Effect of phase of feed current	163
5.5.4	Endfire array	166
5.5.5	Yagi–Uda antenna	166
5.5.6	Log-periodic dipole array	167
5.5.7	Television transmitting antennas	169
5.6	Polynomials in Linear Array Theory	171
5.6.1	The E-field equation	172
5.6.2	The unit circle	172
5.6.3	The binomial antenna array	175
5.6.4	Constant current array	175
5.6.5	Gaussian array	176
	Problems	178
	References	179

6 Statistical Distributions and Diversity Principles — **180**

6.1	Useful Parameters	180
6.2	Shapes of Distributions	184
6.2.1	The basic equations	185
6.2.2	Normalized equations	185
6.3	Gamma functions	186
6.4	Specific Types of Distributions	187
6.4.1	Exponential distribution	187
6.4.2	Rayleigh distribution	191

6.4.3   Rayleigh probability paper   193
6.4.4   Gamma distributions   195
6.4.5   Other related distributions   199
6.4.6   The normal and log-normal distributions   200
6.5   Diversity   202
6.5.1   Basic concepts and definitions   202
6.5.2   Signal-to-noise ratio   206
6.6   Multiple Propagation Paths   207
6.6.1   Some diversity combinations   208
6.7   Selection Diversity   209
6.7.1   Graphical approach   209
6.8   Maximal Ratio Diversity   211
Problems   213
References   215

7   Tropospheric Scatter Radio Links   216
7.1   Simple Model of Troposcatter Link   216
7.2   Path Analysis   218
7.3   Long-term and Short-term Variations   222
7.3.1   Slow variation   223
7.3.2   Fast variation   224
7.3.3   Combined slow and fast variations   224
7.3.4   Effective scattering cross-section   227
7.4   Other Scatter Propagation Systems   228
7.4.1   Ionoscatter radio systems   228
7.4.2   Propagation equations   230
7.4.3   Automatic repeat request   232
7.4.4   Meteor burst communication   234
Problems   236
References   237

8   Microwave Radio Relay Systems   238
8.1   Route Planning   238
8.2   The FDM/FM Radio Relay System   241
8.3   Circuit Specifications   244
8.3.1   Specification of hourly mean   244
8.3.2   Fading   247
8.3.3   Short-term specifications for the HRC   249
8.3.4   Specifications for a real radio relay system   251
8.4   Circuit Calculations with Fast Fading   252
8.4.1   The Rayleigh and inverse Rayleigh distributions   252
8.4.2   Signal-to-thermal noise ratio with Rayleigh fading   254
8.5   Digital Radio Relay Systems   256
Problems   261
References   262

9   Satellite and Space Communication   *gerenl reading*   **263**
9.1   Satellite Orbits   263
   9.1.1   Orbital period of satellite   263
   9.1.2   Satellite range and relative velocity 265
   9.1.3   Doppler frequency shift   268
   9.1.4   Geostationary orbit   268
9.2   Propagation Equations   272
   9.2.1   Carrier-to-noise ratio   273
   9.2.2   Carrier-to-noise-plus-interference ratio 275
9.3   Noise in Satellite Communication Links   278
   9.3.1   Cosmic noise and sky noise   278
   9.3.2   Antenna noise and interference   280
9.4   Some Satellite Systems   283
   9.4.1   Commercial satellites   283
   9.4.2   Weather satellites   286
   9.4.3   Broadcasting from satellites   289
   9.4.4   Marine communication satellites   290
9.5   Deep Space Communication   292
   Problems   294
   References   295

10   Radar   **296**
10.1   Pulse Radar   296
   10.1.1   The radar equation   299
   10.1.2   Noise   301
   10.1.3   Signal-to-noise ratio   305
   10.1.4   Radar cross-section   308
10.2   Search and Tracking Radar   310
   10.2.1   Integration of pulse trains   311
   10.2.2   Search radar equation   312
   10.2.3   Tracking radar   316
10.3   Beacon and Bistatic Radars   316
10.4   Doppler Radars   318
   10.4.1   Doppler principle   319
   10.4.2   Simple applications   320
   10.4.3   Pulse Doppler radar   322
   10.4.4   FM-CW radar   324
   Problems   326
   References   327

A1   Revision of Electromagnetic Principles   **328**
A1.1   Relationships Between Parameters   328
   A1.1.1   Volumetric parameters   328
   A1.1.2   Surface parameters   328
   A1.1.3   Linear parameters   329

	A1.1.4	Stokes's and Green's theorems	329
A1.2	E-field and H-field Equations		333
A1.3	Wave Equations		334
A1.4	Net-Charge-Free Media		334
	A1.4.1	Wave impedance and power flow	336
A1.5	Magnetic Vector Potential		337
	A1.5.1	Solutions for magnetic vector potential	338
A1.6	Boundary Conditions		341
	A1.6.1	Special case: free space to perfect dielectric	342
	A1.6.2	Laws of reflection and refraction	343
	A1.6.3	Fresnel's reflection coefficients	344
A1.7	Complex Dielectric		346

A2 Electrical Noise in Radio Systems   **348**

A2.1	Noise Voltage		348
A2.2	Thermal Noise		350
	A2.2.1	Nyquist's and Planck's equations	352
A2.3	Noise Figure and Noise Temperature		354
	A2.3.1	Effective noise temperature of a matched attenuator	355
	A2.3.2	Cascaded networks	359
	A2.3.3	Noise field	362
A2.4	Measurement of Noise Figure and Noise Temperature		363
A2.5	Noise Sources		366
A2.6	Noise in Digital Systems		367
	References		370

A3 Answers to Problems   **371**

A4 Bibliography   **374**

INDEX   378

Preface

This book is intended as an introduction to the topics of radio wave propagation and associated antennas, somewhere in between the more broadly based texts in communications which cover each of these two topics in single chapters and the very specialized texts on either antennas or propagation which cover in separate books many of the subjects which have been summarized here in chapters.

An attempt has been made to set the level of presentation appropriate to that of university undergraduates in electrical communication subjects and to students on similar courses at polytechnics, colleges and radio-communication training establishments. Some emphasis has therefore been placed on numerical examples and illustrations throughout the text, with further problems at the end of each chapter for students to attempt themselves. To this end, some 93 worked examples and 109 problems will be found throughout the book, and a teacher's manual of solutions to the problems is obtainable from the publishers.

The work is primarily concerned with the basic principles of radio wave propagation, but in order to have a fuller understanding of the propagation principles and equations it is useful to have a brief background knowledge of general electromagnetic theory related to this topic and also of electrical or radio noise. Both topics are included as appendices.

Elsewhere within the text, some brief ideas of statistical distributions are made available as part of a single chapter (Chap. 6) in order to appreciate the order that exists within the randomness associated with the fading and scattering of radio waves and the improvements which can be achieved with diversity arrangements. The approach to statistical distributions is a little unorthodox – more intuitive than mathematically rigorous – but it is intended to show that many of the numerous named varieties of statistical distributions which are encountered in radio wave propagation and radar are related variations of a simple exponential function.

Except for these associated topics, the text follows the principles of radio wave propagation almost on a frequency scale, starting with the fundamentals of ground wave propagation at the lower frequencies, then following on with an explanation of sky wave propagation mainly in the HF band, and finally considering the various line-of-sight and scatter forms of communication at the higher-frequency ranges. Two extreme topics, on this basis, have been omitted: propagation at ELF and EHF. It is considered that these are more appropriate to a postgraduate course. In addition, there are chapters on antenna arrays and on the basic principles of radar.

The book is neither fundamental nor pragmatic in the true senses of these adjectives. It does not concentrate on the basic laws of electromagnetism and the rigorous derivations of all the propagation equations, nor on the practical aspects of the antennas and associated communication systems. Instead the aim has been to form a bridge between the two with the aid of idealized models of the more common applications of radio wave propagation.

At the end of each chapter there is a brief list of books, reports or published papers specifically referred to within the text. These are not intended to be anything other than references. For students who wish to read further about the contents of each chapter, a brief bibliography is included at the end of the book. It is assumed that the reader has a basic knowledge of mathematics and communications, as well as the basics of statistics.

I would like to thank my wife, Mary, for all the time, patience and effort spent in transposing my rough drafts into typescript, often several times over as ideas change or modifications are necessary, and for her constant encouragement over such a long period of time from start to finish. My thanks, also, to the anonymous previewers whose detailed and critical comments have been much appreciated, and to my colleagues at the University of Bath for many useful discussions.

Principal Symbols

A	magnetic vector potential
A	magnitude of magnetic vector potential; attenuation factor; coefficient; area; angle
AF	array factor
B	magnetic flux density
B	magnitude of magnetic flux density; angle
BER	bit error ratio
BF	log-periodic antenna bandwidth factor
BW	beamwidth
C	capacitance; angle
CF	correction factor; cumulative frequency
CIR	carrier-to-interference ratio
CNR	carrier-to-noise ratio
$CN_{th}R$	carrier-to-thermal-noise ratio
CNIR	carrier-to-noise-plus-interference ratio
D	electric flux density
D	magnitude of electric flux density; directivity; divergence factor; diameter
E	electric field strength
E	magnitude of electric field strength; noise field E_n
E_b	bit energy
E_b/N_0	ratio of bit energy to noise spectral density
EF	excitation factor
EIRP	effective radiated power from isotrope
ERPD	effective radiated power from halfwave dipole
F	force
F	magnitude of force; noise factor or noise figure of receiver F_r or system F_s
F_a	antenna noise factor and four-hour median F_{am}
FM	figure of merit; frequency modulation
G	gravitational constant; power gain of amplifier or antenna
G/T	ratio of antenna gain to noise temperature
H	magnetic field strength
H	magnitude of magnetic field strength; height associated with refractive bending within the troposphere
I	electric current
J	electric current density

J	magnitude of electric current density; conduction current density J_c or displacement current density J_d
K	refractivity factor; constant; paraboloidal reflector antenna design parameters K_a and K_b
L	inductance; length; latitude; focal length; attenuation loss; spatial loss L_s; miscellaneous losses L_{misc}; scatter loss L_{sc}; antenna-to-medium coupling loss L_c; etc.
LE	loss effect
LL	land loss
LUF	lowest usable frequency
M	mass of earth; modified refractivity; noise measure
MUF	maximum usable frequency; maximum usable frequency factor MUFF
N	noise power; number of elements in linear arrays; density of electrons in ionosphere; refractivity; number of class intervals in a histogram; number of radar pulses in integration process
N_0	noise spectral density
N_s	surface refractivity
OTF	optimum traffic frequency; optimum working frequency OWF
$\mathbf{P}_a$	power flux density
P	signal power; magnitude of power flux density P_a; atmospheric pressure; probability; probability of detection P_d; probability of false alarms P_{fa}
Q	Q-factor; sky wave availability factor
R	radius; reflection coefficients R_H and R_V; sunspot number; network parameter R_{11}, etc.; resistance, radiation resistance R_{rad}; loop resistance R_L
RE	recovery effect
S	surface area; signal; incremental surface area dS
SG	sea-gain
SNR	signal-to-noise ratio
$SN_{th}R$	signal-to-thermal-noise ratio
T	temperature; time; time between false alarms T_{fa}
U	statistical variable
V	voltage; receiver voltage V_r; threshold voltage V_{th}; noise voltage V_n
W	energy
X	reactance; statistical variable; network parameter X_{12}, etc.
Y	statistical variable; noise parameter
Z	impedance; intrinsic impedance Z_i; network parameter Z_{12}, etc.
a	radius of earth; waveguide dimension; weighting factor a_j; polynomial coefficient
b	waveguide dimension; parameter in gamma integral; bandwidth of voice channel; auxiliary parameter b^0
c	universal constant
cmf	cymomotive force
d	distance; displacement of elements in antenna array
e	electronic charge; exponential; instantaneous emf

emf	electromotive force
f	frequency; beat frequency f_b; critical frequency f_c; frequency deviation f_d(rms); Doppler frequency shift f_D; gyromagnetic frequency f_H; modulation frequency f_m; critical frequency of ionospheric E-layer f_0R; cross-over frequency f_q; pulse repetition frequency f_R; etc.
h	height; Planck's constant; Fresnel zone clearance h_F
$\mathbf{i}$	unit vector
i	angle of incidence; instantaneous current
$\mathbf{j}$	unit vector
j	complex operator
$\mathbf{k}$	unit vector
k	Boltzmann's constant; constants
m	mass of electron; statistical parameter; statistical moments m_1 and m_2; mass of satellite; mode
n	refractive index; statistical parameter; noise; number of fading hops; number of turns in loop antenna
p	numerical distance; probability density function; path length; water vapour pressure
q	q-factor; electric charge; number of hops in a microwave radio relay link
$\mathbf{r}$	vector distance in spherical polar coordinates
r	spherical polar coordinate; angle of reflection; root-mean-square; mean-square $\bar{r}^2$
s	signal; incremental length in loop d$\mathbf{s}$; dimension of loop antenna; Fresnel parameter
t	time; statistical parameter
u	statistical variable
$\mathbf{v}$	velocity
v	magnitude of velocity; volume; Fresnel parameter; instantaneous voltage
var	variance
x	cartesian coordinate; statistical variable; fading attenuation
y	cartesian coordinate; statistical variable
z	cartesian coordinate; statistical varible; polynomial
α	attenuation coefficient; angle of diffraction; azimuth
β	phase-change coefficient
Γ	Γ-function (gamma function)
γ	propagation coefficient; angle; integration efficiency
Δ	attenuation rate; angle of elevation or launch; excess path length
Δf	jammer frequency
ΔP	diversity correction factor
Δp	path difference
δ	incremental value
ε	permittivity; relative ε_r; effective ε'_r; complex ε_r^*
η	efficiency; illumination efficiency; radiation efficiency
θ	spherical polar coordinate; angle; scatter angle

λ wavelength; in guide λ_g; cut-off λ_c

μ permeability; relative μ_r

v electron collision frequency

π universal constant

ρ cylindrical coordinate; charge density

$\sum$ summation

σ conductivity; standard deviation; radar cross-section; forward-scatter cross-section σ_s; noise parameter σ_n

τ mean time between collisions; pulse width; log-periodic antenna design parameter; orbital period of satellite

ϕ spherical polar coordinate; phase angle; angle; magnetic flux

χ sun's zenith angle; statistical parameter

ψ angle; latitude; longitude; phase-angle of antenna array currents; satellite inclination; solid angles; electric flux

ω angular frequency (see f)

∇ Laplace operator

1

Antennas and Propagation

Although a radio communication system, such as that illustrated in Fig. 1.1, consists of a transmitter, a transmitting antenna, the propagation medium, a receiving antenna and a receiver, the scope of this book is confined mainly to those conditions which arise in and between the transmitting antenna and the receiving antenna, with only occasional excursions outside these topics into such related matters as the noise performance of receivers when appropriate. This chapter is intended as a brief introduction to the two topics, antennas and propagation, defining certain fundamental features of each, deriving some of the basic equations, and summarizing the principal mechanisms and applications of radio wave propagation.

The coordinate systems used most frequently in the study of antennas and propagation are cartesian (x, y, z) and spherical (r, θ, ϕ), though cylindrical coordinates (ρ, z, ϕ) are sometimes useful with simple linear antennas. We shall confine ourselves mainly to the first two sets of coordinates. Throughout this book we shall use a conventional notation for the components of vectors in the various coordinate systems, as shown by the following example: using spherical polar coordinates (r, θ, ϕ), a vector **H** at a given point in space is described by its components H_r, H_θ and H_ϕ, where H_r is the component in the direction of increasing r, etc.

1.1 ELECTROMAGNETIC WAVES IN SPACE

A propagating electromagnetic field is a flow of energy through a medium in the form of an electric field **E** and a magnetic field **H**. For a plane electromagnetic wave, the electric field **E**, magnetic field **H** and the velocity of propagation **v** are mutually perpendicular and obey the 'right hand' or vector product rule such that **E** × **H** points in the direction of **v**. In the simplest case of linear polarization (see later) the vectors **E** and **H** are constant in direction. Suitably oriented cartesian axes can then be chosen so that the x, y, z axes are aligned with **E**, **H** and **v**, respectively, and only the components E_x, H_y and v_z are non-zero. For a spherical electromagnetic wave radiating from a point source at $r = 0$, the right-hand rule also applies for E_θ, H_ϕ and v_r at a specified location in space.

The orientation of the electric field component of the electromagnetic field is called its *polarization*. If the direction remains constant with time at a fixed point in space, the

1

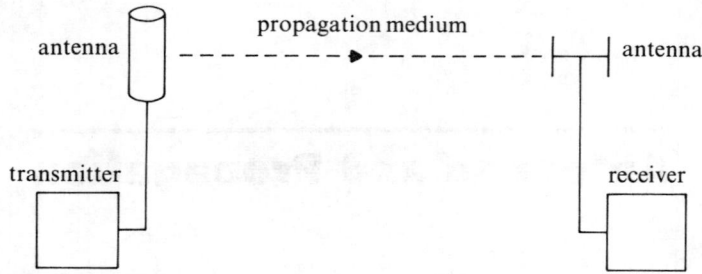

Fig. 1.1 A radio communication system consists of a transmitter, antenna, propagation medium, antenna and receiver.

field is said to be linearly polarized. For wave propagation near the earth's surface, the terms vertical, horizontal and slant polarization are frequently used to denote linear polarizations with the appropriate orientations.

  If the direction of the electric field component at a fixed point in space varies with time, the electromagnetic field may be circularly or elliptically polarized. For circular polarization the E-field vector has constant magnitude but its direction, in the plane perpendicular to the velocity of propagation, rotates smoothly through 360° per wavelength of propagation. The wave is said to have right-handed circular polarization if the electric field vector at a fixed point rotates in the clockwise direction when the wave is receding from the observer. Note that this is contrary to the classical definition found in books on optics and some books on radio waves. For elliptical polarization the tip of the electric field vector traces an ellipse as it rotates 360° per wavelength of propagation. Figure 1.2 illustrates three of the definitions of polarization.

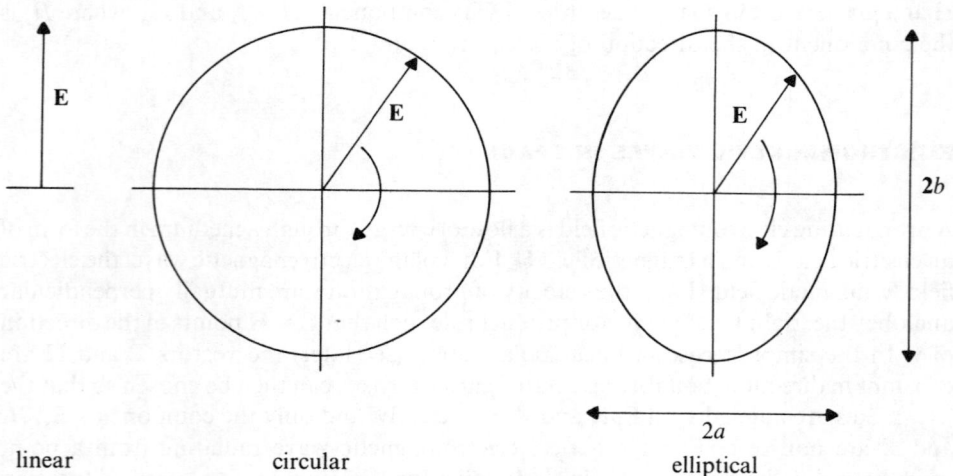

linear   circular   elliptical

Fig. 1.2 Three types of polarization of electromagnetic waves receding from the observer. Clockwise rotation is called right-circular of right-elliptical polarization and the ratio b/a is called the axial ratio of the ellipse.

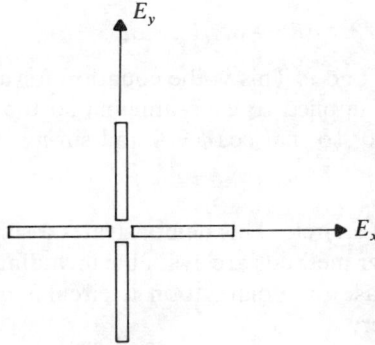

Fig. 1.3 Linear, circular and elliptical polarization can be achieved with two mutually perpendicular linear antennas and suitable phasing between the two signals. The process is similar to that used for obtaining Lissajous figures on a cathode-ray oscilloscope. The z axis is directed out of the page.

One method of producing elliptical polarization is to position two linear antennas at right angles to each other and to supply them with currents which are ϕ radians out of phase, as shown in Fig. 1.3. If the powers supplied to each antenna are different, the corresponding E-fields at a point on the z axis (Fig. 1.3) are

$$\mathbf{E}_1 = E_x\mathbf{i} = E_A \sin(\omega t - \beta z)\mathbf{i} = E_A \sin(\omega t')\mathbf{i}$$

and

$$\mathbf{E}_2 = E_y\mathbf{j} = E_B \sin(\omega t - \beta z + \phi)\mathbf{j} = E_B \sin(\omega t' + \phi)\mathbf{j}$$

where E_x and E_y are the magnitudes of vectors $\mathbf{E}_1$ and $\mathbf{E}_2$ and $\mathbf{i}, \mathbf{j}$ (and $\mathbf{k}$, not used) are the unit vectors of the x, y, z coordinate system. From

$$E_x = E_A \sin \omega t'$$

we obtain

$$\sin \omega t' = E_x/E_A$$

and

$$\cos \omega t' = \sqrt{1 - \sin^2 \omega t'} = \sqrt{1 - (E_x/E_A)^2}$$

By expanding the expression for E_y and substituting the values for $\sin \omega t'$ and $\cos \omega t'$, we can write

$$E_y = E_B(\sin \omega t' \cos \phi + \cos \omega t' \sin \phi)$$

and

$$\frac{E_y}{E_B} = \frac{E_x}{E_A} \cos \phi + \sqrt{1 - \left(\frac{E_x}{E_A}\right)^2} \sin \phi$$

With a little mathematical manipulation this becomes

$$\left(\frac{E_y}{E_B}\right)^2 - \frac{2E_x E_y \cos \phi}{E_A E_B} + \left(\frac{E_x}{E_A}\right)^2 = \sin^2 \phi$$

or

$$aE_x^2 - bE_xE_y + cE_y^2 = 1$$

for suitable constants a, b and c. This is the equation for an ellipse.

If equal powers are supplied to each antenna so that $E_A = E_B$, and the phase difference is adjusted to $90°$ so that $\cos\phi = 0$ and $\sin\phi = 1$, then

$$E_y^2 + E_x^2 = E_A^2$$

represents the equation of a circle. This demonstrates a simple method of producing circular polarization. Other methods are available including, for example, the use of a helical antenna. In this case the polarization is circular only along the axis and it becomes elliptical elsewhere.

The velocity of propagation of an electromagnetic wave in free space is equal to the universal constant c. The latter is related to the permeability μ_0 and permittivity ε_0 of free space via the expression

$$c = \frac{1}{\sqrt{\mu_0\varepsilon_0}} \approx 3 \times 10^8 \, \text{m s}^{-1} \tag{1.1}$$

A more accurate value of c is $2.997925 \times 10^8 \, \text{m s}^{-1}$, which is determined by direct measurement. μ_0 is defined as $4\pi \times 10^{-7} = 1.256637 \times 10^{-6} \, \text{H m}^{-1}$, so that ε_0 is $8.854185 \times 10^{-12} \, \text{F m}^{-1}$ as a result of the equation for c. Thus c and ε_0 have values which are defined by experiment with finite accuracy.

If the propagation medium has a permeability and/or permittivity in excess of their free-space values, the velocity of propagation is less than that of free space. Although c is approximated as $3 \times 10^8 \, \text{m s}^{-1}$ for most purposes, the more precise value is occasionally needed. For time division multiple access (TDMA) satellite communication, which involves distances of the order 40,000 km between earth transmitter and orbiting satellite and synchronized data transmission 'bursts' of about 10 microseconds, the calculated propagation times differ by about 92 microseconds, depending upon which value of c is used.

Example 1.1 Propagation delay over satellite link

What is the approximate propagation delay between two earth stations linked via a geostationary satellite if the up-path and down-path distances are both about 40,000 km?

Ignoring any delays within the systems, it takes $8 \times 10^7/(3 \times 10^8)$ seconds, or roughly one-quarter of a second, for a signal to travel between the two earth stations via the satellite. There is a similar delay in the return signal. This explains the pauses in response when making a telephone call via a satellite link.

For an electromagnetic field varying sinusoidally with time, the E-field at a point in space may be represented by

$$E(t) = E_0 \cos\omega t = E_0 e^{j\omega t}$$

where it is understood that use is being made of the real component of the exponential term. If we consider plane wave propagation from this origin, with the field orientated in the x direction and the velocity of propagation in the z direction, the electric field at distance z from the origin is subject to a propagation delay of z/v seconds or a phase retardation of βz radians, such that

$$E_x(z,t) = E_{x0}\cos(\omega t - \beta z) = E_{x0}e^{j(\omega t - \beta z)}$$

In this equation $v = \lambda f$ is the *velocity of propagation* and $\beta = 2\pi/\lambda = \omega/v$ is the *phase change coefficient*. If the medium is lossy, the attenuation of the propagating field is represented by an *attenuation coefficient* α nepers per metre and the previous equation is modified to

$$E_x(z,t) = E_{x0}e^{-\alpha z}e^{j(\omega t - \beta z)} = E_{x0}e^{j\omega t}e^{-(\alpha + j\beta)z} \qquad (1.2)$$

A wave is said to have been attenuated by 1 neper if its field strength has decayed by $1/e$ (and its power density by $1/e^2$), where $e = 2.7183$.

Equations for the secondary parameters v, α and β in terms of the primary parameters σ, ε_0 and μ_0 and frequency f are given in Appendix 1.

The relationship between power flow and the electric field strength of an electromagnetic wave is obtained via the energy equations. The average energies stored in an incremental volume δv of free space due to the electric field component E (rms) and the magnetic field component H (rms) are given by

$$\delta W_E = \tfrac{1}{2}\varepsilon_0 E^2 \delta v$$

and

$$\delta W_H = \tfrac{1}{2}\mu_0 H^2 \delta v$$

In free space the total energy of an electromagnetic wave is contained equally by the electric and magnetic fields (averaged over a cycle) so that

$$\tfrac{1}{2}\varepsilon_0 E^2 \delta v = \tfrac{1}{2}\mu_0 H^2 \delta v$$

This equation is used to obtain the ratio of E to H, which is called the *intrinsic*

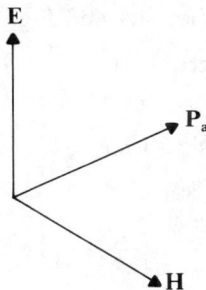

Fig. 1.4 The phasors representing electric field strength **E**, magnetic field strength **H** and the power flux $\mathbf{P}_a$ are mutually perpendicular. They are related via Poynting's theorem, $\mathbf{P}_a = \mathbf{E} \times \mathbf{H}$, so that in terms of magnitudes the mean power flux equals the product of E_{rms} and H_{rms}.

impedance Z_i of free space:

$$Z_i = \frac{E}{H} = \sqrt{\frac{\mu_0}{\varepsilon_0}} \approx 120\pi \approx 377\,\Omega \tag{1.3}$$

The corresponding expressions for Z_i of a lossy or complex medium are given in Appendix 1 and the vector orientations in space are shown in Fig. 1.4.

Poynting's theorem is discussed in Sec. A1.4.1. It relates the power flux P_a (expressed in W/m^2) to the electric and magnetic field components of an electromagnetic wave. In its simplest form, the theorem relates these parameters in free space via

$$P_a = EH = \frac{E^2}{120\pi} = 120\pi H^2 \tag{1.4}$$

if we make use of Eqs. (A1.22) and (1.3). This will prove to be a particularly useful relationship later when we wish to relate the power received by an antenna to the incident electromagnetic field components.

1.2 ANTENNA DEFINITIONS

An antenna is a device for radiating and receiving electromagnetic waves. It is the transition between a guided electromagnetic wave and a radiated (or received) spatial electromagnetic wave. There are many types of antenna, the simpler ones including:

(1) the hypothetical isotrope or isotropic radiator;
(2) the linear antenna such as a dipole or monopole;
(3) the loop antenna;
(4) the aperture antenna such as a slot or horn; and
(5) the lens antenna.

These are described later in the text.

Maxwell's equations, when applied to a linear radiator of incremental length δz, produce the following equations (Ref. 1.1 and Fig. 1.5):

$$E_r = \frac{\hat{I}\delta z \cos\theta}{2\pi\varepsilon_0}\left\{\frac{1}{cr^2} + \frac{1}{j\omega r^3}\right\}e^{j\omega(t - r/c)}$$

$$E_\theta = \frac{\hat{I}\delta z \sin\theta}{4\pi\varepsilon_0}\left\{\frac{j\omega}{c^2 r} + \frac{1}{cr^2} + \frac{1}{j\omega r^3}\right\}e^{j\omega(t - r/c)}$$

$$H_\phi = \frac{\hat{I}\delta z \sin\theta}{4\pi}\left\{\frac{j\omega}{cr} + \frac{1}{r^2}\right\}e^{j\omega(t - r/c)}$$

with $E_\phi = H_r = H_\theta = 0$. The terms involving $1/r^2$ and $1/r^3$ will diminish more rapidly with distance than the $1/r$ term, and substitution of $r = \lambda/2\pi$ into the above equations produces an interesting result.

In most applications of these or similar equations in radio wave propagation, the

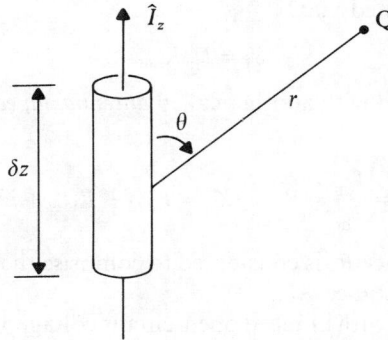

Fig. 1.5 An incremental linear radiator is an element of length δz carrying current of amplitude $\hat{I}_z$.

distance r between the transmitter and receiver is very much larger than a wavelength, and the terms involving $1/r^2$ and $1/r^3$ are negligible with respect to the $1/r$ term. The receiving antenna is then said to be in the *far field* of the transmitting antenna.

An exception to this situation can arise when the radio wave propagation is at very low frequencies. For example, the Omega navigation system operates close to 10 kHz, where the wavelength is about 30 km. Although it is used mainly when $d \gg \lambda$, it is possible for a receiver to be within a few wavelengths of the transmitter where the received field will include contributions from $1/r^2$ and $1/r^3$ terms which are not negligible. The receiving antenna is then said to be within the *near field* of the transmitting antenna. A similar situation can arise even at higher frequencies when an antenna is tested within the confines of a laboratory or anechoic chamber and it may then be necessary to determine a distance from the transmitting antenna beyond which the $1/r^2$ and $1/r^3$ terms can be considered as negligible.

For an antenna of finite physical extent, the far-field condition is that the relative phasing between field contributions from current elements in different parts of the antenna is dependent only on angle (θ, ϕ) and not on distance (r). With such a definition there are two common approximations to the boundary between the near field and the far field: for a linear element of length L the boundary occurs at $r \approx 2L^2/\lambda$, and for an aperture antenna with largest dimension D the boundary occurs at $r \approx 2D^2/\lambda$. In the remainder of this book we shall consider only far-field components and shall revert to using d for distance.

The input power P_{in} to an antenna provides the mean radiated power P_t and the antenna losses L_a such that

$$P_t = \frac{P_{in}}{L_a} \tag{1.5}$$

Many antennas have well-defined 'terminals' at which an input current I_{in} and an input voltage V_{in} can be measured. For these antennas the radiated power is related to I_{in} (rms) in terms of a parameter of proportionality which has the dimension of resistance. Although this parameter is not a physical dissipative resistance, it is called *radiation resistance* R_{rad} and is defined via:

$$P_t = I_{\text{in}}^2 R_{\text{rad}} \tag{1.6}$$

This term can be compared with another called *antenna impedance* Z_a which is defined by

$$Z_a = \frac{V_{\text{in}}}{I_{\text{in}}} = R_a + jX_a = R_{\text{rad}} + R_{\text{loss}} + jX_a \tag{1.7}$$

where the antenna resistance R_a is considered to comprise the radiation resistance R_{rad} and the antenna loss resistance R_{loss}.

If a receiving antenna with induced open-circuit voltage V (rms) and impedance Z_a is terminated by a receiver with impedance $Z_t = R_t + jX_t$, the maximum power transfer between antenna and receiver occurs when the two impedances are matched. This condition is achieved when Z_t is the complex conjugate of Z_a, or $X_t = -X_a$ and $R_t = R_a$. The circuit is illustrated in Fig. 1.6.

For many antennas the radiation resistance is much greater than the loss resistance and the appropriate terminating impedance for matching becomes $Z_t = R_{\text{rad}} - jX_a$. The power received is then

$$P_r = \frac{V^2}{4R_{\text{rad}}} \tag{1.8}$$

for zero antenna loss. In other cases, where the antenna loss is not negligible, a more detailed analysis is needed.

The *maximum effective aperture* A_e of the receiving antenna can be defined via the relationship

$$P_a A_e = P_r = \frac{V^2}{4R_{\text{rad}}} \tag{1.9}$$

where P_a is the power flux in W/m^2 of the incident electromagnetic field at the location of the antenna. Note that the *maximum* effective aperture occurs when the incident

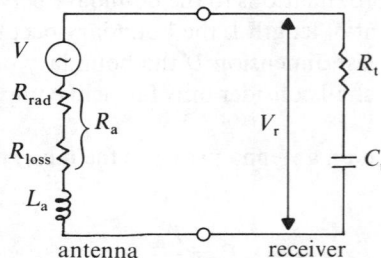

Fig. 1.6 The equivalent circuit of an antenna matched to a receiver. V is the induced voltage, R_{rad} is the radiation resistance, R_{loss} is the antenna loss and R_t is the receiver input resistance. The reactances are represented here by L_a and C_t as examples, and these should cancel for matching. Note that for a matched system the receiver input voltage V_r is exactly half the induced voltage.

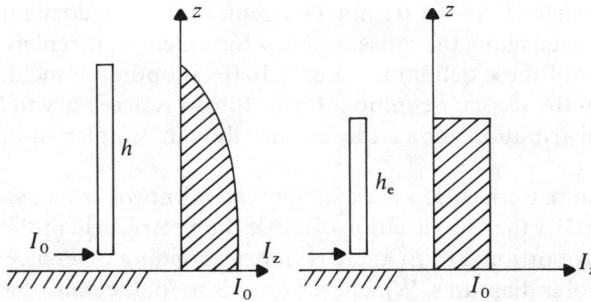

Fig. 1.7 An antenna of actual height h and current distribution I_z which varies with z as shown is equivalent to an antenna of effective height h_e and current $I_z = I_0$ which is independent of z. The areas shaded are equal.

direction of P_a is chosen to maximize P_r. In general, the effective aperture would be a function of the incident direction.

The electrical aperture A_e can be related to the physical aperture A of the antenna by a constant η called the *illumination efficiency*, provided the antenna has a recognizable physical aperture. The derivation of expressions for A_e is often complicated, but it is shown later that the commonly used reference antenna, the isotrope, has $A_{ei} = \lambda^2/4\pi$.

A similar term is sometimes preferred for linear antennas. These are said to have an effective length L_e or an effective height h_e (Fig. 1.7), whichever is the more appropriate description, defined via

$$V = Eh_e \tag{1.10a}$$

in reception, where V is the emf induced in the antenna by field E, or equivalently

$$I_0 L_e = \int_0^L I_z \, dz \tag{1.10b}$$

Example 1.2 Effective length of antenna

If a certain linear antenna has an assumed current distribution $I_z = I_0 \cos \beta z$ along its length from $z = -\lambda/4$ to $z = +\lambda/4$, determine its effective length.

Using the definition of effective length we can write in this case

$$I_0 L_e = \int_{-\lambda/4}^{+\lambda/4} I_0 \cos \beta z \, dz \tag{1.11}$$

from which

$$L_e = 2\left[\frac{\sin \beta z}{\beta}\right]_0^{\lambda/4} = \lambda/\pi$$

This is, in fact, the effective length of a halfwave dipole, described later.

---

in transmission, where I_0 is the terminal current, dz is an incremental length of the linear antenna aligned along the z axis, and I_z is the antenna current at point z from the origin. In the first of these definitions, Eq. (1.10a), an optimum incidence direction is again assumed. In the second definition, Eq. (1.10b), it is necessary to have knowledge of the current distribution along the z axis, though simpler approximations are sometimes used.

The radiation pattern (Fig. 1.8) of an antenna is known as its *polar diagram*. This may be represented by the θ, ϕ variation of either the electric field or the power intensity P_a, but as P_a is proportional to E^2 there is a corresponding difference in the shapes of their respective polar diagrams. Whichever form is used, the half-power *beamwidth* in any specified plane is unambiguous; it is the angle within which the power intensity exceeds half the maximum power intensity. In terms of the electric field strength it becomes the angle within which E exceeds $0.7071\,E_{max}$.

Beamwidth is thus a useful numerical parameter with which to characterize an antenna. Another is called either antenna *gain G* or *directivity D*. These similar terms have slightly different meanings though they are commonly intermixed. Both are defined as the ratio of maximum radiation intensity from the specified antenna to the maximum radiation intensity from a loss-free isotrope, but for gain G there is the proviso that the same value of input power P_{in} is used in each case, whereas for directivity D the ratio requires as reference the same radiated power P_t. The simple relationship is $G = \eta_a D$ where η_a is the *radiation efficiency factor*, $0 < \eta_a < 1$. A η_a value less than 1 denotes energy dissipation in the antenna itself.

For most antennas the definitions we have just considered are reciprocal in the sense that they are the same whether the antenna is used for transmitting or for receiving signals.

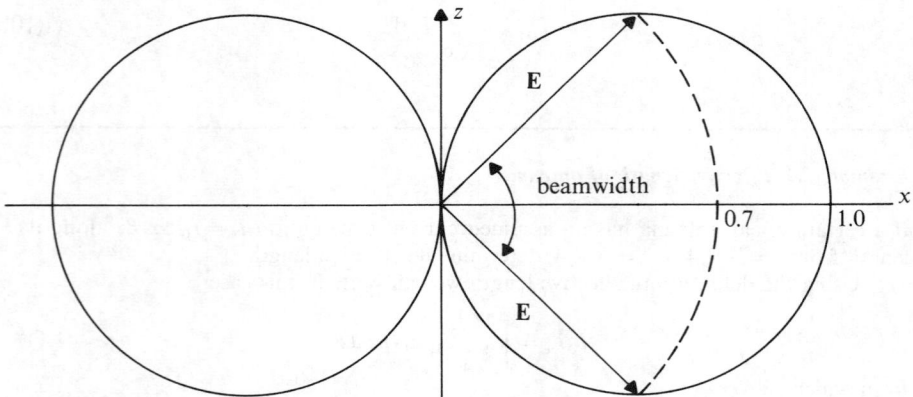

Fig. 1.8 The E-field polar diagram of an incremental linear radiator in the zx plane. The half-power beamwidth is the angle within which E exceeds 0.707 times its maximum value. For the ILR (incremental linear radiator), which has a circular cross-section polar diagram, the beamwidth is 90°.

1.3 PROPAGATION EQUATIONS

In this section we consider some of the very basic forms of the propagation equations which will be used and developed more fully in later chapters of this book. Three types of equations will be considered: those involving the variation of electric field strength with distance d from the transmitting antenna; those involving the corresponding power levels; and the logarithmic forms of these equations (Fig. 1.9).

It is found that in certain propagation circumstances the electric field at distance d from the antenna is proportional to the square root of the radiated power P_t and inversely proportional to distance, provided that d is in the far-field region. Thus we have

$$E = k\sqrt{P_t}/d \qquad (1.12)$$

in SI units, where k is the constant of proportionality which has to be determined for each type of transmitting antenna. If we use the more 'practical' units of mV/m, kW and km, and select the direction in which the radiation is maximum, the constant k is replaced by the term *figure of merit* FM (not to be confused with frequency modulation). Typical values of FM for ideal antennas are 173 for the isotrope, 212 for the incremental linear radiator (ILR), 300 for the short vertical monopole (SVM) and 222 for the halfwave dipole, as shown later.

The corresponding power equations can be derived by considering a receiving

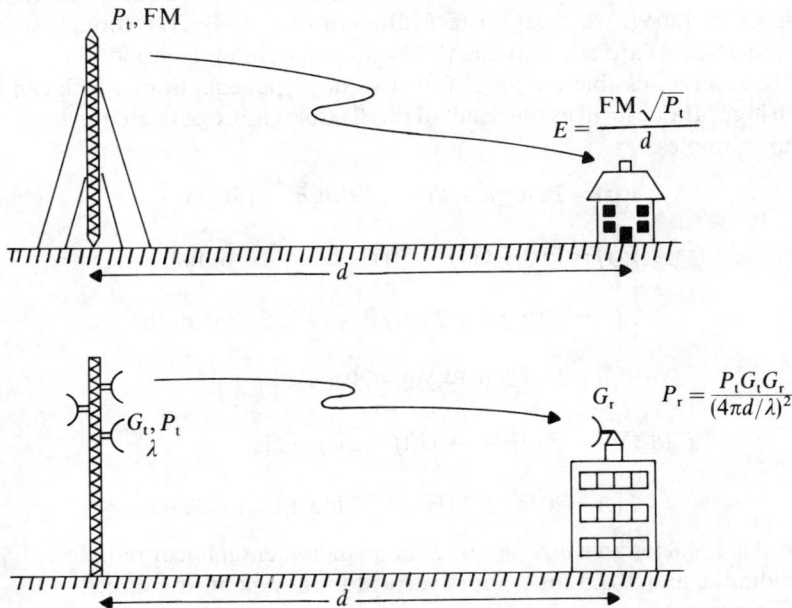

Fig. 1.9 There are two commonly used equations in radio wave propagation, one relating the electric field strength to distance and the other relating power received to distance.

antenna (gain G_r) displaced distance d from a transmitting antenna (gain G_t) which operates with radiated power P_t at wavelength λ. If it is initially assumed that both antennas are isotropes, we can write

$$P_a \times 4\pi d^2 = P_t$$

and

$$P_r = P_a A_{ei} = (P_t/4\pi d^2)(\lambda^2/4\pi)$$

where $4\pi d^2$ is the surface area of a sphere of *radius d* surrounding the isotrope, and $\lambda^2/4\pi$ is the maximum effective aperture of each isotrope. The antenna gains enhance P_a by G_t and A_{ei} by G_r, and hence we obtain the *Friis free-space equation*:

$$P_r = P_t \frac{G_t G_r}{(4\pi d/\lambda)^2} \tag{1.13}$$

in SI units for antennas in a loss-free medium. The larger the denominator, the smaller the value of received power P_r, and hence we describe

$$L_s = (4\pi d/\lambda)^2 \tag{1.14}$$

as the *spatial attenuation*, which is distinct from dissipative or absorption losses.

The two principal propagation equations just considered are commonly expressed in logarithmic form in order to make use of the decibel notation. The reference units for the decibel notation are either the appropriate SI unit or some preferred arbitrary value. In the latter case the reference must be specified clearly. For example, if we wish to express electric field strength in decibels with respect to $1\,\mu V/m$, or power in decibels with respect to $1\,mW$, we must write $E(dB\mu V/m)$ or $P(dBmW)$, though sometimes $E(dB\mu)$ and $P(dBm)$ are adequate as the basic units are understood.

Of the several possible versions of the two principal equations which can now be written in logarithmic form as the result of this flexible choice of reference, the following are useful examples:

$$E(r) = 20 \log k + P(t) - 20 \log d \quad \text{(SI units)} \tag{1.15a}$$

and

$$P(r) = P(t) + G(t) + G(r) - L(s) \quad \text{(SI units)} \tag{1.15b}$$

with

$$L(s) = 20 \log f + 20 \log d - 147.55 \quad \text{(SI units)} \tag{1.15c}$$

or

$$E_r(dB\mu) = k' + P_t(dBkW) - 20 \log d(km) \tag{1.15d}$$

and

$$P_r(dBW) = P_t(dBW) + G(t) + G(r) - L(s) \tag{1.15e}$$

with

$$L(s) = 20 \log f(MHz) + 20 \log d(km) + 32.45 \tag{1.15f}$$

The term k' is 104.8 for an isotrope, 106.5 for an incremental linear radiator, 106.9 for a halfwave dipole, and 109.5 for a short vertical monopole in the direction $\theta = 90°$. A graphical representation is given in Fig. 1.10.

The use of parentheses implies decibels in $G(t)$, $G(r)$ and $L(s)$, or that standard SI units are used as reference in $E(r)$ or $P(t)$. Otherwise the flexible references must be

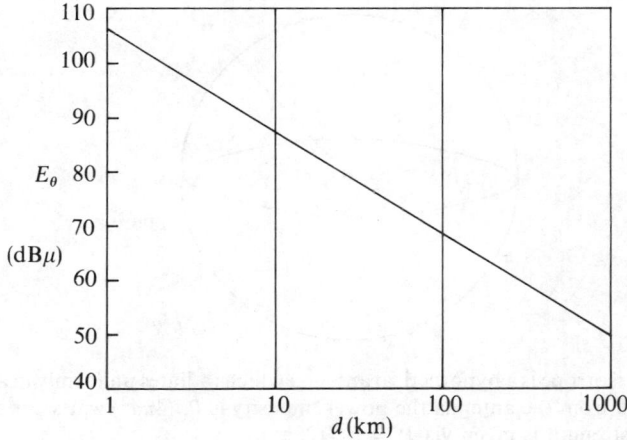

Fig. 1.10 The logarithmic form of the propagation equations lends itself to the use of decibels and logarithmic graph paper. The inverse-distance relationship appears linearly on log-log paper, the above example applying specifically to 1 kW radiated from a halfwave dipole in the direction $\theta = 90°$.

specified. Note that it is incorrect to express logarithmic ratios of frequency or distance as decibels, and the expressions should always be written in full as $20 \log f$, etc. Decibels are defined in terms of power ratios.

1.4 SOME BASIC ANTENNAS

A detailed analysis of antennas is outside the scope of this text but a brief review of some of the essential features of the more basic antenna types is appropriate at this stage. Two of these, the isotrope and the incremental linear radiator, are essentially theoretical in nature, whereas the others have practical equivalents.

1.4.1 Isotrope and Incremental Linear Radiator

The *isotrope* is the hypothetical antenna which radiates energy uniformly in all directions. We can obtain its propagation equation by noting that the power intensity at *radius d* from the isotrope, shown in Fig. 1.11, is related to the electric field, Eq. (1.4), via:

$$P_a = \frac{P_t}{4\pi d^2} = \frac{E_\theta^2}{120\pi}$$

and hence

$$E_\theta = \frac{\sqrt{30 P_t}}{d} \qquad (1.16a)$$

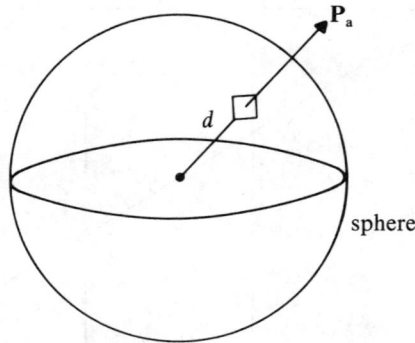

Fig. 1.11 The isotrope is a hypothetical antenna which radiates uniformly in all directions. At a distance d from the antenna the power intensity is $P_t/(4\pi d^2)$ watts per square meter and the field strength is given via $P_a = E_\theta^2/120\pi$.

In alternative forms this equation may be written as either

$$E_\theta(\text{mV/m}) = 173\sqrt{P_t(\text{kW})}/d(\text{km}) \tag{1.16b}$$

or

$$E_\theta(\text{dB}\mu) = 104.8 + P_t(\text{dBkW}) - 20\log d(\text{km}) \tag{1.16c}$$

and $\quad E_0(dBn) = 104.8 + p_t(dB_n) - 20\log d(n)$

The isotrope is the ideal lossless *reference* antenna having gain $G = 1$ and directivity $D = 1$ (or 0 dB in each case), but no parameter R_{rad}.

The *incremental linear radiator* (ILR) is a short linear element of incremental length δz (or equivalent electrical length $2\pi\delta z/\lambda$) carrying oscillating current of constant amplitude. For such an element, it can be shown (Ref. 1.1) that at any point Q in the far-field region (Fig. 1.5), distance d from a reference point midway along the element, the instantaneous field is

$$E_\theta = j120\pi\frac{\delta z}{2\lambda d}\sin\theta \hat{I}_z\cos\omega(t - d/c) \tag{1.17a}$$

or

$$E_\theta(\text{rms}) = 30\frac{\beta\delta z}{d}\sin\theta I_z(\text{rms}) \tag{1.17b}$$

where $\hat{I}_z$ represents current amplitude (i.e. peak) and I_z represents the rms value. The corresponding expression for power can be obtained by first using instantaneous power intensity,

$$\text{inst. } P_a = \frac{E_\theta^2}{120\pi} = 120\pi\left[\frac{\hat{I}_z\delta z\sin\theta}{2\lambda d}\right]^2\cos^2\{\omega(t - d/c) + \pi/2\}$$

and then substituting the mean of $\cos^2\{\omega(t - d/c) + \pi/2\} = \frac{1}{2}$ to obtain the mean power flux:

$$P_a = 30\pi\left[\frac{I_z\delta z\sin\theta}{\lambda d}\right]^2$$

We can now derive an expression for the mean radiated power P_t by integrating P_a over the surrounding sphere:

$$P_t = \int P_a \, dA'$$

where $dA' = 2\pi d^2 \sin\theta \, d\theta$ (from Fig. 1.12). Hence:

$$P_t = \int_0^\pi 30\pi \left[\frac{I_z \delta z \sin\theta}{\lambda d} \right]^2 2\pi d^2 \sin\theta \, d\theta$$

$$P_t = 60\pi^2 I_z^2 (\delta z/\lambda)^2 \int_0^\pi \sin^3\theta \, d\theta$$

$$P_t = 20 I_z^2 (\beta \delta z)^2 \tag{1.18}$$

This analysis permits the radiation resistance R_{rad} to be determined via $P_t = I_z^2 R_{rad}$, giving

$$R_{rad} = 20(\beta\delta z)^2 \tag{1.19}$$

and the electric field strength to be obtained by equating I_z in Eqs. (1.17b) and (1.18):

$$\frac{E_\theta d}{30\beta\delta z \sin\theta} = \frac{1}{\beta\delta z} \sqrt{\frac{P_t}{20}}$$

or

$$E_\theta = \frac{\sqrt{45 P_t}}{d} \sin\theta \tag{1.20}$$

The field strength is therefore maximum when $\sin\theta = 1$ and the (power) gain G of the incremental linear radiator with respect to the reference isotrope is $G = 45/30 = 1.5$,

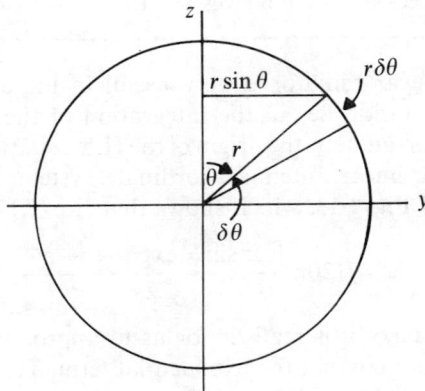

Fig. 1.12 The incremental area dA' is the product of $2\pi r \sin\theta$ and $r\delta\theta$ for each increment $\delta\theta$ in θ, using conventional spherical polar coordinates. In this text we use d (instead of r) to represent distance.

from Eqs. (1.20) and (1.16a), as gains are proportional to the square of the field strengths.

Example 1.3 Maximum effective apertures and antenna gain

Determine the maximum effective apertures of an incremental linear radiator and an isotrope. An antenna has a maximum effective aperture of $0.1\,\mathrm{m}^2$ when $f = 1$ GHz. What is its gain?

We are able to derive an expression for maximum effective aperture by noting that

$$P_a A_e = \frac{V^2}{4R_{rad}}$$

for the incremental linear radiator. Substituting $P_a = E^2/120\pi$ with $V = E\delta z$ because of the constant current amplitude along element δz gives

$$\frac{E^2}{120\pi} A_e = \frac{(E\delta z)^2}{80(\beta\delta z)^2}$$

or

$$A_e = 3\lambda^2/8\pi$$

as the maximum effective aperture of an incremental linear radiator. We have just shown that the ILR has a gain of $\frac{3}{2}$ with respect to the reference isotrope and so:

$$A_{ei} = 2A_e/3 = \lambda^2/4\pi$$

is the maximum effective aperture of the isotrope.

The gain of an antenna with respect to an isotrope is given by

$$G = \frac{P_{ra}}{P_{ri}} = \frac{P_a A_e}{P_a A_{ei}} = \frac{A_e}{A_{ei}} = \frac{4\pi A_e}{\lambda^2}$$

where the suffices a and i refer to the antenna and isotrope, respectively. When $f = 1$ GHz, $\lambda = 0.3$ m and hence direct substitution gives $G = 13.96$ linearly or 11.45 dB.

The incremental linear radiator is very useful in the analysis of larger linear antennas because the solution lies in the integration of the ILR equation over an appropriate length. Unfortunately, the origin of the ILR equation is at *its* midpoint and not at the origin of the linear antenna coordinate system. We can overcome this problem with the aid of Fig. 1.13, which shows that Eq. (1.17a) may be written as

$$E_\theta = j120\pi\, \frac{\hat{I}_z\delta z \sin\theta' \exp j\omega(t - s/c)}{2\lambda s}$$

When the distance d is large it is realistic for us to approximate $\theta' = \theta$, $s = d$ in the denominator and $s = d - z\cos\theta$ in the exponential term. This modifies the equation into the new d, θ, ϕ coordinate system as

$$E_\theta = j120\pi\, \frac{\hat{I}_z \sin\theta \exp j\omega(t - d/c + \{z\cos\theta\}/c)}{2\lambda d}\delta z \qquad (1.21)$$

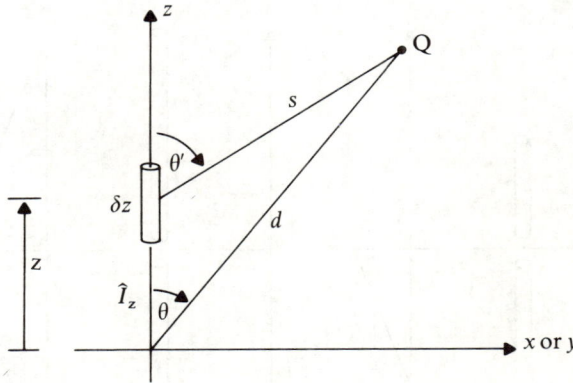

Fig. 1.13 The coordinates of the ILR equation, Eq. (1.17), have their origin midway along the ILR. When this equation is used as part of an integration it is useful to change the origin to some other point on the z axis.

and permits integration over appropriate limits of z with d and θ specified in the new coordinate system.

1.4.2 Short Dipole and Halfwave Dipole

A dipole is a centre-fed linear antenna of length L. Two particular types of interest here are the *short dipole*, which is defined with length $L \ll \lambda$ ($L < 0.2\lambda$ is a good approximation), and the *halfwave dipole*, which is defined with $L = \lambda/2$, though in practice it is often slightly less than half a wavelength to produce a non-reactive input impedance.

As previously mentioned, the derivation of equations associated with linear antennas involves the integration of the ILR expression, and this requires knowledge of the current amplitude distribution along the antenna. For the two antennas just mentioned, the current distributions are approximated by simple linear and cosine functions (along, say, the z axis) of the forms illustrated in Fig. 1.14.

We can calculate the effective length of the short dipole from Eq. (1.10b),

$$I_0 L_e = \int_{-L/2}^{+L/2} I_z \, dz$$

and obtain the obvious solution that $L_e = L/2$, while the corresponding value for the halfwave dipole was obtained in a similar manner in Ex. 1.2 as $L_e = \lambda/\pi$.

The calculation of the radiation resistance is mathematically complicated, but in the case of the short dipole, where L is sufficiently small for the approximation $\sin(\beta L) \approx \beta L$, the solution tends to

$$R_{\text{rad}} \approx 5(\beta L)^2 \approx 20(\beta L_e)^2 \tag{1.22}$$

It can be seen that this is identical to the expression for the radiation resistance of the

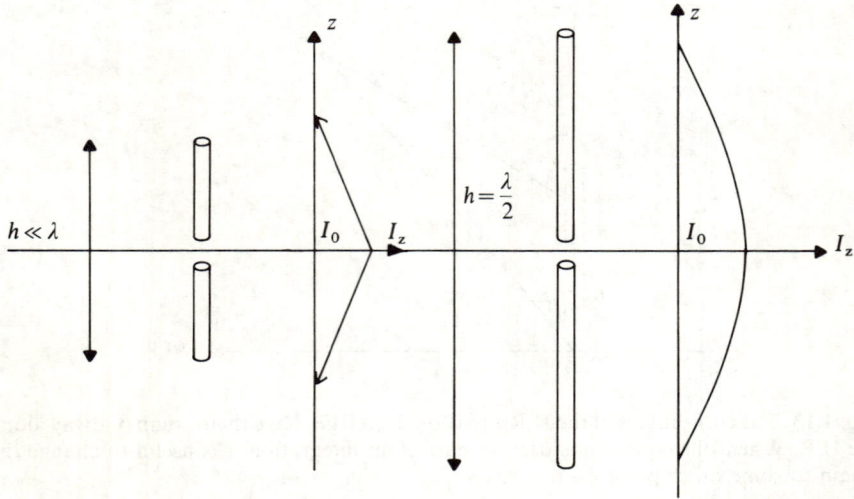

Fig. 1.14 The current distribution along dipole antennas is very complicated, but two useful approximations are illustrated for the short dipole (in which the current is assumed to vary linearly with z) and for the halfwave dipole (in which the current amplitude is assumed to have a cosinusoidal variation with z).

incremental linear radiator if δz is replaced by $L_e = L/2$. Similar solutions arise in the case of maximum effective aperture, with

$$A_e = 3\lambda^2/(8\pi) \tag{1.23}$$

and directivity, with $D = 1.5$ in the direction $\theta = 90°$.

The calculation of the radiation resistance of a halfwave dipole (Ref. 1.1) produces $R_{rad} = 73.2\,\Omega$. Using this value of R_{rad} with $V = EL_e = E\lambda/\pi$ in Eq. (1.9) gives the maximum effective aperture of the halfwave dipole as

$$A_e = \frac{30\lambda^2}{73.2\pi} \tag{1.24}$$

Directivity is then obtained from

$$D = \frac{A_e}{A_{ei}} = \frac{30\lambda^2}{73.2\pi}\frac{4\pi}{\lambda^2} = 1.64 \text{ in direction } \theta = 90° \tag{1.25}$$

The equation for the electric field E_θ of a halfwave dipole is obtained directly from Eq. (1.21) by assuming that $\hat{I}_z = \hat{I}_0 \cos \beta z$ and then integrating between $-\lambda/4$ and $+\lambda/4$. After some lengthy but straightforward mathematics this produces

$$E_\theta(\text{rms}) = \frac{60I_0}{d}\frac{\cos\{\frac{1}{2}\pi\cos\theta\}}{\sin\theta} \tag{1.26}$$

which is maximum when $\theta = 90°$.

1.4.3 Vertical Monopole

The *vertical monopole* is a linear antenna of height h positioned vertically above an infinite horizontal ground plane. The ground plane is required to be a good conductor and forms one of the two terminals of the antenna. It is the type of antenna used most frequently for low- and medium-frequency ground wave propagation. The variation of current amplitude along the vertical z axis is complicated but for antennas with height h less than about 0.4λ the sinusoidal approximation illustrated in Fig. 1.15 is often

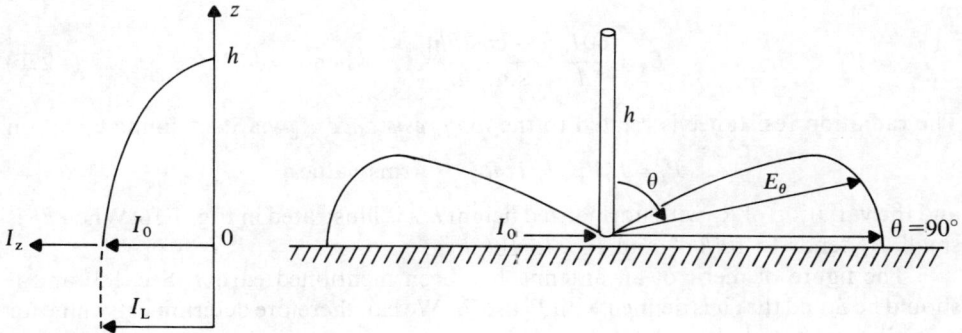

Fig. 1.15 A vertical monopole above a horizontal ground plane with typical E-field polar diagram.

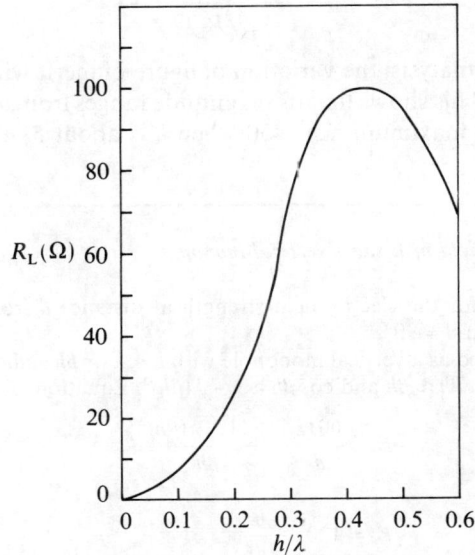

Fig. 1.16 The variation of loop resistance R_L with height h for a vertical monopole.

adequate:

$$\hat{I}_z = \hat{I}_L \sin \beta(h - z)$$
$$\hat{I}_0 = \hat{I}_L \sin \beta h$$

This approximation fails when $h = \lambda/2$ because I_0 is then zero. Using these expressions for I_z and I_0, Eq. (1.21) may be integrated over the range $z = 0$ to $z = h$ to obtain

$$E_\theta = \frac{60 I_L}{d} \frac{\cos(\beta h \cos \theta) - \cos(\beta h)}{\sin \theta} \tag{1.27a}$$

or

$$E_\theta = \frac{60 I_0}{d} \frac{1 - \cos(\beta h)}{\sin(\beta h)} \qquad \text{when } \theta = 90° \tag{1.27b}$$

The radiation resistance is related to the *loop resistance* R_L via its defining equation

$$P_t = I_0^2 R_{rad} = I_L^2 R_L \qquad \text{(rms values)}$$

and the variation of R_L with normalized height h/λ is illustrated in Fig. 1.16. When βh is small, R_L tends to $10(\beta h)^4$ and R_{rad} to $10(\beta h)^2$.

The figure of merit of an antenna has been mentioned earlier (Sec. 1.3) and it should be noted that it is defined with P_t in kW. We can therefore determine its value for a vertical monopole by substituting I_L from

$$1000 P_t(\text{kW}) = I_L^2 R_L$$

into Eq. (1.27) to obtain

$$\left(\frac{\lambda/4}{\text{Gain}} \right) = 3\text{dB above dipole}$$

$$\text{FM} = \frac{60\sqrt{1000}(1 - \cos \beta h)}{\sqrt{R_L}} \tag{1.28}$$

Using a more detailed analysis, the variation of figure of merit with antenna height is illustrated in Fig. 1.17. This shows that its magnitude ranges from about 300 for a short vertical monopole to a maximum near 430 when h is about $5\lambda/8$.

Example 1.4 Electric field of a short vertical monopole

Derive an expression for the electric field strength at distance d from a short vertical monopole (SVM) when $\theta = 90°$.

The SVM is defined as a vertical monopole with $h \ll \lambda$ or βh sufficiently small to use the approximations $\sin \beta h \approx \beta h$ and $\cos \beta h \approx 1 - \frac{1}{2}(\beta h)^2$. Equation (1.27b) then becomes

$$E_\theta = \frac{60 I_0}{d} \frac{1 - \{1 - \frac{1}{2}(\beta h)^2\}}{\beta h}$$

or

$$E_\theta = 120\pi \frac{I_0 h_e}{\lambda d}$$

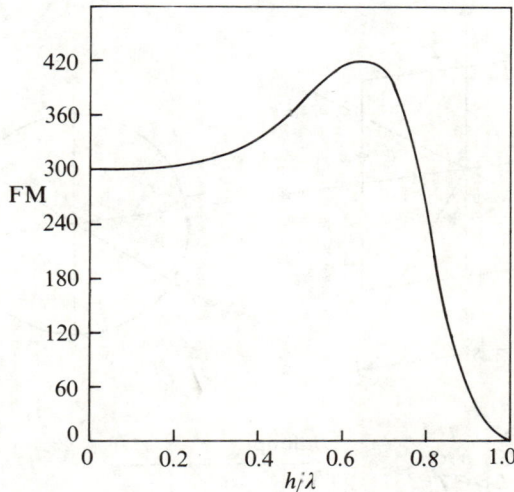

Fig. 1.17 The variation of the figure of merit FM with height *h* of a vertical monopole above a horizontal ground plane.

The electric field equation for a short vertical monopole is identical to that of an incremental linear radiator given in Eq. (1.17b) if $2h_e$ replaces δz, except that only half the power is required for the SVM as its field is confined to the region above the infinite horizontal ground plane. For a given I_0 and P_t, this implies that the radiation resistance of the SVM is only half that of the ILR or short dipole:

$$R_{\text{rad}} = \tfrac{1}{2} \times 20(\beta h)^2 = 160\pi^2 (h_e/\lambda)^2 \tag{1.29}$$

By similar reasoning, the maximum effective aperture and directivity of an SVM are twice those of a short dipole: $A_e = 3\lambda^2/4\pi$ and $D = 3.0$ in the direction $\theta = 90°$. This is equivalent to a gain of 4.77 dB in the horizontal plane.

1.4.4 Small Loop Antenna

A *small loop antenna* is one in which the area of the loop, which may have any shape such as a square or circle, is less than about $0.01\lambda^2$. The small dimension allows us to assume that the current is constant in amplitude around the loop and equal to the current injected at the terminals.

If a square loop of dimension *s* is oriented as shown in Fig. 1.18, the polar pattern of E_θ in the *xy* plane can be obtained by noting that the symmetrical elements 2 and 4 produce fields which cancel, while the fields due to elements 1 and 3 do not quite cancel. When $\theta = 90°$ (in the *xy* plane), Eq. (1.17a) gives the field due to one (ILR) element as

$$E_\theta(\text{ILR}) = j\frac{120\pi s}{2\lambda d}\,\hat{I}_z \cos\omega(t - d/c)$$

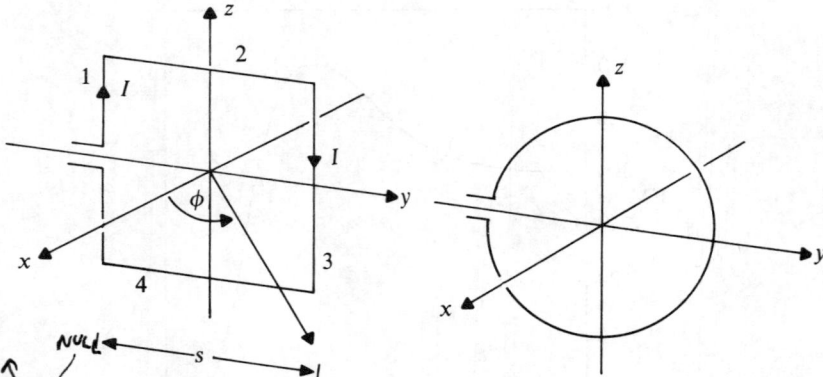

Handwritten annotations: ELECTRICALLY SMALL; NULL; MAX. IN PLANE OF LOOP; NULL; Overall circumference C :- MAX o/p; C ⇒ λ; MAX o/p

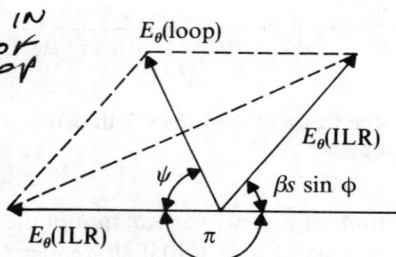

Fig. 1.18 Small loop antennas. The polar plot in the xy plane ($\theta = 90°$) can be obtained by the phasor addition of the E-field contributions of elements 1 and 3 in any direction ϕ. Note that $\cos\psi = \cos((\pi/2) - (\beta s \sin\phi/2)) = \sin((\pi s/\lambda)\sin\phi) \approx (\pi s/\lambda)\sin\phi$ when $\pi s/\lambda$ is very small.

where dimension s is illustrated in Fig. 1.18, but at any distant point in the xy plane there is a phase difference of $\beta s \cos\phi$ radians due to the different path lengths from elements 1 and 3, and a further π radians due the fact that the two element currents flow in opposite directions. Hence the resultant field is the phasor addition of the two contributions (using rms values) as shown in Fig. 1.18:

$$E_\theta(\text{loop}) = E_\theta(\text{ILR}) \times 2\cos\psi$$

or

$$E_\theta(\text{loop}) = 120\pi^2 \frac{IA\sin\phi}{\lambda^2 d} \tag{1.30}$$

where $A = s^2$ is the cross-sectional area of the loop. The same answer arises irrespective of the shape of the loop.

It can also be shown that the radiation resistance of a *small* loop antenna is

$$R_{\text{rad}} = 20(\beta^2 A)^2 = 320\pi^4 A^2/\lambda^4 \tag{1.31}$$

and that the directivity $D = 1.5$ when $\theta = 90°$. If $A = 0.01\lambda^2$ the radiation resistance is $3.1\,\Omega$, but a square loop $0.5\,$m by $0.5\,$m will have $R_{\text{rad}} = 2.4 \times 10^{-7}\,\Omega$ at $f = 1$ MHz.

1.4.5 Aperture-type Antennas

Several types of antenna which we shall describe more fully later in this text may be grouped under the title of aperture antennas. Among others, these include the slot antennas, the flared antennas (or horn antennas) and the reflector antennas, as illustrated in Fig. 1.19.

The *slot antenna*, as its name suggests, is simply an opening cut in a sheet of conductor which is energized in some appropriate manner, such as via a coaxial line or waveguide. One simple type of slot antenna is half a wavelength long with narrow width and excited via a $50\,\Omega$ coaxial cable commonly connected about 0.05λ from one end of the slot to achieve reasonable matching conditions. A horizontal slot so energized produces vertical polarization in the direction normal to the slot, and a vertical slot produces horizontal polarization. Radiation occurs from both sides of the conductive sheet but if the slot is 'boxed' with internal dimension of depth $d_s = \lambda/4$, the radiation is outwards from the opening of the box.

A single half-wavelength slot in many ways resembles the halfwave dipole in terms of gain and radiation pattern except that there is a difference in polarization, described earlier. To enhance the gain and directive properties of the basic slot antenna it is common to have arrays of slots in a manner similar to the arrays of dipoles. Some VHF transmitters employ cylindrical arrays of slots to produce omnidirectional radiation in the horizontal plane with horizontal polarization.

An open-ended rectangular waveguide is a kind of slot radiator, a guide with its broad dimension in the horizontal plane producing vertical polarization. The intrinsic impedance of the space outside the opening of the guide and the waveguide impedance within the guide do not match, and so reflections arise at the opening. In addition, in accordance with diffraction theory, the narrow aperture dimensions result in a wide beamwidth. To overcome the mismatch and reduce the beamwidth in order to increase the gain, the waveguide opening is flared in one or both dimensions to produce the family of *E*-type, *H*-type or *pyramidal horn antennas*. A similar effect is achieved with cylindrical guides (producing conical antennas) and coaxial lines (producing biconical antennas).

An antenna can have its radiation pattern changed by means of a *reflector* so that the beamwidth is reduced and the gain increased, or the reverse can be achieved if required. The reflector may be a large plane sheet of conducting material, a simple reflecting element such as shorted dipole, a 90° corner reflector, a paraboloidal reflector or a variety of other shapes.

In particular, the paraboloidal reflector is designed to produce a plane wavefront from a point source located at its focal length. The (parabolic) curvature of the reflector is such that, irrespective of the angle of incidence, the total distance ABC or AB′C′ or AB″C″ (in Fig. 1.19) between the source and the plane wavefront is equal to a constant value. In reality the source does not radiate uniformly but the paraboloid's geometry helps to reduce the beamwidth and enhance the gain.

For each of the various types of flared and reflector antennas, the gain G at wavelength λ in a direction normal to the physical aperture of area A is given by

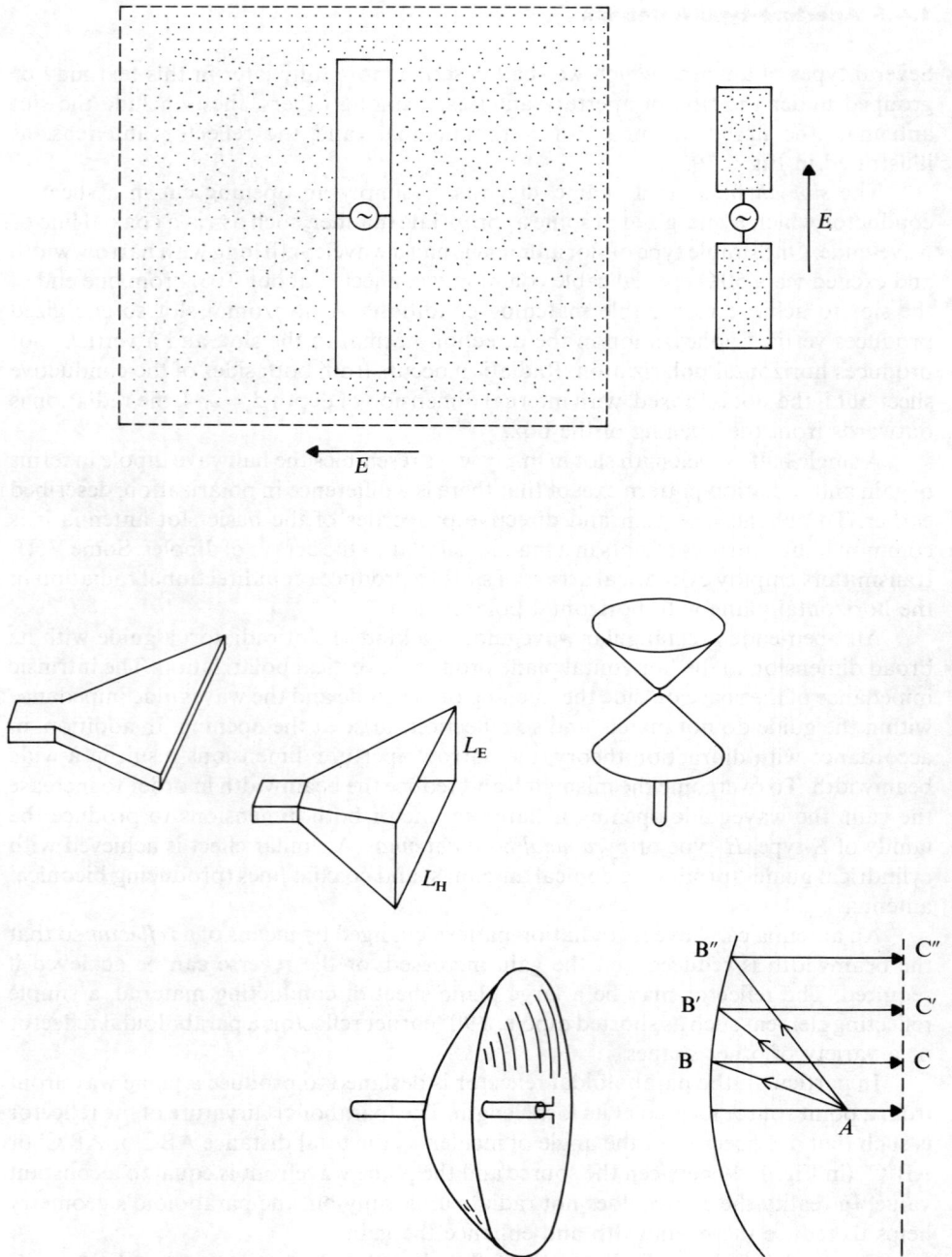

Fig. 1.19 Examples of the slot antenna, flared antenna and the paraboloidal antenna.

$$G = \frac{A_e}{A_{ei}} = \frac{\eta A}{\lambda^2/(4\pi)} = \frac{4\pi\eta A}{\lambda^2} \qquad A = \pi r^2, \; r = \text{rad}^{i}v) \qquad (1.32)$$

where η is the illumination efficiency. We can see that by constructing a paraboloidal reflector with large diameter (relative to a wavelength) the overall gain becomes considerable. However, because of the difficulty of achieving uniform illumination and some problems with side-lobe levels, the illumination efficiency is commonly of the order of 55%. In some special cases this can be increased to about 80%.

In the case of the pyramidal horn antenna it can also be seen that the larger flared antenna would seem to have larger gain, but the difficulty of achieving uniform illumination from a rectangular waveguide means that other conditions need to be satisfied for optimum design. However, the gain of the antenna is a function of the physical dimensions of the pyramid and the waveguide and, as these can be measured accurately, the magnitude of the antenna's gain is known to within a small fraction of a decibel. It can therefore be used as a standard against which the gains of other antennas can be compared.

The beamwidths of the flared and reflector types of antennas are sometimes very small. Diffraction theory indicates that if a uniformly illuminated aperture has a maximum dimension L in any specified plane, the minimum half-power beamwidth within which the radiation can be concentrated in that particular plane is of the order $BW = \lambda/L$ radians. If uniform illumination cannot be achieved, the angle is larger. Thus for uniformly illuminated rectangular apertures and circular apertures (of length L or diameter D), the beamwidths are approximately $51\lambda/L$ and $58\lambda/D$ degrees, respectively. For the pyramidal and sectoral horns, the corresponding beamwidths are about $56\lambda/L_E$ and $67\lambda/L_H$ degrees in the E- and H-planes. For the paraboloidal reflector antenna, the beamwidth depends on the feed antenna at its focal point, but it is commonly about $(70 \pm 10)\lambda/D$ degrees. Thus large diameter paraboloidal reflector antennas have large gains and are highly directive.

Example 1.5 Gain of a paraboloidal reflector antenna

Estimate the approximate gain and beamwidth of a paraboloidal reflector antenna at $f = 4\,\text{GHz}$ if its diameter $D = 20\,\text{m}$ and its illumination efficiency $\eta = 55\%$.

The physical aperture of the paraboloid is $\pi D^2/4 = 314.16\,\text{m}^2$ and the wavelength corresponding to 4 GHz is $\lambda = c/f = 0.075\,\text{m}$. Equation (1.32) may be used to obtain $G = 386,000$ linearly or $G(\text{dB}) = 10 \log G = 55.9\,\text{dB}$. An approximation of the beamwidth is to use $BW = 70\lambda/D$ degrees $= 0.26°$.

1.4.6 Lens Antennas

In Sec. 1.4.5 we noted that a paraboloidal reflector can be used to produce a plane wave from an isotropic source at its focus by arranging that all paths from source to reflector to plane wavefront are of constant magnitude. This is similar to the action of light in a torch where the point source is the filament in the bulb and the reflector is a paraboloidal mirror surface. Another optical technique using the refraction properties of lenses rather than the reflecting properties of mirrors may be adapted to radio frequencies to achieve focusing of the electromagnetic waves.

The *plano-convex lens* illustrated in Fig. 1.20 can be used to achieve this purpose if the relative permittivity of the dielectric exceeds unity and hence the velocity of propagation within the dielectric is less than that in free space. Noting that refractive index $n = c/v$, the geometry requires, in the polar coordinates of Fig. 1.20, that

$$AE/c + EF/v = AB/c + BD/v$$

or

$$r = F + n\text{BC} = F + n(r\cos\theta - F)$$

$$r = \frac{(n-1)F}{n\cos\theta - 1} \tag{1.33}$$

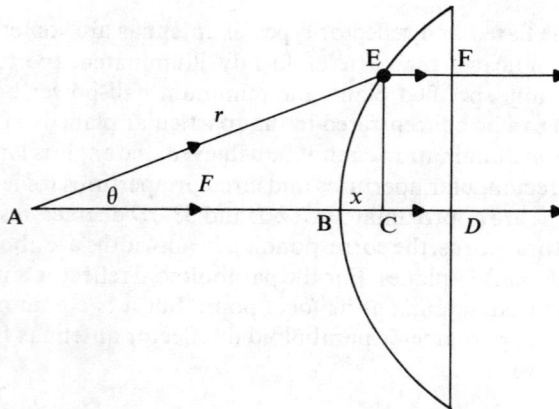

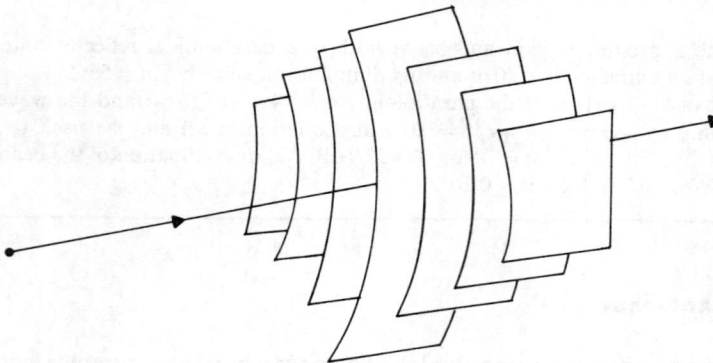

Fig. 1.20 The plano-convex dielectric antenna and the metal plate antenna are typical examples of the lens antenna group.

Thus the profile of the lens in polar coordinates is described for a specified refractive index n and focal length F. In mathematical terminology it describes a hyperbola.

A *metal plate lens* makes use of waveguide theory, which states that the guide wavelength λ_g is related to the free-space wavelength λ via

$$(1/\lambda_g)^2 = (1/\lambda)^2 - (1/2a)^2$$

for the H_{10} mode in a rectangular guide. As the effective refractive index between the plates is given by $n = c/v = \lambda/\lambda_g$,

$$n = \{1 - (\lambda/2a)^2\}^{1/2} \tag{1.34}$$

where a is the wider internal dimension of the rectangular waveguide. Thus n is always less than unity and the phase velocity $v = c\lambda_g/\lambda$ is greater than c. A metal plate lens antenna can be designed on this principle (Fig. 1.20) to produce an equation similar to that of the dielectric lens and by this means produce resultant radiation at a reduced beamwidth and enhanced gain. Such lenses are suitable for use at the higher ranges of radio frequencies.

1.5 RADIO WAVE PROPAGATION

1.5.1 Modes of Propagation

The propagation of radio waves takes place in several different ways but for the sake of simplicity in this introduction we will consider only the following four categories:

1. *Ground wave propagation*, using either the spherical terrestrial waveguide of earth and ionosphere, or the surface wave of Norton and others (Fig. 1.21).
2. *Sky wave propagation*, using the physical conditions of the ionosphere to reflect and refract the radio waves back toward earth (Fig. 1.22).

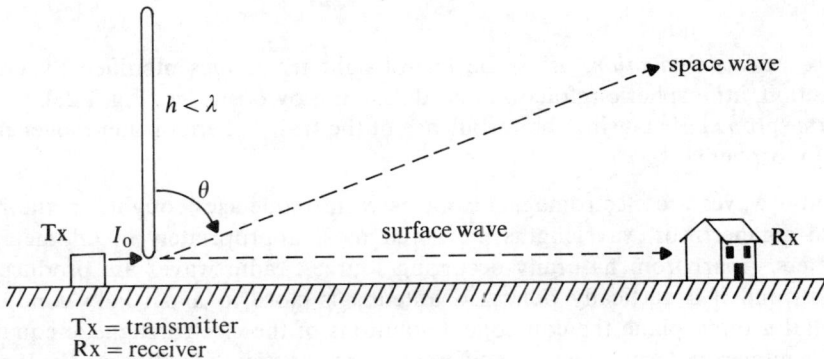

Fig. 1.21 The ground wave consists of the surface wave which is maximum when θ is 90° and the space wave which tends towards zero when θ is 90°. For propagation along the horizontal plane the surface wave is predominant.

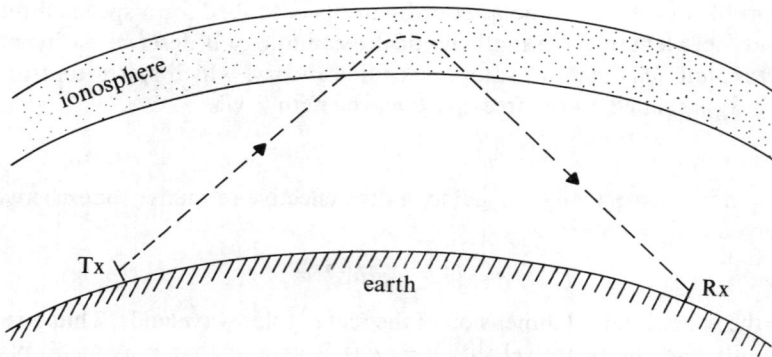

Fig. 1.22 Sky wave propagation occurs when the ionospheric conditions permit the reflection or refraction of the radio wave back towards earth.

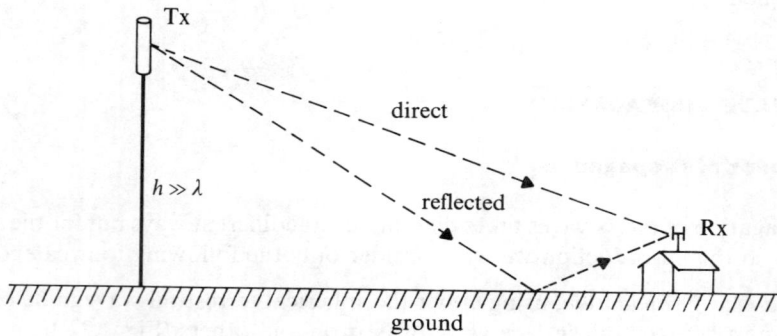

Fig. 1.23 The space wave is predominant when the transmitting antenna is several wavelengths or more above the horizontal ground plane. The signal at the receiving antenna is the resultant of the direct and ground-reflected components.

3. *Space wave propagation*, using the line-of-sight techniques modified by ground reflection, atmospheric refraction and diffraction by obstacles (Fig. 1.23).
4. *Scatter propagation*, using the turbulence of the troposphere or inhomogeneity of the ionosphere.

Radio waves are electromagnetic waves with frequencies roughly in the range hertz to gigahertz or wavelengths measured most appropriately in kilometers to millimeters. Apart from naturally occurring sources, radio waves are produced by antennas, and if a vertically polarized antenna is close to a partially conducting horizontal ground plane the consequent solutions of the electromagnetic equations comprise numerous terms due to the different characteristics of the two media. If all the near-field expressions are ignored, three terms remain, and these are interpreted as the surface wave, direct wave and reflected wave components of the ground wave.

The surface wave is such that its magnitude is greatest when $\theta = 90°$ (along the plane of the horizontal boundary) and diminishes rapidly for $\theta < 90°$. By contrast, the sum of the direct and reflected waves tends to zero when $\theta = 90°$, particularly when the antenna is electrically close to the ground plane, but increases in magnitude for $\theta < 90°$. The combination of the direct and reflected waves is the space wave. If the antenna is displaced a few wavelengths or more above the ground plane, the space wave predominates.

Sky wave propagation is a special case of space wave propagation which occurs at frequencies usually less than about 30 MHz, though the mechanism is occasionally observed well above this limit. The space waves are directed towards the ionosphere, an ionized region of the atmosphere roughly 50–400 km above the surface of earth. This is able to bend the propagation path back toward earth under certain circumstances, which arise mainly in the medium frequency (MF) and high frequency (HF) bands.

Sometimes the propagation medium has scattering mechanisms due to the turbulence of the troposphere, the patchiness of the ionosphere or the ionization trails of meteors. The equivalent is common in optics. For example, in a smokey cinema the scattering of light by the dust particles shows up the beam between the projector and the screen.

1.5.2 General Applications

Radio wave transmissions at very low frequencies (VLF), 3–30 kHz, propagate as vertically polarized waves using earth's surface and the ionosphere as two boundaries of the spherical terrestrial waveguide. Because of the diurnal variation of the D-layer, the effective height of the terrestrial waveguide also varies around the earth, depending on the local time of day. The attenuation at VLF is comparatively low and hence the uses of VLF propagation include long-distance worldwide telegraphy and navigation (such as the Omega navigation system). The bandwidth is very small and the data transmission rate is extremely low. The antennas are physically large but electrically small.

Radio waves in the low frequency (LF) band, 30–300 kHz, also propagate as vertically polarized surface waves with fairly low attenuation and are used for long-distance communication and navigation. The availability of increased bandwidth permits radio broadcasting. Such transmitters broadcast over fairly extensive regions and incorporate antennas which are physically large (commonly 100–200 metres high) and often radiate several hundreds of kilowatts.

The medium frequency (MF) band, 300–3000 kHz, makes use of both the surface wave and the sky wave. The surface wave attenuation is higher than that of LF and hence MF surface waves are used mainly for comparatively small regional broadcasting and communication links. Sky wave propagation at medium frequencies is most effective at night when the attenuating D-layer of the ionosphere is at its minimum. Under such circumstances it is possible for an antenna to receive a surface wave from its local transmitter and interference in the form of a sky wave from the same transmitter or a distant transmitter operating at the same frequency. Whether or not the combination affects reception may be estimated via several alternative methods.

The short wave or high frequency (HF) band, 3–30 MHz, is used for beamed communication services via sky wave reflection and refraction by the ionosphere and for long-range broadcasting. Single and multiple hops permit almost worldwide coverage with both vertical and horizontal polarization but fluctuations in the ionosphere cause problems which can impair reception. This can be improved with the aid of diversity techniques. Basic problems include the estimation of the received field strength, the determination of optimum working frequency, the prediction of diurnal variation, etc. The HF band also makes use of ground wave propagation though the attenuation is fairly restrictive. Some coastal communication is possible due to the reduced attenuation over a sea path.

The very high frequency (VHF) band, 30–300 MHz, is usually too high a frequency for ionospheric refraction and use is now made of the line-of-sight space wave propagation. The available bandwidth is adequate for good quality FM radio and some television channels but the overland range of roughly 100 km restricts its use to local coverage. The VHF band is also used for radio communication between vehicles, aircraft and ships, and for medium-range radar and navigation systems. The analysis of the space wave propagation at VHF and for ultra-high frequencies (UHF), 300–3000 MHz, needs to take into account the problems of reflection from ground and buildings, the refraction by the troposphere, the diffraction over hill tops and buildings, and the scattering or multipath effects of buildings and trees. The UHF band is used for color television, radar and navigation, communication systems, etc. (Fig. 1.24).

Besides the refraction caused by the troposphere, a scattering mechanism is also observed and put to good use in tropospheric scatter radio links across land paths or sea paths, commonly around 700–1000 MHz, sometimes at higher frequencies, and usually over paths which are beyond line-of-sight. The problems involve the estimation of path loss and consequent reliability, involving the statistical analysis both of the propagation path and the necessary diversity arrangements for high reliability.

The super high frequency (SHF) band, 3–30 GHz, is used for high rate digital links, commonly with spacecraft, microwave radio relays, radars, and also for such comparatively short-range land-based applications as collision-avoidance radars for

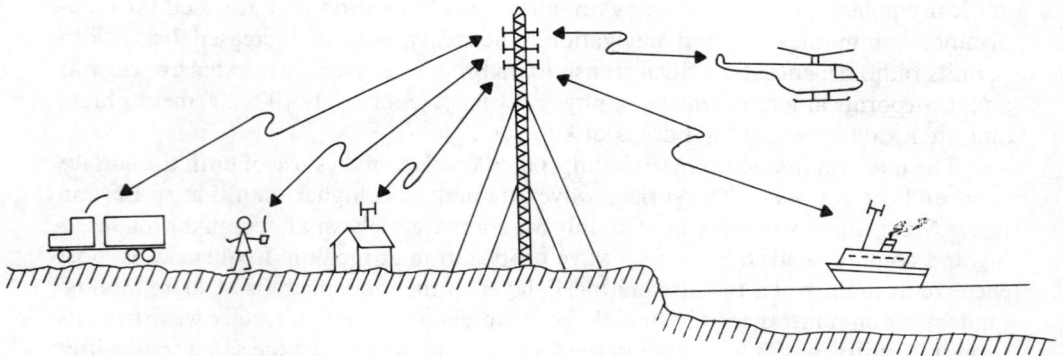

Fig. 1.24 Fixed and mobile radio communication links commonly make use of space wave propagation in the VHF and UHF bands.

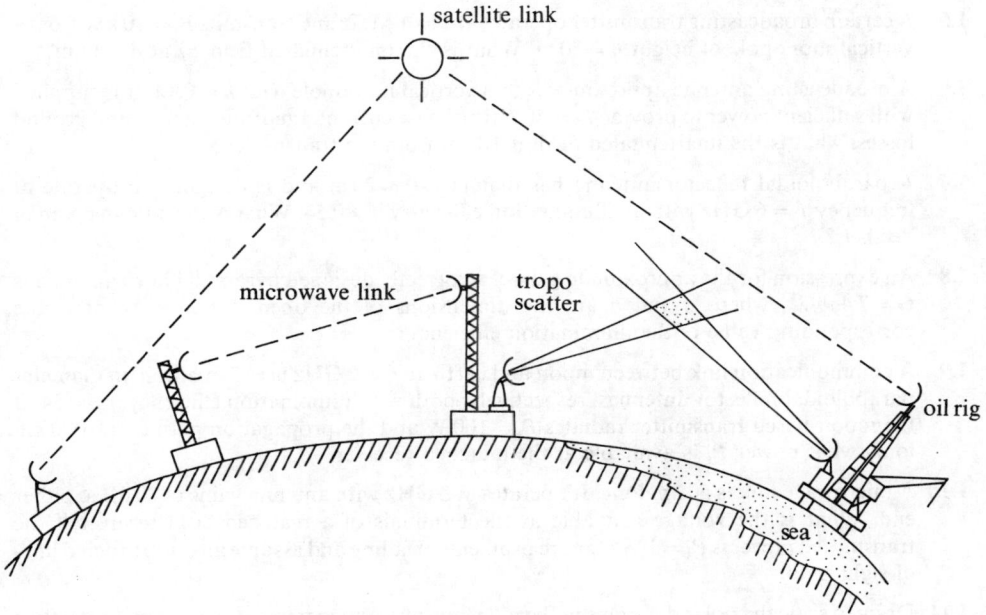

Fig. 1.25 Other examples of the uses of space wave or scatter mechanisms include satellite communication systems, troposcatter radio links, and microwave links.

automobiles, for railway control and communication systems, etc. The problems involve the atmospheric absorption rate at these frequencies, and the effect of rain, snow and clouds on the reliability of propagation.

Three typical applications of radio wave propagation in the UHF and SHF bands are illustrated in Fig. 1.25.

PROBLEMS

1.1 Elliptical polarization is achieved with two halfwave dipoles aligned at 90° to each other to produce $E_x = 20 \cos \omega t$ and $E_y = 40 \cos(\omega t + \pi/2)$. Determine the axial ratio of the ellipse. Is the polarization left- or right-handed for propagation along the z axis?

1.2 What are the electric field strength E_θ, magnetic field strength H_ϕ, electric flux density D_θ, magnetic flux density B_ϕ, and the energies per unit volume stored electrically and magnetically at $d = 1$ km from an isotrope radiating 1 kW of power?

1.3 In a laboratory experiment at 10.0 GHz it is necessary to ensure that a receiving antenna is in the far field of the transmitting antenna. The latter is a pyramidal horn antenna with 10 cm by 10 cm aperture. What is the minimum distance at which the receiving antenna may be placed?

1.4 A halfwave dipole is matched to a receiver at $f = 10$ MHz. What is the voltage V_r at the receiver when the dipole samples an electric field $E = 1 \mu V/m$?

1.5 A certain broadcasting transmitter operates at $f = 1\,\text{MHz}$ and transmits $P_t = 10\,\text{kW}$ from a vertical monopole of height $h = 50\,\text{m}$. What is the unattenuated field E_θ at $d = 1\,\text{km}$?

1.6 A broadcasting antenna approximates to a vertical monopole with $h = 0.1\lambda$. It is supplied with sufficient power to provide $I_0 = 50\,\text{A (rms)}$ base current. Ignoring antenna and ground losses, what is the unattenuated field at $10\,\text{km}$ from the transmitter?

1.7 A paraboloidal reflector antenna has diameter $D = 20\,\text{m}$ and is designed to operate at frequency $f = 6\,\text{GHz}$ with an illumination efficiency $\eta = 0.54$. What is the antenna gain in decibels?

1.8 An expression for the approximate gain of a correctly designed pyramidal horn antenna is $G = 7.4ab/\lambda^2$, where a and b are the dimensions of the open aperture. What is the corresponding value of the illumination efficiency?

1.9 A communication link between moon and earth at $f = 2\,\text{GHz}$ uses $2\,\text{m}$ and $20\,\text{m}$ diameter paraboloidal reflector antennas, respectively, both with illumination efficiency $\eta = 0.54$. If the moon-based transmitter radiates $P_t = 100\,\text{W}$ and the propagation path is $400,000\,\text{km}$ long, what power P_r is available at the matched receiver?

1.10 A microwave link is $50\,\text{km}$ long. It operates at $2\,\text{GHz}$ with antenna gains of $20\,\text{dB}$ at either end. What is the voltage available at the terminals of a matched $50\,\Omega$ receiver if the transmitted power is $P_t = 1\,\text{W}$? Ignore problems of fading and assume an unobstructed line-of-sight path.

1.11 Draw to scale the polar diagrams in the r, θ plane of: (a) an incremental linear radiator; (b) a halfwave dipole; and (c) an ideal vertical monopole of height $h = 0.4\lambda$, and determine the half-power beamwidth of each.

1.12 Derive an expression for the power flux P_a at distance d from a halfwave dipole.

REFERENCE

1.1 Kraus, J. D., *Antennas*, New York: McGraw-Hill Book Company, Inc., 1950.

2
Ground Wave Propagation

Radio wave propagation at the lower end of the radio spectrum is described by several models. Apart from consideration of the near field which is needed when the wavelength is very large, the process involves three types of propagation: ground wave, sky wave and guided electromagnetic wave. The ground wave theory assumes that the propagation path is around a spherical earth with homogeneous parameters, and ignores the effect of the ionosphere. Sky wave theory takes into consideration the effect of the ionosphere where it is appropriate, mainly in the LF and MF bands, but at very low frequencies and low launch angles the similarity between the effective height of the ionosphere and the wavelength of propagation makes the theory of guided waves in spherical waveguides more realistic.

Ground wave propagation operates mainly in the three frequency bands, VLF, LF and MF, or 3 kHz to 3 MHz, with some applications extending into the ELF and HF bands. The range over which ground waves may be used with realistic transmitter powers is determined by the attenuation of the fields, and as attenuation tends to increase with frequency, the ranges of the lower frequencies are likely to be longer than those of the higher frequencies. VLF propagation is virtually world wide, LF propagation is fairly extensive, MF propagation more limited and HF (ground wave) propagation is fairly confined. Communication of information is bandwidth dependent and hence VLF is restricted to low rate telegraphy and narrowband navigation techniques, while LF and MF can be used for radio broadcasting and communication as well as navigation aids. Finally, antenna dimensions are commonly measured in units of wavelength and it should be noted that 3 kHz and 3 MHz are equivalent to wavelengths of 100 km and 100 m, respectively. It is therefore not surprising that ground wave antennas are often physically very large and yet electrically very short.

In this chapter we discuss ground wave propagation more in application than in basic theory, because the surface wave theories of Norton (Ref. 2.1) and others, the guided wave theories of Wait (Ref. 2.2) and others, and the more recent ideas of Rotheram (Ref. 2.3) all have a mathematical content outside the scope of this book.

2.1 GROUND PARAMETERS

When a radio wave is propagated from an antenna close to the ground, such as from a short vertical monopole, the conditions associated with any point on the wave depend on the specific location. The ground absorbs some of the electromagnetic energy, which causes a downward flow of energy from the rest of the wave, and there are boundary conditions to consider. Reflection, refraction and diffraction effects must also be noted.

One of the ground wave propagation theories is based on the assumption of a spherical earth propagation path which has homogeneous electromagnetic parameters and ignores the presence of the surrounding ionosphere. The electric field strength decreases in magnitude because of the spatial spreading of the energy, the absorption by the surface of the earth due to its finite conductivity, and the diffraction loss. The two media involved in this model of ground wave propagation are the atmosphere and earth. These may be represented by the primary parameters (permeability μ, permittivity ε and conductivity σ); the secondary parameters (such as wave impedance Z_i, velocity of propagation v, propagation coefficient γ, attenuation coefficient α, phase coefficient β and refractive index n), which are usually related to the primary parameters via frequency f; and sometimes certain auxiliary parameters which are introduced later. Table 2.1 lists a selection of typical ground parameters.

Within the ground itself, electromagnetic theory indicates two components of current flow. In a homogeneous, lossy dielectric medium having conductivity σ and permittivity ε there is a conduction current density given by

$$J_c = \sigma E$$

and a displacement current density given by

$$J_d = \varepsilon \, dE/dt$$

Converting from time domain to frequency domain with $dE/dt = j\omega E$ gives

$$J = \sigma E + j\omega\varepsilon E \tag{2.1}$$

An auxiliary parameter q is defined as the ratio of the magnitudes of the two component

Table 2.1   Some typical parameters for a selection of ground paths

Ground path	ε_r	σ(mS/m)	f_q(MHz)	q^*
sea water	80	4000	900.0	600.0
fresh water	80	5	1.1	0.8
moist land	15	5	6.0	4.0
	15	20	24.0	16.0
	30	5	2.5	1.7
	30	20	12.0	8.0
rocky land	7	1	2.6	1.7
	7	5	12.9	8.6
dry land	4	1	4.5	3.0
	4	10	45.0	30.0

* The q-factor is calculated at f = 1.5 MHz

current densities:

$$q = |J_c|/|J_d| = \sigma/(\omega \varepsilon_r \varepsilon_0) \tag{2.2}$$

If $q \gg 1$ we can consider the ground to behave mainly as a good conductive medium, but if $q \ll 1$ we must consider it mainly as a good dielectric medium. In between, it is a complex dielectric and the full expressions derived in Appendix 1 must be used.

The frequency at which $q = 1$, the *cross-over frequency* f_q, is a useful parameter to know for a particular ground path and it can be obtained directly from the previous equation in practical units:

$$f_q(\text{MHz}) = 18\sigma(\text{mS/m})/\varepsilon_r \tag{2.3}$$

Table 2.1 lists typical values for f_q for a selection of ground paths and these may be compared with the upper end of the MF broadcasting band, around $f = 1.5\,\text{MHz}$. If the ground path parameters are such that $q \gg 1$ (or equivalently $f_q \gg f$) it is possible to make the approximation that the ground permittivity ε_r does not make any appreciable contribution to the calculation. Otherwise a more detailed analysis is necessary.

From Table 2.1 we see that whereas sea water may be considered highly conductive, with $f_q = 900\,\text{MHz}$ and hence well above the MF band, fresh water is poorly conductive in the MF band. Ground wave propagation over certain moist lands tends to be a little better than over dry land or rocky land, but there is considerable variability and each propagation path should be checked in detail.

2.2 PROPAGATION EQUATIONS

An electromagnetic wave radiated by a vertically polarized incremental linear radiator of height $h \ll \lambda$ above a ground plane contains an electric field (at distance d) which has a vertical component E_z and a horizontal component E_ρ. The expressions for E_z and E_ρ (Ref. 2.1) each contain several terms representing the direct wave, the reflected wave and the surface wave components in terms of the radiation field $(1/d)$, the induction field $(1/d^2)$ and the electrostatic field $(1/d^3)$.

Even at LF and MF the wavelength is still small enough that at the practically important ranges of distance the near-field components can be ignored. For such realistic distances with $d \gg h$ it is possible to simplify the remaining equations by assuming that θ, the angle of the direct wave with respect to the vertical, is 90° and that the reflection coefficient relating the reflected wave to the direct wave is -1. The effect of this is to reduce all the complexity to a very simple expression for the field strength at distance d in Fig. 2.1:

$$E = Z_i \frac{I_0 h_e}{\lambda d} A \tag{2.4}$$

with the retarded function $\exp j(\omega t - \beta d)$ taken as understood, h_e as the effective height of the vertical monopole with base current I_0, and A as an attenuation factor. This is equivalent to the equation in Ex. 1.4 modified by the dimensionless attenuation factor A.

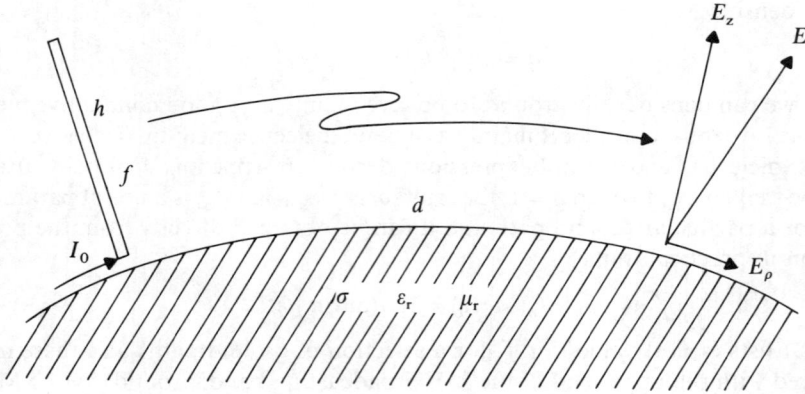

Fig. 2.1 Ground wave propagation depends on the conductivity σ, permittivity ε_r and permeability μ_r of the ground, though the latter is commonly assumed to equal unity. In the far field the electric field has two components, E_z and E_ρ, with E_ρ often neglected.

In the commonly used units of mV/m, kW and km, an alternative expression for the electric field strength at distance d is given by:

$$E = \frac{FM\sqrt{P_t}}{d} A \tag{2.5}$$

as described in Sec. 1.3. When $A = 1$ and $d = 1$ km, $E_1 = FM\sqrt{P_t}$ is called the *unattenuated field at 1 km*, and then Eq. (2.5) becomes:

$$E = E_1 A/d \tag{2.6}$$

with units of mV/m and km. An alternative method is to use the term *cymomotive force* cmf, which is the product of E and d and has the dimensions of volts.

Example 2.1 Electric field at 1 km

Two vertical monopoles are 100 m high and are supplied with 50 A (rms) base current. They operate at 300 kHz and 1 MHz, respectively. What are the radiated powers and the unattenuated fields at 1 km in each case?

The first antenna is a short vertical monopole with $h/\lambda = 0.1$, radiation resistance $R_{rad} = 10(\beta h)^2 = 3.95\,\Omega$ and figure of merit $FM = 300$. The radiated power $P_t = I_0^2 R_{rad} = 9870$ W and the unattenuated field at 1 km $E_1 = FM\sqrt{P_t(kW)} = 943$ mV/m.

The second antenna has $h/\lambda = 0.333$ with $\beta h = 2\pi/3$ radians or 120°. From Sec. 1.4.3 we can deduce loop current $I_L = I_0/\sin 120° = 57.74$ A, and from Fig. 1.16 we obtain the loop resistance as about 80 Ω. Hence the radiated power $P_t = I_L^2 R_L = 266.7$ kW. We can use Eq. (1.28) to obtain the figure of merit as 318 and hence deduce the field at 1 km as $E_1 = FM\sqrt{P_t(kW)} = 5193$ mV/m.

2.2.1 Attenuation Factor *A*

It has been shown in Sec. 2.1 that if the auxiliary parameter q is sufficiently large for the effect of ε_r to be negligible, the ground may be considered as being primarily a conductor. Under these circumstances it has been shown in Ref. 2.1 that the *attenuation factor* may be determined via the expressions

$$A = \frac{(2 + 0.3p)}{(2 + p + 0.6p^2)} \tag{2.7}$$

and

$$p = 0.582d(\text{km})f^2(\text{MHz})/\sigma(\text{mS/m}) \tag{2.8}$$

where the auxiliary parameter p is sometimes called the *numerical distance*. This value of the plane earth ground wave attenuation factor (Fig. 2.2) ignores the effect of diffraction and the ground permittivity. However, it gives realistic answers for distances

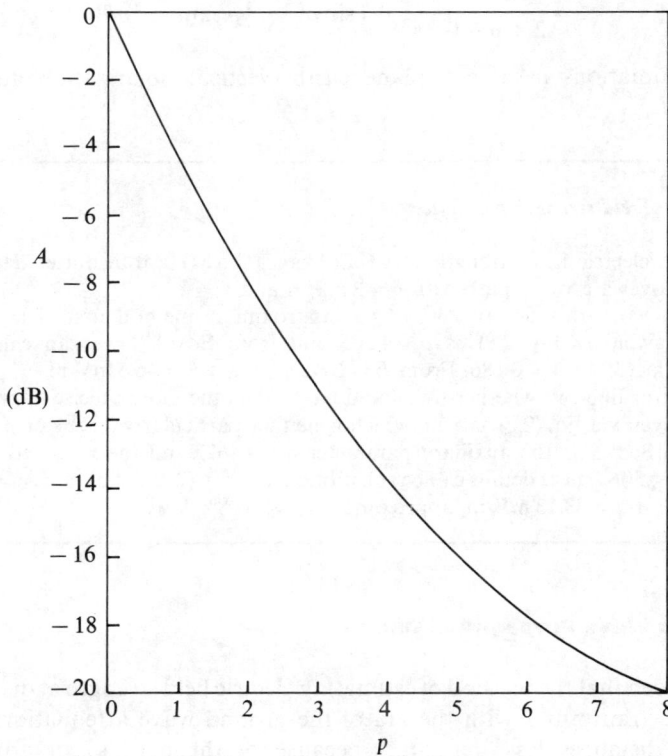

Fig. 2.2 A graph of the approximate values of ground wave attenuation factor A against numerical distance p based on Eq. (2.7). For $p < 3, A = -3.7p$ dB is a good approximation.

less than d_{max}, where

$$d_{max} = 100\, f^{-1/3} \qquad (2.9)$$

and d_{max} is in km if f is in MHz. For $f = 0.5$ to 1.5 MHz, the maximum distance is typically 125 to 90 km.

A slightly more accurate solution (Ref. 2.1) can be obtained within the same limit of distance by incorporating the relative permittivity ε_r of the ground path. This is achieved via yet another auxiliary parameter $b°$ defined by

$$\tan b° = \frac{(\varepsilon_r - 1)f}{18\sigma}$$

with the auxiliary parameter for numerical distance reduced to

$$p = (0.582 d f^2 \cos b°)/\sigma$$

where the practical units of km, MHz and mS/m are used throughout. The expression for A is then changed to

$$A = \frac{2 + 0.3p}{2 + p + 0.6p^2} - [\sin b°]\sqrt{\tfrac{1}{2}p}\,\exp(-5p/8) \qquad (2.10)$$

These approximations assume a plane earth, vertical polarization and distances restricted to d_{max}.

Example 2.2 Field strength prediction

Estimate the electric field strength at 10 km from a 1.5 MHz transmitter having $E_1 = 2500$ mV/m over a ground path with $\sigma = 3$ mS/m and $\varepsilon_r = 7$.

We can work this out two ways, the approximate method first. The numerical distance is given by Eq. (2.8) as $p = 4.365$ and from Eq. (2.7) we can calculate the attenuation factor as $A = 0.186$. From $E = E_1 A/d$ we get $E = 46.5$ mV/m.

Now let us find out whether we should have taken the more precise approach. The q-factor is given via Eq. (2.2) as 5.14, which is neither particularly large nor particularly small. From Sec. 2.2.1 the auxiliary parameter $b = 9.462°$ and the modified numerical distance $p = 4.306$. These details can be substituted into Eq. (2.10) to obtain $A = 0.172$ and hence $E = E_1 A/d = 43.13$ mV/m, approximately 7% to 8% lower.

2.2.2 Ground Wave Propagation Charts

We have just seen that the method of estimating electric field strength at distance d from an LF or MF transmitter with the aid of the ground wave attenuation factor A is restricted to distances less than d_{max} because of the earth's curvature and the consequent diffraction effects. In his original publication (Ref. 2.1), Norton modified his plane earth equation with Watson's diffraction formula. Such is the complication of the resulting expression that it is common practice to use instead charts of field strength E

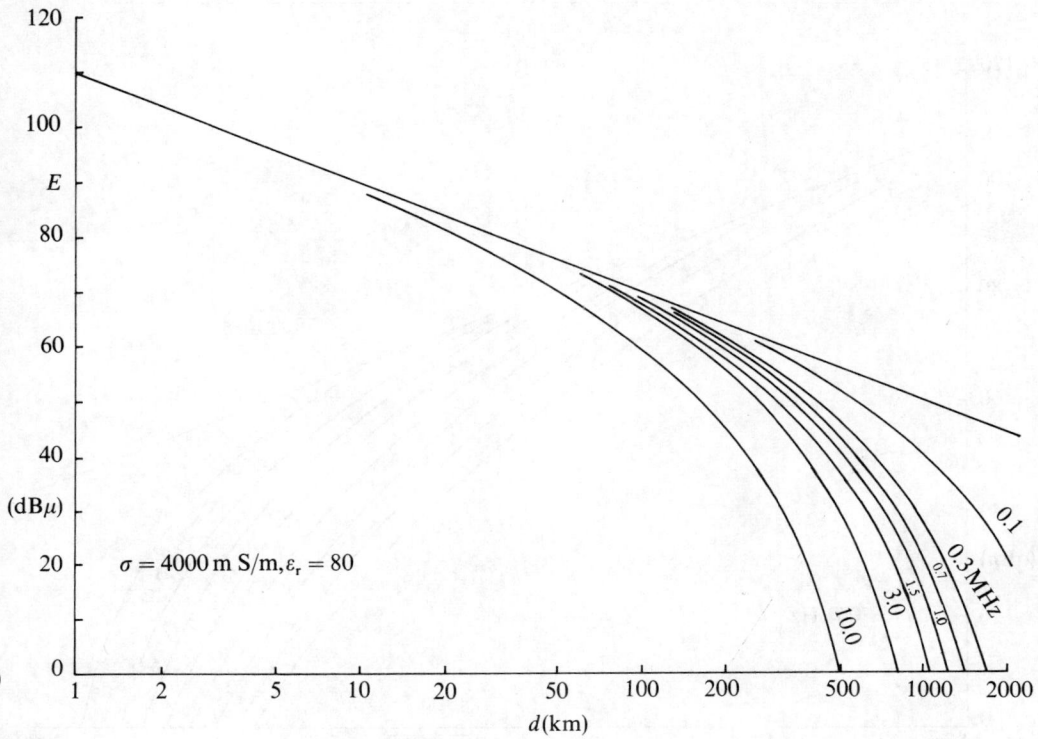

(a)

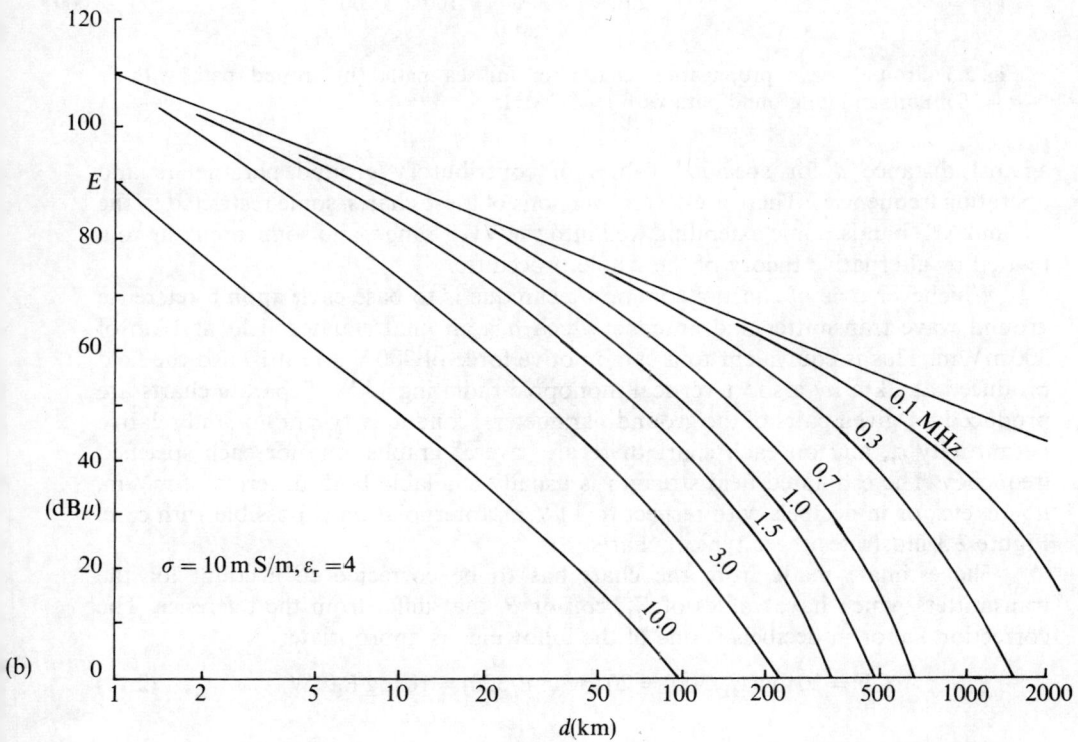

(b)

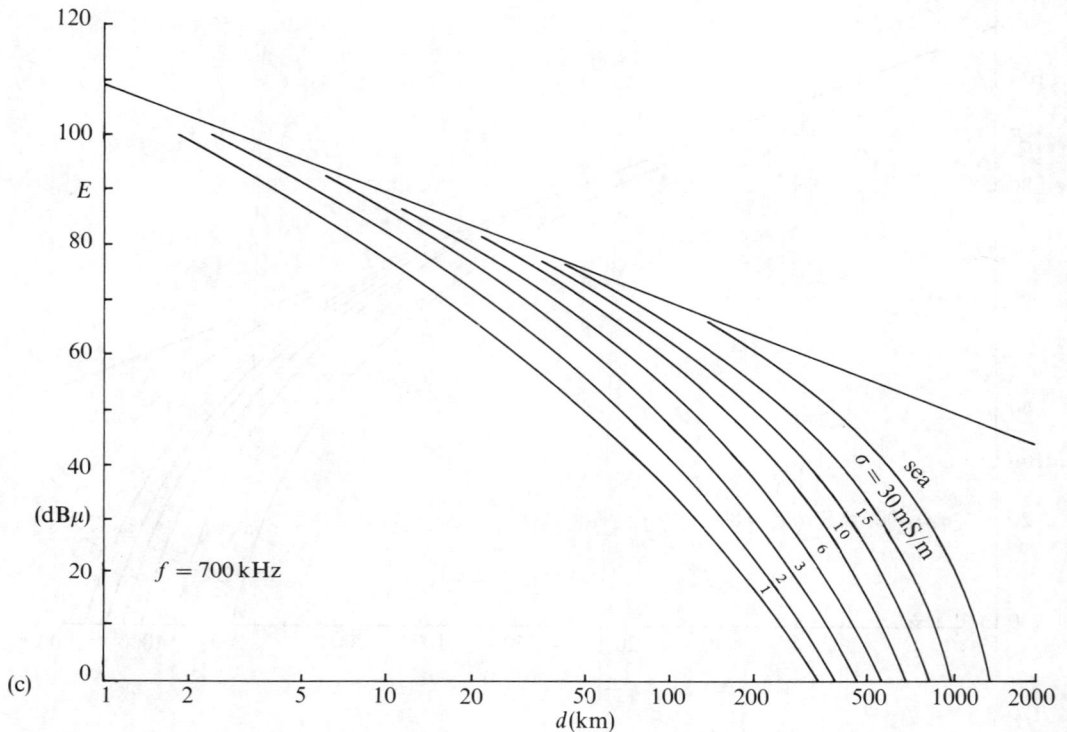

Fig. 2.3 Ground wave propagation charts for (a) sea path, (b) ground path with $\sigma = 10\,\text{mS/m}$, and (c) ground path with $f = 0.7\,\text{MHz}$.

against distance d for specified values of contributory ground parameters and operating frequencies. There are several versions of these charts, some restricted to the LF and MF bands, some extending well into the VHF range, and some applying to a revised or alternative theory of the whole procedure.

Whichever type of chart, a common technique is to base each upon a reference ground wave transmitter and antenna which has an unattenuated field at 1 km of 300 mV/m. This is equivalent to a cymomotive force of 300 V and it is also the field produced at 1 km by a short vertical monopole radiating 1 kW. Separate charts are produced for given pairs of the ground parameters, conductivity σ mS/m and relative permittivity ε_r, and on each chart there are several graphs, one for each specified frequency. The estimated field strength is usually available both in terms of mV/m, μV/m, etc., or in decibels with respect to $1\,\mu$V/m. Interpolation is possible with care. Figure 2.3 illustrates three typical charts.

The estimate made from the chart has to be corrected to account for the transmitters which have values of E_1, cmf or P_t that differ from the reference. The correction factor in decibels is one of the following, as appropriate:

$$\text{CF} = 20\log(E_1/300) = 20\log(\text{cmf}/300) = 10\log P_t(\text{kW}) \tag{2.11}$$

with E_1 in mV/m and cmf in V. The result is a prediction of the ground wave field strength at distance d from the transmitter over a ground path with constant uniform values of conductivity and permittivity at a specified frequency f. The transmitter parameters are usually available in published sources (E_1 and f, for example), distance can be measured on a map, and conductivity is sometimes available from a geological map. The propagation curves give typical values of permittivity appropriate to their use but otherwise a useful approximation is

$$\varepsilon_r = 20(1 - e^{-\sigma/4}) \tag{2.12}$$

with conductivity in mS/m.

Example 2.3 Use of propagation chart

What is the field strength at 100 km from a transmitter having $E_1 = 4500$ mV/m and $f = 1$ MHz if the path parameters are $\sigma = 10$ mS/m and $\varepsilon_r = 4$?

First select the propagation chart for the correct ground parameters (Fig. 2.3b) and note the curve corresponding to a frequency of 1 MHz. When distance is 100 km (on the horizontal axis), the predicted electric field strength is given as $E = 52$ dBμ (on the vertical axis) at the intersection of the $d = 100$ km line and the $f = 1$ MHz curve.

A correction factor is necessary because the chart assumes $E_1 = 300$ mV/m whereas the actual transmitter has $E_1 = 4500$ mV/m. The correction factor is CF = $20 \log(4500/300) = 23.5$ dB. Hence the predicted field is about 75 dBμ at $d = 100$ km.

2.3 PROPAGATION PROBLEMS

2.3.1 Land–Sea Boundary

The use of ground wave propagation charts in the estimation of field strength at distance d assumes constant values of σ and ε_r. In practice this is not realistic, but the effect of slight variations over the total path length is normally not too great. For instance, two estimates could be made, one with slightly optimistic parameters and one with slightly pessimistic parameters, and a rough indication of the possible variation in field strength levels can be obtained.

However, in the case of a mixed land–sea path there is a very abrupt and drastic change in both σ and ε_r at the boundary between land and sea. The propagating wave, adjusted to a downflow of electromagnetic energy to account for ground losses associated with (say) $\sigma = 10$ mS/m, is not immediately able to readjust itself to the very much lower 'ground' losses associated with the $\sigma = 4000$ mS/m of the sea, and a temporary build-up of energy is observed near the surface of the sea. This enhances the electric field strength and is called the *recovery effect*. For ground wave propagation in the reverse direction across a sea–land boundary, the opposite situation exists and the consequent reduction in the electric field immediately after the boundary is called the *loss effect*.

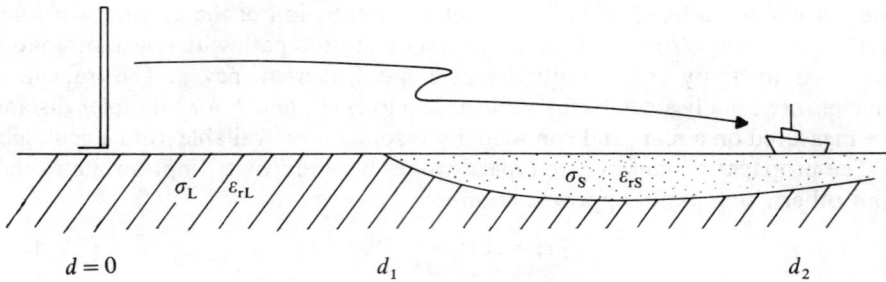

Fig. 2.4 A ground wave communication link across a land–sea path with a distinct change in ground parameters at the boundary.

A simple technique proposed by Millington (Ref. 2.4) may be used in the analysis of radio wave propagation over mixed paths. For the case of the land–sea path illustrated in Fig. 2.4, select the two propagation charts for σ_L and σ_S and note on each the particular propagation curve for frequency f, initially assuming $E_1 = 300 \, \mathrm{mV/m}$ and leaving the correction until later. Millington suggested that we first make an estimate of the field at the receiver in the *forward* direction, as illustrated, ignoring the recovery effect. This will give $E_f \, \mathrm{dB}\mu$. Then we reverse the locations of the transmitter and receiver and make an estimate of the field at the receiver in the *reverse* direction, this time ignoring the loss effect. The result is $E_r \, \mathrm{dB}\mu$. Finally, a good estimate of the actual field at the receiver is the arithmetic mean of the two predicted values as the loss effect is then likely to cancel the recovery effect over two identical paths. In other words,

$$E = (E_f + E_r)/2 \tag{2.13}$$

in decibels with respect to $1 \, \mu\mathrm{V/m}$. If the estimates are made in mV/m, the resultant is the geometric mean of E_f and E_r. The correction factor $\mathrm{CF} = 20 \log \, (E_1/300) \, \mathrm{dB}$ is added to the predicted value to account for a transmitter with parameter other than $300 \, \mathrm{mV/m}$.

When taking a reading from the propagation chart for the sea path in Fig. 2.4, the origin is at the transmitter and not at the boundary between land and sea. The inverse distance law still applies to $d = 0$ at the transmitter. The technique can be extended to, say, a land–sea–land path but it is obviously one step more complicated.

Figure 2.5 illustrates the two propagation curves for the land path with σ_L and the sea path with σ_S for a specified frequency f. The length of the land path is d_1 km and the length of the sea path is $d_2 - d_1$ km. If the entire propagation path had been across d_2 km of sea, the predicted field strength would be $E_S \, \mathrm{dB}\mu$. However, because the first d_1 km of the path was actually across land and not across sea, there must be a reduction of *land loss* LL in the prediction to account for this fact. Hence a better estimate of the received field strength is

$$E_f = E_S - \mathrm{LL} + \mathrm{RE} \tag{2.14}$$

The reverse path is illustrated in Fig. 2.5 and this shows that there is now a

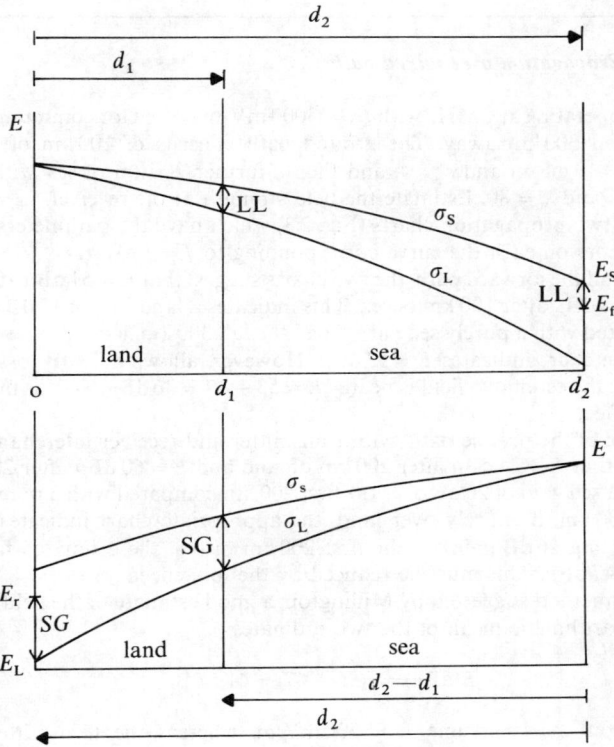

Fig. 2.5 Sketches of the ground wave propagation curves for σ_s and σ_L at frequency f in forward and reverse directions. The forward prediction is $E_f = E_s - LL + RE$ and the reverse prediction is $E_r = E_L + SG - LE$. If $RE \approx LE$, the best estimate of the field is $E = \frac{1}{2}(E_s + E_L + SG - LL)\,dB\mu$.

corresponding *sea gain* SG which must be added to the original prediction E_L and that a better estimate of the field strength is

$$E_r = E_L + SG - LE \qquad (2.15)$$

In these expressions, RE is the unknown recovery effect and LE is the unknown loss effect.

2.3.2 Transmitter Coverage for Broadcasting

One use of ground wave propagation at LF and MF is radio broadcasting, and it is useful to be able to estimate the range of each transmitter under specified conditions. No precise answer can be obtained as the receivers vary in quality and the noise levels

Example 2.4 Propagation over mixed paths

A transmitter operating at 1 MHz with $E_1 = 3000\,\text{mV/m}$ is used for communication with a receiver situated 300 km away. The ground path consists of 100 km of ground with parameters $\sigma = 10\,\text{mS/m}$ and $\varepsilon_r = 4$, and then a further 200 km of sea with parameters $\sigma = 4000\,\text{mS/m}$ and $\varepsilon_r = 80$. Estimate the field strength at the receiver.

Select the two propagation charts (Fig. 2.3) which have the parameters given in the question and note on each the curve corresponding to $f = 1\,\text{MHz}$.

In the case of the forward path, the two charts suggest that $E = 51\,\text{dB}\mu$ after 100 km of land and $E = 68\,\text{dB}\mu$ after 100 km of sea. This indicates a land loss of 17 dB over the first 100 km compared with a purely sea path. Over the full 300 km, if entirely over a sea path, the appropriate chart indicates $E = 53\,\text{dB}\mu$. However, allowing 17 dB loss for the first 100 km of land, the estimated field is nearer $E = 53 - 17 = 36\,\text{dB}\mu$. To this must be added the recovery effect.

In the case of the reverse path, with transmitter and receiver interchanged, the two charts suggest that $E = 34\,\text{dB}\mu$ after 200 km of land and $E = 60\,\text{dB}\mu$ after 200 km of sea. This indicates a sea gain of 26 dB over the first 200 km compared with a purely land path. For the full 300 km, if entirely over land, the appropriate chart indicates $E = 21\,\text{dB}\mu$. However, allowing 26 dB gain for the first 200 km of sea, the estimated field is nearer $E = 21 + 26 = 47\,\text{dB}\mu$. This must be reduced by the loss effect.

Using the method suggested by Millington, a good estimate of the field at the actual receiver is the arithmetic mean of the two estimates:

$$E(\text{dB}\mu) = \tfrac{1}{2}(36 + 47) = 41.5\,\text{dB}\mu$$

The propagation charts assume $E_1 = 300\,\text{mV/m}$, whereas the transmitter had $E_1 = 3000\,\text{mV/m}$, which is 20 dB higher. Hence the final correction gives the estimated field at the receiver as $61.5\,\text{dB}\mu$ or 1.2 mV/m.

differ considerably with environment. In addition, good reception is a matter of subjective assessment.

The reception of radio signals is affected by receiver noise, natural noise and man-made noise. Most practical receivers are adequately designed so that receiver noise is no problem except for the smaller and cheaper portable models. Natural noise or atmospheric noise varies greatly with time, season and location. It is caused mainly by the radiation from terrestrial lightning discharges and other natural electrical disturbances, and is guided by the ionosphere over considerable distances. For this reason it tends to be worse at night than during the day and a few decibels higher in summer than in winter. It is also much greater in tropical areas than in mid-northern latitudes because of the larger number of thunderstorms in those climates, and very much less in the Arctic and Antarctic regions. Man-made electrical noise in the LF and MF bands arises from the use of numerous forms of electrical equipment and machinery. In industrial areas and large cities it is at its greatest but its mode of propagation is such that it does not extend over very long distances. Consequently the reception of radio signals in rural areas is much easier.

Example 2.5 Electrical field strength and noise level

A receiver is connected to a short vertical monopole and is matched over the MF range. If there is zero incident field, the receiver output is due to its own thermal noise level. If this arrangement is equated to a noise-free receiver and an equivalent noise field strength E_n at the antenna, what value of E_n would you expect at 1 MHz for a 5 kHz bandwidth receiver with 10 dB noise figure at reference temperature $T_0 = 290$ K?

Appendix 2 shows that the thermal noise power of a receiver is given in terms of Boltzmann's constant k (J/K), the reference temperature T_0 (K), the noise bandwidth B (Hz) and the noise figure F as:

$$N = kT_0BF \tag{2.16}$$

in watts. If this is equated to the input power of a noise-free receiver, Eq. (1.8) may be used to obtain:

$$kT_0BF = \frac{V^2}{4R_{\text{rad}}} = \frac{(Eh_e)^2}{4R_{\text{rad}}}$$

We have seen in Sec. 1.4.3 that the effective height of a short vertical monopole is $h_e = h/2$ and that its radiation resistance is $R_{\text{rad}} = 10(\beta h)^2$. Simple substitution results in:

$$E^2 = 640\pi^2 \frac{kT_0BF}{c^2} f^2 \tag{2.17}$$

where $k = 1.38 \times 10^{-23}$ J/K. For this particular example only, direct substitution of data gives:

$$E^2 = 1.404 \times 10^{-29} f^2$$

or

$$E(\text{dB}\mu) = -48.5 + 20 \log f(\text{MHz})$$

and the numerical solution, when $f = 1$ MHz, is $E_n = -48.5$ dBμ.

We have seen in Ex. 2.5 that the effect of thermal noise is normally insignificant in the reception of LF and MF radio waves. At these frequencies it is the natural and man-made noise levels which are the more important, and Figs. 2.6 and 2.7 illustrate the order of magnitude of each. In mid-northern latitudes, for example, it may be necessary to provide field strengths of the order 0.1 to 50 mV/m to obtain very good reception, depending on the frequency and on the noise environment of the reception area.

Thus if we are to make a prediction for transmitter coverage, it is first necessary to decide upon the minimum field strength which is required to provide a specified standard of reception. If the transmitter is aimed at a sparsely populated area, the minimum field strength required is likely to be low, but if intended to broadcast primarily to a city or industrial region, the minimum field strength needed is likely to be much higher. Whatever the level, let us call it E_m. We can make a rough estimate of the transmitter coverage by using Eqs. (2.6) and (2.8) to obtain:

$$\frac{p}{A} = \frac{0.582 f^2 E_1}{\sigma E_m} \tag{2.18}$$

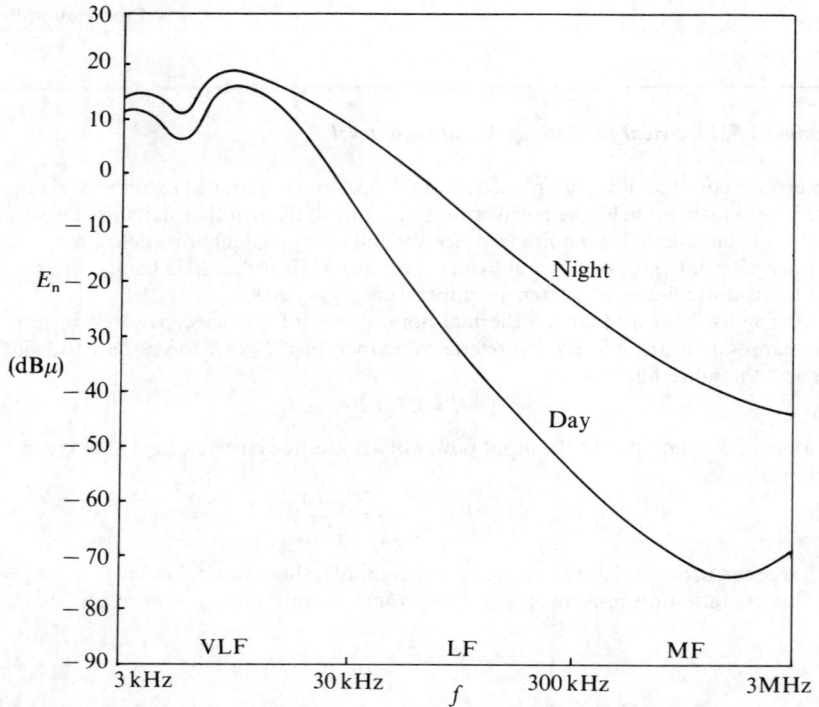

Fig. 2.6 Sketch of the typical shape of the atmospheric noise field E_n per hertz bandwidth at a specified location in winter. If the receiver has noise bandwidth B Hz, the noise field strength is $E_n + 10 \log B$. This is in addition to the receiver noise which is commonly negligible by comparison at VLF-LF-MF.

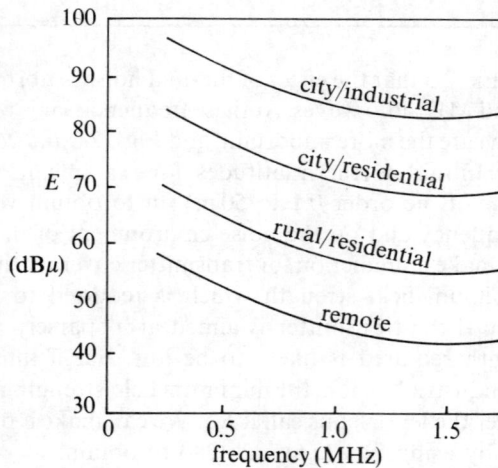

Fig. 2.7 The order of magnitude of field strength needed for good quality reception of radio broadcasts in mid-latitudes in the presence of receiver, atmospheric and man-made noises.

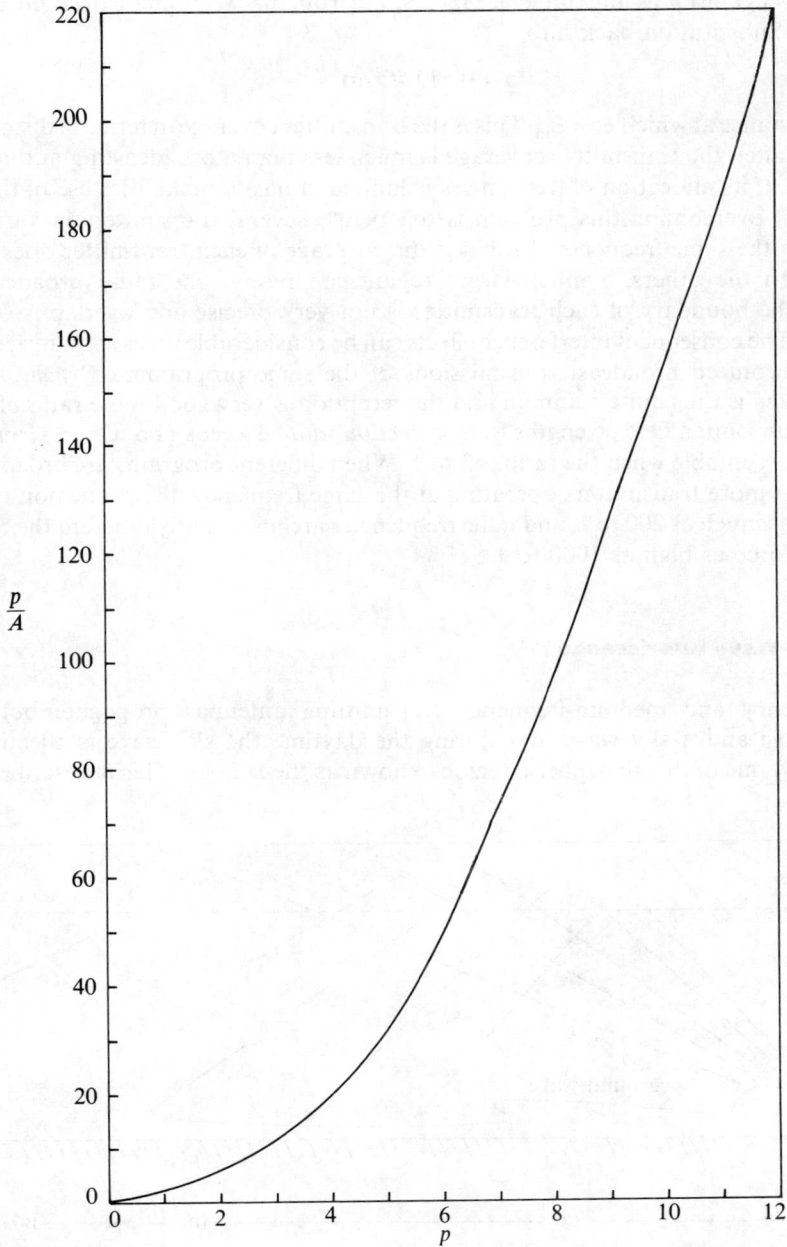

Fig. 2.8 Graph of the ratio p/A against numerical distance p for use with Eq. (2.18) in estimating transmitter coverage.

in appropriate units of MHz, mV/m and mS/m. With the aid of Eq. (2.7) we can plot a graph of p/A against p, as illustrated in Fig. 2.8, and from this we estimate the numerical distance p. Substitution back into

$$p = 0.582 \, df^2/\sigma$$

gives the distance at which $E = E_m$. This is the transmitter coverage in terms of distance.

Quite often the transmitter coverage is much less than a broadcasting authority might wish. If its allocation of frequencies is limited, it has to make best use of them. One way of overcoming this problem is to operate several transmitters in various locations at the same frequency. Provided the coverage of each transmitter does not overlap with the others, a much larger region can receive the radio broadcasts. However, the boundary of each transmitter is not very precise and overlap is often inevitable. The consequent interference effects can be considerable unless minimized by using synchronized broadcast transmissions of the same programme. *Synchronous group working* is thus quite common and the reception is very good if the ratio of the wanted to unwanted field strengths (the *protection ratio*) exceeds about 5 to 1, and is even quite acceptable when the ratio is 3 to 1. When different programs are broadcast from two or more transmitters operating at the same frequency, the protection ratio can rise to as much as 200 to 1, and if the frequencies are just slightly different the ratio may need to be as high as 1000 to 1.

2.3.3 Sky-Wave Interference

Low-frequency and medium-frequency transmitting antennas propagate both a ground wave and a sky wave, but during the daytime the sky wave is effectively absorbed by one of the ionospheric regions known as the *D-layer*. This is described in

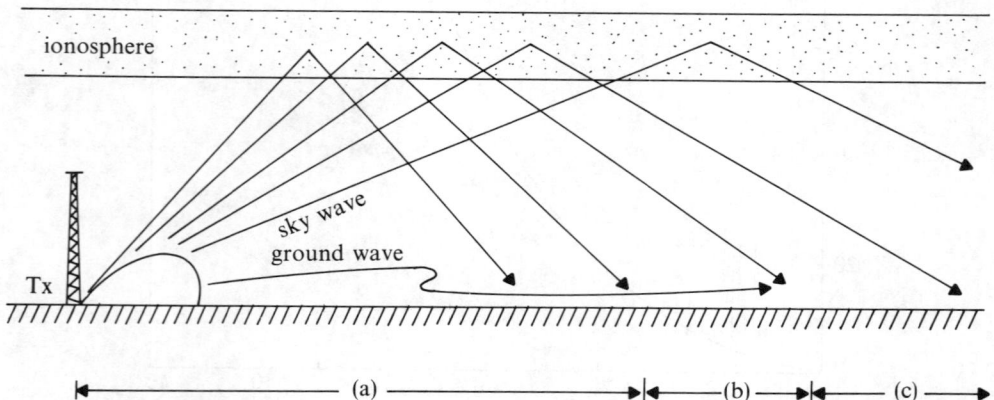

Fig. 2.9 The interference between sky wave and ground wave from the same transmitter. The reception is principally that of ground wave in zone (a) and of sky wave in zone (c), but the two modes produce interference in zone (b). A flat earth model is used for illustration.

more detail in Chap. 3 but it is sufficient here to describe it as an ionized region in the upper atmosphere which occurs mainly during the daylight hours and decays rapidly after sunset. The level of ionization is such that it behaves very effectively as an attenuator of LF and MF radio waves. Thus, after sunset, when the D-layer has diminished, the transmitting antenna produces two modes of propagation, as illustrated in Fig. 2.9, and the received field strength is the combination of the two.

The polar diagram of the transmitting antenna indicates a maximum value of the launch angle and, for a given virtual height of the reflecting point in the ionosphere, there is a corresponding minimum distance known as the *skip distance* before which the sky wave is received back at ground level. Between the transmitter and the skip distance the ground wave reception predominates. Thereafter there is interference between the two propagating modes which can drastically impair reception quality, until eventually the level of the sky wave mode exceeds the level of the ground wave mode by the protection ratio necessary for good reception. For distances greater than this value the sky wave mode gives good reception until its field strength becomes too low.

Such a condition obviously refers to synchronous signals as they actually come from the same transmitting antenna. It was mentioned in Sec. 2.3.2 that a broadcasting authority may use several synchronized transmitters in different locations to achieve wider coverage. These may cause similar interference both between their ground waves and their sky wave at night. The situation becomes worse when the interference is from a transmitter belonging to a different broadcasting authority operating at almost the same frequency and with a different programe. The consequence can then be quite drastic. Figure 2.10 illustrates the situation.

The calculations involving sky wave propagation are described in more detail in the next chapter, but one very simple method can be described here. For those antennas which approximate to a short vertical monopole, the field strength at a distant location can be predicted if the ionospheric height and the transmitter power are known and the skywave propagation is assumed to suffer no attenuation. Experience has shown that the field strength should be reduced to about one-quarter this level to account for

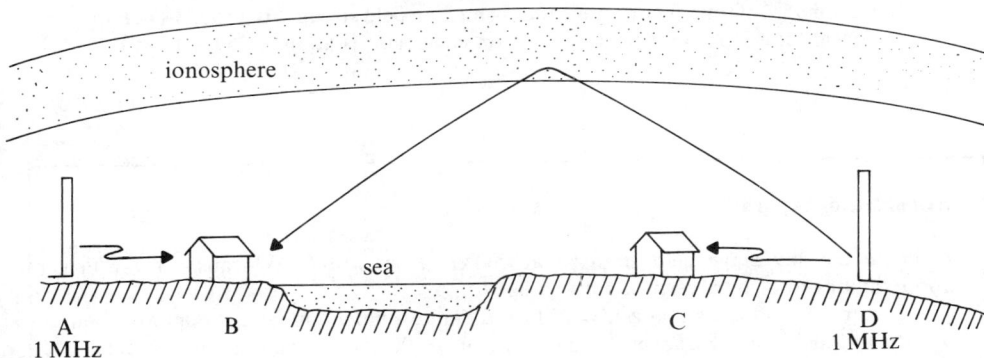

Fig. 2.10 Night-time interference from the sky wave produced by a non-synchronized transmitter (D) operating at the same frequency as the wanted transmitter (A) but broadcasting a different program.

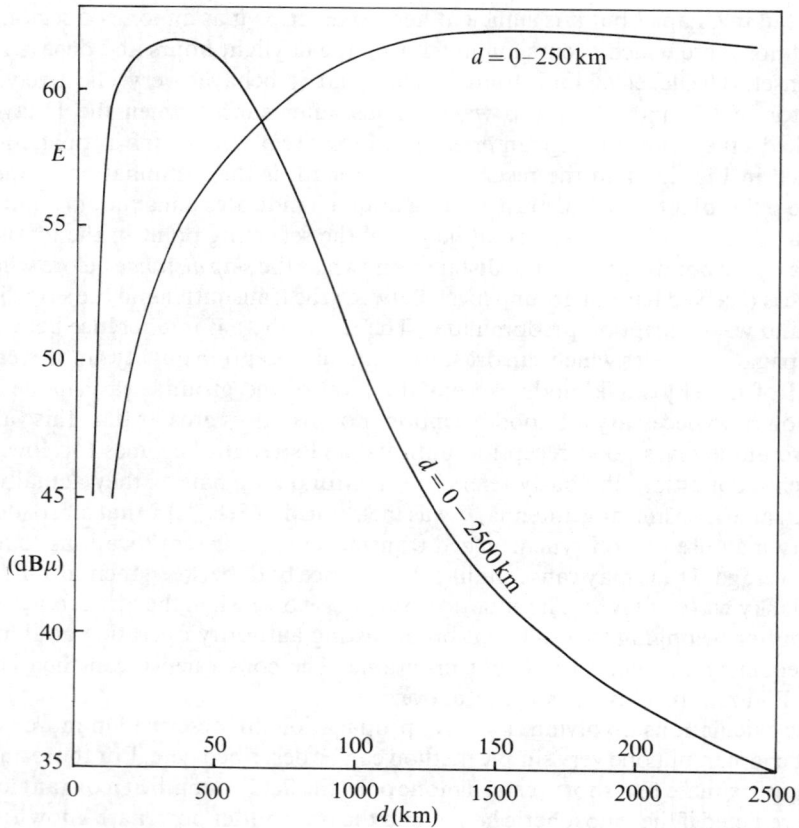

Fig. 2.11 A curve of quasi-maximum sky wave field strength against distance assuming 1 kW radiated from a short vertical monopole and unity ionospheric reflection coefficient. The signal is greater than the quasi-maximum value for 5% of time, or less than the quasi-maximum for 95% of time. The corresponding median field strength is about 8–9 dB lower and a typical long-term ionospheric reflection coefficient is about 0.25 (-12 dB).

Example 2.6 Sky wave interference

Estimate the sky wave field strength at 700 km from a 100 kW transmitter which approximates to a short vertical monopole.

Figure 2.11 indicates that at $d = 700$ km the quasi-maximum field strength is about 57 dBμ for each kW of transmitted power. For a 100 kW transmitter the estimate is therefore 77 dBμ. Typically the ionospheric reflection coefficient is 0.25 (-12 dB) and hence the estimate is now reduced to 65 dBμ exceeded for 5% of time, or 56 dBμ exceeded for 50% of time with the 9 dB difference between the quasi-maximum and median levels.

typical path losses, and that further modifications are necessary to account for the rapid variations of field with time that is characteristic of a fluctuating medium. Figure 2.11 represents this simplified approach. The curve gives the quasi-maximum field strength, the value exceeded for only 5% of time, at distance d from a short vertical monopole (SVM) transmitting 1 kW with unity ionospheric reflection coefficient. The median level of field strength is found to be about 8–9 dB lower.

2.4 VLF PROPAGATION

We have shown in the case of MF propagation that the received signal is the resultant of the ground wave and sky wave components. At shorter distances the ground wave predominates, and then there is a region of interference until eventually the ground wave attenuation reduces its contribution to well below that of the sky wave. When applied to lower frequencies in the LF and VLF bands the ground wave attenuation is much smaller and the interference pattern extends over a greater distance. Calculations can be made of the combined effect out to a distance of perhaps one or two megameters. As the propagation frequency is even further reduced and the wavelengths can be measured in tens of kilometers, the ionospheric height becomes of the order of a wavelength or so. Under these circumstances the situation is very similar to the propagation of signals along a waveguide with the SVM as the probe. It is a little more complicated in that the waveguide is spherical and its height varies with time and distance due to the diurnal variation of ionospheric height. Thus the theory of waveguide propagation, moding and attenuation can be applied to ground waves in the VLF and ELF frequency bands.

2.4.1 Attenuation Rates

In Sec. 2.2 we saw that the electric field strength at distance d due to ground wave propagation is given by

$$E = \frac{\text{FM}\sqrt{P_t}}{d} A$$

in appropriate units of mV/m, kW and km. The same equation can be written in a different way by using E' as the unattenuated field at distance d and $A' = 1/A$. This gives

$$E = E'/A'$$

or, in logarithmic form:

$$A'(\text{dB}) = E'(\text{dB}\mu) - E(\text{dB}\mu) \tag{2.19}$$

With the propagation equation in this form, the problem now reduces to the calculation of the unattenuated field at distance d (E') and the determination of the path attenuation (A').

The calculation of A' is not simple. The waveguide mode theory indicates that A' depends on such factors as frequency, time of day, time of year, solar activity, path conductivity, location of transmitter and receiver, direction of propagation, latitude, and so on. It also depends upon the waveguide moding. To ease the problem, graphs are available (for example in Ref. 2.5) of attenuation rates Δ_α in dB/Mm for each of the contributory factors and for the principal waveguide modes. The attenuation A'(dB) is then given by

$$A'(\text{dB}) = \alpha d = \sum \Delta_\alpha d \tag{2.20}$$

for each waveguide mode, where the summation is taken for all contributory factors.

Phase velocity is also an important parameter in VLF propagation as the accuracy of certain navigational systems depends on a precise knowledge of phase velocity. This too can be determined from graphs such as those given in Ref. 2.5.

2.4.2 Electric Field Strength

The electric field strength equation for VLF propagation is a little different from the surface wave expressions used at LF and MF. To have some idea of the content of the equation, consider first three particular examples. For an isotrope in free space:

$$P_a = \frac{P_t}{4\pi d^2} = \frac{E^2}{120\pi}$$

or

$$20 \log E = 10 \log 30 + 10 \log P_t - 20 \log d \tag{2.21a}$$

where $20 \log d$ can be interpreted as the spatial loss due to the spreading of the electromagnetic energy uniformly in all directions from the antenna. If the space is confined between two parallel plates separated by a small dimension h_i (Fig. 2.12a):

$$P_a = \frac{P_t}{2\pi d h_i} = \frac{E^2}{120\pi}$$

or

$$20 \log E = 10 \log 60 + 10 \log P_t - 10 \log d - 10 \log h_i \tag{2.21b}$$

with $10 \log d + 10 \log h_i$ now representing the spatial loss within the confines of the parallel plates. Finally, for concentric spheres with radii a and $a + h_i$ ($a \gg h_i$), the cross-sectional area is $2\pi r' h_i$ (Fig. 2.12b), where $r' = a \sin \theta$, and hence

$$P_a = \frac{P_t}{2\pi r' h_i} = \frac{P_t}{2\pi a \sin(d/a) h_i} = \frac{E^2}{120\pi}$$

or

$$20 \log E = 10 \log 60 + 10 \log P_t - 10 \log\{a \sin(d/a)\} - 10 \log h_i \tag{2.21c}$$

with $10 \log\{a \sin(d/a)\} + 10 \log h_i$ representing spatial loss due to dispersal within the confines of two concentric spheres. Figure 2.13 shows how the electromagnetic energy

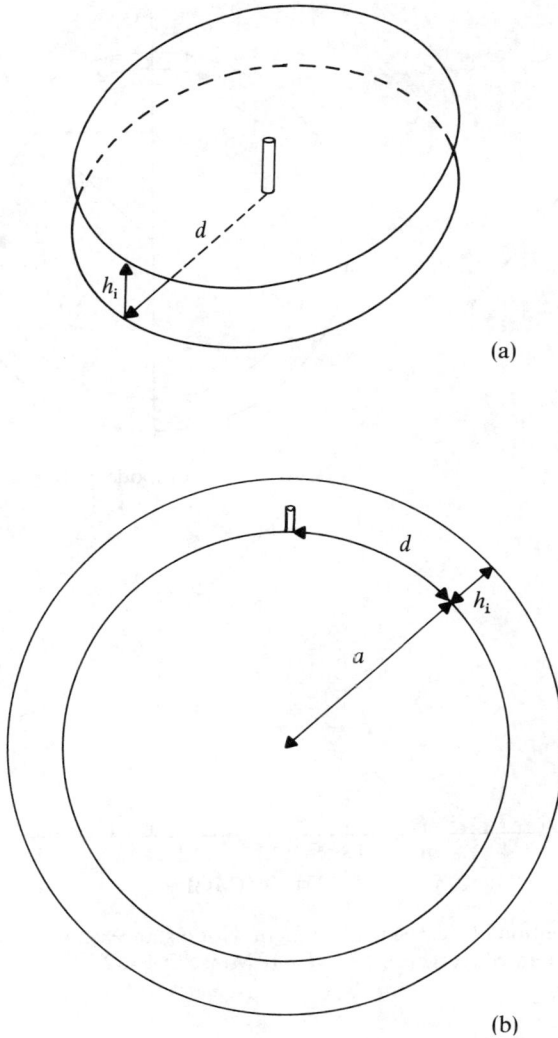

(a)

(b)

Fig. 2.12 Radio wave propagation at VLF (a) between two parallel conductive bound-aries displaced by h_i and (b) between two spherical conductive boundaries (earth and ionosphere) displaced by h_i, where a is the radius of earth. The E-field expressions are given by Eqs. (2.21b) and (2.21c).

initially spreads around half the spherical waveguide before converging again around the other half towards the antipode.

As dimension h_i increases, the explanation is not so simple and the analysis is more complicated. The theory of spherical waveguide moding (Ref. 2.5 for example) is needed to show that the electric field E_{wn} at distance d around a spherical earth of radius

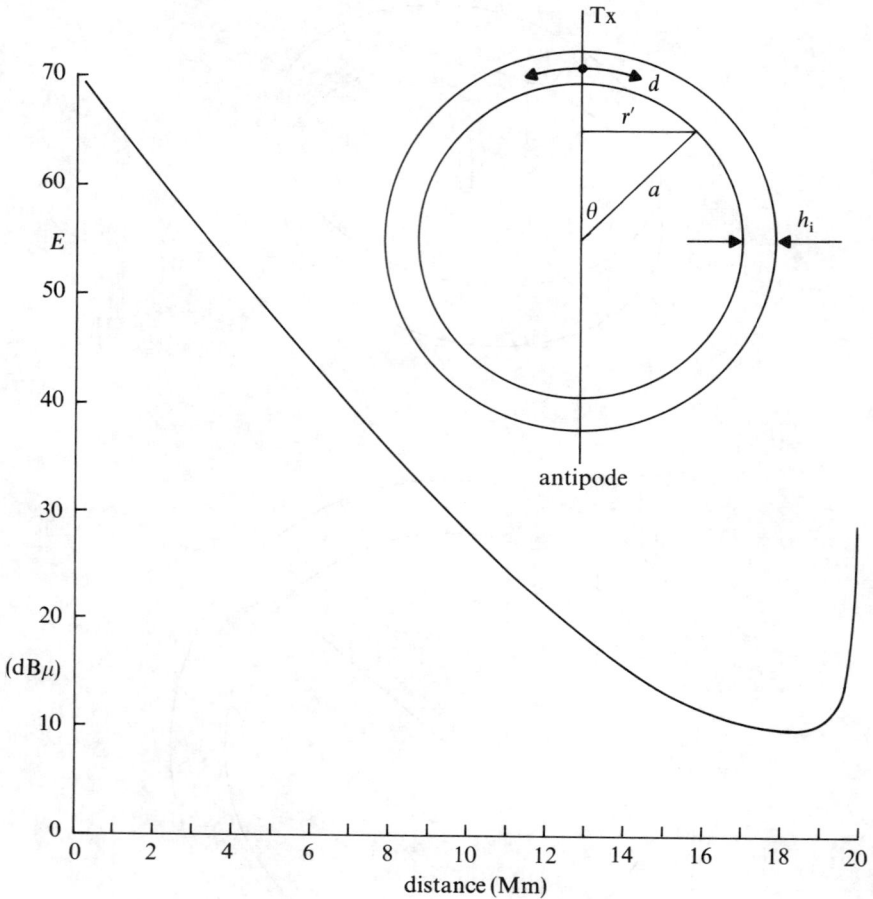

Fig. 2.13 The variation of electric field strength with distance, showing spatial dispersal up to $d = 10$ Mm and convergence after $d = 10$ Mm. See Ex. 2.7.

$a = 6370$ km, for a specific waveguide mode $n = 1, 2$ or 3, is given by:

$$20 \log E_{wn} = 104.3 + 10 \log P_t - 10 \log f - 20 \log h_i + 20 \log EF_n$$
$$- 10 \log \{a \sin(d/a)\} - \alpha d/10^6 \tag{2.22}$$

in SI units throughout, except for α which is in decibels per megameter. The waveguide moding theory introduces the frequency term and an additional component for ionospheric height h_i, and $20 \log EF_n$ is the excitation factor for each waveguide mode. The latter is a measure of the ability of the antenna to excite each component mode, and typical values are available in graphical form in Ref. 2.5.

Example 2.7 VLF field strength prediction around earth

To demonstrate the effect of spatial loss within the confines of two concentric spheres (earth and ionosphere), determine the electric field at distances $d = 0$ Mm to $d = 20$ Mm for a transmitter radiating 10 kW at 10 kHz with $20 \log \text{EF}_n = 1.0$ dB. Although unrealistic over such great distances (due to diurnal variation), assume $h_i = 70$ km and $\alpha = 3$ dB/Mm.

The field strength can be determined from Eq. (2.22) with appropriate substitution of data. If we use E_{wn} in μV/m and distance d in Mm, the logarithmic form of the equation reduces to:

$$E(\text{dB}\mu) = 60.36 - 10 \log \sin(d/6.37) - 3d$$

This equation is plotted as a graph in Fig. 2.13 and this clearly shows the dispersal towards $d = 10$ Mm and the convergence towards $d = 20$ Mm.

2.5 LOW-FREQUENCY ANTENNAS

2.5.1 Transmitting Antennas

Low-frequency transmitting antennas are usually vertical monopoles of large physical dimensions though electrically only a small fraction of a wavelength in height. By using Eq. (1.29), $R_{rad} = 10(\beta h)^2$, we can see that an antenna of height $h = \lambda/2\pi$ has a radiation resistance of 10 Ω while a 10 kHz transmitting antenna, 250 m high, has a radiation resistance of only 0.03 Ω. This gives rise to the definition of *radiation efficiency* η due to

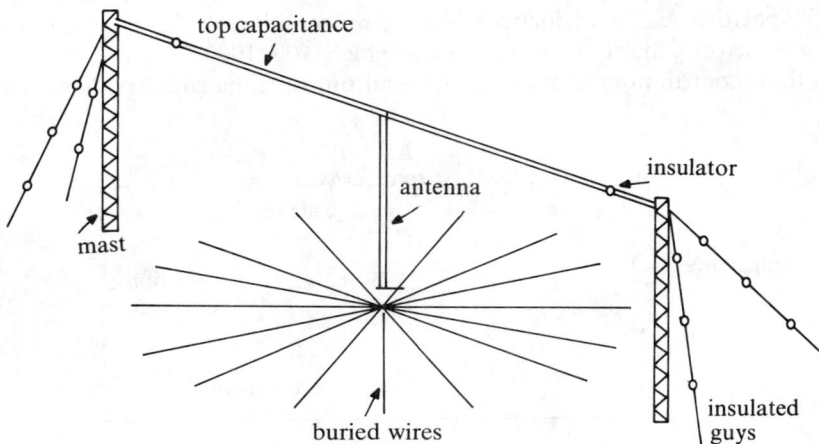

Fig. 2.14 A short vertical monopole transmitting antenna with buried conductors to reduce ground losses and top capacitance to increase the effective height.

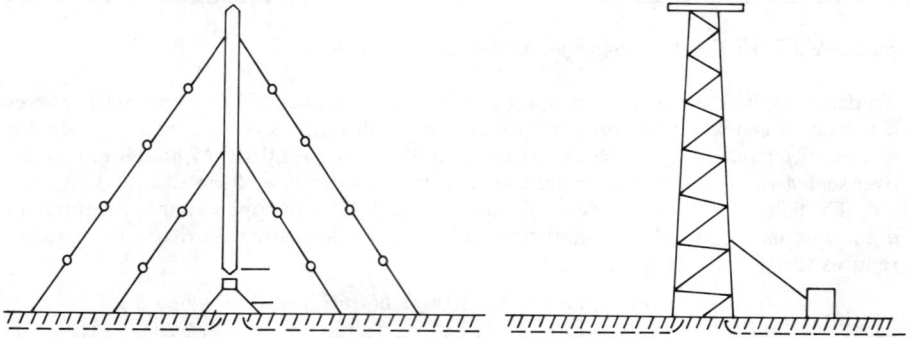

Fig. 2.15 Transmitting antennas at MF include the vertical mast supported by insulated guys or by a self-supporting tower with some top capacitance.

antenna losses as

$$\eta = \frac{R_{rad}}{R_{rad} + R_{loss}} \tag{2.23}$$

and the antenna losses are produced by the vertical antenna itself, the ground return path and the components needed for matching. To reduce the ground loss, a large number of wires are buried in the ground and oriented radially away from the base of the antenna up to a distance at least equal to the antenna height. The structure holding up the antenna, such as guys and non-radiating masts, must also be carefully insulated to reduce their contribution to R_{loss} (Figs. 2.14 and 2.15).

The input impedance of a short vertical monopole is capacitive and this is much larger a reactance than that of the small inductance L_a of the mast or wire. To tune out the net capacitive reactance for matching, a series inductor (L, r) is used, and the resulting Q-factor of the circuit is usually quite high (100–1000). Figure 2.16 illustrates some of these contributory losses in the transmitting antenna circuitry. Note that as a

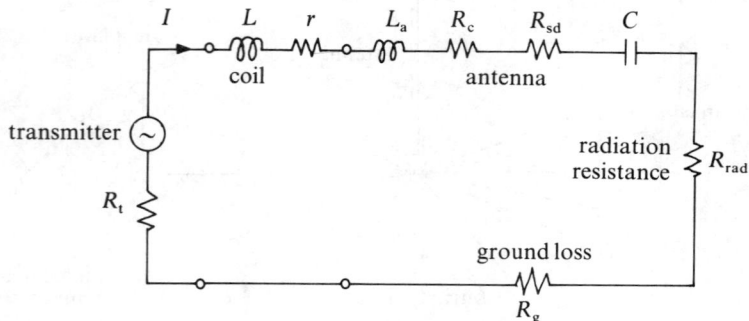

Fig. 2.16 Equivalent circuit of a transmitting antenna with tuning coil (L, r), antenna inductance (L_a), copper loss (R_c), capacitance (C), series dielectric loss (R_{sd}), radiation resistance (R_{rad}) and ground loss (R_g).

result of this high Q-factor the resonant voltages can become sufficiently high to break down the insulation or cause corona effects. There is thus a voltage limit to such antennas. For example, the voltage across the inductance is

$$V_{\mathrm{L}} = \omega L I = \frac{1}{\omega C} I = \sqrt{\frac{L}{C}} I \qquad (2.24)$$

and its magnitude can be reduced if capacitance C is made larger and inductance L is then automatically smaller for resonance.

The electric field strength at distance d is given by Eq. (2.4) and this indicates that the field is directly dependent upon the product of I and h_{e}. For a short vertical monopole this product is represented by the area of the shaded triangle in Fig. 2.17a. However, if some top capacitance loading is added, by extending the wire in the horizontal plane as shown in Fig. 2.17b, the effective product of I and h_{e} is much improved. *Top capacitance* may take the form of wires radiating from the top of the mast or some configuration of loops attached to the top of the vertical antenna.

At medium frequencies, 300 kHz to 3 MHz, where the corresponding wavelengths are 1000 m to 100 m, vertical monopoles can have heights which are appreciable fractions of a wavelength. Equation (2.5) illustrates that the field at distance d for such an antenna is dependent on its figure of merit and Fig. 1.17 shows that this is maximum when h is about five-eighths of a wavelength. However, the polar diagram also varies with antenna height and enhances the unwanted sky wave component. A compromise is generally taken as $h \approx 0.53\lambda$ and this improves the fringe area reception. Simple parallel-structured masts with insulated guys may be used, as well as self-supporting tapered towers, though the latter produce some undesirable electrical problems.

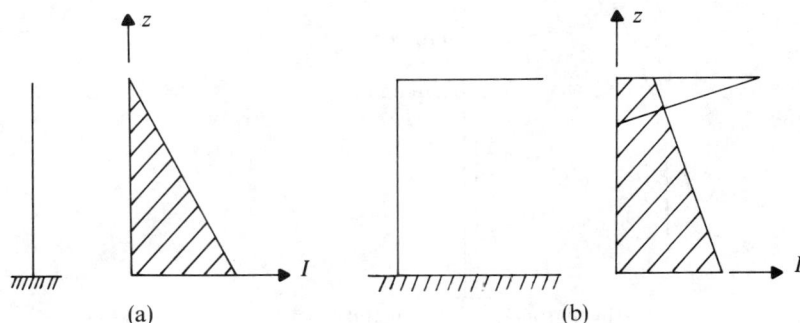

Fig. 2.17 The short vertical monopole (a) has a tapered current distribution, but if the antenna is extended and then bent to form top capacitance, the current distribution in the vertical portion is changed to that illustrated in (b). The far-field component of electric field strength is proportional to the magnitude of the shaded areas and hence increases when top capacitance is added.

2.5.2 Receiving Antennas

A propagating ground wave may be said to have vertical polarization, but close to the surface of earth the *E*-field has a tilt of perhaps 2–3° to the vertical due to the finite conductivity of the ground. Thus the electric field has two components, the ratio of the horizontal to vertical components being about 1 or 2% for good ground and up to 6–8% when the ground conductivity is poor. If a horizontal antenna several wavelengths long is placed about 10 metres above ground level, supported by insulators on poles, the electric field of the ground wave induces small voltages along the line which add cumulatively as the field passes along the length of the wire. The ends of the line are terminated by a matched impedance at one end and a matched receiver at the other, as shown in Fig. 2.18. The traveling wave antenna so produced, sometimes called the

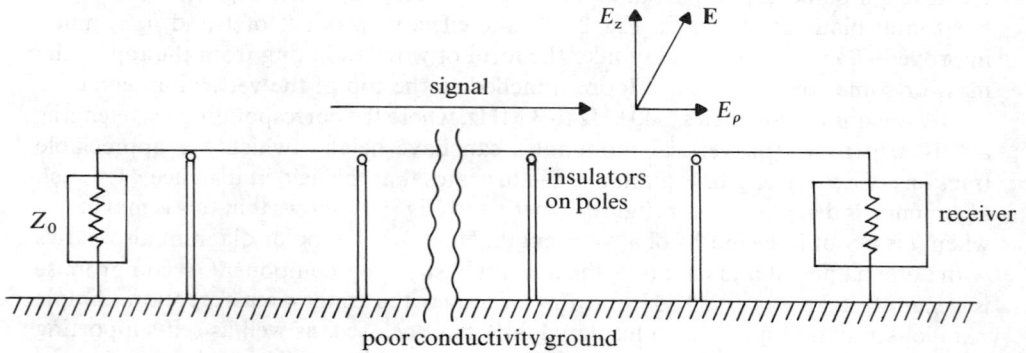

Fig. 2.18 A Beverage antenna is a horizontal wire supported on insulated poles and several wavelengths long. A signal is obtained from the small E_ρ component of the incident electric field.

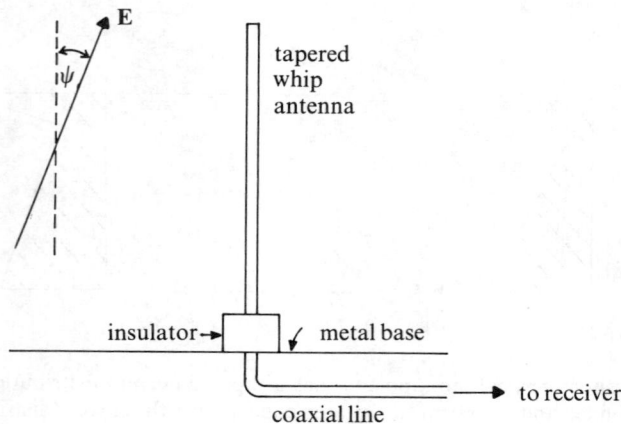

Fig. 2.19 The whip antenna is a tapered short vertical monopole above a metal plane, commonly connected by a coaxial cable to the receiver.

Beverage antenna, is not as good as a tall vertical antenna but its simplicity and directional properties make it attractive for reception at LF when a large area of poor soil conductivity is available.

On quite a different scale, the *whip antenna* is a simple small or mobile version of the short vertical monopole, illustrated in Fig. 2.19, in which the induced voltage is

$$V = Eh_e \cos \psi \tag{2.25}$$

where ψ represents the alignment of the antenna with the incident electric field. In theory the antenna has a linear gain of 3 with respect to the isotrope in the direction $\theta = 90°$.

Another simple type of low-frequency receiving antenna is the *loop antenna,* sometimes modified with a ferromagnetic core to enhance its performance (Fig. 2.20). The *n*-turn antenna couples with the magnetic field to produce an induced voltage in accordance with Faraday's law, $V = -n \, d\phi/dt$. For a simple sinusoidal variation in magnetic flux ϕ,

$$d\phi/dt = j\omega\phi = j\omega\mu H A \cos \psi$$

where flux ϕ is replaced by $\mu H A \cos \psi$ and $A \cos \psi$ is the area perpendicular to the flux. Hence for an air-cored loop antenna the magnitude of the induced voltage is

$$V = \frac{\mu_0}{60} E A n f \cos \psi = \frac{2\pi}{\lambda} E A n \cos \psi \tag{2.26}$$

If the loop is wound around a ferromagnetic core there is an increase in ϕ due to the addition of the core. This increase depends upon the core material, cross-sectional shape, and its length-to-diameter ratio. When the latter is between 10 and 50, the corresponding increase in V is typically 50 to 100.

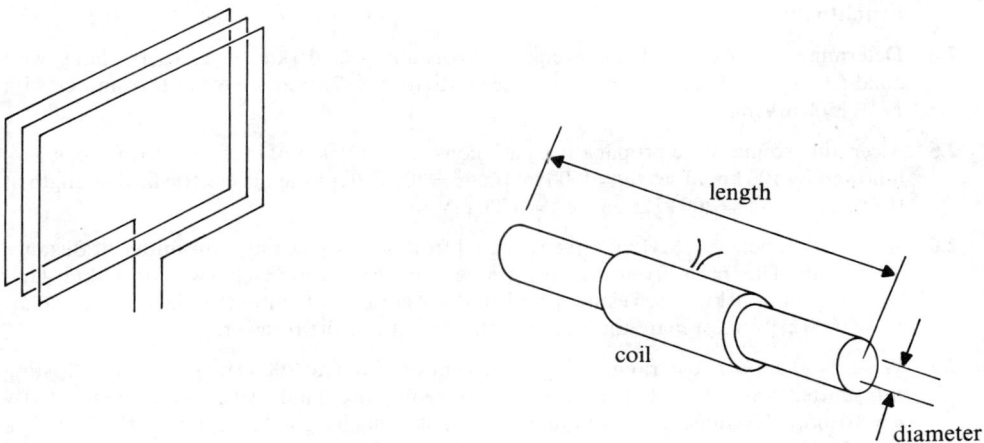

Fig. 2.20 Loop antennas may have square or circular cross-section with air core or ferromagnetic core.

Example 2.8 Loop antenna

An MF field strength indicator uses a circular loop antenna of diameter 600 mm with 5 turns. Its inductance is 40 μH and resistance is about 1 Ω. The antenna is tuned with a capacitance and the voltage across the capacitance is taken to a high-impedance receiver input. If the loop samples an electromagnetic field with $E = 100\,\mu$V/m (rms) at 1.2 MHz, estimate the voltage across the capacitor.

The antenna is tuned with the capacitor to give maximum V_c and then turned so that $\cos\psi = 1$. Hence

$$V_c = QV = \frac{\omega_0 L}{r}\frac{2\pi}{\lambda}E\left(\frac{\pi D^2}{4}\right)n = 1.1\,\text{mV}$$

PROBLEMS

2.1 Estimate the electric field strength at $d = 50$ km from a short vertical monopole radiating $P_t = 10$ kW at $f = 900$ kHz over ground with mean conductivity $\sigma = 10$ mS/m.

2.2 An antenna operating at $f = 200$ kHz approximates to a vertical wire of height $h = 100$ m connected at the top to a horizontal wire of length 100 m. It is fed with base current $I_0 = 100$ A (rms) and has an ideal ground plane. Making a very simplified estimate of the effective height h_e of the antenna, estimate the electric field strength at $d = 100$ km over ground with mean conductivity $\sigma = 5$ mS/m.

2.3 Estimate from appropriate propagation charts the electric field strength at $d = 150$ km from: (a) a 1 MHz transmitter with $E_1 = 4500$ mV/m over land with mean conductivity $\sigma = 10$ mS/m and permittivity $\varepsilon_r = 4$; and (b) a vertical monopole of height $h = 0.5\lambda$ radiating $P_t = 50$ kW over a sea path with conductivity $\sigma = 4000$ mS/m and relative permittivity $\varepsilon_r = 80$.

2.4 Determine the electric field strength at distance $d = 10$ km over rocky land with conductivity $\sigma = 1$ mS/m and relative permittivity $\varepsilon_r = 7$ from a 3 MHz transmitter with $E_1 = 1500$ mV/m.

2.5 A certain ground wave propagation path consists of 150 km of land ($\sigma = 10$ mS/m, $\varepsilon_r = 4$) followed by 100 km of sea ($\sigma = 4000$ mS/m, $\varepsilon_r = 80$). Determine the electric field strength at the receiver if $f = 700$ kHz and $E_1 = 4500$ mV/m.

2.6 A receiver tuned to 1 MHz receives a signal from a broadcasting transmitter at distance $d = 80$ km. The path from the transmitter to the receiver consists of 40 km land ($\sigma = 10$ mS/m), 20 km of sea estuary, and 20 km of land ($\sigma = 10$ mS/m). If the transmitter has $E_1 = 3000$ mV/m, estimate the electric field strength at the receiver.

2.7 What is the probable range of good reception from a 700 kHz radio broadcasting transmitter with $E_1 = 1500$ mV/m over surrounding land with mean conductivity $\sigma = 10$ mS/m? Assume that the minimum level of signal for good reception is 10 mV/m in a city, 2 mV/m in residential suburbia, and 0.5 mV/m in rural areas.

2.8 A 50 kW transmitter operating at 1 MHz with $E_1 = 3000$ mV/m excites both a ground wave and a sky wave at night. Assume: (a) that the curves of Fig. 2.3 and Fig. 2.11 apply for the ground wave and sky wave, respectively; (b) that a good signal is defined as one in which the ratio of wanted to unwanted signals is at least 4:1 for 95% of time; (c) that the 5% level of the sky wave is 17 dB below the quasi-maximum 95% curve; and (d) that $E_{min} = 30$ dBμ. Under these conditions determine the boundaries between the ground wave zone, the interference zone, the good sky wave zone and the limit of good sky wave reception.

2.9 Two transmitters, 1500 km apart, transmit different programs at the same frequency $f = 1$ MHz and with the same radiated powers. What is the difference between the daytime and the night-time coverage of each transmitter if a protection ratio of 200:1 is required for 95% of time with $E_{min} = 0.1$ mV/m? Assume $E_1 = 3000$ mV/m and $\sigma = 10$ mS/m.

2.10 The CCIR have published world maps and charts of atmospheric radio noise in the form of a noise factor $F_a = 10 \log f_a$ defined for a short vertical monopole (with $T_a = 0$) via

$$kT_0 B f_a = \frac{V_n^2}{4R_{rad}}$$

Show that if E_n is the equivalent rms noise field strength in dBμ in 1 kHz bandwidth and f is the frequency in MHz,

$$E_n = F_a - 65.54 + 20 \log f$$

2.11 A CCIR Report (Ref. 2.6, pp. 20–21) indicates that during the period 0000–0400 hours in mid-winter the median atmospheric radio noise figure F_{am} for the UK is 154 dB at 10 kHz; 108 dB at 100 kHz; 60 dB at 1 MHz; and 30 dB at 10 MHz. What are the corresponding median values of the rms noise field strength E_n in dBμ per 1 Hz bandwidth? Repeat the question for the daytime 1200–1600 hours period in which the corresponding values of F_{am} are 152 dB, 81 dB, 25 dB and 33 dB. Plot the results on a graph.

2.12 A small loop antenna is 0.4 by 0.4 m in size. It has 12 turns and an inductance of 120 μH. If the Q-factor is 100 at 1250 kHz and the untuned antenna is coupled to a 50 Ω input resistance receiver, what is the electric field strength when the output of the 50 dB receiver is 250 mV? Assume the loop is aligned for maximum reception.

REFERENCES

2.1 Norton, K. A., 'The propagation of radio waves over the surface of the earth and in the upper atmosphere,' *Proc. IRE*, 24 (October 1936), 1367–87; 25 (September 1937), 1203–36.

2.2 Wait, J. R., 'The mode theory of VLF ionospheric propagation for finite ground conductivity,' *Proc. IRE*, 45 (June 1957), 760–67.

2.3 Rotheram, S., 'Ground-wave propagation,' *Proc. IEE*, 128, Part F (October 1981), 285–95.

2.4 Millington, G., 'Ground-wave propagation over an inhomogeneous smooth earth,' *Proc. IEE*, 96, Part III (1949), 53–64.

2.5 Watt, A. D., *VLF Radio Engineering*. London: Pergamon Press, 1967.

2.6 CCIR, 'World distribution and characteristics of atmospheric radio noise,' *Report* 322–1, International Telecommunication Union, Geneva, 1974.

3

Sky Wave Propagation

Some early radio wave propagation experiments by Marconi and others at the turn of the century included the reception of signals at Newfoundland from a transmitter in Cornwall. To explain this long-range radio link, Heaviside and Kennelly suggested that the upper regions of the atmosphere may have conductive properties which help to reflect the signal back to earth. Two decades later, Appleton confirmed the existence of such a region which he and Watson-Watt named the *ionosphere*. It was found that the ionosphere was a little more complex than previously thought and Appleton recorded the presence of at least two regions which he named the *E-layer* and the *F-layer*. His initial techniques were similar to those used in optics, such as Lloyd's single-mirror interferometer, and he measured part of the interference pattern caused by the ground wave and night-time sky wave at Oxford from the BBC transmitter at Bournemouth. It was deduced from this experiment, carried out on 11 December 1924, that the reflection occurred at a height of about 100 km. In 1929 the experiments detected subsidiary fringes which, upon analysis, indicated reflections from a higher layer at about twice the height of the original E-layer. He called this the F-layer though occasionally others referred to it as the Appleton layer.

Sky wave, high-frequency or short wave radio communication over the succeeding years has made use of the reflecting or refracting properties of the ionosphere, sometimes using single hops or various combinations of multiple hops from the same or different layers (Fig. 3.1). Long-range point-to-point communications of telegraphy, telephony and broadcasting have made extensive use of the rather limited HF band for

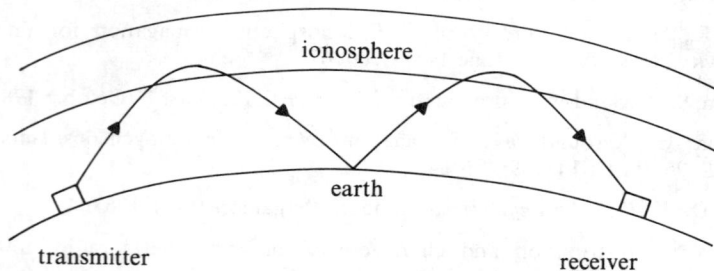

Fig. 3.1 Sky wave propagation between transmitter and receiver via two reflections by the ionosphere and one reflection by the earth.

commercial, amateur, diplomatic, civil and military applications, including shore–ship and ground–air links as well as reliable data communications.

3.1 THE IONOSPHERE

Since the discoveries by Appleton and others, much research has been undertaken into the structure of the ionosphere. Very generally it may be said that the ionosphere is that region of the atmosphere in which the pressure is sufficiently low for free electrons and positive ions to exist for a comparatively long period of time before recombination. Because of this, and the several available sources of ionizing energy from the sun (ultraviolet and x-rays) and from cosmic rays, the amount of ionization which occurs in

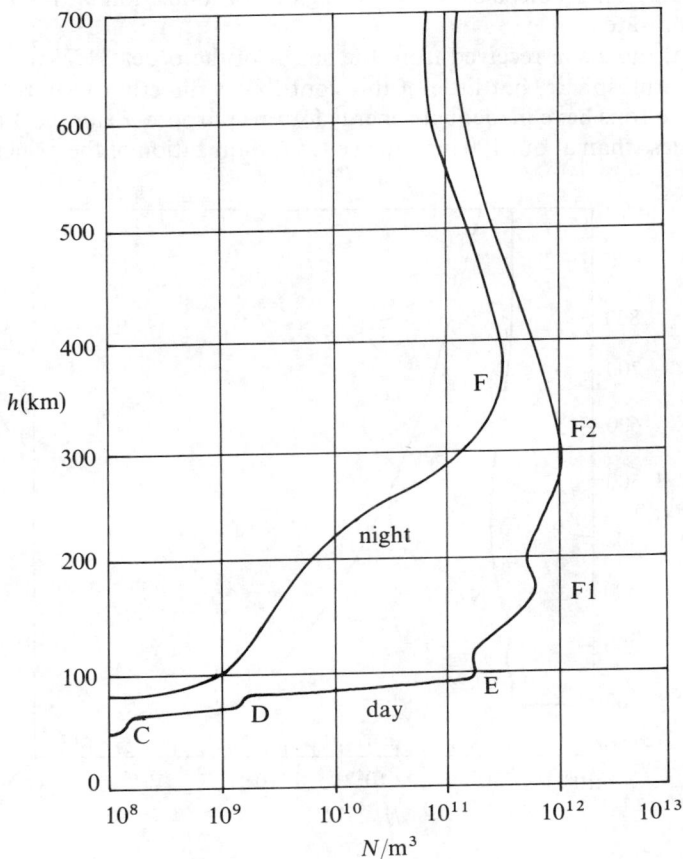

Fig. 3.2 A sketch of typical variation of free electron density with height above ground level.

the upper atmosphere varies in a non-uniform manner with height h above ground level. It is found that there are several regions of localized higher ionization densities which are now commonly labeled C, D, E and F, or sometimes D1, D2, E1, E2, F1 and F2. Both the height and density of these layers vary with time of day, season of year, sunspot number and many other factors. A typical variation of the free electron density N with height h is shown in Fig. 3.2, and the patterns of diurnal variations are illustrated later in Fig. 3.11.

The mechanism of the ionization is complicated. The ionizing radiation is maximum in the outer atmosphere and is gradually absorbed as it approaches ground level. The greatest molecular density is at ground level and reduces with height, as shown in Fig. 3.3. At great heights there are comparatively few molecules or atoms to ionize despite the abundance of energy and low recombination rate, and at ground level there is little energy left to ionize the abundance of molecules and the recombination rate is high. Somewhere in between there are optimum states with sufficient remaining radiation energy and molecules to set up regions of ionization without too rapid a recombination rate.

The total radiation received from the sun is of the order $3\,kW/m^2$ in the outer limits of the atmosphere, but little of this contributes directly to ionization because most of the photons have inadequate energy for this purpose. Example 3.1 shows that wavelengths less than about 100 nm are needed for ionization of the principal gases in

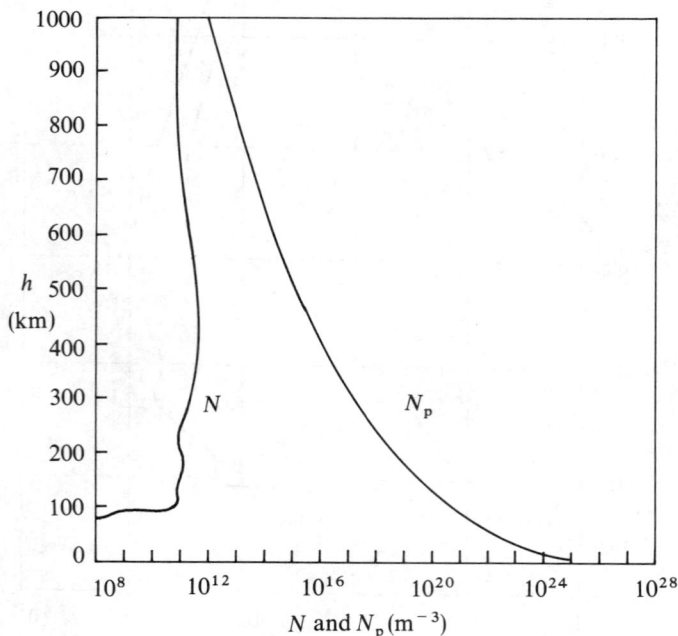

Fig. 3.3 A comparison between the number of free electrons N and the number of gas molecules N_p per cubic meter at various heights above ground level. Note that the origin is not at zero density.

the atmosphere. Radiation at 80 nm penetrates down to about $h = 150$ km; that at 20 nm penetrates to about 125 km; that at 2 nm to about 100 km; and that at 0.3 nm to about 75 km. Their radiation intensities are of the order mW/m^2.

Example 3.1 Ionization radiation

The atmosphere around the earth contains something like 10^{44} molecules, about 90% of them below a height of 20 km. The most common molecules are nitrogen (78%), oxygen (21%), argon (1%) and carbon dioxide (1%). Determine the maximum wavelengths of incident radiation which can ionize these molecules and show that visible light makes no contribution.

A photon of incident energy hf joules must at least equal the ionization energy of the molecule E_i such that:

$$E_i(\text{eV}) = \frac{hf}{e} = \frac{hc}{e\lambda} = \frac{1240}{\lambda(\text{nm})}$$

where eV = electron-volts, h = Planck's constant and e = electronic charge. If the ionization energies are 15.8 eV (N_2), 12.5 eV (O_2), 12.8 eV (O_3), 15.8 eV (A) and 13.8 eV (CO_2), the corresponding wavelengths are 78 nm, 99 nm, 97 nm, 78 nm and 90 nm. Visible light, 400 nm $< \lambda <$ 800 nm, is therefore insufficient to produce ionization of these molecules.

The maximum ionization wavelengths are in the ultraviolet band of radiation, but the atmospheric molecules are also ionized by x-rays (10 nm $> \lambda > 10^{-3}$ nm) with photon energies in keV. Several processes are available: (1) the kinetic energy of the incident photon is divided between an electron ejected from the outer shell of the molecule and a secondary photon; (2) an electron is lifted out of an inner shell and an electron in an outer shell falls into the vacated energy state, emitting a secondary photon; (3) a secondary photon may strip an additional electron from the same molecule; and (4) the secondary photons pass on to ionize other molecules and produce further electrons and photons until the energy levels are too low for the process to continue. The x-ray yield is of the order 1 electron for every 30–35 eV of incident energy.

At the highest levels, roughly above 250 km, it is believed that the oxygen atom is the main constituent, producing an O^+ ion and an e^- electron. At lower levels, around 150–200 km, the positive ions of nitric oxide (NO^+), oxygen atoms (O^+) and oxygen molecules (O_2^+) are found, having been ionized by the ultraviolet radiations and x-ray emissions from the sun. The E-layer region, approximately 80–110 km, consists of more oxygen and nitrogen ions, this time with soft x-rays (1–3 nm) as the major ionizing agent as well as the Lyman-β line. The even lower *D-layer* region involves combinations of oxygen and nitrogen which have been ionized by x-rays (0.2 to 0.5 nm) and the Lyman-α line. Below this is an occasional C-layer region. In both latter cases an elaborate series of reactions take place.

The *D-layer* has comparatively low free electron density ($N = 10^9/m^3$) compared with a molecular density of the order $10^{20}/m^3$. It occurs during daylight hours at a height of about 60–90 km and responds very quickly to the diurnal variation of the sun's movement. It has little effect on the bending of high-frequency radio waves but produces considerable attenuation effects at lower sky wave frequencies, particularly in the medium-frequency band.

The *E-layer* occurs mainly during daylight hours and only weakly at night. It is located about 90–130 km above ground level with a free electron density near $10^{11}/m^3$. It causes some bending and attenuation but mainly at the lower sky wave frequencies.

The *F-layer* is in many respects the most useful part of the ionosphere, particularly at night. At such times it may exhibit two regions of localized maximum free electron density and these are then labeled F1 and F2.

There are occasional and unpredictable appearances in the E-layer region of drifting electron clouds in which the ionization is much higher than is normal for the layer. These clouds are quite variable in size and form what is known as the *sporadic E-layer*. Such layers occur at heights of about 90–150 km and are caused by various complex mechanisms including the equatorial electrojet during daytime at low latitudes, precipitating auroral electrons at night at high latitudes, and the neutral wind shear at middle latitudes. There are seasonal and diurnal variations in its occurrence.

3.2 PRIMARY AND SECONDARY PARAMETERS

An ionized and therefore partially conducting medium such as the ionosphere may be defined in terms of primary and secondary parameters. In this text we shall restrict the primary parameters to permittivity ε and conductivity σ, thereby requiring the medium to be non-magnetic and isotropic at any given point.

Alternatively, for sinusoidally varying electromagnetic waves, we may derive certain secondary parameters in terms of ε, σ and μ, together with frequency in the form $\omega = 2\pi f$. The most common secondary parameters include attenuation coefficient α, phase coefficient β, velocity of propagation v, refractive index n and intrinsic impedance Z_i. These are defined in Appendix 1 and their equations deduced from the fundamental laws of electromagnetic theory.

3.2.1 Attenuation

A propagating radio wave may be represented in frequency domain by its electric field strength

$$E = E \exp[-(\alpha + j\beta)z] \exp(j\omega t) \tag{3.1}$$

where α nepers per meter is the attenuation coefficient. It is shown in Appendix 1 that

$$\alpha = \omega \sqrt{\frac{\mu\varepsilon}{2}\{\sqrt{1 + (\sigma/\omega\varepsilon)^2} - 1\}} \tag{3.2}$$

In the particular situation where $\mu_r = 1$ and the conductivity is sufficiently small relative to ε and ω to write

$$(\sigma/\omega\varepsilon)^2 \ll 1$$

and

$$\sqrt{1 + (\sigma/\omega\varepsilon)^2} = 1 + \tfrac{1}{2}(\sigma/\omega\varepsilon)^2$$

direct substitution into the previous equation for attenuation gives

$$\alpha = \tfrac{1}{2}\sigma\sqrt{\mu/\varepsilon} = 60\pi\sigma(\varepsilon_r)^{-1/2} \qquad \text{Np/m} \tag{3.3a}$$

$$\alpha = 1.637 \times 10^6 \sigma/(\varepsilon_r)^{1/2} \qquad \text{dB/km} \tag{3.3b}$$

with conductivity in S/m. This has used the relationship that 1 Np/m = 8.686 dB/m. We may use either equation as required for a partially conducting medium or otherwise revert to the full equation for attenuation.

3.2.2 Refractive Index

Maxwell's equations indicate two types of current: conduction current density J_c and displacement current density J_d, such that

$$J = J_c + J_d = \sigma E + \varepsilon \partial E/\partial t \tag{3.4}$$

If a plane wave $E = \hat{E}\sin\omega t$ passes through an ionized medium having N free electrons per cubic metre available for conduction, the force F on an electron in some specified direction x is given by

$$F_x = (-)eE_x$$

or

$$m\frac{d^2 x}{dt^2} = (-)e\hat{E}_x \sin\omega t$$

or

$$\frac{dx}{dt} = (+)\frac{e\hat{E}_x \cos\omega t}{\omega m} = v_x$$

where m is the mass of the electron and v_x is its velocity in the specified direction. With this relationship we can obtain

$$J_{cx} = N(-e)v_x = \frac{N(-e)e\hat{E}_x \cos\omega t}{\omega m}$$

and

$$J_{dx} = \varepsilon_0 \frac{dE_x}{dx} = \varepsilon_0 \omega\hat{E}_x \cos\omega t$$

Combining the two equations gives:

$$J_x = \varepsilon_0\{1 - Ne^2/(\omega^2 m\varepsilon_0)\}\omega\hat{E}_x \cos\omega t$$

or

$$J_x = \varepsilon_0\{\varepsilon_r'\}\omega\hat{E}_x \cos\omega t$$

This is the equation we would expect for a dielectric with *effective relative permittivity* ε_r' and hence the ionosphere can be considered as a dielectric with refractive index $n = \sqrt{\varepsilon_r'}$ given by:

$$n = \left\{ 1 - \frac{Ne^2}{\varepsilon_0 m \omega^2} \right\}^{1/2} = \left\{ 1 - \left(\frac{\omega_c}{\omega} \right)^2 \right\}^{1/2} = (1 - 81 N/f^2)^{1/2} \qquad (3.5)$$

where 81 is a common approximation of $e^2/(4\pi^2 \varepsilon_0 m) = 80.5 \, \text{m}^3\text{s}^{-2}$. Note that when $\omega^2 = \omega_c^2 = Ne^2/(\varepsilon_0 m)$, the refractive index $n = 0$.

3.2.3 Conductivity and Permittivity

A little more precisely, the laws of motion indicate that force is proportional to the rate of change of momentum. For an electron (with mass assumed constant) subjected to an electric field, the force $F = eE$ is proportional to the product of mass and the rate of change of velocity:

$$eE = m \frac{dv}{dt}$$

in the appropriate direction. In the event of collisions between electrons and gas molecules, the energy is lost in the form of heat. The frictional, retarding or viscous force representing collisions is mv/τ or mvv, where τ is the mean time between collisions and v is the *frequency of collisions* ($v = 1/\tau$). Thus a better equation for force is

$$Ee = m \, dv/dt + mvv$$

In time domain, with $E = E_0 e^{j\omega t}$ and $v = v_0 e^{j\omega t}$, the solution of the equation is

$$v = \frac{Ee}{mv + j\omega m}$$

Hence we can substitute this value into

$$J_c = Nev = \frac{Ne^2 v E}{m(v^2 + \omega^2)} - j \frac{\omega Ne^2 E}{m(v^2 + \omega^2)}$$

and compare the result with

$$J = J_c + J_d = \sigma E + j\omega \varepsilon_0 E$$

$$= \frac{Ne^2 v}{m(v^2 + \omega^2)} E + j\omega \varepsilon_0 \left[1 - \frac{Ne^2}{\varepsilon_0 m(v^2 + \omega^2)} \right] E$$

to obtain expressions for conductivity and relative permittivity

$$\sigma = \frac{Ne^2 v}{m(v^2 + \omega^2)} \qquad (3.6)$$

and

$$\varepsilon_r' = 1 - \frac{Ne^2}{\varepsilon_0 m(v^2 + \omega^2)} \qquad (3.7)$$

Thus the ionized medium can be represented by the primary parameters σ, ε'_r and $\mu_r = 1$, or by any secondary parameter related to these via frequency f.

Example 3.2 Complex dielectric constant

Derive an expression for the complex dielectric constant of the ionosphere in terms of frequency (via $\omega = 2\pi f$), critical frequency (via $\omega_c = 2\pi f_c$) and electron collision frequency v. Ignore the effect of earth's magnetic field.

It was stated in Sec. 3.2.3 that the solution of the equation of motion is

$$v = \frac{Ee}{mv + j\omega m}$$

If we now replace the three current densities by $J_c = Nev$, $J_d = j\omega\varepsilon_0 E$, and the total current by $J = j\omega\varepsilon_0\varepsilon_r^* E$, where ε_r^* is called the *complex dielectric constant* of the partially conducting medium as described in Appendix 1, then

$$j\omega\varepsilon_0\varepsilon_r^* = j\omega\varepsilon_0 + \frac{Ne^2}{mv + j\omega m}$$

Replacing $Ne^2/(\varepsilon_0 m)$ by ω_c^2 as in Eq. (3.5) gives us:

$$\varepsilon_r^* = 1 + \frac{\omega_c^2}{j\omega(v + j\omega)} = 1 - \frac{\omega_c^2}{\omega(\omega - jv)} \tag{3.8}$$

which is the expression for the complex dielectric constant of the ionized medium using ω_c and v as its defining parameters.

3.2.4 Earth's Magnetic Field

If an electron with velocity **v** is in the presence of a magnetic field it experiences a force $e\mathbf{v} \times \mathbf{B}$ in a direction mutually perpendicular to **v** and **B**. The effect is to superimpose circular motion upon the electron's path of radius r such that $v = 2\pi r/(1/f_H) = \omega_H r$, where f_H is the frequency of rotation known as the *gyromagnetic frequency*. Balancing the magnitudes of the centrifugal force and the magnetic force gives:

$$mv^2/r = evB$$
$$m\omega_H = eB$$
$$\omega_H = \mu_0 He/m = 2\pi f_H \tag{3.9}$$

For typical values of earth's magnetic field, the gyromagnetic frequency is of the order 0.7 to 1.9 MHz, depending on the location. World maps of f_H are available (see, for example, Ref. 3.1).

If the ideas of Sec. 3.2.2 and Sec. 3.2.3 are taken one stage further, we must include four components in the equation of motion of a free electron in the ionosphere. These are: (1) the force due to the electric field $e\mathbf{E}$; (2) the inertia force represented by $m\,d\mathbf{v}/dt$; (3) the frictional term $mv\mathbf{v}$; and now (4) the force due to the earth's magnetic field $e\mathbf{v} \times \mathbf{B}$, all in appropriate directions. The *vector* equation is:

$$e\mathbf{E} + e\mathbf{v} \times \mathbf{B} = m\,d\mathbf{v}/dt + m\nu\mathbf{v}$$

and the solution of this equation in terms of the complex dielectric constant is (Ref. 3.1):

$$\varepsilon_r^* = 1 - \frac{\omega_c^2}{\omega(\omega_c \pm \omega_H - j\nu)} \tag{3.10}$$

where $\omega_c^2 = Ne^2/\varepsilon_0 m$ from Eq. (3.5), $\omega_H = \mu_0 He/m$ from Eq. (3.9), and ν is the electron collision frequency.

Thus the complex dielectric constant has two values, one for each of the plus and minus signs. This is interpreted as signifying that an incident sky wave excites two components when entering the ionized medium, one called the *ordinary wave* (positive sign) and one called the *extraordinary wave* (negative sign). It is found that the ordinary wave has smaller phase velocity, greater group velocity and lower attenuation rate, and that the polarizations of the two waves are different. The two waves recombine when leaving the ionosphere but because of the different amplitudes and phases they produce net elliptical polarization.

An even more precise analysis results in the Appleton–Hartree equation (see Ref. 3.1), which is not considered here.

3.3 ELECTRON COLLISION FREQUENCY

The effect of collisions between electrons and heavy particles in the ionosphere is taken into account by adding a frictional term into the equation of motion of an electron in an

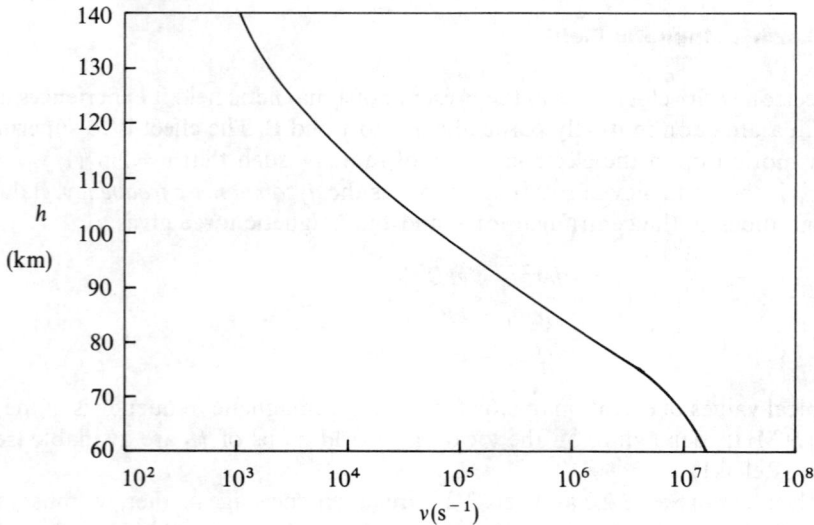

Fig. 3.4 The variation of effective electron collision frequency with height.

Table 3.1   Numerical data for Ex. 3.3

	1 MHz day	*1 MHz night*	*10 MHz day*	*10 MHz night*
σ(S/m)	6.96×10^{-7}	6.96×10^{-9}	7.12×10^{-9}	7.12×10^{-11}
ε_r'	0.92	1.0	1.0	1.0
α(dB/km)	1.19	0.01	0.01	0.0001
n	0.96	1.0	1.0	1.0

electromagnetic field. This requires a knowledge of v, the frequency of collisions, which includes two contributory factors, v_i and v_n, the frequencies of collisions between electrons and ions or neutral particles, respectively. These depend on temperature and pressure and thus vary with height above ground level, season of year, etc. Figure 3.4 illustrates a typical relationship between v and h. Note that the equations for conductivity σ and effective permittivity ε_r' involve the term $(v^2 + \omega^2)$ and that there are cases in which the approximations $v^2 \gg \omega^2$ or $\omega^2 \gg v^2$, as appropriate, simplify the expressions. In the case of the D-layer an appropriate equation for collision frequency

---

Example 3.3 D-layer ionospheric parameters

Determine the values of conductivity σ, relative permittivity ε_r', attenuation rate α and the refractive index n at a certain level in the D-layer if the corresponding free electron density is $N = 10^9/\text{m}^3$ during the day and $N = 10^7/\text{m}^3$ during the night, and the effective electron collision frequency is $v = 10^6/\text{s}$. Assume $f = 1$ MHz and $f = 10$ MHz for comparison.

    The daytime solutions for $f = 1$ MHz are given via four equations. Firstly, for conductivity, we use Eq. (3.6):

$$\sigma = \frac{Ne^2v}{m(v^2 + \omega^2)} = 0.28 \times 10^{-7} \times \frac{Nv}{v^2 + \omega^2} = 6.96 \times 10^{-7} \, \text{S/m}$$

Knowing the numerical value of conductivity, a slightly shorter expression for relative permittivity combining Eqs. (3.6) and (3.7) is:

$$\varepsilon_r' = 1 - \frac{\sigma}{\varepsilon_0 v} = 1 - 0.11 \times 10^6 \times \sigma = 0.92$$

The attenuation rate can be obtained from Eq. (3.3b):

$$\alpha = 1.64 \times 10^6 \sigma/(\varepsilon_r')^{1/2} = 1.19 \, \text{dB/km}$$

and the refractive index n is simply

$$n = \sqrt{\varepsilon_r'} = 0.96$$

The numerical results for each of the four cases specified in the example are given in Table 3.1, the purpose of the example being to demonstrate that the D-layer is essentially a daytime attenuator of medium frequencies. During the night-time and at high frequencies the D-layer has comparatively little attenuation.

is

$$v = 7.5 \times 10^5 p \tag{3.11}$$

where p is the atmospheric pressure measured in pascals or newtons per square meter (Ref. 3.2).

3.4 D-LAYER ATTENUATION

If the night-time values of the electron collision frequency and the free electron density in the vicinity of the D-layer are as illustrated in Figs. 3.4 and 3.5, it is possible to calculate the attenuation rate at any height by the method outlined in Ex. 3.3. At a frequency of 1 MHz the resulting curve is as shown in Fig. 3.6 over the range 60–90 km above ground level.

If the permittivity is close to unity and $\omega^2 \gg v^2$, an approximate expression for attenuation rate is

$$\alpha = 1.637 \times 10^6 \frac{Ne^2 v}{m\omega^2} \approx \frac{Nv}{f^2} \times 10^{-15} \, \text{dB/km} \tag{3.12}$$

with f in MHz. Thus a similar curve to Fig. 3.6 with $f = 10$ MHz would be 100 times less in magnitude. These all assume vertical incidence.

The daytime effect of the D-layer may be estimated by assuming that the free electron density N increases from about $10^7/\text{m}^3$ to about $10^9/\text{m}^3$. If the other factors are constant there is thus about a 100-fold increase in the attenuation rate during the day.

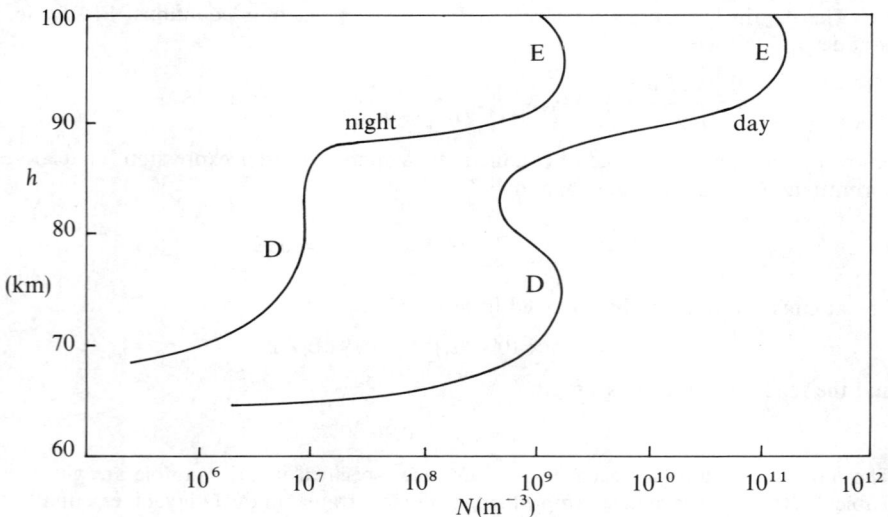

Fig. 3.5 Sketches of the variation of free electron density per cubic meter N in the vicinity of the D-layer for use in examples and problems.

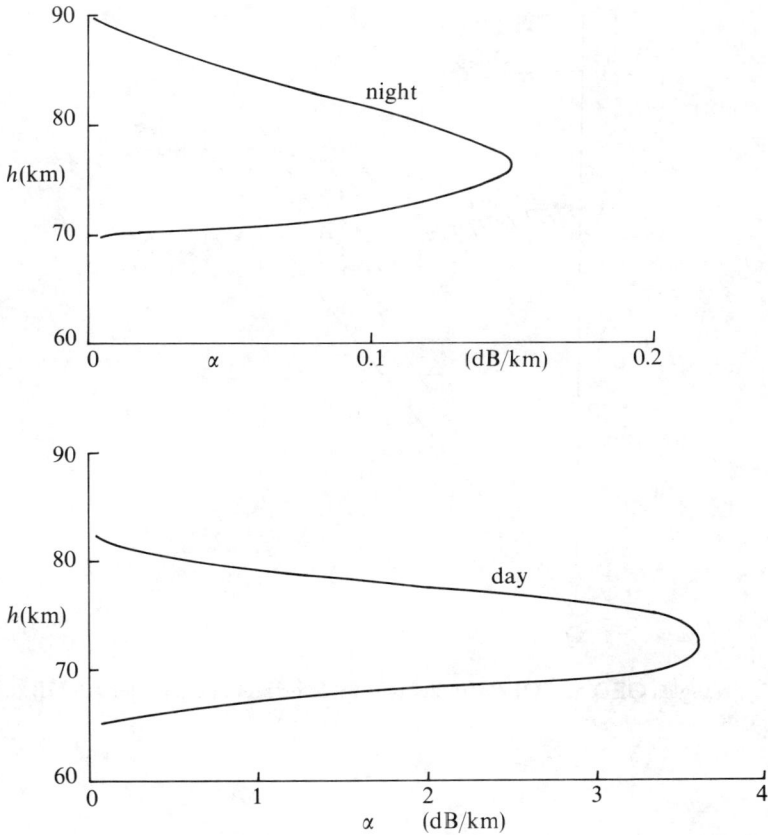

Fig. 3.6 The attenuation rate of the D-layer with height at a frequency of 1 MHz.

At 1 MHz the attenuation is then considerable and this effectively prevents sky wave interference at medium frequencies. At 10 MHz, the 100-fold increase in attenuation due to N is canceled by the 100-fold decrease due to f and so the daytime attenuation is quite small.

For sky wave communication links the propagation path through the D-layer is not vertical. The path length is consequently much greater and the total path loss is higher. Figure 3.7 illustrates the difference. For the values labeled in the illustration, distance DB can be obtained via the trigonometrical sine rule applied to triangles EOD and EOB.

The path length and attenuation is thus dependent upon the launch angle Δ and the ground range is also dependent upon Δ. Hence graphs of attenuation loss through the D-layer against ground range can be calculated for specific frequencies and reflection heights (Fig. 3.8).

Similar analyses can be made for the E-layer, and the overall attenuation is the integration of the α–h curve between appropriate limits.

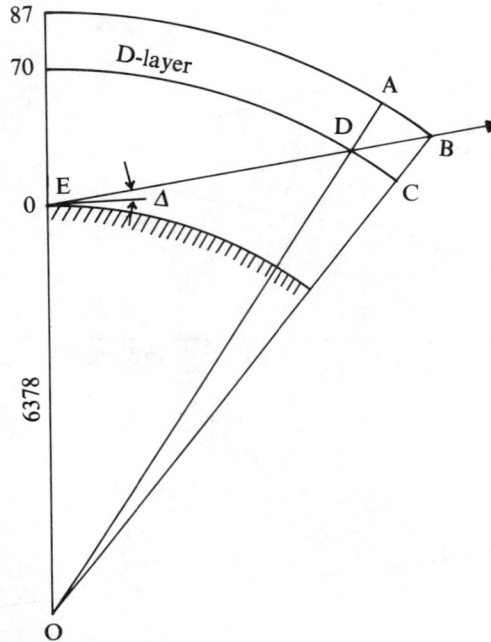

Fig. 3.7 Triangles OED and OEB can be used to determine the path length DB through the D-layer.

3.5 THE PROPAGATION PATH

The sky wave propagation path consists of the up-path from transmitter to ionosphere, the refraction and reflection by the ionosphere, and the down-path from ionosphere to receiver. In the case of multiple hops this pattern is repeated each hop. To analyze the propagation path, we first consider an ideal horizontally stratified ionosphere as illustrated in Fig. 3.9a.

3.5.1 Snell's Law of Refraction

Snell's law is derived from first principles in Appendix 1, and is illustrated in Fig. 3.9. This indicates that

$$n_0 \sin i_0 = n_1 \sin r_1$$
$$n_1 \sin i_1 = n_2 \sin r_2$$

and so on for each incremental level of the ionosphere. However, because of the linear

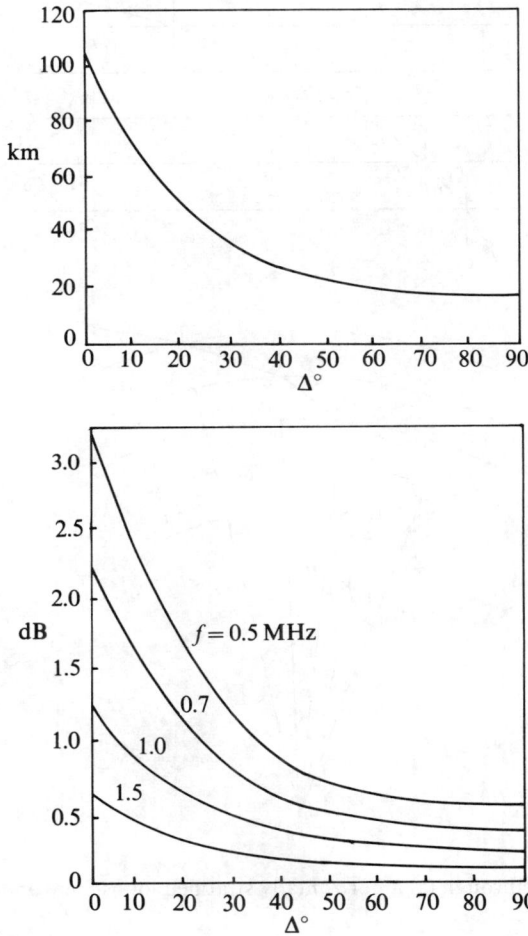

Fig. 3.8 Path length DB through D-layer and night-time attenuation against launch angle Δ.

stratification in this idealized example, $i_x = r_x$ and hence we can write

$$n_0 \sin i_0 = n_x \sin r_x \qquad (3.13)$$

where $x = 1, 2, 3, \ldots$ to an appropriate limit.

Let us consider two special cases. Firstly, if $n_0 = 1.000$ and reflection occurs when angle $r_x = 90°$, we can write

$$\sin i_0 = n_x$$

Secondly, if $n_0 = 1.000$ and $i_0 = 0°$, reflection from vertical incidence will occur when

$$n_x = \sin 0° = 0$$

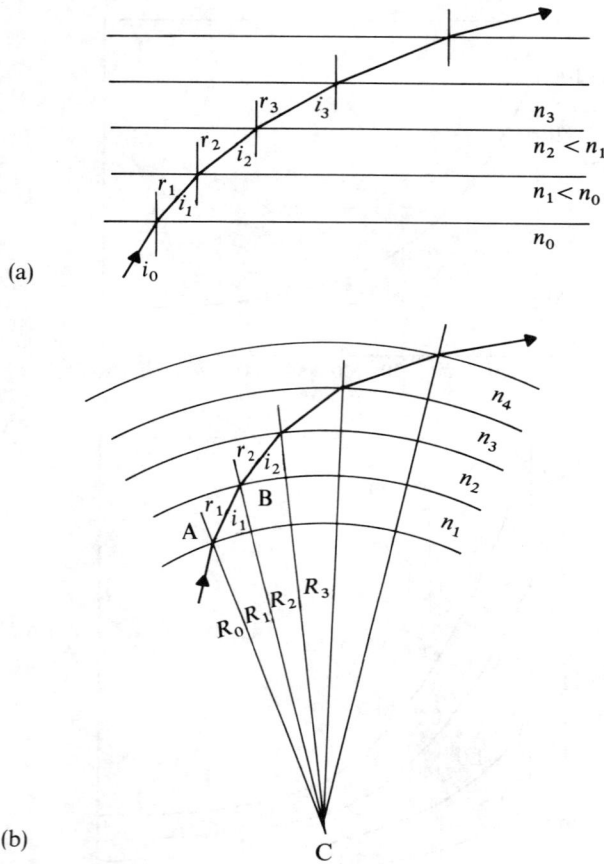

(a)

(b)

Fig. 3.9 Refraction through (a) a horizontally stratified ionosphere and (b) a spherically stratified ionosphere.

In words, if a sky wave is incident upon the ionosphere at angle i_0 with respect to the normal then it will be reflected by the ionosphere at the height at which n_x is numerically equal to $\sin i_0$. This means that if the angle of incidence is vertical ($i_0 = 0°$) the reflection will occur at the height at which $n_x = 0$. If such conditions cannot be met, the wave is not reflected by the ionosphere but passes through.

3.5.2 Critical Frequency and Maximum Usable Frequency

If an electromagnetic wave is launched vertically into the ionosphere, reflection will occur when (and if) $n_x = 0$. We can write Eq. (3.5) as:

$$n_x = \sqrt{1 - \frac{81 N_x}{f^2}} = 0 \tag{3.14}$$

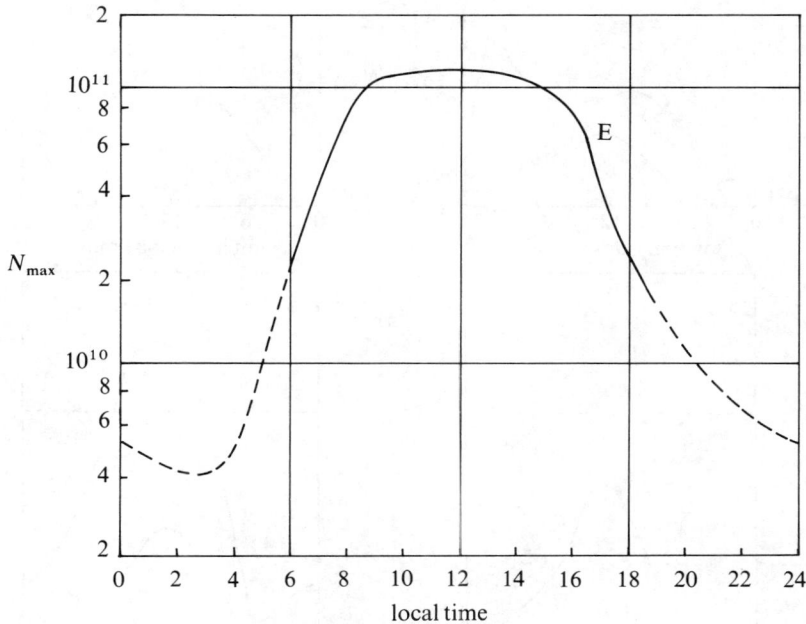

Fig. 3.10 Sketch of a typical variation of N_{max} of the E-layer during the course of one day.

and from this obtain the highest frequency at which reflection will occur when N_x is a maximum:

$$f_c = 9\sqrt{N_{max}} = \frac{\omega_c}{2\pi} \tag{3.15}$$

This term is usually called the *critical frequency* or sometimes the *plasma frequency*. Each of the ionospheric layers has a specific N_{max} at a given location and time, and the maximum frequencies at which reflection is made by the various layers are often labeled f_0E, f_0F_1 and f_0F_2. These are not constant but vary diurnally, seasonally and with other solar patterns. Figures 3.10 and 3.11 illustrate typical variations. Two estimates of these critical frequencies, which are valid for any time of day and any season, are given in Ref. 3.3:

$$f_0E = 0.9[(180 + 1.44R)\cos\chi]^{0.25} \text{ MHz} \tag{3.16a}$$

and

$$f_0F_1 = (4.3 + 0.01R)\cos^{0.2}\chi \qquad \text{MHz} \tag{3.16b}$$

where R is the sunspot number and χ is the zenith angle of the sun.

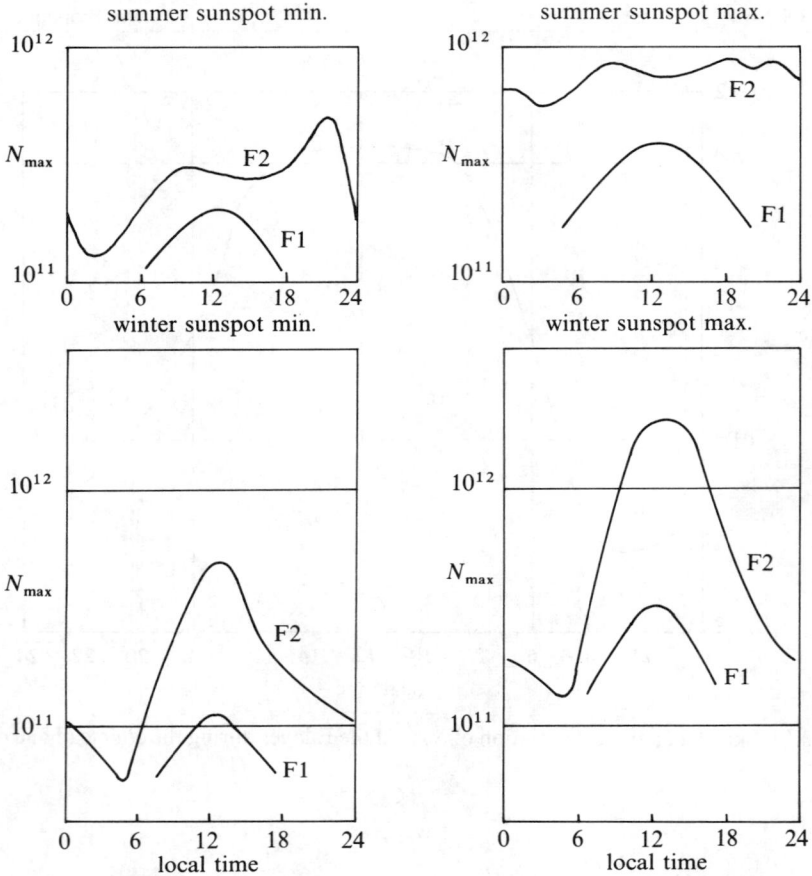

Fig. 3.11 Sketches of typical variations of N_{max} in the vicinity of the F_1 and F_2 layers.

Example 3.4 Diurnal variation of critical frequency

Sketch the predicted diurnal variation of $f_0 E$ in mid-January at a latitude of 40°N when the sunspot number is 100.

The sun's zenith angle χ is the angle between the perpendicular to the earth at the observing point and the direction of the sun. If the observing point is at latitude L_1, the sun's subpolar point is at latitude L_2 and the angle of the sun referred to the observing point is L_3, spherical geometry gives

$$\cos \chi = \sin L_1 \sin L_2 + \cos L_1 \cos L_2 \cos L_3$$

Charts of χ are commonly available (for example, Ref. 3.1), usually for mid-months. From these it is possible to obtain χ against local time for a specified latitude. Noting that sunrise occurs when χ is about 93° for the E-layer, the following data have been obtained from such a chart for mid-January at latitude 40°N. Equation (3.16a) has been used with $R = 100$ to estimate $f_0 E$.

zenith angle	90	80	70	65	65	70	80	90
local time	07.20	08.20	09.40	10.25	13.30	14.30	15.45	16.50
$f_0 E$(MHz)	——	2.46	2.92	3.08	3.08	2.92	2.46	——

These are sufficient points to make a sketch of $f_0 E$ against local time.

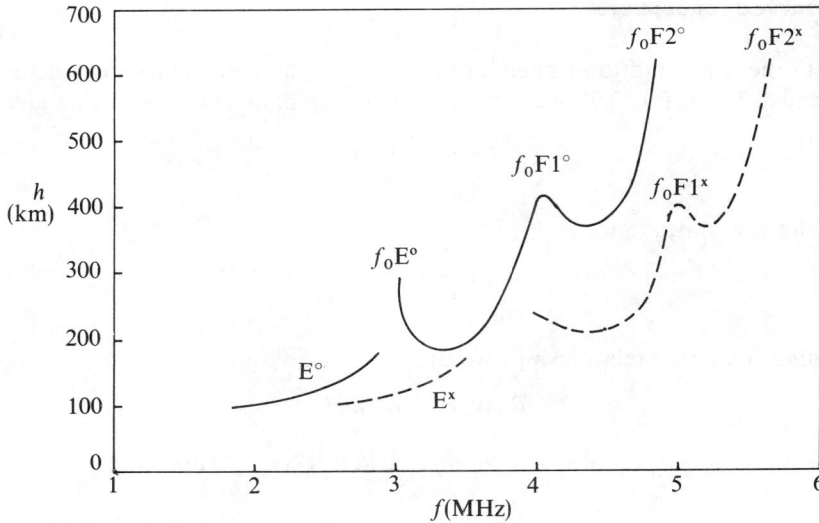

Fig. 3.12 A sketch of an ionogram. Note that o and x refer to the ordinary and extraordinary components of the waves.

An instrument called an *ionosonde* is used to measure the variation of critical frequency with ionospheric height and the results are used to determine the corresponding free electron density at each level of height. The instrument sends a succession of short pulses of radio waves vertically into the ionosphere and records the time delay between transmission and reception. Knowing the velocity of propagation, the range of the point of reflection can easily be determined. At first the frequency of the radio wave pulses is relatively low and the tests are repeated for progressively increasing frequencies so that a plot of reflection height h against frequency f can be plotted. This is called an *ionogram* and is illustrated in Fig. 3.12. There may be very little apparent reflection at the lower frequency range because the signals are attenuated by the D-layer, but the critical frequencies of the E- and F-layers are clearly observed.

If an electromagnetic wave is launched at an angle of incidence i_0 into the ionosphere, reflection can now occur at a higher frequency because

$$n_x = \sin i_0 = \sqrt{1 - 81 N/f^2}$$

and

$$\sin i_0 = \sqrt{1 - 81 N_{max}/MUF^2}$$

or

$$\sin i_0 = \sqrt{1 - (f_c/MUF)^2} \qquad (3.17)$$

where MUF is called the *maximum usable frequency*. This equation can be rewritten as

$$MUF = f_c \sec i_0 \qquad (3.18)$$

but it is correct only for the idealized horizontally stratified ionosphere.

3.5.3 Curved Ionosphere

In reality the earth and ionosphere are curved and the mathematics is a little more complicated. From Fig. 3.9b we can see that the trigonometrical sine rule gives

$$\frac{\sin i_1}{R_0} = \frac{\sin r_1}{R_1}$$

and Snell's law states that

$$\frac{\sin r_1}{\sin i_0} = \frac{n_0}{n_1} \qquad \text{etc.}$$

Combining these two relationships we obtain

$$R_0 \sin r_1 = R_1 \sin i_1$$

or

$$n_0 R_0 \sin r_0 = n_1 R_1 \sin r_1 = n_2 R_2 \sin r_2 \qquad \text{etc.}$$

or

$$n_x R_x \sin r_x = \text{constant} \tag{3.19}$$

This is the modified version of Snell's law.

The skip distance or ground range d can be determined from Fig. 3.13 where Δ is the launch angle, R is the radius of earth and h is known as the *virtual height*.

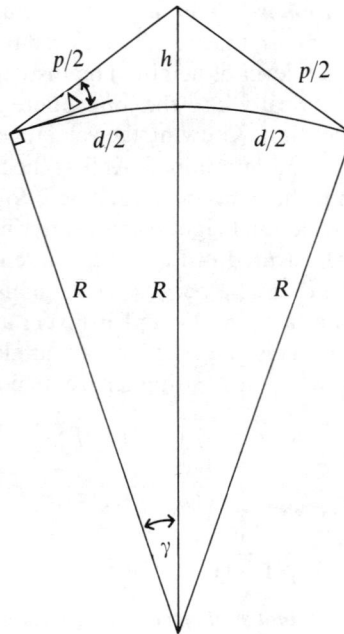

Fig. 3.13 The trigonometry for determining path length p and launch angle Δ.

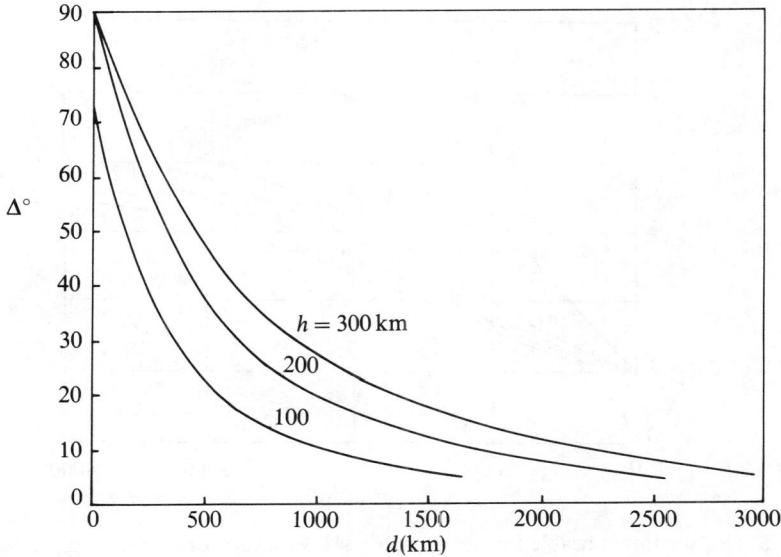

Fig. 3.14 The relationship between launch angle Δ and propagation range d for several values of virtual height of the ionosphere.

Unfortunately, the virtual height varies both with frequency and time (due to many factors) and so graphs relating range and launch angle usually refer to specified typical values of virtual height. Figure 3.14 illustrates such a chart.

3.5.4 Maximum Usable Frequency Factor and Optimum Working Frequency

Because the earth and ionosphere are spherical, the maximum usable frequency calculations must be appropriately modified with the aid of spherical geometry. If the overall effect of the modification can be represented by k_e then

$$\text{MUF} = k_e f_c \sec i_0 = \text{MUFF} f_c \tag{3.20}$$

where MUFF combines both k_e and $\sec i_0$ and is known as the *maximum usable frequency factor*. The MUFF also depends on the virtual heights of the various layers and consequently is available in graphical form for certain typical values for each (Fig. 3.15).

The techniques used for ionospheric forecasting may be read elsewhere (see Ref. 3.1). With the aid of such techniques, charts are produced each month of the median MUF for each high-frequency operational route. To account for day-to-day variations within the month the *optimum traffic frequency* OTF or *optimum working frequency* OWF is set at 85% of the MUF. Because the attenuation is lower at the higher frequencies and the atmospheric noise levels tend to be lower, the higher the operational frequency the better the likely performance. Lower frequencies are thus

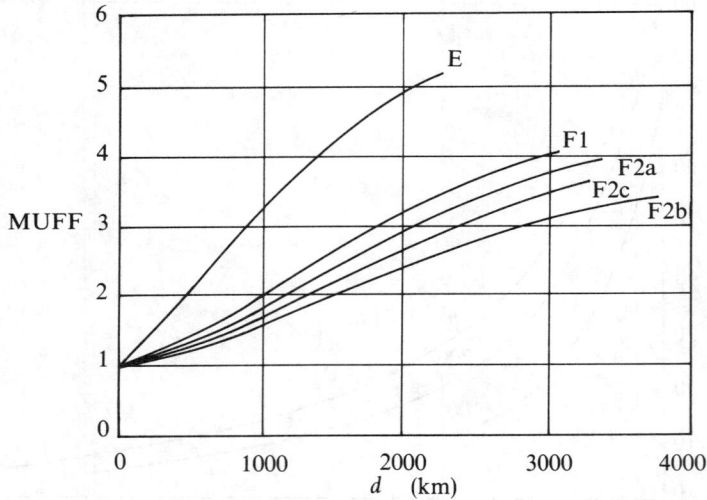

Fig. 3.15 The maximum usable frequency factor MUFF against range for typical heights of ionospheric layers. F2a is for a winter day, F2b is for a winter night and F2c is for summer day or night.

avoided except by high-power commercial links. There is a limit in the lower-frequency region when the signal-to-noise ratio becomes inadequate for the type of communication, and this is known as the *lower usable frequency* LUF limit. Sketches of these predictions are shown in Fig. 3.16.

Example 3.5 Optimum working frequency

If it is assumed that the monthly median estimate of the free electron density of the F2 layer in winter is as shown in Fig. 3.16, determine the monthly median MUF curve and the corresponding OWF curve for a distance of 2000 km.

'Monthly median' implies that propagation is satisfactory for fifteen days of the month. The OWF corresponds to the frequency limit for satisfactory propagation during all undisturbed days of the month. Frequencies in excess of the median MUF will propagate satisfactorily but only for less than fifteen days.

For each point on the local time scale we can record $N(F2)$. For example, at 1200 hours $N(F2) = 2 \times 10^{12}/m^3$. The corresponding critical frequency is $f_c = 9\sqrt{N} = 12.7$ MHz. The MUFF chart in Fig. 3.15 indicates that for a distance of 2000 km the MUFF varies diurnally between 2.4 and 2.9 for F2 winter, day and night. This gives $MUF = MUFF \times f_c = 2.9 \times 12.7 = 36.8$ MHz at 1200 hours, and the OWF as $0.85 \times 36.8 = 31.3$ MHz.

In predicting MUF and OWF charts, the starting point is more likely to be an estimate of the $f_0 F2$ frequency rather than the free electron density, but otherwise the method is the same.

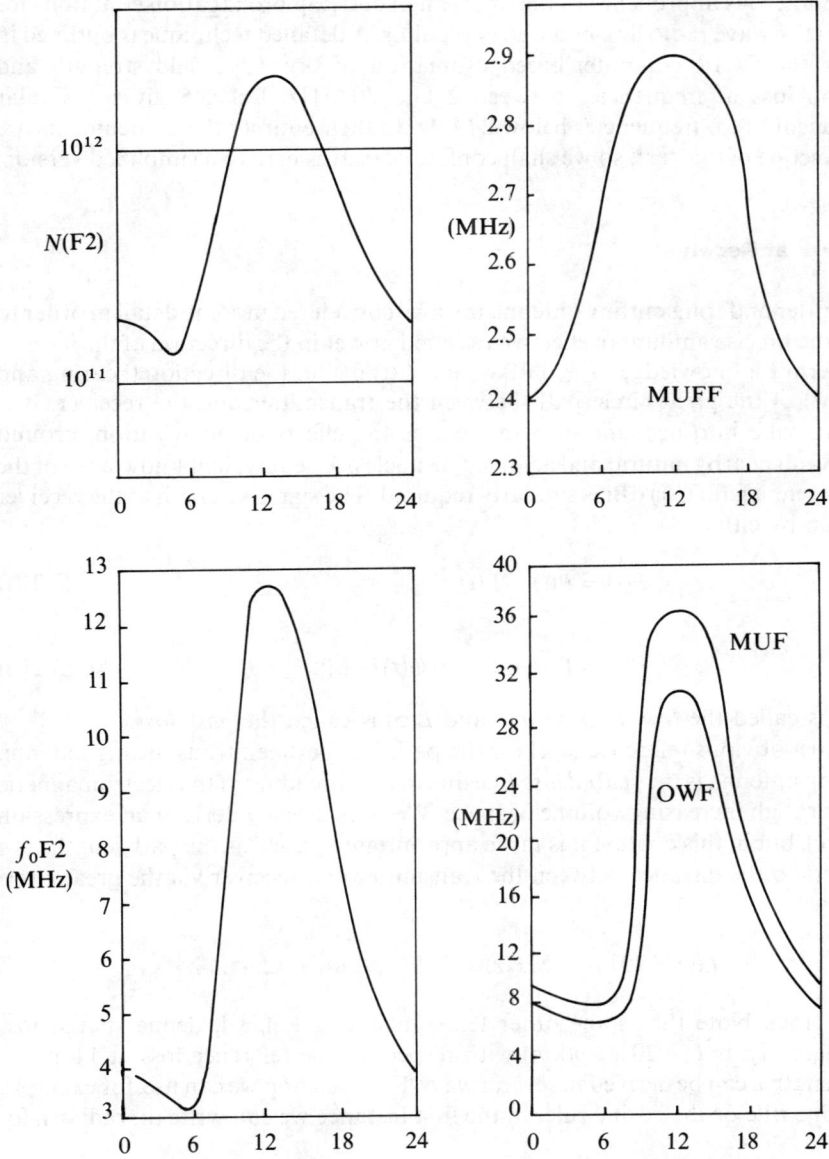

Fig. 3.16 The calculation of optimum working frequency (OWF).

3.6 PROPAGATION EQUATIONS

There are numerous approaches to the problem of deriving propagation equations for a multi-hop sky wave radio link between two points. A detailed technique is outlined in Ref. 3.4 for the CCIR computer-based estimation of sky wave field strength and transmission loss at frequencies between 2 and 30 MHz. Ref. 3.5 gives a similar analytical account for frequencies below 2 MHz. In their entirety these documents are outside the scope of this text, so we shall confine ourselves here to a simplified version.

3.6.1 Power at Receiver

The transmitter and transmitting antenna must be considered in some detail in order to determine the precise amount of effective radiated power in the direction of the launch. This will permit a knowledge of $P(t)$ dBW and $G(t)$ dBi in the direction (bearing and elevation $\Delta°$) of the great circle path between the transmitter and the receiver. It is necessary to take into account such matters as the effects of polarization, ground parameters and nearby natural and artificial obstacles. An equivalent knowledge of the receiving antenna gain $G(r)$ dBi is similarly required. The signal strength at the receiver is then given by either

$$P(r) = P(t) - L(t) \tag{3.21a}$$

or

$$P(r) = P(t) + G(t) + G(r) - L(p) \tag{3.21b}$$

where $L(t)$ is called the *transmission loss* and $L(p)$ is called the *path loss.*

The most obvious reference level for the path loss between transmitting antenna and receiving antenna is the *spatial loss* $L(s)$ due to the spreading of the electromagnetic radiation through increasing volume of space. We have already derived an expression in Eq. (1.15c), but in this context it is more appropriate to use p as the path length, as d usually refers to the distance between the transmitter and receiver via the great circle path. Thus:

$$L(s) = 20 \log f(\text{MHz}) + 20 \log p(\text{km}) + 32.45 \tag{3.22}$$

from Eq. (1.15c). Note that some other texts, including Ref. 3.4, define spatial loss slightly differently as $L = 20 \log p(\text{km})$ with respect to the reference loss at 1 km.

Path length p can be derived in several ways. For each hop we can use, for example, either the sine rule or the cosine rule. In the first instance we can write (in radians) for Fig. 3.13:

$$\frac{R}{\sin(\pi/2 - \Delta - \gamma)} = \frac{p/2}{\sin \gamma}$$

where $\gamma = d/2R$ radians and $\sin(\pi/2 - \Delta - \gamma) = \cos(\Delta + \gamma) = \cos(\Delta + d/2R)$, and hence (still in radians):

$$p = 2R \frac{\sin(d/2R)}{\cos(\Delta + d/2R)} \tag{3.23}$$

In the second instance, the cosine rule gives us:

$$(p/2)^2 = R^2 + (R + h)^2 - 2R(R + h)\cos\gamma$$
$$p = \{8R(R + h)[1 - \cos(d/2R)] + 4h^2\}^{1/2}$$

or, replacing $\cos(d/2R)$ by the first two terms of its expansion, $1 - \frac{1}{2}(d/2R)^2$,

$$p = [d^2(1 + h/R) + 4h^2]^{1/2} \tag{3.24}$$

In the first case p is defined via launch angle Δ, but Δ depends upon the virtual height h of the ionosphere and the great circle distance d.

Thus, so far, the received power would be $P(\text{r}) = P(\text{t}) + G(\text{t}) - L(\text{s}) + G(\text{r})$, but several other losses and gains need still to be considered. One obvious path loss is the absorption of the electromagnetic waves as they pass through the ionosphere. As mentioned in Sec. 3.2.4, the incident ray splits into two components, the ordinary ray O and the extraordinary ray X, which have different paths, attenuation and polarization effects. The corresponding *daytime ionospheric absorption* can be estimated via a long and complicated procedure outlined in Ref. 3.4, designed for computer analysis but outside the scope of this text. An alternative version, taken from Refs. 3.6 and 3.7, describes the total daytime ionospheric absorption at frequency f (MHz) as

$$L(\text{i}) = \frac{677.2 \sec i}{1.98(f + f_{\text{H}}) + 10.2} \sum_{j=1}^{n} I_j \tag{3.25}$$

where i is the angle of incidence at $h = 100$ km, n is the number of hops, f_{H} is the gyromagnetic frequency and

$$I_n = (1 + 0.0037R_{12})(\cos 0.881\chi)^{1.3} \tag{3.26}$$

for sunspot number R_{12} averaged over twelve months and zenith angle χ at the penetration point of the absorbing region. World maps of f_{H} and χ are available in various texts, such as Ref. 3.1.

The ionospheric absorption loss varies diurnally as a result of the zenith angle term in Eq. (3.26) and the application of Eq. (3.25) becomes invalid towards night-time as I_n approaches roughly one-tenth. Thereafter an alternative expression, taken from Ref. 3.8, becomes more appropriate. This suggests that the residual *night-time absorption* of the ordinary ray is approximately

$$L(\text{i}) = \frac{(7 + 0.019d)(1 + 0.015R_{12})}{f^2 + 10} \tag{3.27}$$

with f in MHz and d in km. This equation should be used only when its magnitude is larger than that of Eq. (3.25).

A possibly unexpected path gain occurs as the result of the ionosphere behaving as a concave reflector which helps to focus the reflected waves back to earth, particularly at the lower angles near grazing. The effect enhances the received field strength by a factor known as the *ionospheric convergence gain* $G(f)$ which depends on the angle of launch Δ. Graphs are available in Ref. 3.4.

There is, similarly, a possibly unexpected path loss which arises because the sky wave incident upon the ionosphere is split into two components, O and X, which propagate with different attenuations and polarizations. The amount of power coupled to the O and X waves and the proportion of the downward waves which couples with the receiving antenna (as a result of its polarization characteristics) depend upon the limiting polarizations of the O and X waves at the receiver. The combined effects are all represented by a loss known as the *polarization coupling loss* $L(c)$.

If the sky wave propagation involves more than one hop it is necessary to include the loss produced by the ground reflection between each hop. It is customary to approximate the resulting polarization of the downwave as circular and to define ground reflection loss as

$$L(g) = 10 \log \frac{|R_H|^2 + |R_V|^2}{2} \tag{3.28}$$

where R_H and R_V are the Fresnel reflection coefficients defined and derived in Appendix 1.

There are other factors to be taken into consideration, but those described so far are sufficient to illustrate briefly some of the problems involved with sky wave propagation. For a single hop the monthly median equation becomes

$$P(r) = P(t) + G(t) + G(r) + G(f) - L(s) - L(i) - L(c) \tag{3.29}$$

Converting from power levels into root-mean-square electric field strength via $P_r = A_e P_a = G_r(\lambda^2/4\pi)E^2/(120\pi)$ results in:

$$E(dB\mu) = 107.2 + P(t) + G(t) + 20 \log f \,(\text{MHz}) - L(p) \tag{3.30}$$

if the transmitted power is recorded in dBW and $E(dB\mu)$ is the monthly median field strength of the down-coming fading sky wave.

An alternative approach to the problem is to replace Eq. (3.29) by

$$P(r) = P(t) + G(t) + G(r) - L(s) - L(i) - L(b) \tag{3.31}$$

for a single hop, where $L(b)$ is called the excess system loss (Ref. 3.9). This is available in tabular form in Ref. 3.6 but an approximate estimate which ignores the slight diurnal variation is given by either

$$L(b) = 9 + 7 \exp\left\{ -\frac{(\psi - 68)^2}{112} \right\} \tag{3.32}$$

during the summer or winter season or

$$L(b) = 9 + 10.6 \exp\left\{ -\frac{(\psi - 68)^2}{145} \right\} \tag{3.33}$$

during the spring and autumn seasons, where ψ is the latitude in degrees.

Example 3.6 High-frequency radio link

Estimate the path loss, power received, field strength at the receiver and signal-to-noise ratio for an HF radio link with the following parameters: $P(t) = 30\,dBW$, $f = 10\,MHz$, $d = 2000\,km$, $h = 300\,km$, $R_{12} = 110$, $f_H = 1.25\,MHz$, latitude $\psi = 50°$, December noon with $\chi = 77°$, $G(t) = G(r) = 10\,dB$, $B = 3.4\,kHz$ and $F_{am} = 33\,dB$.

Using radius of earth $R = 6370\,km$, great circle distance $d = 2000\,km$ and virtual height $h = 300\,km$, we can determine the total path length as $p = 2133\,km$ from Eq. (3.24). This means that at a frequency $f = 10\,MHz$ the spatial loss from Eq. (3.22) is $L(s) = 119.1\,dB$.

We can apply the cosine rule to a triangle with sides $R = 6370\,km$, $p/2 = 1066.5\,km$ and $R + h = 6670\,km$, to obtain $\gamma = 9°$ subtended at the centre of the earth and hence $\Delta = 12°$ = angle of elevation. Note that $i = 74.2°$ = angle of incidence at the ionosphere. The ionospheric absorption loss during the daytime, at a time when the 12-month moving average of the sunspot number is $R_{12} = 110$, and at a location where $f_H = 1.25\,MHz$, is given via Eq. (3.25) as $L(i) = 30.3\,dB$.

Finally, the excess system loss makes use of Eq. (3.33) with latitude $\psi = 50°\,N$. This gives $L(b) = 9.4\,dB$.

The path loss is the sum of these component losses, $L(p) = 158.8\,dB$, and the received power from Eq. (3.31) with $G(t) = G(r) = 10\,dB$ is $P(r) = -108.8\,dBW$. Equation (3.30) gives the electric field strength as $E(r) = 8.4\,dB\mu$.

The atmospheric noise factor at winter noon is given as 33 dB and the noise power in 3.4 kHz bandwidth is $N(r) = -204 + 10\log 3400 + F_{am}$, or $N(r) = -135.7\,dBW$. The signal-to-noise ratio of the channel (ignoring other factors) is $SNR = P(r) - N(r) = 27\,dB$.

3.6.2 Reception Levels

The minimum field strength required for good reception of radio telephony via a high-frequency sky wave communication link in middle latitudes may resemble the sketch shown in Fig. 3.17 under certain atmospheric conditions. The principal cause of atmospheric noise is the electromagnetic radiation from lightning strokes. Local thunderstorms propagate some energy via the ground wave mechanism but more distant storms use the sky wave mode. The ability of the ionosphere to propagate the sky wave mode varies with frequency and diurnally, some of the upper frequencies being cut off as the sky waves escape through the ionosphere.

For short-wave broadcasting the reception level needs to be about 15–25 dB higher than that needed for radio telephony, whereas for manual telegraphy, where the bandwidth is much reduced, the minimum field strength is about 15 dB lower than that for telephony.

The ionosphere is an inhomogeneous and unstable medium, continually changing, and the electric field strength at the receiver is the net result of 'rays' which have traveled from the transmitter via numerous slightly different sky wave paths. One effect is that the phases of each component vary randomly and the amplitude of the resultant varies with time. Another cause of the consequent *fading* of the signal is the variation of the polarization of the received signals brought about by the superposition of the ordinary

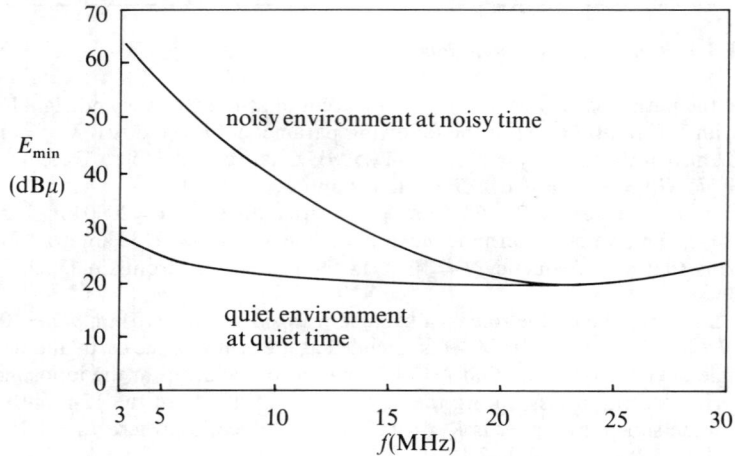

Fig. 3.17   The equivalent atmospheric noise field in the HF frequency range.

and extraordinary rays which are differently polarized. As the polarization of the receiving antenna is usually fixed, polarization fading occurs. The propagation parameters are also frequency dependent and sometimes this is sufficient to affect signal components within the communication bandwidth and thereby cause distortion of the signal via selective fading.

These variations occur continuously, some effects taking place quickly and producing rapid or deep fading, and other bulk effects such as the attenuation by the D-layer changing more slowly and producing slow or shallow fading.

If the electric field strength (measured linearly as volts per metre and not in the more usual logarithmic form of dBμ) is sampled rapidly and the measurements plotted as a statistical histogram, it is found that the resultant variation approximates to the Rayleigh distribution when the fast variation is observed over a sufficiently short period of time (such as a few minutes) for the slow variation to be considered as constant.

Measurements of the slow variation of the electric field strength approximate to the log-normal distribution when observed over a long period of time.

These best-fitting statistical distributions enable some estimates to be made of the quality or reliability of the received signal in a high-frequency communication link. The effect of fading is to require either the use of greater transmitter power to achieve reliable reception, an increase of 13–14 dB over the monthly median being common, or the use of diversity techniques. Some introductory comments on statistical distributions and the use of diversity systems to combat fading are described in Chap. 6.

3.6.3 Noise Levels

To obtain a numerical estimate of the externally produced noise level in a given HF radio link, use may be made of CCIR Report 322–1 (Ref. 3.10), which is also reproduced in Ref. 3.1. This report contains twenty-four pages of maps and charts, firstly divided into four sections to account for each of the four seasons, and then further divided into six sections per season to account for local time in four-hour divisions throughout a day.

Each of the twenty-four pages contains a map giving worldwide values of F_{am} at 1 MHz. If F_a is the equivalent antenna noise factor resulting from the external noise power available from a loss-free antenna, measured in decibels relative to thermal noise level, then F_{am} is defined as the median value of F_a over the given four-hour period.

A second chart on each page indicates how it is possible to transfer from F_{am} at 1 MHz to F_{am} at the operating frequency, if this is not 1 MHz. A third chart includes the decile-to-median ratios of F_{am} for each frequency, also in decibels.

By selecting as reference a bandwidth of 1 Hz, the reference thermal noise level is simply $N = kTB = kT$ watts, or $N(\text{dBW}) = -204$. The noise level per 1 Hz bandwidth in the HF radio link is then given by

$$N(\text{r}) = F_{am} - 204 \quad \text{dBW} \tag{3.34}$$

and the signal-to-noise ratio per 1 Hz bandwidth can be determined from Eqs. (3.29) and (3.30), with due allowance for bandwidth in Eq. (3.29). See also Ex. 2.5 and Prob. 2.10.

3.6.4 Circuit Reliability

We can make an estimate of the probable reliability of a high-frequency communication link over a particular path during a specified month and at a given local time. Such a path will have several predicted parameters and these will include the diurnal graph of the median maximum usable frequency for the given month against local time. The monthly median MUF is the maximum predicted frequency at which sky waves are expected to be available at the receiving antenna for 50% of the days of the month at the given local time of day. Thus, for example, between 1100 and 1200 hours the median MUF is expected to be (say) 7.2 MHz. On 15 days of the month it will probably exceed this value and on 15 days of the month it will probably be less.

The approximate statistical variation of MUF during the course of a month is known, but calculations can be simplified by using MUF_{90}, MUF_{50} and MUF_{10} to represent the maximum frequencies at which sky waves can be expected for 90%, 50% and 10% of the days of the specified month for given local time. In statistical terminology these terms are known as the *lower decile, median* and *upper decile,* respectively.

On the basis that the actual statistical distribution of day-to-day MUF follows the chi-squared distribution (Chap. 6), the following equations can be derived for the *sky wave availability factor* $Q\%$ (based on Refs. 3.11 and 3.12):

$$Q\% = \frac{80(\text{MUF}_{10} - \text{MUF}_{50})}{(\text{MUF}_{10} - \text{MUF}_{50}) + (f - \text{MUF}_{50})} - 30 \qquad \text{if } f > \text{MUF}_{50} \tag{3.35}$$

$$Q\% = 130 - \frac{80(\text{MUF}_{50} - \text{MUF}_{90})}{(\text{MUF}_{50} - \text{MUF}_{90}) + (\text{MUF}_{50} - f)} \qquad \text{if } f \leqslant \text{MUF}_{50} \tag{3.36}$$

providing the limiting values of $Q\%$ are 0% and 100%. Thus $Q\%$ is the sky wave availability at a given local time on a specific sky wave link for a transmission at frequency f (MHz).

Example 3.7 The sky wave availability factor

If some hypothetical values of MUF for a given HF radio link and specified local time are $\text{MUF}_{90} = 10\,\text{MHz}$, $\text{MUF}_{50} = 12\,\text{MHz}$ and $\text{MUF}_{10} = 14\,\text{MHz}$, draw a graph of sky wave availability against frequency.

    For frequencies above 12 MHz we can use Eq. (3.35) to obtain $Q\%$ for $f = 13, 14, 15$ and 16 MHz as 23%, 10%, 2% and the unrealistic -3%. For frequencies below 12 MHz we can use Eq. (3.36) to obtain $Q\%$ for $f = 11, 10, 9$ and 8 MHz as 77%, 90%, 98% and the unrealistic 103%. These can be plotted on a graph as in Fig. 3.18 to illustrate the sky wave availability variation with frequency for the specified conditions.

    The reliability of an HF radio link depends not only on the sky wave availability but also on the signal-to-noise ratio (SNR) at the receiver. The link may exist but the transmission losses and noise levels could produce a signal which is at an unacceptably low level. To simplify the mathematics a little we can use SNR in terms of a 1 Hz reference bandwidth and make the necessary corrections later.

    As with MUF, described earlier, transmission loss $L(t)$ and noise power in 1 Hz bandwidth N(dBW) can be represented by approximate statistical distributions, but again we can make good estimates from the upper deciles L_{90} and N_{90}, the medians L_{50} and N_{50}, and the lower deciles L_{10} and N_{10}. Thus, for example, L_{90} is the maximum transmission loss we would expect for 90% of days in the month for a given local time and N_{10} is the maximum noise level per 1 Hz bandwidth we would expect for 10% of the days of the month at the same local time.

    The transmission loss and noise levels contribute to the overall signal-to-noise ratio per 1 Hz bandwidth and this too can be represented by similarly defined SNR_{90}, SNR_{50} and SNR_{10}. Using the earlier references it can be shown that

$$(\text{SNR}_{10} - \text{SNR}_{50})^2 = (L_{50} - L_{90})^2 + (N_{50} - N_{90})^2 \qquad (3.37a)$$

$$(\text{SNR}_{50} - \text{SNR}_{90})^2 = (L_{10} - L_{50})^2 + (N_{10} - N_{50})^2 \qquad (3.37b)$$

Details are available in Refs. 3.4 and 3.10 which enable the L and N deciles and medians to be predicted for given paths, locations and times, and hence the SNR deciles can be determined in terms of SNR_{50}.

    Thus the signal-to-noise ratio for a 1 Hz bandwidth at the end of a particular sky wave radio link will vary from day to day (at the given local time), but it will have a monthly median equal to SNR_{50}. The receiver requires a certain *minimum* SNR to produce an acceptable output SNR and this can be labeled SNR_m. The probability of the SNR exceeding this minimum level for the sky wave path just described is given by

$$P\% = \frac{80(\text{SNR}_{10} - \text{SNR}_{50})}{(\text{SNR}_{10} - \text{SNR}_{50}) + (\text{SNR}_m - \text{SNR}_{50})} - 30 \qquad \text{if } \text{SNR}_m \geqslant \text{SNR}_{50} \quad (3.38a)$$

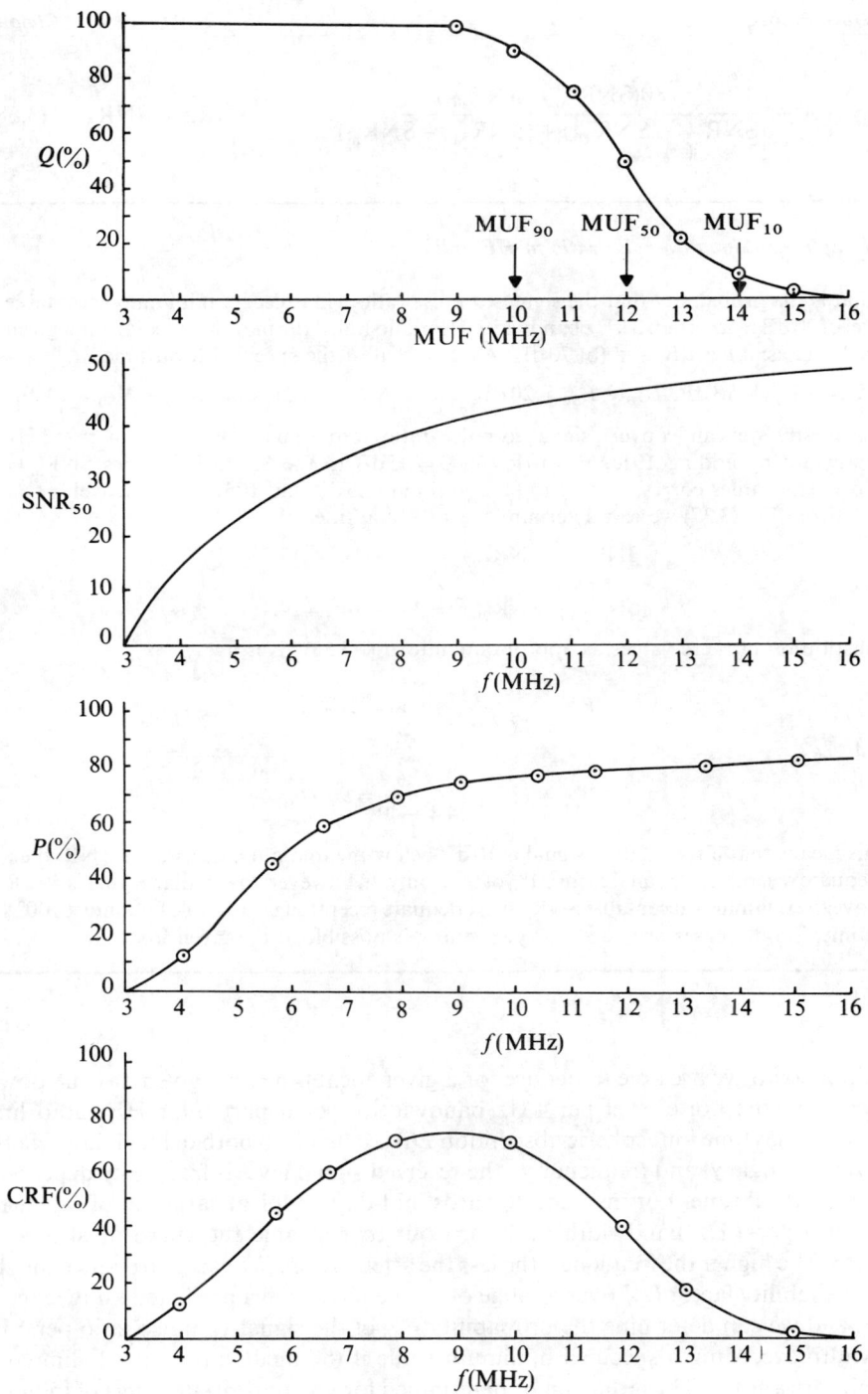

Fig. 3.18 The reliability of an HF radio link.

$$P\% = 130 - \frac{80(\text{SNR}_{50} - \text{SNR}_{90})}{(\text{SNR}_{50} - \text{SNR}_{90}) + (\text{SNR}_{50} - \text{SNR}_{m})} \qquad \text{if } \text{SNR}_m < \text{SNR}_{50} \quad (3.38b)$$

Example 3.8 *Signal-to-noise ratio in HF radio link*

What is the probability that the signal-to-noise ratio will exceed a minimum acceptable level of 30 dB in a certain 3 kHz bandwidth HF radio link if the monthly median for a given local time is: (a) 20 dB, and (b) 70 dB? Assume that at the specified local time

$$L_{50} - L_{90} = 13\,\text{dB}, \quad L_{10} - L_{50} = 20\,\text{dB}, \quad N_{50} - N_{90} = 12\,\text{dB} \quad \text{and} \quad N_{10} - N_{50} = 14\,\text{dB}.$$

Firstly we can convert signal-to-noise ratio into signal-to-noise ratio per 1 Hz bandwidth by adding $10 \log B = 10 \log 3000 = 35\,\text{dB}$ to the SNR. This gives $\text{SNR}_m = 65\,\text{dB}$. The values corresponding to (a) and (b) are 55 dB and 105 dB, respectively.

From Eq. (3.37) we can determine the two quantities

$$(\text{SNR}_{10} - \text{SNR}_{50})^2 = 13^2 + 12^2 = 17.7^2$$

and

$$(\text{SNR}_{50} - \text{SNR}_{90})^2 = 20^2 + 14^2 = 24.4^2$$

Substitution of these values as appropriate into Eq. (3.38) gives us

$$P\% = \frac{80 \times 17.7}{17.7 + 10} - 30 = 21.1\%$$

and

$$P\% = 130 - \frac{80 \times 24.4}{24.4 + 40} = 99.7\%$$

This means that if the median signal is 10 dB below the minimum acceptable SNR, then adequate reception is available for 21% of time only. If, however, the median signal is 40 dB above the minimum acceptable SNR, then adequate reception is available for almost 100% of time, in both cases provided a skywave link is possible at the given frequency.

Thus so far we are able to deduce for a given local time on a given day the power received and the noise level per 1 Hz bandwidth over a particular HF radio link. Because the daytime ionospheric absorption $L(i)$ is a function both of local time (via the sun's zenith angle χ) and frequency f, the received signal level is frequency dependent for a given local time. For instance, towards mid-day a typical variation of signal-to-noise ratio per 1 Hz bandwidth may turn out to resemble the curve illustrated in Fig. 3.18. The higher the frequency, the less the attenuation. We can also deduce the sky wave availability factor $Q\%$ over a range of frequencies from a predicted knowledge of MUF, and we can determine the probability $P\%$ of the signal-to-noise ratio per 1 Hz bandwidth exceeding a specified minimum value if the median (and decile limits) of the SNR are known. The latter can be determined for an appropriate range of frequencies.

Example 3.9 Circuit reliability

Assuming that the signal-to-noise ratio per 1 Hz bandwidth over a certain HF radio link varies with frequency in the manner illustrated in Fig. 3.18, determine the values for $P\%$ if SNR_m per 1 Hz bandwidth = 30 dB and the differences between the deciles and median are those given in Ex. 3.8.

The solution is indicated in Ex. 3.8, but we must now modify the equations for each frequency. Thus, for example, when $f = 4\,\text{MHz}$,

$$P\% = \frac{80 \times 17.7}{17.7 + (30 - 14)} - 30 = 12\%$$

and when $f = 11.2\,\text{MHz}$,

$$P\% = 130 - \frac{80 \times 24.4}{24.4 + (44 - 30)} = 79\%$$

The results of such calculations over the given frequency range are illustrated in Fig. 3.18.

We now have two graphs of $Q\%$ and $P\%$ against frequency f. The overall *circuit reliability factor* is defined as the product of these two quantities and, making appropriate allowances for the percentages, is therefore given by

$$\text{CRF}\% = \frac{Q\% \times P\%}{100} \tag{3.39}$$

For the details given in Ex. 3.7 and Ex. 3.9, the curve of $\text{CRF}\%$ is as illustrated in Fig. 3.18. It is obvious that for the given HF radio link there is an optimum frequency for highest reliability. We have used an arbitrary signal-to-noise ratio per 1 Hz bandwidth of 30 dB to illustrate the technique. If the circuit is to be used for good quality single-sideband voice channel with a bandwidth of about 3.4 kHz, the actual SNR per 1 Hz bandwidth needs to be about 68 dB. Poorer quality reception is available with a SNR per 1 Hz bandwidth down to 50 dB. These figures correspond to actual signal-to-noise ratios of about 33 dB and 15 dB. To provide the necessary SNR the transmitter power level must be increased appropriately.

Because of their different bandwidths, double-sideband radiotelephony needs to have a SNR about 6 dB higher, broadcasting about 20 to 30 dB higher, but telegraphy about 10 dB lower.

3.7 HIGH-FREQUENCY ANTENNAS

Despite the advent of new radio communication systems, including satellite links, there is still a great need for the sky wave mode of radio wave propagation. Consequently there have been continued improvements in the design and construction of high-

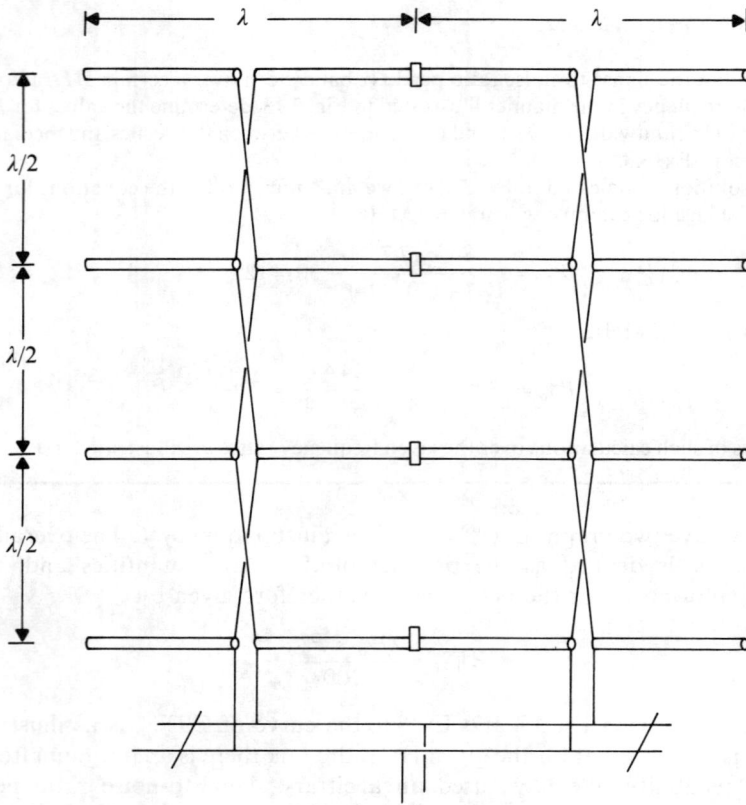

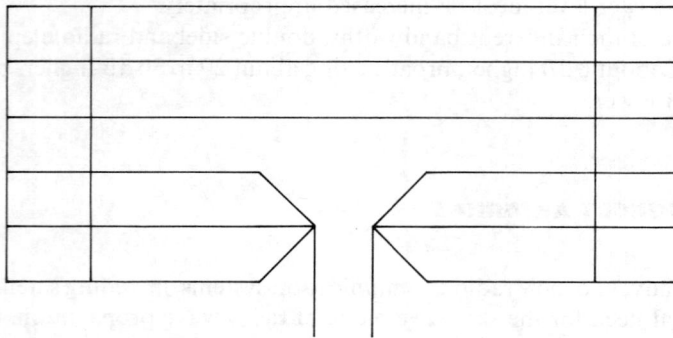

Fig. 3.19 The pine-tree antenna array and the flat folded dipole.

frequency transmitters and receivers and also of high-frequency antennas. A halfwave dipole can be used for sky wave communication. It has low gain and little in the way of directive properties. To improve matters in each case, several dipole elements are combined to form an array, described in more detail in Chap. 5, and the resulting antenna has higher gain in a specific direction. As a general rule, if correctly designed, the greater the number of elements in the array, the higher the gain and the better the directive properties.

A typical early array of this sort is known as the *pine-tree* or *Kooman array*. The curtain array of full-wave dipoles illustrated in Fig. 3.19 has a gain of about 12 dB, increased by about 3 dB when used with parasitic reflectors and by almost 6 dB if used with a perfectly conducting ground plane to produce an image of the antenna. The total gain thus approaches 20 dB in a given direction but the frequency of operation is limited due to the resonant requirements of the full-wave dipoles. If h is the mean height of the array above ground, the angle of launch $\Delta°$ of the main beam is given approximately by $\sin \Delta = \lambda/4h$. Beam slewing up to about $10°$ can be achieved by adjusting the phasing of the feed to one-half of the array or the other.

Halfwave dipoles can be modified to produce broadbanding and to increase the input impedance. Hence curtain arrays of stacked broadband folded dipoles ($Z_{in} \approx 300 \, \Omega$) are more commonly used for short wave broadcasting and these arrays can be designed to cover four adjacent short wave broadcasting bands with a voltage standing wave ratio less than about 1.5:1. One such array uses a flat folded dipole radiating element made out of six wires, as illustrated in Fig. 3.19, with an input impedance of $300 \, \Omega$. An array of such elements can be designed to transmit large peak powers of the order of megawatts and be capable of beam slewing by as much as $\pm 30°$.

Another type of high-frequency antenna that came into use in the 1940s is known as the *rhombic antenna*. The basic traveling wave radiator of the type described by Beverage has been mentioned in Sec. 2.5.2. It consists of a horizontal wire, several wavelengths long, supported about 10 m above the ground. When excited at one end and appropriately terminated at the other, it behaves as a traveling wave radiator. It may be considered as a long array of numerous incremental linear radiators, each fed by the traveling wave. The polar diagram of such an array is like a hollow cone, as illustrated in Fig. 3.20, with various sidelobes (omitted from the sketch). The angle subtended by the major lobe ψ and its gain in that direction depend on the length of the wire in terms of the wavelength.

If the horizontal wires are shaped into a diamond or rhombic form, with angle ϕ (Fig. 3.20) roughly equal to the complement of ψ, the polar patterns reinforce under appropriate conditions in the direction indicated. The small variation of ψ with L/λ indicates why such antennas operate over a 2:1 band of frequencies. There are also unwanted sidelobes to consider.

Combined with the earth plane, the rhombic antenna produces a vertical polar diagram with an angle of elevation Δ and a gain G which are both interrelated to the basic parameters of L (length of antenna arm), ϕ (the semi-angle) and h (the height of the horizontal wires above ground level). Designs for higher gains tend to produce low angles of elevation Δ, but this is a favorable situation for long-distance communication. Similarly, designs involving small angles of elevation are related to larger values of

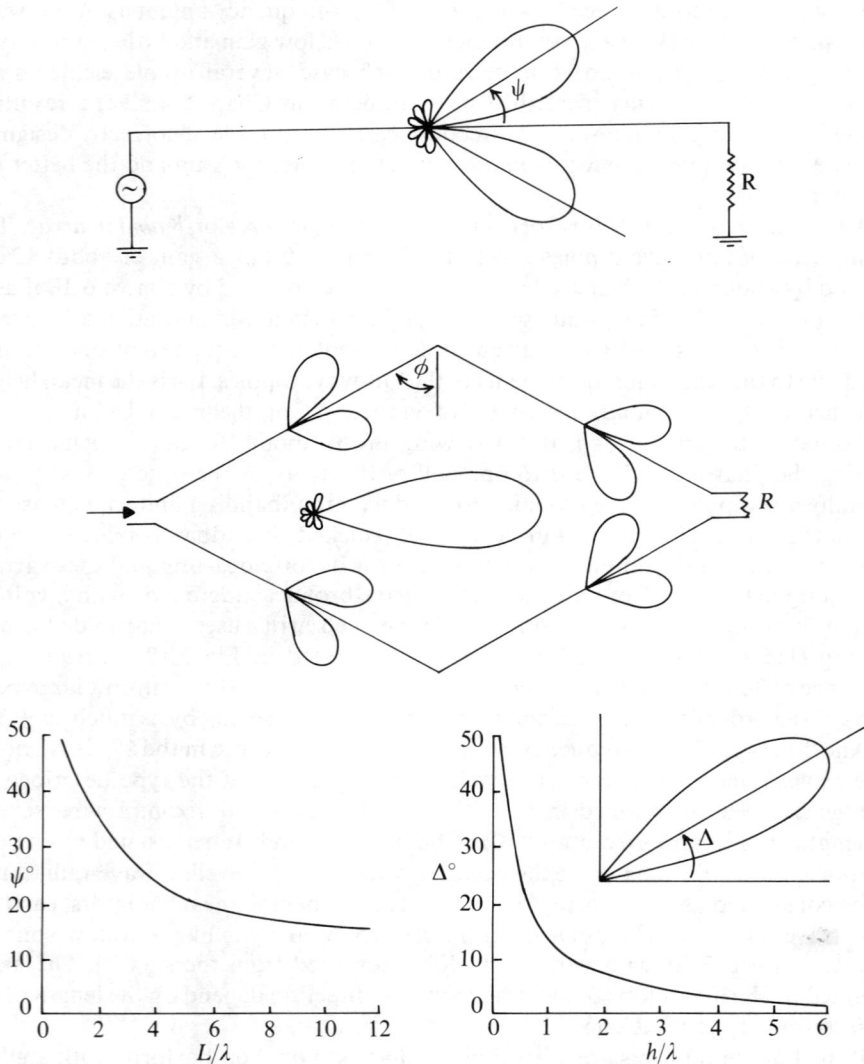

Fig. 3.20 The rhombic antenna and some associated parameters.

height, and typical values of $h = 25$ m and $h = 50$ m are to be found in practice, with $\Delta \approx 5$ to $10°$ or more commonly 7 to 8°.

For distances less than, say, 4000 km the angle of elevation for the most favorable propagation conditions varies widely according to range, time of day, etc. A more flexible system is required than that provided by the rhombic antenna, and this is made possible with the aid of the *logarithmically periodic antenna*, LPA. The LPA is an array of linear elements carefully dimensioned so that its performance remains realistically uniform over a wide range of frequencies. Gains up to about 12 dB can be achieved at high frequencies and by adjusting the alignment of the antenna to the ground a variety of radiation patterns can be obtained which are suitable for short-, medium- or long-range communication.

The LPA is usually smaller than the rhombic antenna and is commonly a double curtain type for use with higher power. Some may be constructed so as to produce vertical polarization and others may be attached horizontally to the top of a rotating mast in order to provide horizontally polarized radiation or reception in any desired direction. Alternatively, for larger versions, six LPAs may be aligned in a circle to provide reasonable all-round coverage. With modifications in their design, it is possible to achieve a given performance with even fewer elements and hence produce antennas with smaller overall dimensions.

The arrays, rhombics and log-periodic antennas can be used for transmitting and receiving high-frequency sky wave signals. Electrically small antennas can also be used for receiving high-frequency communications under appropriate conditions because the received level of sky wave atmospheric radio noise is high in the HF band and the antenna–receiver noise figure has comparatively little effect on the received signal-to-noise ratio.

The omnidirectional *whip antenna* may be used when the angle of elevation is low

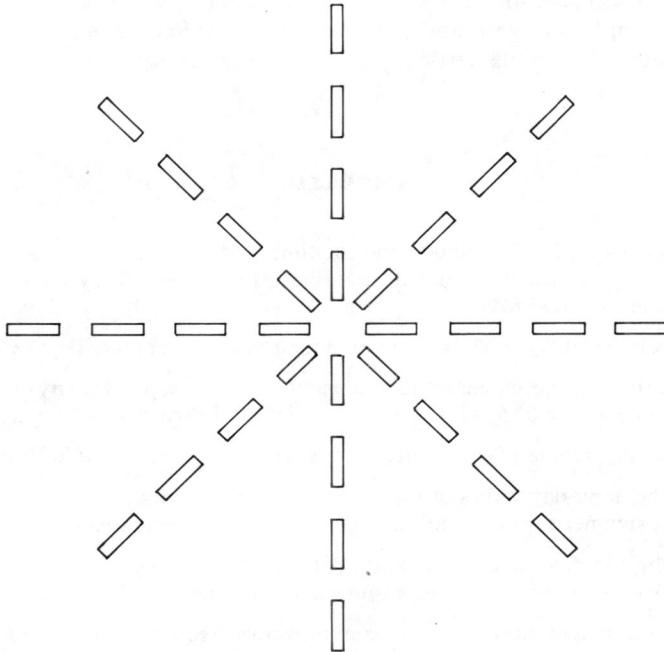

Fig. 3.21 A plan view of 32 small vertical loop antennas arranged so that, with appropriate connections, reception is available from eight different directions in the azimuth plane.

and the polarization is vertical. It needs tuning for maximum response, and self-tuning techniques are available. Even so, the frequency range is restricted so that approximately $0.25 < L/\lambda < 0.625$. *Broadband monopoles* using the principle of the inverted cone have been designed to operate without tuning over a 2.5:1 frequency range.

The *loop antenna* is a useful small receiving antenna at high frequencies. The vertical loop has a figure-of-eight polar pattern and several loops in line can be set up in an array formation with reception bidirectional along the line of the loops. Four sets of eight loops arranged in star formation (four loops at each 45° direction) can be adjusted to have available a choice of eight beams at 45° intervals with adequate overlap, thus forming a multibeam array (Fig. 3.21).

Sometimes the requirement is not for a directional antenna for fixed point-to-point communication but for a more omnidirectional arrangement in order to provide links with mobile receivers on board ships or aircraft. Horizontally polarized wideband omnidirectional antennas are based on the same principle as that of the UHF television transmitting antenna (Fig. 5.10). Horizontal dipoles are arranged in a quadrant to provide almost omnidirectional properties in the horizontal plane when their feeds are correctly phased. To enhance the gain and the vertical polar pattern, vertical stacking of the quadrants provides broadside array improvements in gain and directive properties. For HF antennas the vertical arrangement is designed on the LPA principle for wideband operation, and sometimes two such arrays of quadrants are stacked vertically for even higher gain and power capabilities. Such complex antennas have higher gains than the simple vertically polarized monopole.

PROBLEMS

3.1 Using Figs. 3.4 and 3.5, estimate the daytime and night-time values of ionospheric conductivity σ, permittivity ε_r' and attenuation α dB/km for the D-layer at height $h = 80$ km and frequency $f = 0.85$ MHz.

3.2 Repeat Prob. 3.1 at $f = 8$ MHz to observe the change in magnitudes of σ, ε_r' and α.

3.3 Determine the daytime and night-time values of the complex permittivity of the ionosphere at $h = 80$ km and $f = 0.85$ MHz, ignoring the effect of earth's magnetic field.

3.4 What is the magnitude of earth's magnetic field at a location in which $f_H = 1.5$ MHz?

3.5 Estimate the noon-day values of the zenith angle of the sun at latitude 50°N around the winter and summer solstices and the spring and autumn equinoxes.

3.6 Estimate the mid-day values of f_0E and f_0F1 at latitude 50°N at the time of the previously described solstices and equinoxes. Assume a sunspot number $R_{12} = 110$.

3.7 What are the magnitudes of the maximum usable frequency MUF and the optimum working frequency OWF associated with a high-frequency radio link extending over $d = 2000$ km via reflection from the F1 layer only, taking Fig. 3.2 as representative of the ionosphere?

3.8 Determine the path length p, the spatial loss $L(s)$ at the OWF and the angle of launch Δ of the radio link described in Prob. 3.7.

3.9 Determine the angle of incidence into the ionosphere at height $h = 100$ km for the radio link described in Probs. 3.7 and 3.8.

3.10 Estimate the mid-winter daytime ionospheric loss of the radio link previously described. Assume $f = $ OWF, $f_H = 1.2$ MHz, $R_{12} = 110$ and $\chi = 73.45°$ at $50°$ N.

3.11 What is the received power in the case of the radio link just described if $P_t = 1$ kW, $G(t) = 15$ dB and $G(r) = 10$ dB?

REFERENCES

3.1 Picquenard, A., *Radio Wave Propagation*. London: The Macmillan Press Ltd., 1974.

3.2 Piggott, W. R., 'Collision frequencies in the D-region and stratospheric-mesospheric relations,' in *Progress in Radio Science, 1963–66*, Brussels: U.R.S.I., 1967, pp. 826–54.

3.3 Davies, K., *Ionospheric Radio Propagation*, N.B.S. Monograph 80, Washington, 1965.

3.4 CCIR, 'Second CCIR computer-based interim method for estimating sky-wave field strength and transmission loss at frequencies between 2 and 30 MHz,' Supplement to *Report* 252–2, International Telecommunication Union, Geneva, 1980.

3.5 Knight, P., 'MF propagation: a wave-hop method for ionospheric field-strength prediction,' *BBC Engineering*, 100, (June 1975), 22–33.

3.6 CCIR, 'Interim method for estimating sky-wave field strength and transmission loss at frequencies between the approximate limits of 2 and 30 MHz,' *Report* 252–2, International Telecommunication Union, Geneva, 1970.

3.7 Lucas, D. L., and G. W. Haydon, 'Predicting statistical performance indexes for high frequency ionospheric telecommunication systems,' *ESSA Tech. Report*, IER, 1-ITSA 1, US Government Printing Office, Washington, D.C. 20402.

3.8 Wakai, N., 'Nomogram for easy readout of the night-time absorption,' CCIR IWP 6/1, *Document* 12.

3.9 Barghausen, A. F., 'Predicting long-term operational parameters of high-frequency sky-wave telecommunication systems,' *ESSA Tech. Report*, ERL 110-ITS 78, US Government Printing Office, Washington, D.C. 20402.

3.10 CCIR, 'World distribution and characteristics of atmospheric radio noise,' *Report* 322–1, International Telecommunication Union, Geneva, 1974.

3.11 Bradley, P. A., and C. Bedford, 'Prediction of h.f. circuit availability,' *Electronic Letters*, 12, (January 1976), 32–33.

3.12 Maslin, N. M., 'Assessing the circuit reliability of an h.f. sky-wave air-ground link,' *The Radio & Electronic Engineer*, 49, (October 1978), 493–503.

4

Space Wave Propagation in the Troposphere

In this chapter we restrict ourselves to the consideration of the basic principles of space wave propagation at VHF and UHF within the lower troposphere, leaving to later chapters the related topics of tropospheric scatter, satellite communication links and microwave radio links. Five effects in particular will be covered: *reflection* by the ground, *refraction* by the troposphere, *diffraction* by obstacles along the propagation path, *scattering* by an urban environment, and *fading*.

4.1 LINE-OF-SIGHT RANGE

Because of the curvature of the earth there is a limiting distance at which a receiving antenna has an unobstructed view of the transmitting antenna. This is illustrated in Fig. 4.1 for the case of a smooth earth path. Simple geometry indicates that the maximum line-of-sight range is given by $d = d_1 + d_2$ with:

$$d_1^2 = (h_t + R)^2 - R^2 = h_t^2 + 2h_t R \approx 2h_t R$$
$$d_2^2 = (h_r + R)^2 - R^2 = h_r^2 + 2h_r R \approx 2h_r R$$

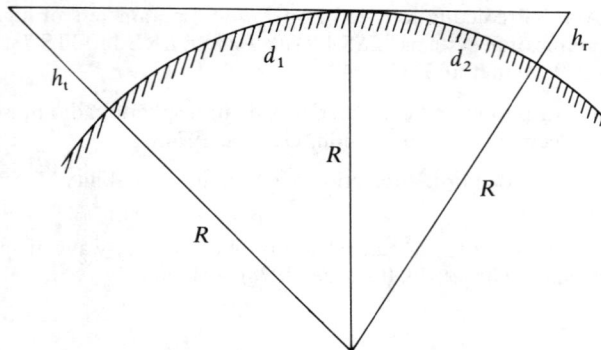

Fig. 4.1 Maximum unobstructed view of the transmitting antenna due to the curvature of a smooth earth.

and

$$d = \sqrt{2R}(\sqrt{h_t} + \sqrt{h_r}) \tag{4.1}$$

where h_t and h_r represent the heights of the transmitting and receiving antennas and R is the effective radius of the earth. It is shown later (Sec. 4.4) that under certain 'standard' atmospheric refraction conditions, $R \approx 8500$ km. Thus if distance is measured in kilometers and heights in meters, the 'standard' atmosphere will produce:

$$d = 4.12(\sqrt{h_t} + \sqrt{h_r}) \tag{4.2}$$

This equation assumes a smooth path between transmitter and receiver but it works reasonably well over slightly undulating ground if the heights are measured with respect to the mean terrain level near the reflection point.

4.2 REFLECTION

The laws of reflection are derived in Appendix 1. When radio waves propagate in the vicinity of two media separated by a plane boundary, the boundary conditions are satisfied when the radio waves are considered in terms of three components of the space wave: the *incident* wave, the *reflected* wave and the *transmitted* wave, as illustrated in Fig. 4.2. Ground wave contributions, usually negligible, are ignored in this description. Although the boundary conditions can be applied to any polarization, it is customary and simpler to consider two specific cases for a plane horizontal boundary: (a) the electric field of the incident wave is horizontal, or (b) the magnetic field of the incident wave is horizontal. In each of these two special cases the reflection coefficient, R_H or R_V, can be derived, as in Appendix 1, to give:

$$R_V = \frac{E_r}{E_i} = \frac{\varepsilon_r^* \cos \theta_i - \{\varepsilon_r^* - \sin^2 \theta_i\}^{1/2}}{\varepsilon_r^* \cos \theta_i + \{\varepsilon_r^* - \sin^2 \theta_i\}^{1/2}} \tag{4.3a}$$

$$R_H = \frac{E_r}{E_i} = \frac{H_r}{H_i} = \frac{\cos \theta_i - \{\varepsilon_r^* - \sin^2 \theta_i\}^{1/2}}{\cos \theta_i + \{\varepsilon_r^* - \sin^2 \theta_i\}^{1/2}} \tag{4.3b}$$

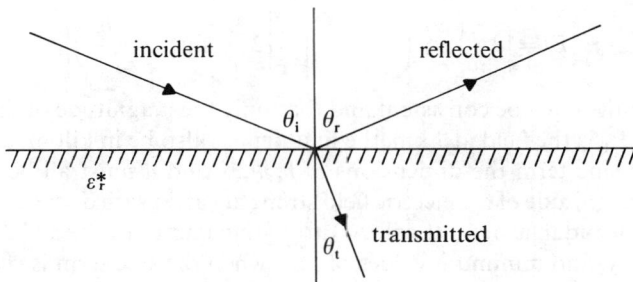

Fig. 4.2 The incident, reflected and transmitted waves needed to satisfy the boundary conditions between two media.

In these equations θ_i represents the angle of incidence illustrated in Fig. 4.2 and ε_r^* is the complex relative permittivity of the ground. It is shown in Appendix 1 that:

$$\varepsilon_r^* = \varepsilon_r(1 - j\sigma/\omega\varepsilon)$$

Both reflection coefficients are complex and are commonly available in graphical form as $|R|\underline{/\psi}$ for specified general conditions. In particular, it is noted that when θ_i approaches 90° the reflection coefficient for horizontal polarization tends towards $R_H = 1\underline{/180°}$.

4.2.1 Flat Earth Reflection

One simple model of space wave propagation is illustrated in Fig. 4.3. The received signal consists of two components, the direct wave and the ground reflected wave, and it is assumed that the boundary is flat and horizontal. For the particular case in which $d \gg h_1$ with horizontal polarization and $R_H = -1$ we can write:

$$d_d = [d^2 + (h_t - h_r)^2]^{1/2} \approx d\left\{1 + \frac{1}{2}\left[\frac{h_t - h_r}{d}\right]^2\right\}$$

$$d_r = [d^2 + (h_t + h_r)^2]^{1/2} \approx d\left\{1 + \frac{1}{2}\left[\frac{h_t + h_r}{d}\right]^2\right\}$$

and

$$d_r - d_d = 2h_t h_r/d$$

Thus the path difference between the reflected wave E_r and the direct wave E_d is $2h_t h_r/d$, which corresponds to a phase lag of

$$\phi = \frac{2\pi}{\lambda}\frac{2h_t h_r}{d} = \frac{4\pi h_t h_r}{\lambda d} \quad \text{radians}$$

The magnitudes of E_d and E_r are almost identical, approximately E_1/d, where E_1 is the direct field at 1 km. In addition, the reflection coefficient $R_H = 1\underline{/180°}$ adds π radians to ϕ. The received field strength is the resultant of E_d and E_r and can be obtained from Fig. 4.3 as:

$$|E| = \left|2\frac{E_1}{d}\cos\left(\frac{\pi - \phi}{2}\right)\right| = \left|2\frac{E_1}{d}\sin\left(\frac{2\pi h_t h_r}{\lambda d}\right)\right| \quad (4.4)$$

Note that the units must be consistent and that only the magnitude of the field is being considered. As E_1 is the field at 1 km, the first d must also be in kilometers. Within the brackets of the sine term the dimensions of h_t, h_r, λ and d must all be consistent.

Thus the magnitude of the electric field strength varies with distance both inversely and in some sinusoidal fashion. It will go through maximum values of $2E_1/d$ when the sine term is unity, and minimum values of zero when the sine term is zero. In practice the reflection coefficient is never exactly -1 and the minimum values of E-field are typically within $-20\,\text{dB}$ of E_1/d, though the maximum values are sometimes almost 6 dB above E_1/d.

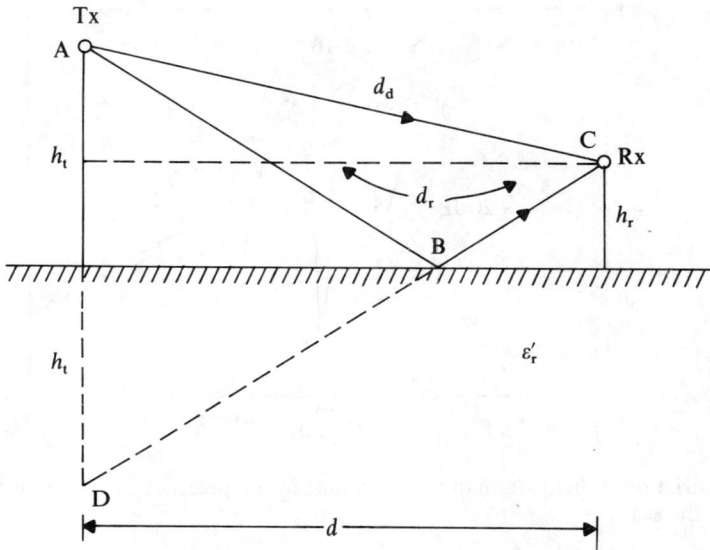

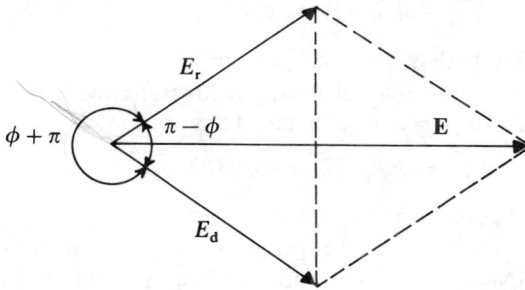

Fig. 4.3 Space wave propagation with ground reflection. The net field is the resultant of the direct and ground-reflected components.

4.2.2 Reflection with Variable Distance

We take as an example the case of an aeroplane traveling over the sea at 1000 m above sea level toward a land-based receiver located on the coast at 25 m above sea level. If the aircraft transmitter operates at 120 MHz with sufficient power to produce $E_1 = 87\,\mathrm{dB}\mu$, we can draw a sketch of the received field strength against the aircraft's range from the receiver.

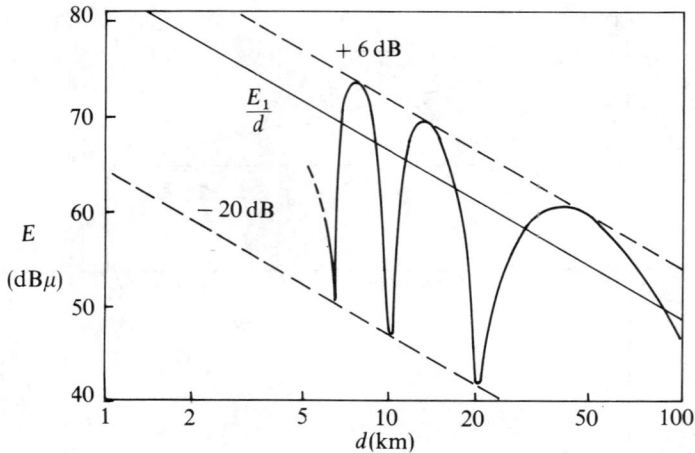

Fig. 4.4 Variation of field strength with distance in the presence of reflection from the surface of the sea.

The successive maximum values of electric field strength given by Eq. (4.4) occur when the term within the brackets is $k\pi/2$, where $k = 1, 3, 5$, etc. Thus

$$d_{max} = 4h_t h_r/(\lambda k) = 40,000/k$$

in meters, giving $d_{max} = 40.0, 13.3, 8.0, 5.7, 4.4, 3.6$ km, etc.

The corresponding minimum values of electric field strength occur when the term within the bracket is $k\pi/2$, with $k = 2, 4, 6, 8$, etc. Thus

$$d_{min} = 4h_t h_r/(\lambda k) = 40,000/k$$

in meters, giving $d_{min} = 20.0, 10.0, 6.7, 5.0, 4.0, 3.3$ km, etc.

These points can be located on the graph in Fig. 4.4 at the appropriate intersections with the free space plus 6 dB line and the free space minus 20 dB line. The sinusoidal variation in between may be sketched or calculated from the logarithmic form of Eq. (4.4),

$$E(\text{dB}\mu) = E_1(\text{dB}\mu) + 6 - 20\log d(\text{km}) + 20\log\left|\sin\left(\frac{2\pi h_t h_r}{\lambda d}\right)\right| \qquad (4.5)$$

It must be noted that the conditions $d \gg h_1$ and $R_H = -1$ do not hold for small values of d.

4.2.3 Reflection with Variable Wavelength

A receiver may be designed to pick up several channels, such as a multi-channel UHF television receiver. For each of, say, four channels the maximum and minimum signal

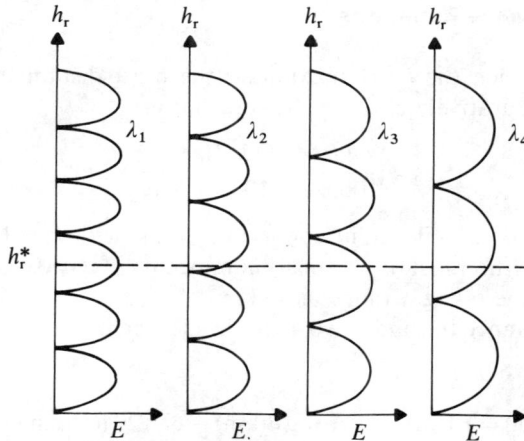

Fig. 4.5 The variation of field strength with receiver height and signal wavelength.

strengths occur when

$$\frac{2\pi h_t h_r}{\lambda d} = \frac{k\pi}{2}$$

with $k = 1, 3, 5$, etc., for maximum, and $k = 2, 4, 6$, etc., for minimum. For a fixed transmitter height h_t and a fixed distance d between transmitter and receiver, the magnitude of the electric field strength will vary with height h_r and wavelength λ as shown in Fig. 4.5. At a given antenna height h_r^* the receiver signal strength varies from channel to channel as shown even though equal powers are transmitted for each channel. In practice this variation may not be noticeable to the television viewer as the receiver automatically compensates for signal variations except when the signal strength is very small near the limits of the transmitter coverage area. Then it will be more apparent.

4.2.4 Reflection with Variable Heights and Wavelengths

A situation can arise when the point of reflection varies with time, such as the propagation of signals across a tidal estuary where the sea level varies by several meters daily, particularly at certain times of the year. This causes changes in the effective values of h_t and h_r, which are measured with respect to the point of reflection. In addition, the variation with wavelength or frequency is similar to that described in Sec. 4.2.3.

Again, in practice this may not be too much of a problem for radio broadcasting or television transmissions, for example, when distance d is small because the nulls are rarely more than about 20 dB below the free-space value and the receivers can easily cope with this amount of variation. However, in the case of low-power microwave links operating at several GHz across larger estuaries or bays, the effect of fading needs to be considered more carefully.

4.2.5 Inverse Distance Equations

In those cases in which the $k = 1$ maximum (the most distant maximum from the transmitter) is comparatively close to the transmitter:

$$d_{k=1} = \frac{4h_t h_r}{1000\lambda} = \frac{4h_t h_r f(\text{MHz})}{1000 \times 300} \quad \text{km} \tag{4.6}$$

For example, in VHF mobile radio communications, with $h_t = 100\,\text{m}$, $h_r = 3\,\text{m}$ and $f = 100\,\text{MHz}$, the farthermost ($k = 1$) maximum occurs at $d = 0.4\,\text{km}$. By comparison, Eq. (4.1) gives the line-of-sight range as 48 km.

For a short distance beyond $d = 0.4\,\text{km}$ the electric field strength is best given by the full equation:

$$E(\text{dB}\mu) = E_1(\text{dB}\mu) - 20\log d(\text{km}) + 6 + 20\log\left|\sin\frac{2\pi h_t h_r}{\lambda d}\right| \tag{4.7}$$

For longer distances towards the line-of-sight limit we can use the approximation $\sin\theta = \theta$ so that

$$\sin\frac{2\pi h_t h_r}{\lambda d} = \frac{2\pi h_t h_r f(\text{MHz})}{300\, d(\text{km})\, 1000}$$

and Eq. (4.7) reduces to:

$$E(\text{dB}\mu) = E_1(\text{dB}\mu) + 20\log(h_t h_r) + 20\log f - 87.6 - 40\log d \tag{4.8}$$

with heights in meters, frequency in megahertz and distance in kilometres. The error involved in this approximation is illustrated in Fig. 4.6 for $h_t = 50\,\text{m}$, $h_r = 3\text{m}$, $f = 450\,\text{MHz}$ and $E_1 = 100\,\text{dB}\mu$, giving $d_{k=1} = 0.9\,\text{km}$. Note also that the $40\log d$ term in Eq. (4.8) implies a $1/d^2$ variation in electric field strength.

Example 4.1 Halfwave dipole radiating 1 kW

Derive an approximate expression for the electric field produced by a halfwave dipole radiating 1 kW at a distance in excess of $d_{k=1}$.

We can use Eq. (1.15d) with $k' = 106.9$, $P_t(\text{dBkW}) = 0$ and $20\log d = 0$ to obtain $E_1(\text{dB}\mu) = 106.9$ for a halfwave dipole. Substitution into Eq. (4.8) gives:

$$E(\text{dB}\mu) = 19.3 + 20\log(h_t h_r) + 20\log f - 40\log d \tag{4.9}$$

with frequency in MHz and distance in km. This is a commonly used reference in mobile radio with which to compare losses due to multipath propagation in urban areas.

An alternative form of the equation can be obtained in terms of signal power levels. Knowing that, in linear terms,

$$P_r = P_a \times A_e = \frac{E^2}{120\pi} \times G_r \times \frac{\lambda^2}{4\pi}$$

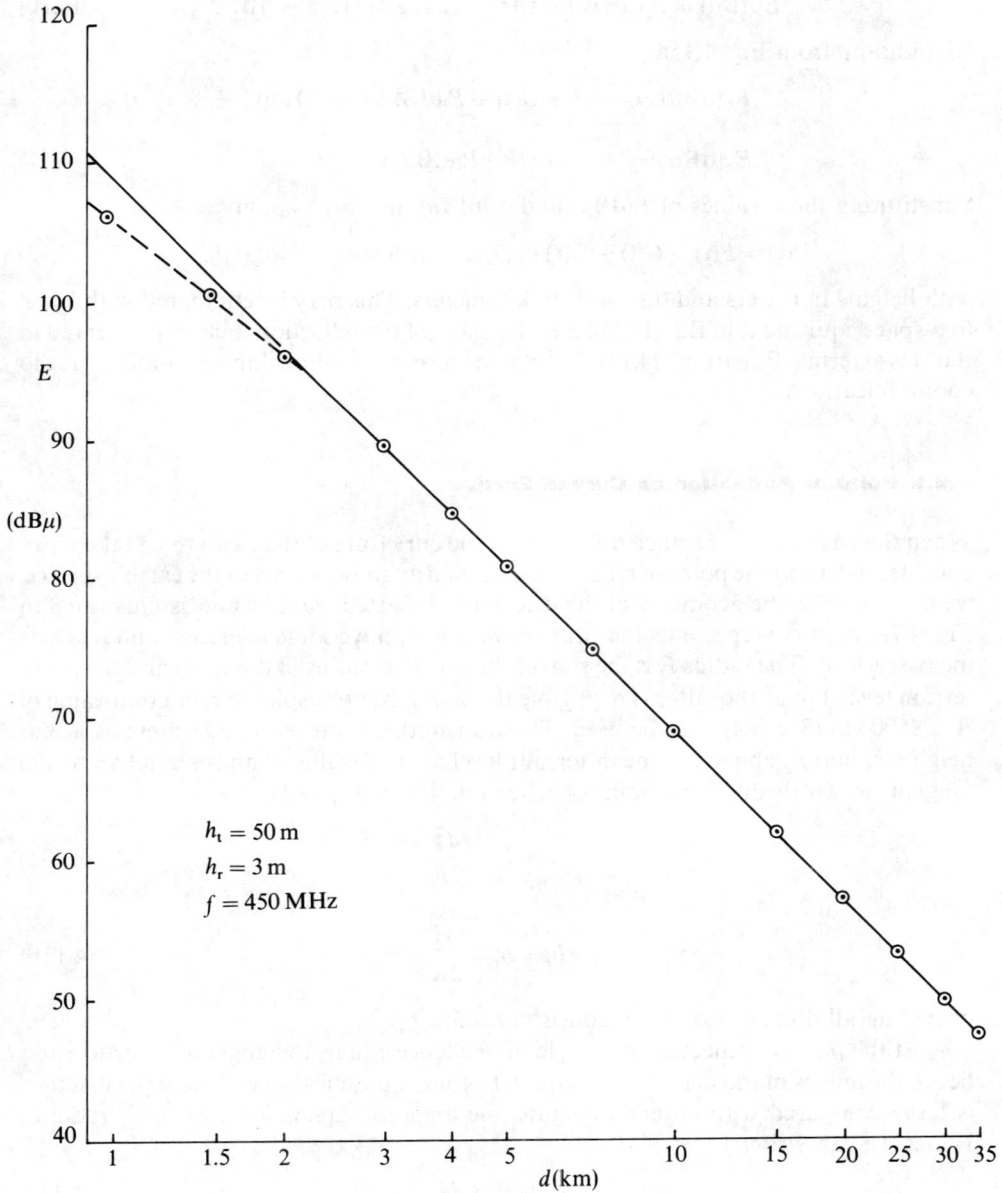

Fig. 4.6 The approximate value of field strength beyond the farthermost ($k = 1$) maximum. The broken line indicates the exact values.

the corresponding logarithmic relationship is

$$E(\mathrm{dB}\mu) = P_r(\mathrm{dBW}) - G(r) + 20\log f(\mathrm{MHz}) + 107.2 \tag{4.10}$$

In addition, from Eq. (1.15a),

$$E_1(\mathrm{dB}\mu) = 104.8 + G(t) + P_t(\mathrm{dBkW}) - 20\log 1$$

or

$$E_1(\mathrm{dB}\mu) = 74.8 + G(t) + P_t(\mathrm{dBW}) \tag{4.11}$$

Substituting these values of $E(\mathrm{dB}\mu)$ and $E_1(\mathrm{dB}\mu)$ into Eq. (4.8) gives:

$$P(r) = P(t) + G(t) + G(r) - \{120 + 40\log d - 20\log(h_t h_r)\} \tag{4.12}$$

with heights in meters and distances in kilometers. This may be compared with Friis' free-space equivalent in Eq. (1.15b) and the effect of the reflection is clearly observed in the loss term. Equation (4.12) is also a useful relationship in mobile radio communications.

4.2.6 Point of Reflection on Curved Earth

When the path length is sufficiently long for the curvature of the earth to be taken into consideration and the point of reflection is on a fairly smooth area of the earth's surface, we can consider the geometry of the direct and reflected waves to be as illustrated in Fig. 4.7. The curve represents the *mean terrain level* above some reference value such as mean sea-level. The radius R is the sum of the effective radius of the earth and the mean terrain level, but as the latter is negligible the standard atmospheric refraction value of $R \approx 8500\,\mathrm{km}$ (Sec. 4.4) may be used. The transmitting and receiving antennas are at heights h'_t and h'_r above the mean terrain level, or at heights h_t and h_r relative to the tangent drawn through the point of reflection. From Eq. (4.1):

$$h_t = h'_t - \frac{d_1^2}{2R} \tag{4.13a}$$

and

$$h_r = h'_r - \frac{d_2^2}{2R} \tag{4.13b}$$

providing all dimensions are in consistent units.

At the point of reflection the angle of incidence equals the angle of reflection and hence the angles marked ψ' are also equal. For most practical cases where the distance is large compared with antenna heights, the angle ψ' equals h_t/d_1 or h_r/d_2 radians. Hence we can write:

$$h_t/h_r = d_1/d_2 \tag{4.13c}$$

Finally, the rather obvious but essential relationship between the distances gives us the fourth equation:

$$d_1 + d_2 = d \tag{4.13d}$$

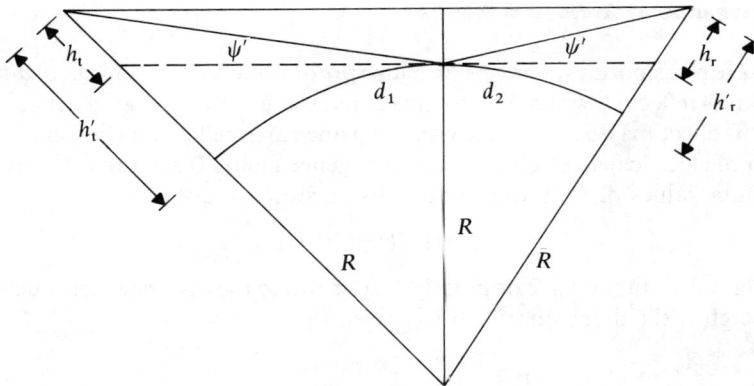

Fig. 4.7 The point of reflection on a smooth curved earth with unequal antenna heights.

For a specific set of circumstances these four equations contain four unknowns, the effective heights h_t and h_r and the distances to the point of reflection d_1 and d_2.

Example 4.2 Point of reflection on curved earth

Find the point of reflection on the curved earth between two antennas of heights $h'_t = 400$ m and $h'_r = 100$ m above mean terrain level and distance $d = 100$ km apart. Assume that the effective radius of earth R and the mean terrain level total 8500 km.

To solve the simultaneous equations we substitute the data into the first two expressions. If we use heights in meters and distances in kilometers:

$$h_t = 400 - d_1^2/17$$

and

$$h_r = 100 - (100 - d_1)^2/17$$

if the fourth expression is used to eliminate d_2. To introduce the third simultaneous equation we must divide h_t by h_r to obtain:

$$\frac{h_t}{h_r} = \frac{400 - d_1^2/17}{100 - (10{,}000 - 200d_1 + d_1^2)/17} = \frac{d_1}{100 - d_1}$$

This is an expression in terms of distance d_1 which reduces to:

$$d_1^3 - 150d_1^2 + 750d_1 + 340{,}000 = 0$$

The cubic equation can be solved by successive approximations, knowing that the point of reflection with plane earth is at $d_1 = 80$ km. The solution is about 70.1 km. Prob. 4.3 gives an alternative approach.

Substituting $d_1 = 70.1$ km into the expressions for the antenna heights gives $h_t = 110.9$ m and $h_r = 47.4$ m. These are the parameters to be used in Eq. (4.4) to obtain the field strength at distance $d = 100$ km over a curved earth.

4.2.7 Divergence of Reflected Waves

The curvature of a smooth spherical earth also produces a certain amount of *divergence* of the ground-reflected wave due to the convex reflection plane, as illustrated in Fig. 4.8. This effect may be included with the plane earth reflection coefficient R in the calculation of electric field strength via a divergence factor D such that the maximum and minimum values of E at appropriate distances d are given by

$$|E| = |E_d|(1 \pm D|R|) \tag{4.14}$$

for the spherical earth model. For practical cases where the distances are much greater than the heights, the divergence factor is given (Ref. 4.1) by:

$$D \approx \left[1 + \frac{2d_1 d_2}{R(h_t + h_r)} \right]^{-1/2} \tag{4.15}$$

in SI units. If d_1 and d_2 are measured in kilometers and h_t and h_r are given in meters, $R = 8.5$ in the equation for standard atmospheric conditions.

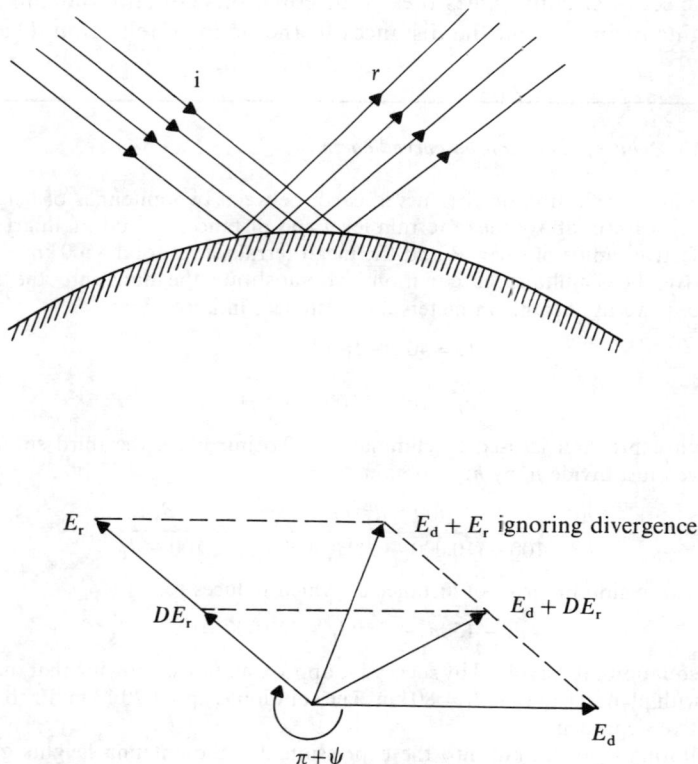

Fig. 4.8 The convex curvature of earth's surface can cause the reflected waves to diverge. The resultant field is then the phasor sum of the direct field and the modified reflected field.

Example 4.3 Effect of curvature and divergence

Compare the field strengths at the receiver for the space wave radio link described in Ex. 4.2 if it is assumed the earth is: (a) flat, and (b) curved with effective radius $R = 8500$ km. The frequency is 100 MHz.

We can use Eq. (4.4) to obtain the solution for the flat-earth case with $h_t = 400$ m, $h_r = 100$ m, $d = 100$ km and $\lambda = 3$ m.

$$E = \frac{2E_1}{d} \sin\left[\frac{2\pi h_t h_r}{\lambda d}\right] = 14.9 \times 10^{-3} E_1$$

Ignoring the divergence factor we can use Eq. (4.4) with effective antenna heights $h_t = 110.9$ m, $h_r = 47.4$ m, distance $d = 100$ km and wavelength $\lambda = 3$ m from Ex. 4.2 to obtain

$$E = \frac{2E_1}{100} \sin\left[\frac{2\pi \times 110.9 \times 47.4}{3 \times 100,000}\right] = 2.2 \times 10^{-3} E_1$$

Including the divergence factor with the plane earth reflection coefficient, we first obtain

$$D = \left[1 + \frac{2d_1 d_2}{R(h_t + h_r)}\right]^{-1/2} = 0.49$$

The direct field component is simply $E_1/d = 10 \times 10^{-3} E_1$ to match the form of the earlier calculations. The reflected field, assuming $R_H = -1$, is only $0.49 \times E_1/100 = 4.9 \times 10^{-3} E_1$. The phase lag of the reflected field component, including the effect of $R_H = 1\underline{/180°}$, is

$$\psi = \frac{4\pi h_t h_r}{\lambda d} + \pi = 0.22 + \pi \quad \text{radians}$$

The resultant field is the vector addition of the direct and reflected components, as illustrated in Fig. 4.8. By this method we can see that (in units of $10^{-3} E_1$):

$$E^2 = (10 - 4.9\cos 0.22)^2 + (4.9\sin 0.22)^2$$

where $E = 5.3 \times 10^{-3} E_1$ is the magnitude of the field at 100 km.

Divergence has the effect of reducing the field strengths above the free-space value, but enhancing the field strengths below the free-space value.

4.2.8 Fresnel's Ellipsoid

We can derive an expression for the locus of the point of reflection on which the path difference between the direct and reflected waves is an integer number of half wavelengths with the aid of Fig. 4.9. This turns out to be the equation of an *ellipsoid*. However, the mathematics can be simplified considerably by assuming $d_1 \gg r$ and $d_2 \gg r$. This is realistic except in the immediate vicinity of the transmitting and receiving antennas. Hence:

$$AB + BC = AC + k\lambda/2$$

$$(d_1^2 + r^2)^{1/2} + (d_2^2 + r^2)^{1/2} = d_1 + d_2 + k\lambda/2$$

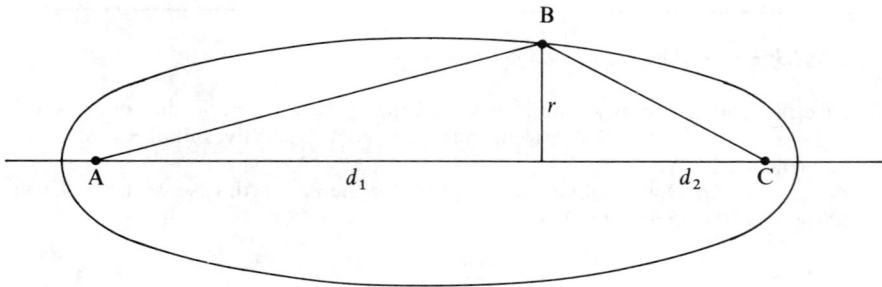

Fig. 4.9 The geometry of the ellipse used to obtain Fresnel's ellipsoid.

$$d_1\left(1 + \frac{r^2}{2d_1^2}\right) + d_2\left(1 + \frac{r^2}{2d_2^2}\right) = d_1 + d_2 + k\lambda/2$$

or

$$r^2 = k\lambda \frac{d_1 d_2}{d_1 + d_2} \tag{4.16}$$

Note that when $d_1 = 0$ or $d_2 = 0$, $r = 0$, which is not correct for an ellipsoid, but for most practical applications in space wave propagation this error is negligible.

If an obstruction or reflecting area is outside the first ($k = 1$) *Fresnel zone* we can consider the direct path between the transmitter to be reasonably clear.

The antenna gains and directivities also help to reduce the effect of reflection and divergence. The reflected wave is often much smaller in initial amplitude due to the transmitting antenna's directivity, and a properly aligned receiving antenna is less sensitive to the reflected wave due to its directivity.

4.3 THE TROPOSPHERE

The troposphere is the lowest region of the atmosphere, about 6 km high at the poles and about 18 km high at the equator. It may be characterized by various parameters including *temperature* (T K), *pressure* (P mB) and *humidity* or water vapour pressure (p mB). Because of the turbulent nature of the atmosphere the parameters do not remain constant and to facilitate some tentative analysis we use simple models of the troposphere (Ref. 4.3).

Within the first 1–2 km of the troposphere the conditions may permit a simple linear relationship between the parameters and height h km such as:

$$T(h) = 290 - 6.5h \qquad \text{K}$$
$$P(h) = 950 - 117h \qquad \text{mB}$$
$$p(h) = 8 - 3h \qquad \text{mB}$$

These are only 'models' of the lower troposphere. $T(0)$, $P(0)$, $p(0)$, dT/dh, dP/dh and dp/dh must actually be measured.

4.3.1 Permittivity of the Troposphere

The permittivity of the troposphere depends on T, P and p, and is therefore itself a function of height. Its value is given by:

$$\varepsilon_r = 1 + \frac{155.1}{T}\left[P + \frac{4810p}{T}\right] \times 10^{-6} \qquad (4.17)$$

If, say, some appropriate values are $T(0) = 17\,°C = 290\,K$, $P(0) = 950\,mB$, and $p(0) = 8\,mB$, then:

$$\varepsilon_r(h = 0) = 1 + (579 \times 10^{-6}) = 1.000\,579$$

At a height $h = 1\,km$, using the linear model parameters listed above, $T(1) = 283.5\,K$, $P(1) = 833\,mB$ and $p(1) = 5\,mB$. Substitution of these values into Eq. (4.17) gives:

$$\varepsilon_r(h = 1) = 1 + (502 \times 10^{-6}) = 1.000\,502$$

The variation of permittivity with height is thus quite small.

4.3.2 Refractivity of the Troposphere

Electromagnetic waves bend when passing through propagation media with different values of permittivity, and because the bending is related directly to changes in the velocity of propagation v, where $v = c/\sqrt{\varepsilon_r}$, a parameter $n = \sqrt{\varepsilon_r}$ may be preferred. This is called the *refractive index* of the medium and is obtained from Eq. (4.17) as

$$n = \sqrt{\varepsilon_r} = \left\{1 + \frac{155.1}{T}\left[P + \frac{4810p}{T}\right]10^{-6}\right\}^{1/2}$$

or

$$n = 1 + \frac{77.6}{T}\left[P + \frac{4810p}{T}\right] \times 10^{-6}$$

using the binomial expansion. This is known as the *Smith–Weintraub relationship* (Ref. 4.2).

Using the data given in the introduction to Sec. 4.3 we obtain the corresponding refractive indexes as $n(0) = 1.000\,289\,5$ and $n(1) = 1.000\,251$. These are such awkward numbers that a simpler term *refractivity* is commonly used instead, where refractivity N is related to refractive index n via

$$N = (n - 1) \times 10^6 = \frac{77.6}{T}\left[P + \frac{4810p}{T}\right] \qquad (4.18)$$

We can now write $N_0 = 289$ and $N(1) = 251$, a much simpler set of data.

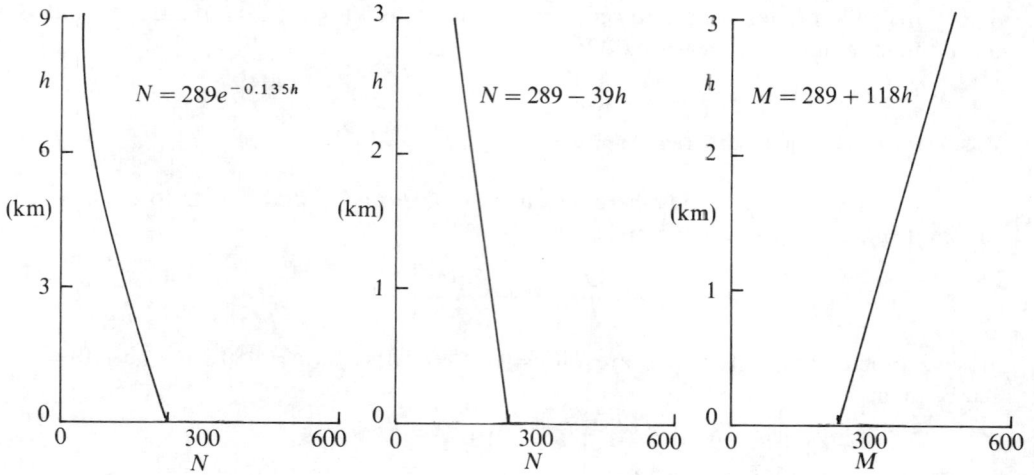

Fig. 4.10 The refractivity and modified refractivity (Sec. 4.5) variations with height at a location where $N_0 = M_0 = 289$.

4.3.3 Standard Models of the Troposphere

Simple mathematical models of the troposphere are usually defined in terms of refractivity N. Three standard versions include the *general model*, defined

$$N = N_0 \exp(qh) \tag{4.19a}$$

with h in km and coefficient q usually in the range $-0.21 < q < +0.12$; the *standard exponential atmosphere*, defined

$$N = 289 \exp(-0.135h) \tag{4.19b}$$

with $dN/dh(0) = -39$ at ground level; and the *standard linear atmosphere*, defined

$$N = 289 - 39h \tag{4.19c}$$

which is applicable for heights of less than about 2 km. These are illustrated in Fig. 4.10.

4.4 CURVATURE OF SPACE WAVE IN THE TROPOSPHERE

The bending of electromagnetic waves as they pass through a medium with varying permittivity or refractivity is governed by *Snell's law*. This is derived in Appendix 1. For a horizontally stratified medium such as that illustrated in Fig. 4.11:

$$n_x \sin r_x = \text{constant}$$

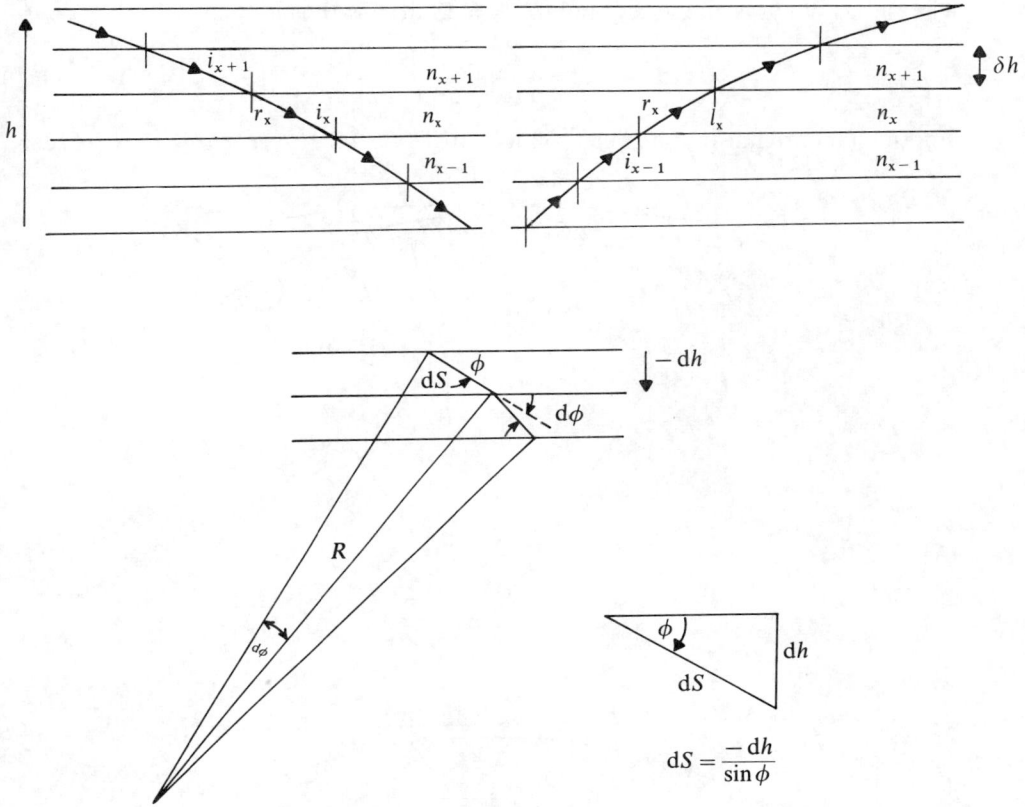

Fig. 4.11 Snell's law applied to a horizontally stratified medium.

The initial launch could be upwards or downwards from an elevated antenna, or even horizontal. For most of the time the refractive index decreases with increasing height, so that $n_x > n_{x+1}$, and hence the curvature is in the direction shown. Occasionally there are atmospheric conditions which result in $n_x < n_{x+1}$ and in such cases the curvature is reversed. Using complementary angle ϕ_x instead of r_x to simplify the mathematics slightly, we can write:

$$\sin(r_x)n_x = \cos(\phi_x)n_x = \text{constant}$$

Differentiating both sides with respect to S results in

$$\cos \phi_x \, dn_x/dS - n_x \sin \phi_x \, d\phi_x/dS = 0$$

and noting from Fig. 4.11 that $dS = - dh/\sin \phi_x$,

$$\frac{d\phi_x}{dS} = \frac{\cos \phi_x \, dn_x/dS}{n_x \sin \phi_x} = -\frac{\cos \phi_x (dn/dh)}{n_x}$$

The radius of curvature at any point, $R_x = dS/d\phi_x$, is thus

$$R_x = \frac{n_x}{\cos \phi_x} \frac{1}{(-dn/dh)} \tag{4.20}$$

For the standard linear atmosphere with $dn/dh = -39 \times 10^{-6}$ per km, the waves bend

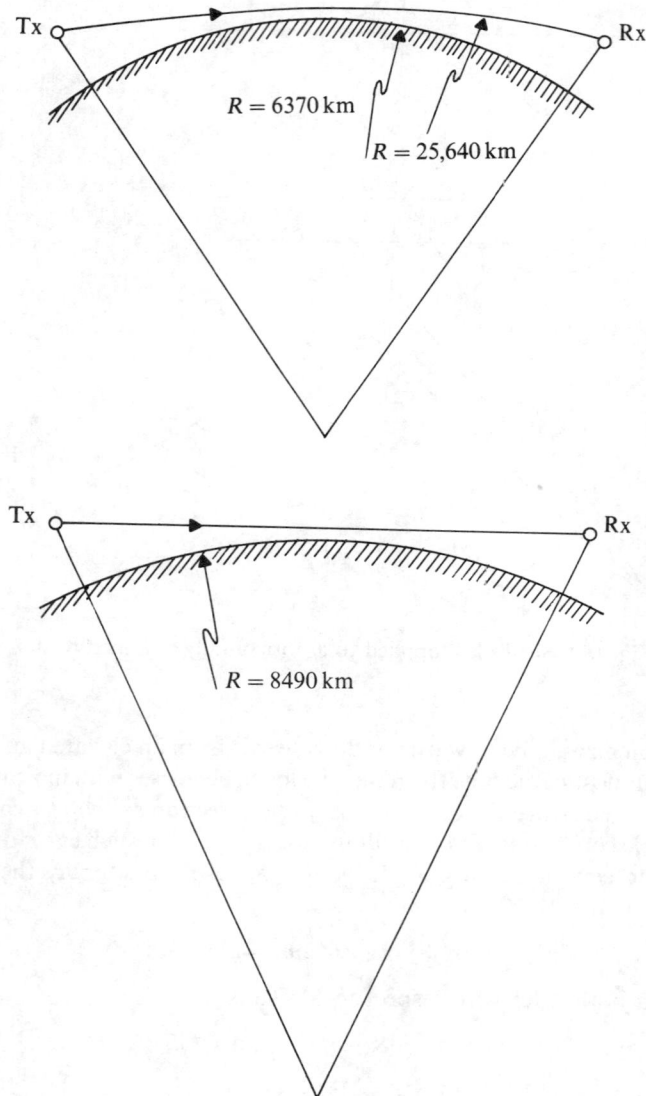

Fig. 4.12 The earth's effective radius of 8470 km permits the use of straight-line propagation paths in a standard atmosphere.

in the troposphere with a radius of curvature given by

$$R_x = 25{,}641 n_x/\cos \phi_x$$

If the launch angle is close to the horizontal, the ratio of n_x to $\cos \phi_x$ is close to unity and within this limitation the propagation path can be described as a circle of radius 25,641 km, or curvature defined $1/R = 1/25{,}641 = 39 \times 10^{-6}$. By comparison, the curvature of the earth is $1/6370 = 157 \times 10^{-6}$, as the radius of earth is 6370 km.

If the curvatures of both the propagation path and of the earth are reduced by 39×10^{-6}, as in Fig. 4.12, the propagation path has an effective curvature of zero (which is a straight line) and the earth has an effective curvature of $(157 - 39) \times 10^{-6} = 118 \times 10^{-6}$ or an effective radius of 8470 km. This is approximately four-thirds as large as the true radius of earth.

It is therefore commonplace to use a propagation profile based upon an earth with effective radius of 8470 km and then to assume that the space wave propagation is linear. Such an idea is illustrated in Fig. 4.13a. The propagation path profile is obtained

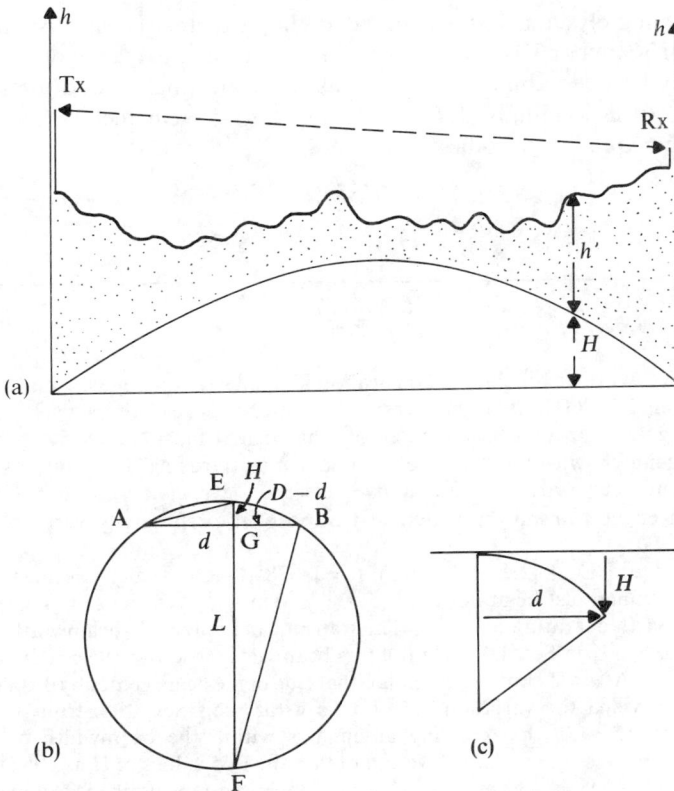

Fig. 4.13 A path profile superimposed on a four-thirds earth radius. Because of a different scaling for heights and distances, H can be determined via the geometry of a circle.

from the contours on a map with heights above some reference datum such as mean sea level. This is superimposed on the curvature of the earth assuming a radius of 8470 km. We are then able to see whether there is an unobstructed path between the transmitter and receiver by simply joining them with a straight radio wave propagation line. It is also possible to construct the appropriate Fresnel zone between transmitter and receiver with the aid of Eq. (4.16).

In reality the vast difference in scales between heights and distances on the propagation path means that such profiles are usually drawn with different horizontal and vertical scales. The dimension H can be obtained from Fig. 4.13b in which AB is perpendicular to EF. As AB is so small compared with the dimensions of the earth, EF is virtually the diameter of the earth. A geometric theorem states that $A\hat{E}F = A\hat{B}F$ and $E\hat{A}B = E\hat{F}B$, which implies that triangles AEG and GBF are similar and hence

$$H \times L = d(D - d)$$

or

$$H = \frac{d(D - d)}{L} = \frac{d(D - d)}{2K6370}$$

as L is approximately equal to the diameter of the earth and the effective radius of earth is some factor K times 6370 km. The factor K has a standard value of 4/3, but this can vary typically between about 0.7 and 2.0 as the atmospheric conditions change and dn/dh takes correspondingly different values. Using H in meters and distances in kilometers the equation becomes:

$$H = 0.0785d(D - d)/K \qquad (4.21a)$$

Example 4.4 Effect of tropospheric refractivity

A 30 km microwave link includes two paraboloidal reflector antennas, each 1 m diameter, and operating at 10 GHz. (a) If the transmitting antenna is 50 m above reference level and aligned to give horizontal beam center, at what height must the receiving antenna be placed to coincide with the beam center under four-thirds earth conditions? (b) If the receiving antenna is also at 50 m above reference level, is it within the half-power beamwidth of the transmitting antenna if tropospheric refractivity variations result in $0.67 < K < 2.0$?

We can use Eq. (4.21b) to obtain $H = 0.0785d^2/K = 53$ m. This means that the receiving antenna must be at height $50 + 53 = 103$ m above reference to compensate for four-thirds earth curvature at 30 km. The transmitting antenna's beamwidth is given by $BW \approx 70\lambda/D = 2.1°$ in Sec. 1.4.5. Half of this beamwidth is above the center of the beam and half below. At $d = 30$ km this extends either side of the beam center by $30,000 \tan 1.05°$, or 550 m. Provided the variation δH of H is within this deviation from 103 m, when $K = 0.67$ and $K = 2.0$, the receiving antenna is within the beamwidth of the transmitting antenna. Using Eq. (4.21b) we can obtain the two values of H as 105 m and 35 m. Thus $\delta H = 53 - 105 = -52$ m and $\delta H = 53 - 35 = 18$ m are unlikely to have a drastic reduction in received field strength for this reason.

The alternative method of displaying path profile, illustrated in Fig. 4.13c, produces

$$H = 0.0785 d^2 / K \qquad (4.21b)$$

In either case, the smaller values of K produce the larger values of H and hence tend to reduce the unobstructed range of a given transmitter.

4.5 MODIFIED REFRACTIVITY

In Sec. 4.4 we considered how it was easier to produce a path profile with an effective earth's radius $R = Ka$, where a is the actual radius of earth, so that the effect of refractive bending of radio waves can be ignored. An alternative approach is to alter the curvature of the radio waves so that the earth may be considered as flat. Such an alternative makes use of a modified refractive index n^* and a *modified refractivity* defined as $M = (n^* - 1) \times 10^6$.

Consider the spherical troposphere illustrated in Fig. 4.14. It was shown in Sec. 3.5.3 that Snell's law for a spherically stratified dielectric medium is given by

$$R_1 n_1 \cos \phi_1 = R_2 n_2 \cos \phi_2 = R_x n_x \cos \phi_x$$

In this case the smallest value of R is a, the radius of earth. At any given height h the corresponding refractive index is n and we can write

$$a n_0 \cos \phi_0 = (a + h) n \cos \phi$$

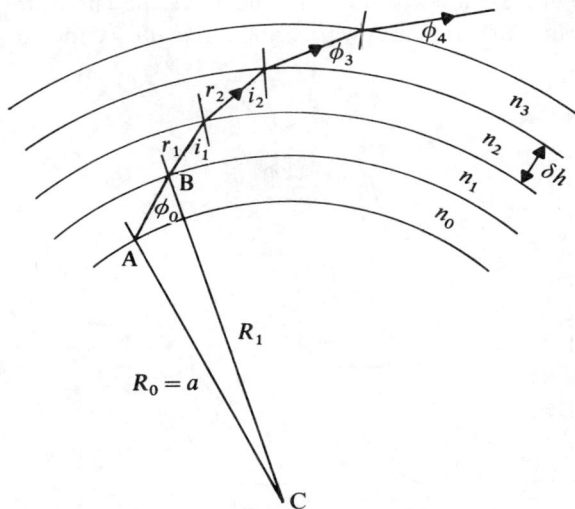

Fig. 4.14 Snell's law applied to a spherically stratified medium.

where n_0 is the refractive index at earth's surface. This can be simplified to

$$n_0 \cos \phi_0 = \{n(a+h)/a\} \cos \phi = n^* \cos \phi$$

This is of identical form to the plane-earth version of Snell's law with

$$n^* = n(a+h)/a \simeq n + h/a \tag{4.22}$$

as the refractive index n is close to unity. In the case of the standard linear troposphere with $N_0 = 289$ and $dn/dh = -39 \times 10^{-6}$ per km:

$$n^* = n + h/a = n_0 + h\,dn/dh + h/a = n_0 + 118h \times 10^{-6}$$

This gives the corresponding values of $M_0 = 289$ and $dM_/dh = +118$ per km.

The curvature of the propagation path of the radio waves is still given by

$$1/r = -dn^*/dh = -118 \times 10^{-6} \quad \text{km}^{-1}$$

which turns out to be a radius of 8490 km as before, but in the upward direction. In other words, if the path profile is drawn with a flat earth reference, the radio waves propagate with an upward radius of 8490 km. This is illustrated in Fig. 4.15.

The variation of the modified refractivity with height for a standard linear atmosphere is shown in Fig. 4.10. If

$$N = N_0 - 39h$$

then

$$M = M_0 + 118h \tag{4.23}$$

with $M_0 = N_0 = 289$ and h measured in kilometers. If the meteorological conditions change so that the variation of M with h is similar to that shown in Fig. 4.16, which is a *surface duct*, then the direction and amount of bending of the radio waves depend on the height above ground level. If dM/dh is negative the curvature is downwards; if dM/dh is positive the curvature is upwards; and if $dM/dh = 0$ the radius is infinite. This

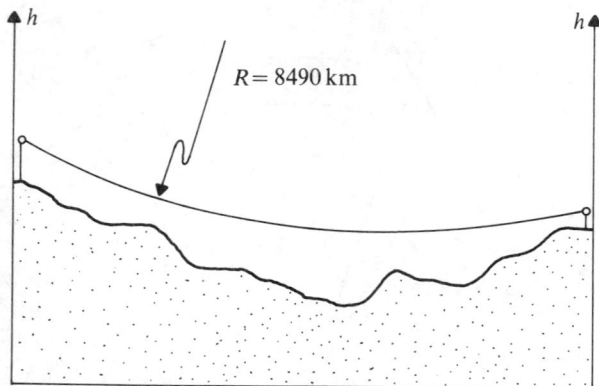

Fig. 4.15 Curved propagation path superimposed on a flat-earth model using the modified refractivity technique.

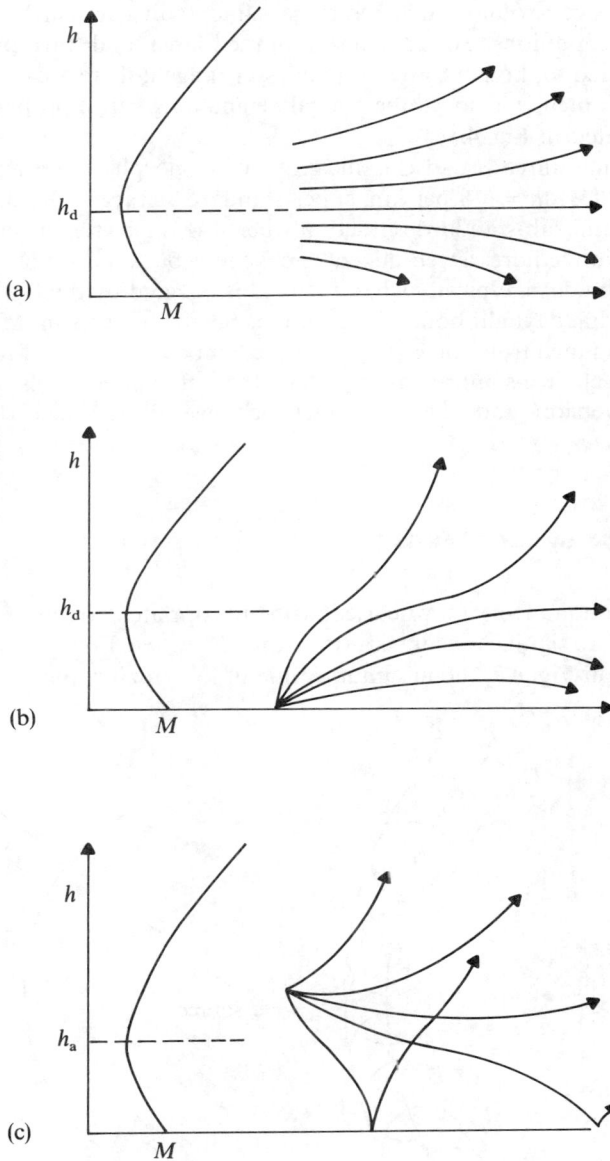

Fig. 4.16 The effect of modified refractive index on (a) horizontally propagating waves at different heights, (b) waves emanating from a low antenna, and (c) waves emanating from an elevated antenna. The variation of M with height indicates a surface duct.

means that if a space wave is launched horizontally at the height at which $dM/dh = 0$ it will propagate at a constant height above the equivalent flat earth, which is the same as saying it will curve around earth with the same radius as earth. In reality the meteorological conditions extend over a limited area and this phenomenon is obviously restricted to the same area or to a lesser range if there are obstacles. Even if the transmitting antenna is not at the critical height, long-range propagation can still occur, as illustrated in Fig. 4.16.

There are thus three broad classifications of tropospheric refraction: standard refraction when $dM/dh = 118$ per km, super-standard refraction when dM/dh is less than this value, and sub-standard refraction when dM/dh is greater than 118 per km. Standard refraction requires a well-mixed troposphere, but super-refracting conditions occur on fine calm days, typically when the land is covered by the trailing edge of an anticyclone. During daylight hours the ground is heated by the sun. After the sun has set, the heat is radiated from the ground and the temperature level of the troposphere near ground level drops more rapidly than that at higher levels. This causes a temperature inversion and the consequent change in the pattern of modified refractivity.

4.6 DIFFRACTION OF SPACE WAVES

Space wave reflection theory may be used, with appropriate corrections for refraction, provided there is no significant contribution from *diffraction*. For the smooth spherical earth illustrated in Fig. 4.7, the minimum value of the grazing angle ψ' for reflection

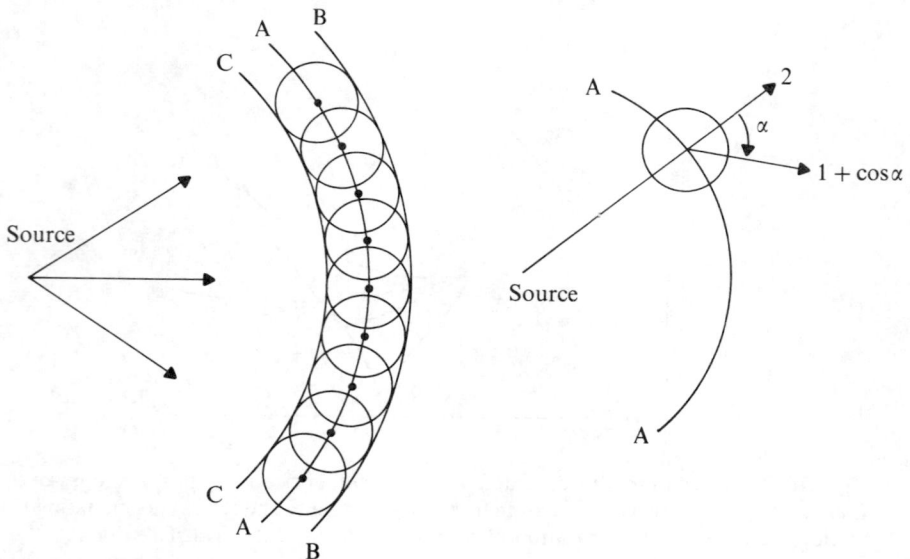

Fig. 4.17 Huygens' principle applied to a spherical wavefront.

theory is given in Ref. 4.4 as:

$$\psi' = \{\lambda/(2Ka)\}^{1/3} \quad \text{radians} \tag{4.24}$$

In the VHF and UHF frequency ranges, with $K = 4/3$, ψ' is much less than half a degree.

4.6.1 Huygens' Principle

Diffraction is the property of all electromagnetic waves by which they curve around the edges of obstacles in their propagation path. The full mathematical treatment of diffraction is outside the scope of this text and our approach shall be more qualitative than quantitative.

Huygens suggested that each point on a wavefront, such as the spherical wavefront AA illustrated in Fig. 4.17, produces a secondary wavelet and that these secondary wavelets combine to produce wavefront BB but not CC. To achieve this effect, the

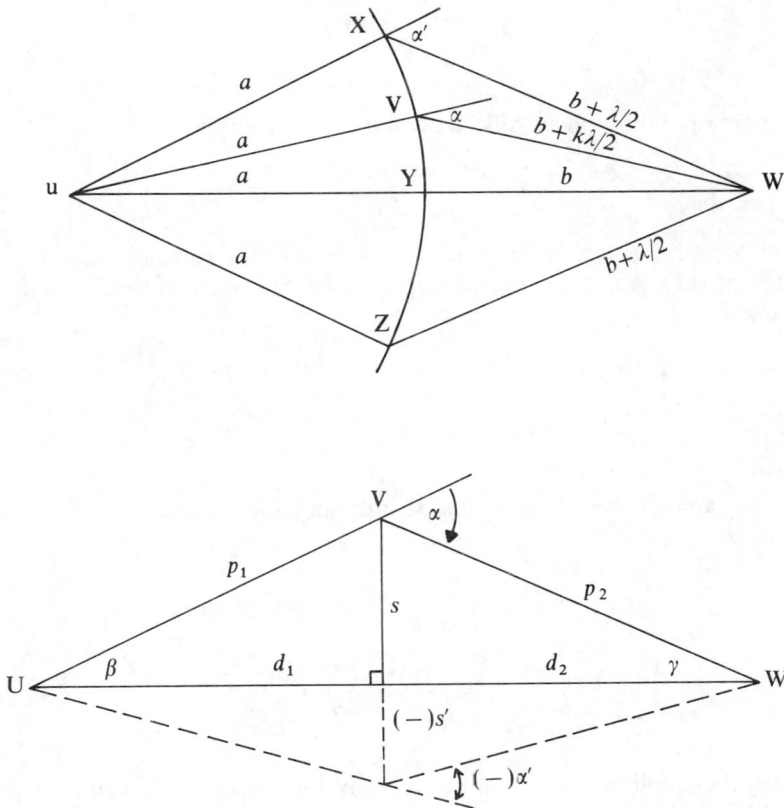

Fig. 4.18 The geometry of diffraction.

secondary wavelets cannot have uniform amplitude in all directions. If angle α represents the direction of interest, the relative amplitude of the secondary wavelets varies as $(1 + \cos \alpha)$. The contribution of wavelet Y at some distant point W (Fig. 4.18) will have an amplitude proportional to $(1 + \cos 0°) = 2$ and a reference phase of $0°$ with respect to itself. Wavelet V will have an amplitude proportional to $(1 + \cos \alpha)$, which is less than 2, and a phase lag of $k\pi$ radians, where $0 < k < 1$. Wavelet X will have an amplitude proportional to $(1 + \cos \alpha')$ and a phase lag of π radians. This procedure is repeated incrementally over the entire wavefront.

   Figure 4.18 illustrates how to obtain a simple geometric relationship between the distances and the heights. The total path length $UV + VW$ is longer than the direct path UW and produces an appropriate phase lag ϕ. The excess path length Δ is given by

$$\Delta = (p_1 + p_2) - (d_1 + d_2)$$

where

$$p_1 = (d_1^2 + s^2)^{1/2} = d_1(1 + s^2/d_1^2)^{1/2} \approx d_1 + s^2/(2d_1)$$

and

$$p_2 \approx d_2 + s^2/(2d_2)$$

Hence

$$\Delta = \frac{s^2}{2} \frac{d_1 + d_2}{d_1 d_2} \tag{4.25a}$$

and the corresponding phase difference $\phi = \beta\Delta$ is given by

$$\phi = \frac{\pi}{2} \frac{2(d_1 + d_2)s^2}{\lambda d_1 d_2} = \frac{\pi}{2} v^2 \tag{4.25b}$$

where v is an auxiliary parameter used to normalize the calculations.

   Alternatively, with the same assumption that s is much smaller than d_1 or d_2 we can write

$$\alpha = \beta + \gamma = s/d_1 + s/d_2$$

or

$$s = \alpha \frac{d_1 d_2}{d_1 + d_2} \tag{4.26}$$

as $\tan x \approx x$ under these conditions. Combining these equations gives the other relationships

$$\phi = \frac{\pi \alpha^2}{\lambda} \frac{d_1 d_2}{d_1 + d_2} \qquad \text{radians} \tag{4.27}$$

and

$$v = s\sqrt{\frac{2(d_1 + d_2)}{d_1 d_2 \lambda}} = \alpha\sqrt{\frac{2d_1 d_2}{(d_1 + d_2)\lambda}} \tag{4.28}$$

Thus both the amplitude and phase of the wavelet contributions can be expressed in terms of the auxiliary parameter v.

4.6.2 Amplitude Diagrams

As a rough qualitative approximation, to simplify the mathematics, we can take the portion YX of the wavefront illustrated in Fig. 4.18 and divide it into nine sections such that the corresponding values of ϕ are conveniently $10°, 30°, 50°, \ldots, 170°$. The diffraction angle α may be obtained from Eq. (4.27) for each value of ϕ and the amplitude of each of the nine contributions can be considered as being proportional to $(1 + \cos\alpha)$. We can now find the total contribution of wavefront YX to the resultant field at distant point W by adding vectorially each of the nine contributions of $|(1 + \cos\alpha)|\underline{/\phi}$. The result is shown in Fig. 4.19.

For two such sections of the wavefront the resultant of the first nine wavelets $(\mathbf{E}_1)$ and the resultant of the second nine wavelets $(\mathbf{E}_2)$ may be added vectorially to produce $\mathbf{E}$.

The *amplitude diagram* so formed, when extended to appropriate limits and containing an infinite number of wavelet contributions per section, becomes *Cornu's spiral*, Fig. 4.20. This makes use of the auxiliary parameter v of Eq. (4.28) to make the spiral independent of d_1, d_2, s, α and λ. Assuming that one of the contributory vectors in the amplitude diagram represents an incremental portion of the wavefront and has amplitude δv and phase $\phi = (\pi/2)v^2$ from Eq. (4.25b), then the resolution into x and y coordinates gives:

$$\mathrm{d}x = \mathrm{d}v\cos\phi = \cos\left(\frac{\pi}{2}v^2\right)\mathrm{d}v$$

$$\mathrm{d}y = \mathrm{d}v\sin\phi = \sin\left(\frac{\pi}{2}v^2\right)\mathrm{d}v$$

Thus the x and y components of the total contribution of a particular wavefront are the

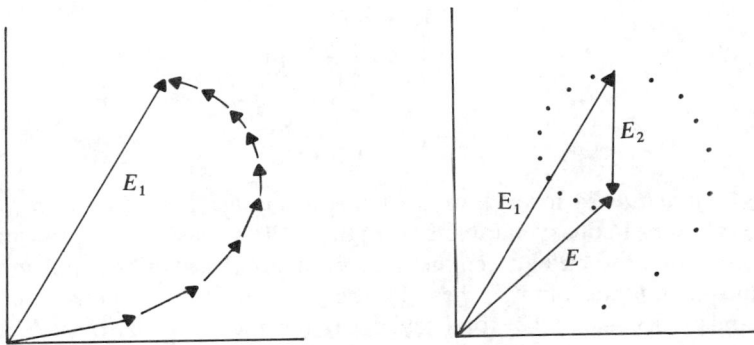

Fig. 4.19 The amplitude diagrams for one or two portions of a spherical wavefront assuming nine segments to each portion.

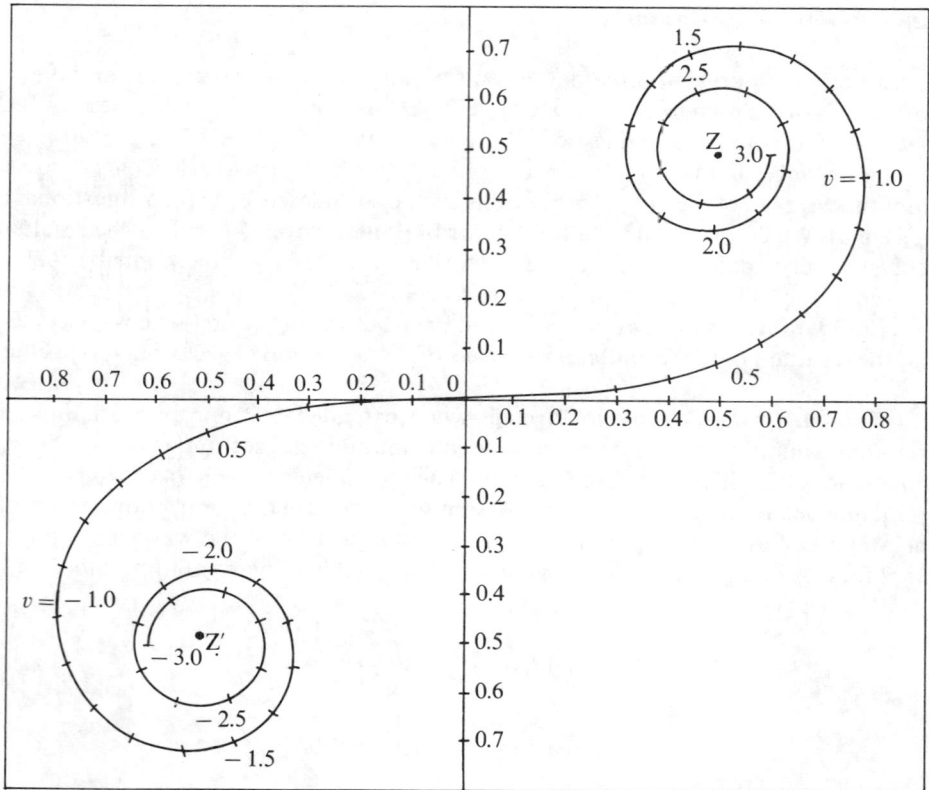

Fig. 4.20 Cornu's spiral for diffraction over a knife-edge obstacle. It is a plot of the Fresnel's integrals in terms of the auxiliary parameter v.

integrals over the appropriate limits of v, namely

$$X = \int_0^v \cos\left(\frac{\pi}{2}v^2\right)dv$$

$$Y = \int_0^v \sin\left(\frac{\pi}{2}v^2\right)dv$$

The *Fresnel integrals* so formed are available in tabular form for given (positive or negative) values of auxiliary parameter v. Figure 4.20 is constructed from such a table.

If there is no obstacle between the transmitter and receiver, the appropriate limits of v are plus and minus infinity. These are the points marked Z and Z' at coordinates $(0.5, 0.5)$ and $(-0.5, -0.5)$. The total field due to the integration between Z' and Z is the vector Z'Z, and this must of course represent the free-space field strength at the receiver. Note that the length of the vector is 1.414 units and this permits us to scale the diagram by saying that 1 Cornu's spiral unit of length corresponds to 0.707 times the magnitude of the free space field at the receiver.

4.6.3 Knife-edge Diffraction

The simplest of diffraction problems arises when the obstacle between the transmitter and the receiver can be represented by an infinitely thin 'knife-edge' as illustrated in Fig. 4.21. Depending on the location of the knife-edge with respect to the direct line between transmitter and receiver, the diffraction angle is either positive, equal to zero, or negative. This will not affect the sign of the amplitudes of the wavelets as $\cos(+\alpha) = \cos(-\alpha) =$ positive for $\alpha < 90°$, the limiting values for diffraction. However, the parameter v takes the same sign as angle α as a result of Eq. (4.28).

Whichever the case, the field strength at the receiver is obtained by determining the auxiliary parameter v corresponding to the location of the knife-edge, locating its position on Cornu's spiral, and drawing the vector from point v to the limiting value Z. The length of this line is then measured in Cornu's spiral units and converted into field strength via the relationship: 1 Cornu's spiral unit equals 0.707 times the free-space field strength at the receiver. This is commonly converted into decibels below (or above) free space value. Figure 4.22 illustrates the logarithmic variation, and Table 4.1 gives some approximate values of *knife-edge loss* L(ke).

In many examples of VHF or UHF broadcasting, the transmitter is described as

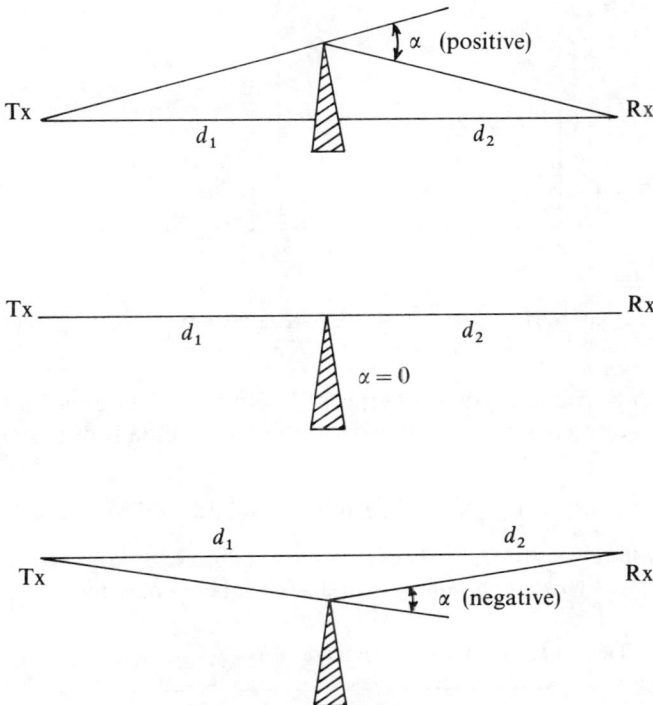

Fig. 4.21 Knife-edge diffraction with positive, zero and negative values of diffraction angle.

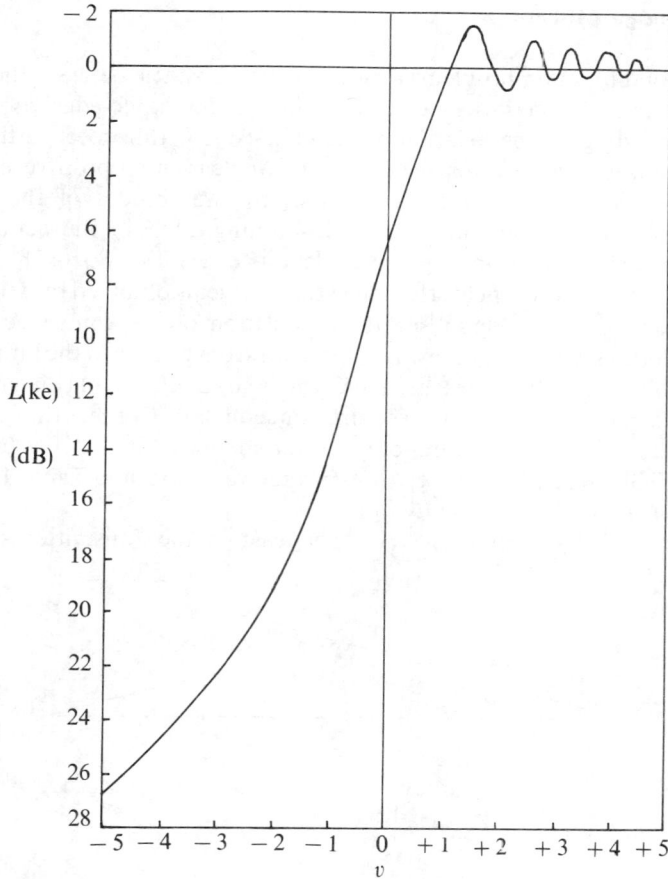

Fig. 4.22 The variation of knife-edge loss $L(ke)$ with Fresnel's parameter v.

having an *effective radiated power* of erpd kW with respect to a halfwave dipole. The field strength at a receiver in the knife-edge diffraction region is thus a modification of Eq. (1.15d):

$$E(\mathrm{dB}\mu) = 106.9 + 10\log \mathrm{erpd(kW)} - 20\log d(\mathrm{km}) - L(\mathrm{ke}) \qquad (4.29)$$

where erpd distinguishes the reference antenna as a dipole. By comparison, eirp implies *effective isotropically radiated power* with an isotrope as reference.

Table 4.1   Knife-edge loss $L(ke)$ dB

v	0	-1	-2	-3	-4	-5	-10	-20
$L(\mathrm{ke})$	6	13	19	22	25	27	33	39

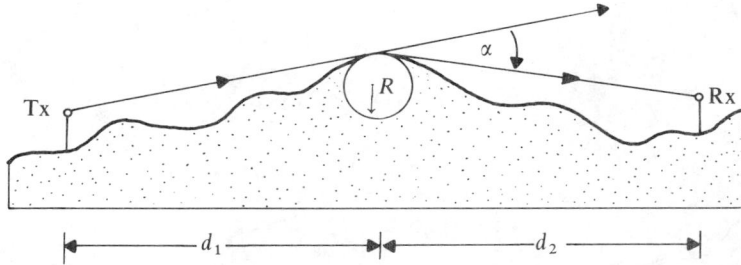

Fig. 4.23 Diffraction over a rounded hill.

4.6.4 UHF Propagation over Rounded Hills

Radio wave propagation is more commonly impeded by rounded hills rather than knife-edge obstructions. The theory of diffraction shows that the diffraction loss around curved obstacles is higher than the knife-edge loss. If a rounded hilltop can be expressed in terms of a circle of radius R, an *excess loss* $L(\text{ex})$ can be added to the knife-edge loss $L(\text{ke})$:

$$L(\text{ex}) = 11.7(\pi R/\lambda)^{1/2}\alpha \qquad \text{dB} \tag{4.30a}$$

where R has the same dimensions as λ, and α is the diffraction angle in radians, as shown in Fig. 4.23. This expression is only a close approximation to theory and applies to horizontally polarized UHF signals.

Example 4.5 Diffraction over rounded hill

What is the prospect of receiving adequate television signals at site X in Fig. 4.24? The UHF television transmitter has 500 kW erpd and operates with $\lambda = 0.4$ m. The rough rounded hill at $d_1 = 22$ km has a radius of about 3 km. The receiving site at $d = 24$ km (or $d_2 = 2$ km) receives television signals via a diffraction angle $\alpha = 0.033$ radians, calculated from detailed maps of the area, assuming four-thirds earth radius.

The free-space field strength without obstructions would have been:

$$E(\text{dB}\mu) = 106.9 + 10\log(500\,\text{kW}) - 20\log(24\,\text{km}) = 106\,\text{dB}\mu$$

The knife-edge diffraction loss can be determined via Eq. (4.28), giving $v = 3.16$, with $L(\text{ke}) = 23$ dB from Table 4.1. The rough rounded hill loss with $R = 3000$ m, $\lambda = 0.4$ m and $\alpha = 0.033$ radians is $L(\text{ex}) = 38$ dB from Eq. (4.30b). The net field at the receiver is likely to be of the order

$$E(\text{dB}\mu) = 106 - 23 - 38 = 45\,\text{dB}\mu$$

This is well below the field strength needed for normal television reception. A typical UHF receiver and antenna requires a signal level of the order of 70 dBμ for reasonable reception. The area within the diffraction shadow is thus unable to receive television signals directly from the main UHF transmitter but will probably be served by a local low-power relay.

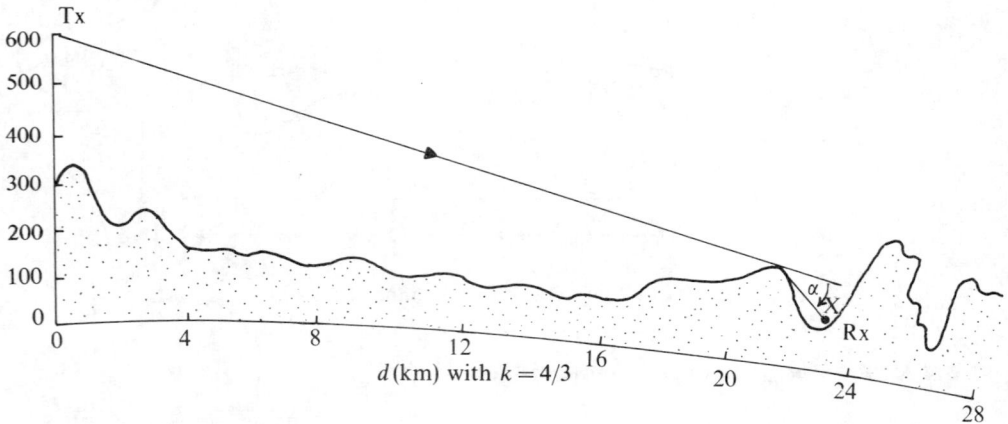

Fig. 4.24 UHF television reception in the diffraction zone.

More often than not the hilltop is covered with trees or buildings, especially in or near urban reception areas. The rougher edges help to enhance the diffraction field and the excess loss $L(\text{ex})$ is a little smaller than that for a smooth rounded hill:

$$L(\text{ex}) = 7.5(\pi R/\lambda)^{1/2}\alpha \qquad \text{dB} \tag{4.30b}$$

With this modification an approximate estimate of the field in the shadow of a rounded hill (from Eq. (4.29)) is:

$$E(\text{dB}\mu) = 106.9 + 10\log \text{erpd}(\text{kW}) - 20\log d(\text{km}) - L(\text{ke}) - L(\text{ex}) \tag{4.31}$$

Ref. 4.5 describes the technique in more mathematical detail.

4.7   MOBILE RADIO IN AN URBAN ENVIRONMENT

A common application of space wave communication is the mobile radio link between a fixed station and a moving vehicle. There are several problems to be overcome, including the shielding, scattering and multipath effects of the urban environment and the Doppler shift due to the vehicle's velocity. The consequence of these effects is a signal of rapidly varying magnitude and phase which can be represented by a suitable statistical distribution and associated parameters.

If we confine ourselves to an estimate of the median field strength, leaving the more detailed analysis to specialized texts such as Ref. 4.6, the matter resolves itself to an extension of Sec. 4.2.5. The electric field at distance $d(\text{km})$ is given by Eq. (4.8), where

$$E_1(\text{mV/m}) = \frac{\sqrt{30G_tP_t}}{d} \qquad \text{(for } d = 1 \text{ km)}$$

or

$$E_1(\text{dB}\mu) = 74.8 + G(\text{t}) + P(\text{t})$$

Substituting this expression into Eq. (4.8) leaves us with

$$E(\text{dB}\mu) = P_t(\text{dBW}) + G(t) + 20\log(h_t h_r) + 20\log f(\text{MHz})$$
$$- 12.8 - 40\log d(\text{km}) \tag{4.32}$$

This assumes that the distance is greater than $d_{k=1}$ given in Eq. (4.6).

Because of the urban and suburban environment the direct and indirect components are attenuated and scattered and are subject to multipath effects. If we assume that as a consequence the path loss is modified by an additional *environmental factor* $A(\text{dB})$ to produce the overall median field strength, where (Ref. 4.7):

$$A(\text{dB}) = A^*(\text{dB}) - 30\log f(\text{MHz}) \tag{4.33}$$

then the median field strength within the urban or suburban environment becomes

$$E(\text{dB}\mu) = \{P(t) + G(t) + 20\log(h_t h_r) - 12.8 - 10\log f\}$$
$$+ A^*(\text{dB}) - 40\log d \tag{4.34a}$$

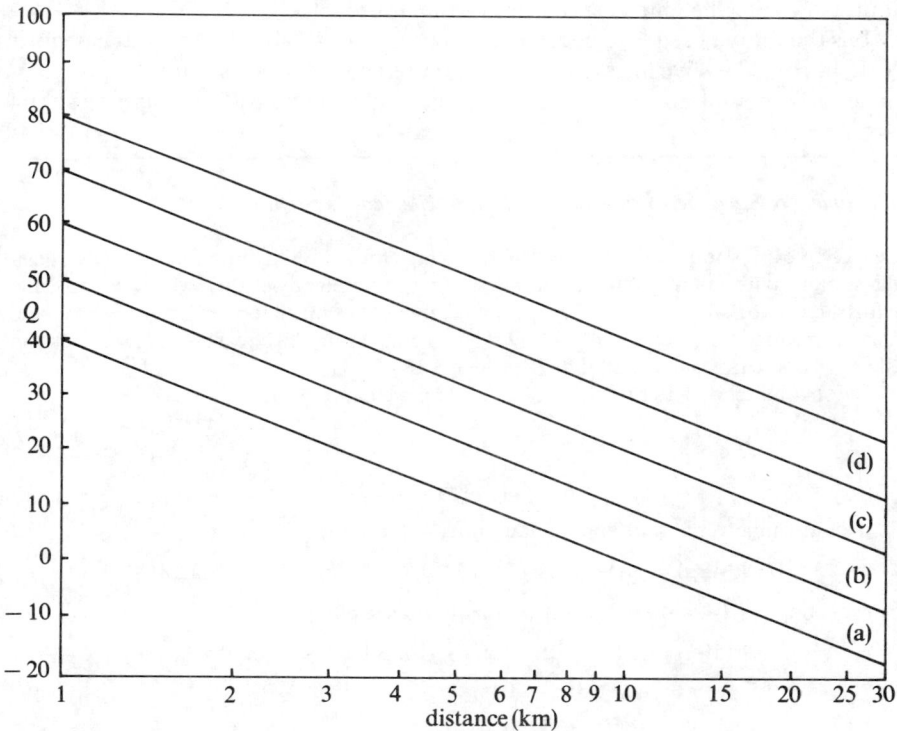

Fig. 4.25 The prediction chart for mobile radio communication in an urban or suburban environment, where $Q = E(\text{dB}\mu) - k(\text{E}) = P_r(\text{dBW}) - k(\text{P}) = V(\text{dB}\mu) - k(\text{V})$ as appropriate. The environment limits are (a) older cities, (b) modern cities, (c) suburbia and (d) open rural, with $\pm 5\,\text{dB}$ limits of the median predictions.

The terms within the brackets are all constants for a given mobile radio system provided the area is quasi-flat. This means we can rewrite Eq. (4.34a) as

$$E(dB\mu) - k(E) = A^*(dB) - 40 \log d \qquad (4.34b)$$

which is a simple linear relationship provided $A^*(dB)$ is constant.

The factor $A^*(dB)$ is obviously not a constant unless the environment is perfectly homogeneous and uniform. However, if the environment is divided into several generalized categories then typical values of $A^*(dB)$ can be determined for each. One possible set of categories and associated parameters is as follows:

$A^*(dB) = 45 \pm 5 \, dB$ for older, densely packed cities with narrow, twisting streets.

$A^*(dB) = 55 \pm 5 \, dB$ for more modern cities with long, straight, wide streets.

$A^*(dB) = 65 \pm 5 \, dB$ for typical suburbia with two-storey houses and gardens; and for thinly tree-clad rural areas.

$A^*(dB) = 75 \pm 5 \, dB$ for unobstructed open areas.

There are, of course, other types of environments, such as thick forests, but the above quoted values of environmental factor $A^*(dB)$ seem to work reasonably well in many instances despite the empirical nature of the approach.

It is therefore possible to construct a chart of Eq. (4.34b) with the four categories of $A^*(dB)$ bordered by five lines (Fig. 4.25), and this chart can be used for the prediction of mobile radio communication field strengths in specific quasi-flat urban or suburban environments.

Example 4.6 Receiver voltage in an urban mobile radio system

Derive an expression for the receiver input voltage for a mobile radio link within a quasi-flat urban environment given that the transmitter radiated power $P(t) = 10 \, dBW$, antenna gains $G(t) = 5 \, dB$ and $G(r) = 2 \, dB$, antenna heights above mean terrain level $h_t = 80 \, m$ and $h_r = 2.5 \, m$, and frequency $f = 450 \, MHz$. The antenna is matched to a 50-ohm receiver and the environment is such that $A^*(dB) = 45 \pm 5 \, dB$.

For a matched antenna-to-receiver system we can rewrite

$$P_r = A_e P_a$$

as

$$V^2/R = G_r(\lambda^2/4\pi)(E^2/120\pi)$$

The logarithmic version in appropriate units reduces to

$$V(dB\mu) = G(r) + E(dB\mu) + 10 \log R - 20 \log f(MHz) + 12.8$$

Substituting Eq. (4.34a) for $E(dB\mu)$ we finally obtain either

$$V(dB\mu) = \{P(t) + G(t) + G(r) + 20 \log(h_t h_r) + 10 \log R - 30 \log f\}$$
$$+ A^*(dB) - 40 \log d \qquad (4.36a)$$

or

$$V(dB\mu) - k(V) = A^*(dB) - 40 \log d \qquad (4.36b)$$

For the numerical solution, appropriate substitution of the data gives $V(dB\mu) = 45.4 - 40 \log d$.

Alternatively, from Eq. (4.12), we can derive an equivalent expression in terms of signal power levels. Modifying Eq. (4.12) by A(dB) gives us

$$P(r) = \{P(t) + G(t) + G(r) + 20\log(h_t h_r) - 30\log f - 120\}$$
$$+ A^*(dB) - 40\log d \qquad (4.35a)$$

Again, the parameters within the brackets are constants of a given mobile radio system and we can write

$$P_r(dBW) - k(P) = A^*(dB) - 40\log d \qquad (4.35b)$$

where $k(P)$ represents the net effect of the contents of the brackets. Thus the same chart can be used for field strength or power received.

4.8   VARIATION OF SPACE WAVE SIGNAL STRENGTH

When space waves propagate through a theoretical standard atmosphere, the signal strength at the receiving antenna should equally theoretically be constant. In reality, the refractivity of the troposphere is neither constant nor uniform, and continuous fluctuations of the signal strength are observed as a consequence (Fig. 4.26). There are

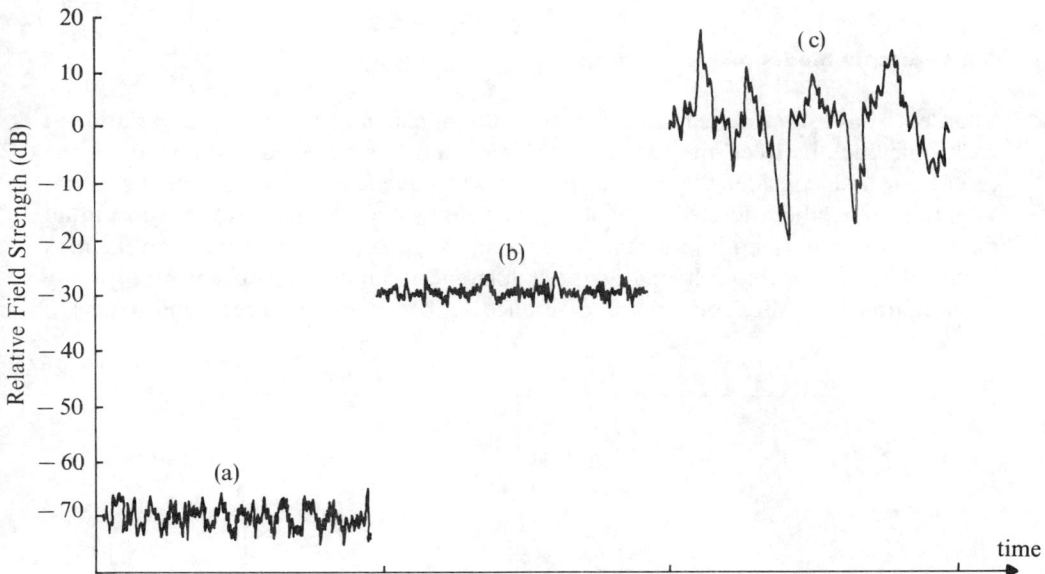

Fig. 4.26 Signal patterns under conditions of (a) sub-standard, (b) standard and (c) super-standard refraction.

many causes of this variation or *fading* of signal strength, some of which are described in Chap. 8. At this stage we restrict our consideration to the effects of refraction on fading caused by simple variations in the K-factor.

The received signal is the resultant of the direct and reflected components of the space wave and every large or small variation in the parameter K produces changes in the effective values of h_t, h_r, d_1, d_2, E_d and E_r via the corresponding changes in effective earth radius R and divergence factor D, with the consequent variation in the resultant E.

The continuous changes in K are considered over different time periods and the consequent variations in E are then called either *fast fading* or *slow fading*. When the meteorological conditions change, the mean value of K can change quite substantially. Small values of mean K cause the effective path profile to rise and sub-standard refraction to take place. The resulting fading may resemble Fig. 4.26a. Large values of K give rise to super-refraction, lowering the effective path profile, and producing fading resembling that in Fig. 4.26c.

The problems of fading concern the amount of power needed to maintain the received signal level above some appropriate minimum level for a specified percentage of time. Abnormal propagation, on the other hand, concerns the interference between two transmissions at the same frequency over long distances. In the latter case we are concerned with the percentage of time that an unwanted signal exceeds some appropriate maximum level and thereby causes interference. These effects are dependent on meteorological conditions, some of which have seasonal and diurnal cycles as well as geographical variation, but otherwise they are fairly random.

4.8.1 Simple Model of Fast Fading

In order to have some numerical idea of the amount of fading caused by the variation in path difference between the direct component and *one* reflected component of the received signal, consider the effect at fairly high frequencies where the small random variations of K about its steady (or slowly varying) value are able to cause substantial differences in path length in terms of wavelengths because of the small wavelengths involved. Over certain path lengths these differences can amount to several or many wavelengths and thus the phase difference ϕ between the direct and reflected

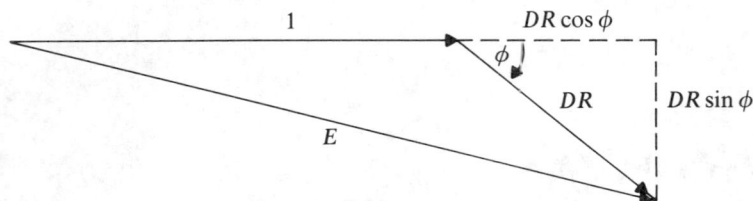

Fig. 4.27 A vector diagram of the sum of the normalized direct component and the normalized reflected component, using the magnitude of the direct component as reference.

components, including the phase change due to ground reflection, can take any value between 0 and 2π. As shown in Fig. 4.27, this means that the *magnitude* of the normalized resultant field strength, denoted by E, can take any value between $E = 1 - DR$ and $E = 1 + DR$, where D is the divergence factor and R is the magnitude of the reflection coefficient.

For our simple model of fast fading let us assume that for a well mixed atmosphere, with a radio link operating in the higher frequency bands, the statistics of ϕ are very simple: all values of ϕ occur with equal probability. As the *magnitude* of E is dependent on $\phi \equiv \phi - k\pi$, with k an integer, the limits of ϕ are 0 and π. Thus we can say that the probability that ϕ exceeds some specified value ϕ^* (within these limits) is:

$$P(\phi) = P(\phi > \phi^*) = \frac{\pi - \phi^*}{\pi} \times 100\% \tag{4.37}$$

Figure 4.27 shows that for each value of ϕ between 0 and π there is a corresponding value of the magnitude of E (normalized with respect to the direct component).

To find an expression relating probability with fading levels, we first note that $\phi^* = 0, 0.5\pi, 0.9\pi, 0.99\pi$ and π radians correspond via Eq. (4.37) to $P(\phi) = 100, 50, 10, 1$ and 0%, respectively. The resultant is:

$$E = \sqrt{(1 + DR \cos \phi)^2 + (DR \sin \phi)^2} = \sqrt{1 + 2DR \cos \phi + (DR)^2}$$

and the median value E_m occurs when $P(\phi) = 50\%$ or $\phi = 0.5\pi$ radians:

$$E_m = \sqrt{1 + (DR)^2}$$

Using decibel notation, we can obtain the depth of fade $L_f(\text{dB})$ as:

$$L_f(\text{dB}) = -20 \log(E/E_m) = 20 \log \sqrt{1 + (DR)^2}$$
$$- 20 \log \sqrt{1 + 2DR \cos \phi + (DR)^2} \tag{4.38}$$

with the ratio being positive if E is below the median and negative if E is above the median.

It is a matter of simple calculation to plot probability against E/E_m dB for specified values of DR. For example, if $DR = 0.9$, the ratio E/E_m dB corresponding to $P = 1\%$ or $\phi = 0.99\pi$ radians is

$$E/E_m = (-)\{10 \log[1 + 1.8 \cos(0.99\pi) + 0.81] - 10 \log[1 + 0.81]\} = +22.2 \, \text{dB}$$

This is interpreted as the probability that the fading will exceed 22.2 dB below the median is 1% when $DR = 0.9$. The results are illustrated in Fig. 4.28.

In practice there are several other variations which affect the fading levels. Firstly, the changes in path difference due to small variations in K are greater when the wavelength is small. Thus there is a tendency for fading to get worse as frequency increases and less when frequency decreases. Secondly, there are often several reflected components of the space wave and the combined effect of many such components will eventually tend toward the statistics of a slowly varying direct component plus a rapidly varying resultant of the reflected components. The effect is that the fading levels are somewhat less than those predicted by the simple model described above, and are

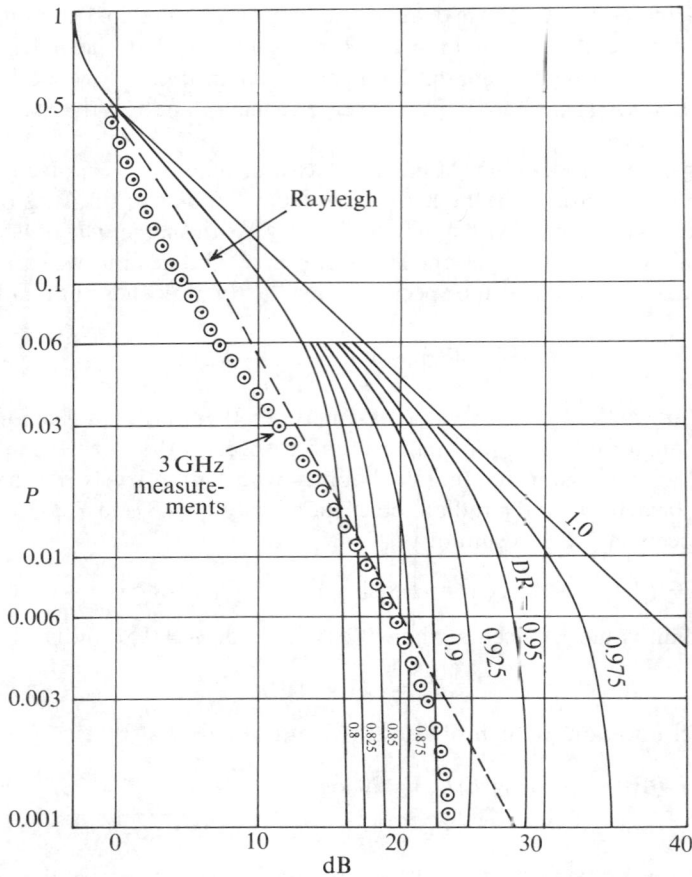

Fig. 4.28 Fading characteristics with one reflected component for various values of *DR* compared with the Rayleigh distribution and a typical curve of measured values.

usually less than a Rayleigh model. For the higher frequencies, the variation within a defined period of time during which the mean is sensibly constant can be taken pessimistically as Rayleigh distributed.

Because of its relationship with meteorological conditions, fading is least on blustery winter days and greatest during calm summer nights; it is less on shorter paths than on long paths.

4.9   ANTENNAS FOR SPACE WAVE PROPAGATION

For the ground-based applications of space wave propagation at VHF and UHF described in this chapter, the most common types of antennas are the monopole, halfwave dipole, slot antenna and their derivatives.

Example 4.7 Fast fading

A radio link 40 km long operates at 3 GHz with 28 dB antennas. The receiver requires a minimum $P(r) = -100$ dBW. The meteorological conditions are such that the median of the received signal is 3 dB below the free-space value and transmitter losses are 2 dB. If the transmitter power before losses is 1 W, during how many seconds per hour is the received signal below the minimum $P(r)$? What transmitter power is needed so that the loss in communication is only one second per hour?

Starting with the basic free-space transmission equation,

$$P(r) = P(t) + G(t) + G(r) - L(s) - L(\text{misc})$$

where

$$L(s) = 20 \log f + 20 \log d + 32.45 = 134 \text{ dB}$$

We can obtain $P(r) = -83$ dBW. This leaves a margin of 17 dB for fast fading. Assuming a Rayleigh distribution, Rayleigh probability paper can be used to show that a fade exceeding 17 dB occurs for 1.37% of time. This corresponds to about 50 seconds per hour.

To obtain an outage of 1 second per hour, or 0.0278%, Rayleigh probability paper again indicates that a fade margin of 34 dB is necessary. This can be achieved by increasing $P(t)$ by $34 - 17 = 17$ dB to 17 dBW (50 W).

The impedance of a *halfwave dipole* is about $73 + j43\,\Omega$, but such is the rate of change of reactance with length that a dipole of length just a little less than half a wavelength resonates with $Z = 73 + j0\,\Omega$. The gain of the resonant halfwave dipole, in the plane perpendicular to the dipole, is 1.64 linearly or 2.16 in decibels, and its beamwidth in the plane of the dipole is 78°.

The bandwidth of the halfwave dipole is of the order $\pm 5\%$ depending on the definition and whether the antenna length-to-diameter ratio is large or small. At 300 MHz the bandwidth may therefore be of the order ± 15 MHz. For a dipole operating at 300 MHz, with length-to-diameter ratio about 60, the resonant length may be 475 mm, compared with half a wavelength of 500 mm, and the diameter about 8 mm. When the dipole operates at $\pm 5\%$ from 300 MHz, i.e. at 285 MHz and 315 MHz, when the resonant lengths should be 500 mm and 450 mm, respectively, the impedances are such that the VSWR (voltage standing wave ratio) is about 2 in each case. This is due mainly to the rapid variation in the dipole reactance with frequency.

An alternative form of dipole as called the *folded dipole*. This may be the conventional folded dipole illustrated in Fig. 4.29b or one with several elements. The input resistance at resonance is about $70N^2\,\Omega$ for an N-element arrangement, or about $280\,\Omega$ for the conventional folded dipole. This is useful when the resonant halfwave dipole is the active element in an antenna array and the coupling of the passive elements tends to reduce the effective input impedance of the active element. A slightly more variable arrangement is to use one of the versions of the shunt-fed dipole in which the input impedance depends on the dimension between the feed points.

There are corresponding versions of the monopole with similar effects on input resistance, etc.

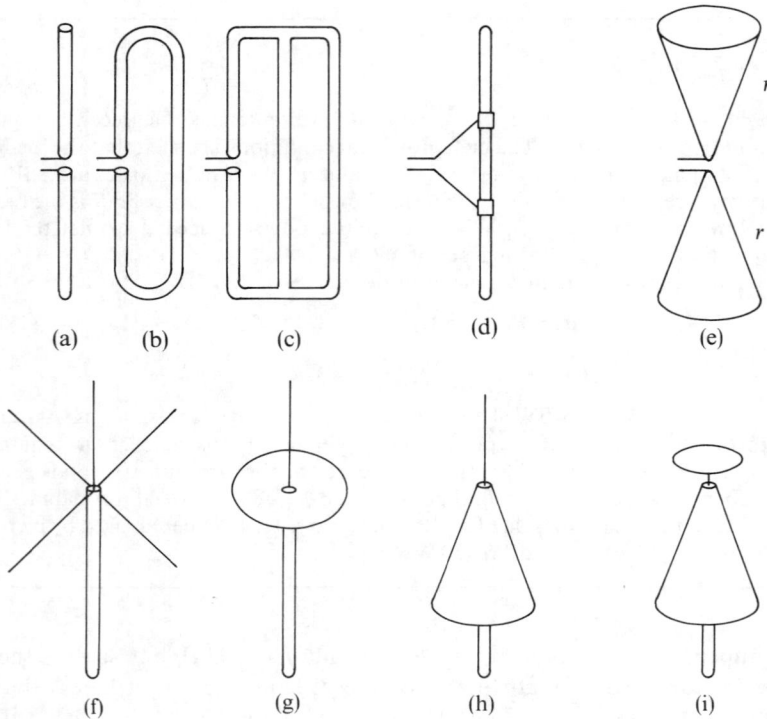

Fig. 4.29 Simple types of space wave antennas include (a) halfwave dipole, (b) folded dipole, (c) multi-element folded dipole, (d) shunt-fed dipole, (e) biconical antenna, (f), (g) & (h) ground-plane monopoles and (i) discone.

Broadbanding a dipole antenna to permit its use over a wider range of frequencies can be achieved by reducing its length-to-diameter ratio. This is sometimes called a *cylindrical antenna* or a *cage antenna* when constructed out of wires instead of being solid. The bandwidth can be doubled in this manner though the antenna input impedance and resonant length are changed.

If the two elements are changed in shape from two cylinders to two cones, as in Fig. 4.29e, the result is known as a *biconical antenna*. The improvement in operational bandwidth is increased, perhaps by a ratio as high as 4:1, varying roughly between the frequencies at which $r = \lambda/4$ to $r = \lambda$ or even more, with a VSWR less than about 2, and input impedance about $150\,\Omega$ when the semi-angle is 30°. Sometimes one of the cones is replaced by a disc to form the discone antenna (Fig. 4.29i) with similar broadband properties and lower input resistance, and sometimes the discone is made up of an open skirt or even wires instead of solid metal.

It is possible to arrive at the discone antenna by considering the monopole above an artificial ground plane, as illustrated in Fig. 4.29. The circular disk is about half a wavelength in diameter. Once again, a wire structure may be used instead of a solid disk, either like a cartwheel or ultimately just four radial wires at 90° to each other.

Alternatively the disk ground plane may be reshaped in the form of a hollow cone, and the $\lambda/4$ stub can be replaced by a disk to produce the discone again.

However, there are differences in the polar diagrams of these antennas. The replacement of the infinite ground plane by the disk results in a radiation mainly at some angle α *above* the plane of the disk. By using a cone instead of the disk the magnitude of α is reduced, and for the discone the radiation is virtually in the horizontal plane.

Another version is to modify the circular disk into a cylinder rather than a cone. This produces a sleeve antenna and radiation is normal to the axis of the antenna.

The halfwave dipole and slot antennas can be combined into the form of arrays to enhance the overall gain in particular directions. Some of the arrays involve driven elements, such as a broadside array, some incorporate passive or parasitic elements, such as a Yagi–Uda antenna, and some have their dimensions carefully arranged to produce uniform characteristics over a very wide frequency range, such as the log-periodic antenna. These and other arrays are described in the next chapter.

Combinations of dipoles with appropriate phasing of their feed currents can be used to produce circular polarization, and other arrangements can be designed to produce almost omnidirectional radiation in the horizontal plane with horizontal polarization (turnstile antenna).

PROBLEMS

4.1 A television transmitting antenna is at height $h_t = 300$ m above mean terrain level (mtl). What is its approximate line-of-sight range if a typical receiving antenna height is taken as $h_r = 10$ m above mtl and tropospheric refraction is that associated with $K = 0.7, 4/3$ or 2.0?

4.2 What is the field strength at a receiving antenna at height $h_r = 30$ m if the transmitting antenna has height $h_t = 90$ m, $E_1 = 1000$ mV/m and $f = 100$ MHz. Assume a flat earth, distance $d = 20$ km and $R_H = -1$.

4.3 Determine the point of reflection on a four-thirds curved earth for a radio link at 900 MHz between two antennas $h'_t = 120$ m and $h'_r = 50$ m above mtl and separated $d = 35$ km, and hence the effective heights of the antennas above the tangent through the point of reflection. Use first the method outlined in Sec. 4.2.6 and then compare with the CCITT solution quoted in Ref. 4.8, viz.

$$T = (d^2/12) + 2.125K(h_t + h_r)$$
$$R = 1.5925Kd(h_r - h_t)$$
$$\psi^\circ = \cos^{-1}(RT^{-3/2})$$
$$d_3 = 2T^{1/2}\cos(240^\circ + \psi^\circ/3)$$
$$d_1 = \frac{d}{2} + d_3 \quad \text{and} \quad d_2 = d - d_1$$
$$h_t = h'_t - 4d_1^2/(51K) \quad \text{and} \quad h_r = h'_r - 4d_2^2/(51K)$$

with d in kilometers and heights in meters.

4.4 Determine the field strength for the space wave path described in Prob. 4.3 if $E_1 = 1000$ mV/m: (a) ignoring divergence factor D, and (b) including divergence factor D.

4.5   Plot graphs of the E-field against K at two receiving antennas, one at $h'_r = 50\,\text{m}$ and the other at $h'_r = 60\,\text{m}$ above sea level, for a 2 GHz sea path radio link over a distance $d = 50\,\text{km}$ with the transmitter at $h'_t = 300\,\text{m}$. Ignore the effects of divergence, tidal variation, wave heights and fast-fading phenomena. If the two receiving antennas are connected to a sensing mechanism which automatically switches the receiver to whichever antenna has the maximum signal, plot a graph of the effective E-field seen by the receiver. Assume the weather conditions are such that for most of the time $0.6 < K < 2.0$ and abnormal propagation does not occur. Normalize $2E_1/d = 1000$.

4.6   In a certain quasi-flat region residents receive television signals from a transmitter operating with four channels, $\lambda = 0.626\,\text{m}$, $\lambda = 0.596\,\text{m}$, $\lambda = 0.569\,\text{m}$ and $\lambda = 0.537\,\text{m}$. If the transmitting antenna is 500 m above mtl and a particular receiving antenna is at 40 m above mtl, with $d = 25\,\text{km}$, estimate the field strength of each channel in decibels with respect to the direct space wave, assuming a flat earth profile. Is there any difference in the quality of reception between channels?

4.7   Direct line-of-sight propagation of UHF television signals between a 800 MHz, 100 kW erpd transmitter with $h_t = 300\,\text{m}$ above mtl is obstructed at $d = 25\,\text{km}$ by a protruding ridge of height 150 m above mtl. A receiving antenna at height 120 m above mtl and at $d = 30\,\text{km}$ is within the radio shadow of the ridge. What is the approximate field strength at the receiving antenna if $0.6 < K < 2.0$ and the ridge is assumed to behave as a knife-edge obstacle?

4.8   If the ridge in Prob. 4.7 is a rough rounded hill of radius $R = 4\,\text{km}$, what further reduction in field strength would you expect at the receiver?

4.9   A receiver operating at 750 MHz is located so that the field strength at its antenna is 75 dBμ. The antenna has a gain $G = 10\,\text{dB}$ and is correctly matched to the receiver. What is the voltage across the receiver terminals?

4.10   A microwave link has 2 m diameter paraboloidal antennas at each end of the propagation path and operates at 9 GHz. The antennas are aligned beam-center when $K = 4/3$. Show that if $0.6 < K < 2.0$ the receiving antenna is unlikely to deviate out of the beam of the transmitting antenna due to variation in refraction for $d < 100\,\text{km}$.

4.11   A certain single-frequency experimental space wave link has a median received signal which is 5 dB below the free-space value. The variation about the mean follows a Rayleigh distribution which requires a margin of 18 dB (28 dB) for 99% (99.9%) reliability. Determine the corresponding transmitter powers necessary to achieve these limits if the minimum receiver power level is $-100\,\text{dBW}$, $d = 40\,\text{km}$, $f = 9\,\text{GHz}$, transmitter losses are 5 dB, and antennas are 1 m diameter paraboloidal reflector types with 0.54 illumination efficiency. Ignore other factors.

4.12   A mobile radio system has $h_t = 60\,\text{m}$, $h_r = 3\,\text{m}$, $G(t) = 8\,\text{dB}$ and $G(r) = 2\,\text{dB}$. The environment is a modern city with $A^* = 55 \pm 5\,\text{dB}$. Estimate the median transmission loss at $d = 15\,\text{km}$ and frequencies $f = 150\,\text{MHz}$, 450 MHz and 900 MHz. Assume antenna heights are measured with respect to the mean terrain level and the area is quasi-flat.

REFERENCES

4.1 Fagot, J., P. Magne and R. Aubert, *Frequency Modulation Theory*. Oxford: Pergamon Press, 1961, p. 36.

4.2 Smith, E. K., Jr. and S. Weintraub, 'The constants in the equation for atmospheric refractive index at radio frequencies,' *Proc. IRE*, 41 (August 1953), 1035–7.

4.3 David, P., and J. Voge, *Propagation of Radio Waves*. Oxford: Pergamon Press, 1969.

4.4 Picquenard, A. *Radio Wave Propagation*. London: Macmillan Press Ltd., 1974, p. 66.

4.5 Hacking, K. 'Propagation over rounded hills,' *BBC Research Report RA-21*, 1968.

4.6 Lee, W. C. Y., *Mobile Communications Engineering*. New York: McGraw-Hill Book Company, 1982.

4.7 McGeehan, J. P., and J. Griffiths, 'Normalized prediction chart for mobile-radio reception,' *Fourth International Conference on Antennas and Propagation*, IEE Conference Publication No. 248, April 1985, pp. 395–9.

4.8 Henk, A. J., 'Optimum diversity separation for over-sea line-of-sight radio links,' *The Radio and Electronic Engineer*, 51 (November/December 1981), 561–70.

5

Antenna Arrays

In the middle of the VHF and UHF frequency bands, where $10 > \lambda > 0.01$ m, the halfwave dipole has very manageable dimensions. It is a simple antenna to construct and its radiation resistance is easily matched to a receiver. However, the maximum gain of 2.16 dB and effective aperture of $0.131\lambda^2$ are rather small. For those applications where a larger gain is required in a specified direction, several techniques can be employed to achieve this objective with dipoles. Of these, one method is to use a large paraboloidal reflector; another is to set up an array of halfwave dipoles. The enhancement of gain in specified directions gives not only better directive properties but also improves the signal-to-noise ratio of the radio link as the energy is not dispersed unnecessarily into space and, when the array is used as a receiving antenna, the interference is reduced if it arrives in a direction other than that of the main lobe.

An array of dipoles, or other types of antenna, is a collection of such radiating elements arranged in a geometric pattern and driven in some suitable electrical manner by adjusting the antenna current amplitude and phase. In addition, such improvements in antenna gain and associated directivity are reciprocal, applying equally to both transmitting and receiving antennas. In these notes we shall consider mainly linear arrays of halfwave dipoles with variables N (number of elements in the array), d (the spacing of the linear array), I (the feed current amplitude), ψ (the phase of the feed current with respect to some reference value) and r, θ, ϕ (the spherical coordinate system). Note the new meanings of d and r.

5.1 THE HALFWAVE DIPOLE

The halfwave dipole has been considered earlier but we recap briefly at this point. It is shown in Appendix 1 that if an element of length dz carrying current I is aligned along the z axis in free space, the magnitude of the retarded vector magnetic potential dA_z is:

$$dA_z = \frac{\mu_0}{4\pi} \frac{[I]\,dz}{r}$$

where $[I]$ denotes the retarded function $\hat{I} \exp j\omega(t - r/c)$, r is distance in spherical polar coordinates, and $A_x = A_y = 0$. For the radiation field component it can be shown that

$$dB_\phi = -j\beta \sin\theta \, dA_z$$

where $\beta = 2\pi/\lambda$, and hence we can write:

$$dE_\theta = \frac{Z_i}{\mu_0}dB_\phi = \frac{j60\pi[I]\,dz\sin\theta}{\lambda r}$$

using $Z_i \approx 120\pi\,\Omega$. By changing the origin of the coordinate system, as shown previously in Fig. 1.13, we can write:

$$dE_\theta = \frac{j60\pi\hat{I}\exp j\omega(t - s/c)\sin\theta'}{\lambda s}dz$$

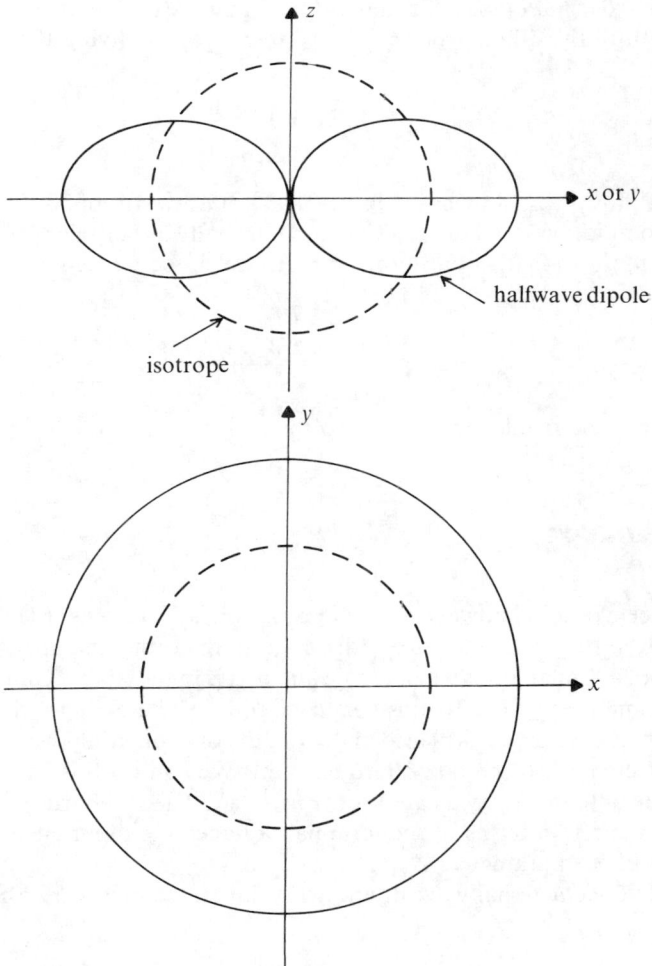

Fig. 5.1 Radiation patterns of a halfwave dipole in the xz, yz and xy planes compared with those of an isotrope (broken lines). The maximum gain is 2.16 dB.

or

$$dE_\theta = \frac{j60\pi \hat{I} \exp j\omega\{t - (r - z\cos\theta)/c\}\sin\theta}{\lambda r} dz$$

using the approximation for large r discussed in Sec. 1.4.1. For a halfwave dipole, with assumed current distribution $\hat{I} = \hat{I}_0 \cos\beta z$,

$$dE_\theta = \frac{j60\pi \hat{I}_0 \sin\theta \exp j\omega(t - r/c)}{\lambda r} \exp\left\{\frac{j\omega z \cos\theta}{c}\right\} \cos\beta z \, dz$$

or

$$dE_\theta = E_0 e^{az} \cos\beta z \, dz$$

where $a = j\beta\cos\theta = j(\omega/c)\cos\theta$. To find E_θ for a halfwave dipole we must integrate dE_θ over the length of the dipole from $z = -\lambda/4$ to $z = +\lambda/4$, giving the rms value:

$$E_\theta = \frac{60I_0}{r} \frac{\cos\left(\dfrac{\pi}{2}\cos\theta\right)}{\sin\theta} \tag{5.1a}$$

The resulting plot of E_θ, shown in Fig. 5.1, has a beamwidth of 78° in the yz and xz planes but is omnidirectional in the xy plane. If the halfwave dipole is not orientated in the z direction, the E-field equation may alternatively be written as:

$$E_d = \frac{60I_0}{r} \frac{\cos\left(\dfrac{\pi}{2}\cos\alpha\right)}{\sin\alpha} \tag{5.1b}$$

where α is the angle made with respect to the axis of the dipole.

5.2   DIPOLE IMPEDANCE

The impedance of a thin halfwave dipole in isolation is $73.13 + j42.5\,\Omega$. This value will change if: (1) the length-to-diameter ratio is reduced; (2) the antenna is placed in the close vicinity of a similar antenna; or (3) the antenna is in the close vicinity of a reflector, etc. For the time being we will consider the second of these alternatives.

When a halfwave dipole is placed in the vicinity of another halfwave dipole, *mutual impedance* effects need to be taken into account. We can do this by treating the N-element dipole antenna array as an N-port network. The standard N-port equations may then be written in terms of the terminal voltages V_N, input currents I_N and the network impedance parameters Z_{11}, Z_{12}, etc.

For the two-element halfwave dipole array illustrated in Fig. 5.2, the two network equations are:

$$V_1 = Z_{11}I_1 + Z_{12}I_2$$

and

$$V_2 = Z_{21}I_1 + Z_{22}I_2$$

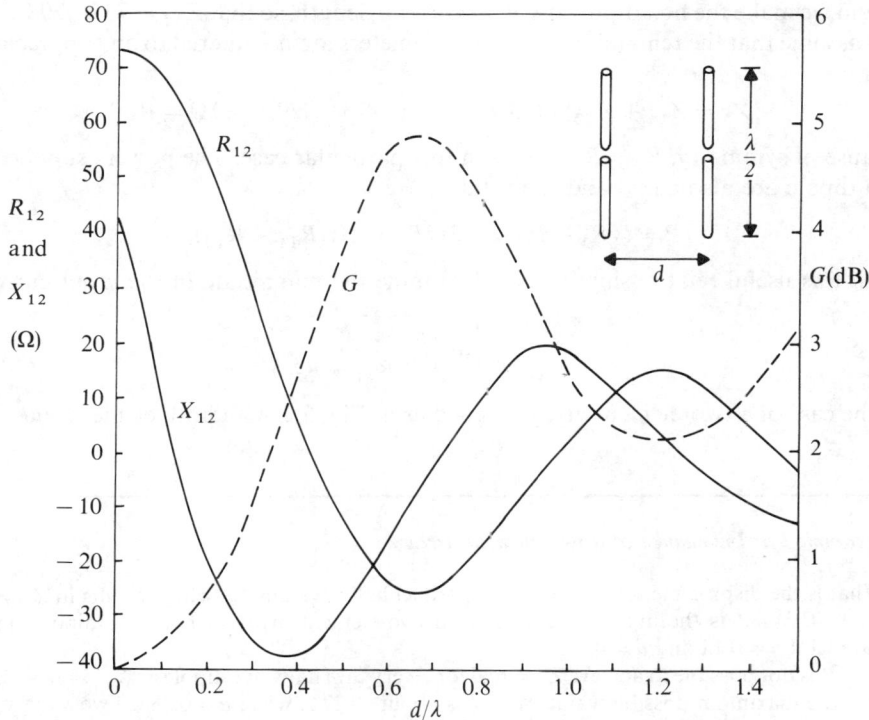

Fig. 5.2 An array of two halfwave dipoles arranged in parallel broadside configuration. The coupling parameters R_{12} and X_{12} depend on the displacement d/λ. The gain G is given in decibels with respect to a single halfwave dipole.

The network impedance parameters Z_{11}, etc., include resistive and reactive components R_{11} and X_{11}, etc.

We can make Z_{12} equal to zero by removing the second halfwave dipole from our two-element array and then $Z_{11} = 73 + j43\,\Omega$ as before, with $R_{11} = 73\,\Omega$ and $R_{12} = 0$. It can be shown (Ref. 5.1) that when the second halfwave dipole is displaced distance d from the first halfwave dipole, R_{12} and X_{12} will vary with d/λ in the manner shown in Fig. 5.2 for *thin* dipoles.

If the elements are displaced by $d = \lambda/2$, for example, the approximate values of the network parameters are $R_{11} = 73\,\Omega$, $R_{12} = -13\,\Omega$, $X_{11} = 43\,\Omega$ and $X_{12} = -29\,\Omega$. If we chose to design an array with the two input currents equal in amplitude and phase, then $I_2 = I_1$ and:

$$V_1 = I_1 Z_{11} + I_1 Z_{12}$$

with

$$Z_1 = V_1/I_1 = Z_{11} + Z_{12}$$

or

$$Z_1 = (73 + j43) + (-13 - j29) = 60 + j14\,\Omega$$

Should we make the first dipole slightly shorter in length, so that $Z_{11} = 73 + j29\,\Omega$, and then assume that the remaining network parameters are not altered to any appreciable extent:

$$Z_1 = Z_{11} + Z_{12} = (73 + j29) + (-13 - j29) = 60\,\Omega = R_1$$

Because of symmetry, $R_2 = R_1 = 60\,\Omega$ in this particular case. The powers supplied to each dipole are also equal and the total power is:

$$P = I_1^2 R_1 + I_2^2 R_2 = 2I_1^2 R_1 = 2I_1^2(R_{11} + R_{12})$$

From this useful relationship we can determine the magnitude of the input current:

$$I_1 = \sqrt{\frac{P}{2R_1}} = \sqrt{\frac{P}{2(R_{11} + R_{12})}}$$

for the case of a two-element array, aligned as in Fig. 5.2, with both elements *tuned* to real Z_1.

Example 5.1 Impedance of a two-element array

What is the displacement d between two parallel halfwave dipoles which results in $Z_1 = R_1 + j0$? What is the input impedance of a two-element array of halfwave dipoles in parallel if $d = 0.1\lambda$ and $d = 0.01\lambda$?

It is not possible to achieve $Z_1 = $ real for a very thin halfwave dipole as $X_{11} = +43\,\Omega$ and the maximum possible value of X_{12} is about $-37\,\Omega$ when $d = 0.38\lambda$. Two ways of obtaining a real Z_1 include the use of thicker dipoles or slightly shorter dipoles.

For small displacements, say $0 < d < 0.1\lambda$, the input impedance increases as the displacement decreases. Thus, when $d = 0.1\lambda$,

$$Z_1 = Z_{11} + Z_{12} = 73 + j43 + 67 + j8 = 140 + j51\,\Omega$$

and when d is only 0.01λ, Z_1 is practically twice Z_{11}.

5.3 TWO-ELEMENT ARRAY

Although a two-element array can be analyzed via the generalized array theory discussed later, it is useful to introduce the basic ideas and definitions of arrays via this simplest of types. Firstly we consider an array of two isotropes and then we can derive an expression for an array with two halfwave dipoles by pattern multiplication.

5.3.1 Array of Two Isotropes

The electric field strength at distance r from a loss-free isotrope is given by:

$$E_i = \frac{\text{constant}}{r} I$$

provided r is sufficiently large to be in the far field of the antenna. The total power required to provide this field is $P = I^2 R$, where R is the radiation resistance of the antenna when isolated from the influence of other antennas, etc.

When two identical loss-free antennas are displaced distance d along (say) the z axis, the electric field strengths at distance r are:

$$E_{i1} = \frac{\text{constant}}{r} I_1$$

and

$$E_{i2} = \frac{\text{constant}}{r} I_2$$

However, because of the displacement between the two antennas there is a phase difference of $\phi = \beta d \cos \theta$ between E_{i1} and E_{i2} and the resultant is easily shown to be:

$$E_\theta = 2E_{i1} \cos \left(\frac{\pi d}{\lambda} \cos \theta \right) \tag{5.2}$$

This is illustrated in Fig. 5.3.

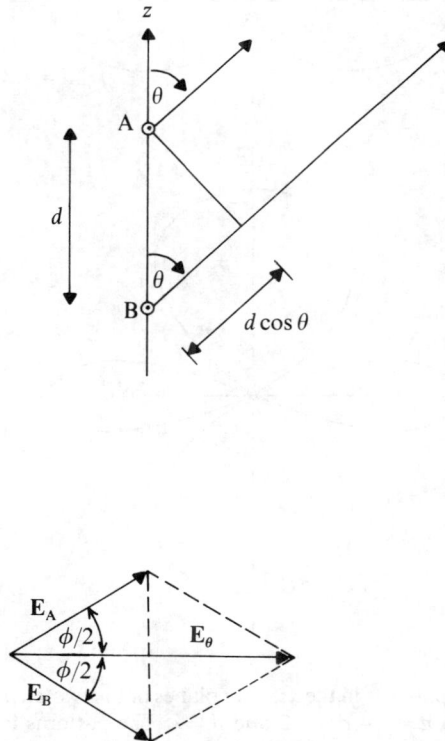

Fig. 5.3 An array of two isotropes displaced distance d along the z axis. The resultant field strength $\mathbf{E}_\theta$ is the vector sum of $\mathbf{E}_\mathbf{A}$ and $\mathbf{E}_\mathbf{B}$ and has a magnitude twice as long as $E_\mathbf{A} \cos(\phi/2)$.

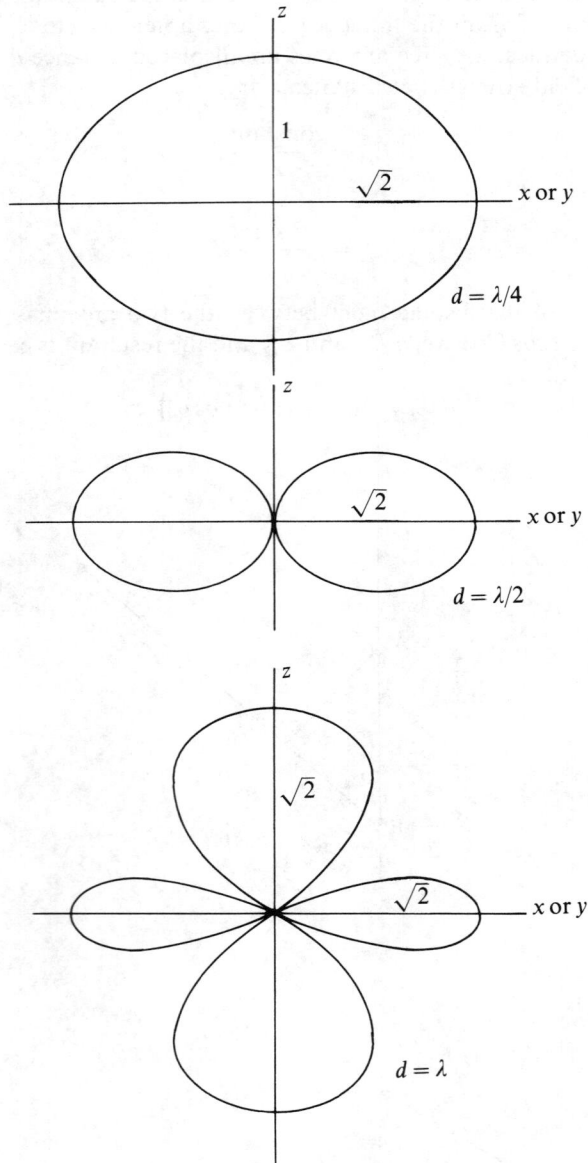

Fig. 5.4 The radiation patterns in the xz or yz planes of the two-element array of isotropes shown in Fig. 5.3, with $d = \lambda/4, d = \lambda/2$ and $d = \lambda$. The patterns in the xy plane are all omnidirectional.

The total power required by the array to provide E_θ is $P = 2I_1^2 R_1$, where R_1 is the resistance of each antenna in the presence and under the influence of the other. Noting that:

$$\frac{E_{i1}}{E_i} = \frac{I_1}{I} = \sqrt{\frac{P}{2R_1}}\sqrt{\frac{R}{P}} = \sqrt{\frac{R}{2R_1}} \tag{5.3}$$

we finally obtain the relationship:

$$\frac{E_\theta}{E_i} = \sqrt{2}\sqrt{\frac{R}{R_1}}\cos\left(\frac{\pi d}{\lambda}\cos\theta\right) \tag{5.4}$$

The ratio E_θ/E_i is called the *array factor* AF, sometimes normalized so that its maximum value is unity.

The array factor clearly depends on the displacement between the two antennas, not only in the term d/λ but also in the ratio R/R_1. Figure 5.4 illustrates the array factor patterns for $d = \lambda/4$, $d = \lambda/2$ and $d = \lambda$, ignoring the term $\sqrt{R/R_1}$. For isotropic antennas there are no specific values of R and R_1, but the law of conservation of energy requires that $0.707 < \sqrt{R/R_1} < 1$ as displacement d varies from zero to infinity, though in practice a value near 1 is achieved quickly.

Note that the array patterns in Fig. 5.4 have a maximum value of $\sqrt{2}$ in the directions $\theta = 90°$ and $270°$, though with certain values of d/λ the gains in directions $\theta = 0°$ and $180°$ can also reach a similar value. The need for $\sqrt{R/R_1}$ becomes particularly apparent when $d < \lambda/4$ as the array factor without $\sqrt{R/R_1}$ implies that $E_\theta > E_i$ in *every* direction. If $d \approx 0$, the array pattern ignoring $\sqrt{R/R_1}$ would be a sphere with $E_\theta = \sqrt{2}E_i$, which is, of course, impossible.

5.3.2 Arrays of Two Halfwave Dipoles

Halfwave dipoles can be arranged to produce several types of two-element antenna arrays, the three most common being the collinear, broadside and endfire arrays.

The polar pattern of an array of two *collinear* halfwave dipole antennas can be obtained by *pattern multiplication* of the polar diagram of the halfwave dipole and the array factor of the two-element array in each of the three cartesian planes. In this case the displacement d between the feed points must obviously be greater than half a wavelength, with perhaps an insulator isolating the tips of each antenna when $d = \lambda/2$. With this restriction we can write Eqs. (5.1) and (5.4) for both the xz and yz planes:

$$E_d = \frac{60I_0}{r}\frac{\cos\left(\dfrac{\pi}{2}\cos\theta\right)}{\sin\theta}$$

and

$$AF = \sqrt{2}\sqrt{\frac{R}{R_1}}\cos\left(\frac{\pi d}{\lambda}\cos\theta\right)$$

The field strength of the antenna array $E_a = \text{AF} \times E_d$, or

$$E_a = \sqrt{2}\sqrt{\frac{R}{R_1}}\cos\left(\frac{\pi d}{\lambda}\cos\theta\right)\frac{\cos\left(\frac{\pi}{2}\cos\theta\right)}{\sin\theta}E \qquad (5.5)$$

where $E = 60I_0/r$ is the field strength at distance r when $\theta = 90°$.

If displacement $d = 0.7\lambda$ or spacing $S = 0.2\lambda$, and we again make the assumption that slightly shortened dipoles produce resonant conditions without affecting the other parameters too greatly, the approximate network parameters are $R_{11} = 73\,\Omega$ and $R_{12} = 7\,\Omega$, from Fig. 5.5. Hence $\sqrt{R/R_1} = \sqrt{73/80} = 0.955$ and the polar diagrams

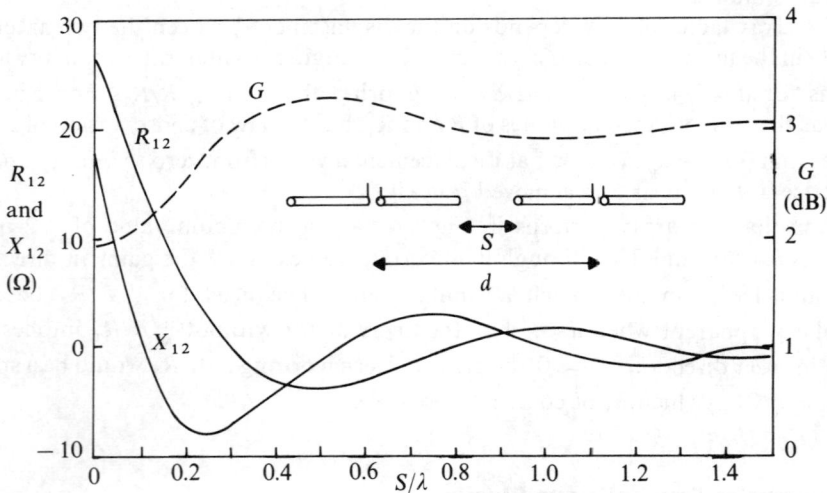

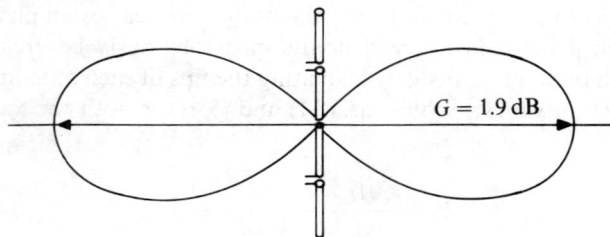

Fig. 5.5 The coupling components R_{12} and X_{12} of an array of two halfwave dipoles in collinear configuration. The gain is given in decibels with respect to a single halfwave dipole. The polar plot is for such a collinear array when $d = \lambda/2$ or $S = 0$, and includes the effect of R_{12} and X_{12}.

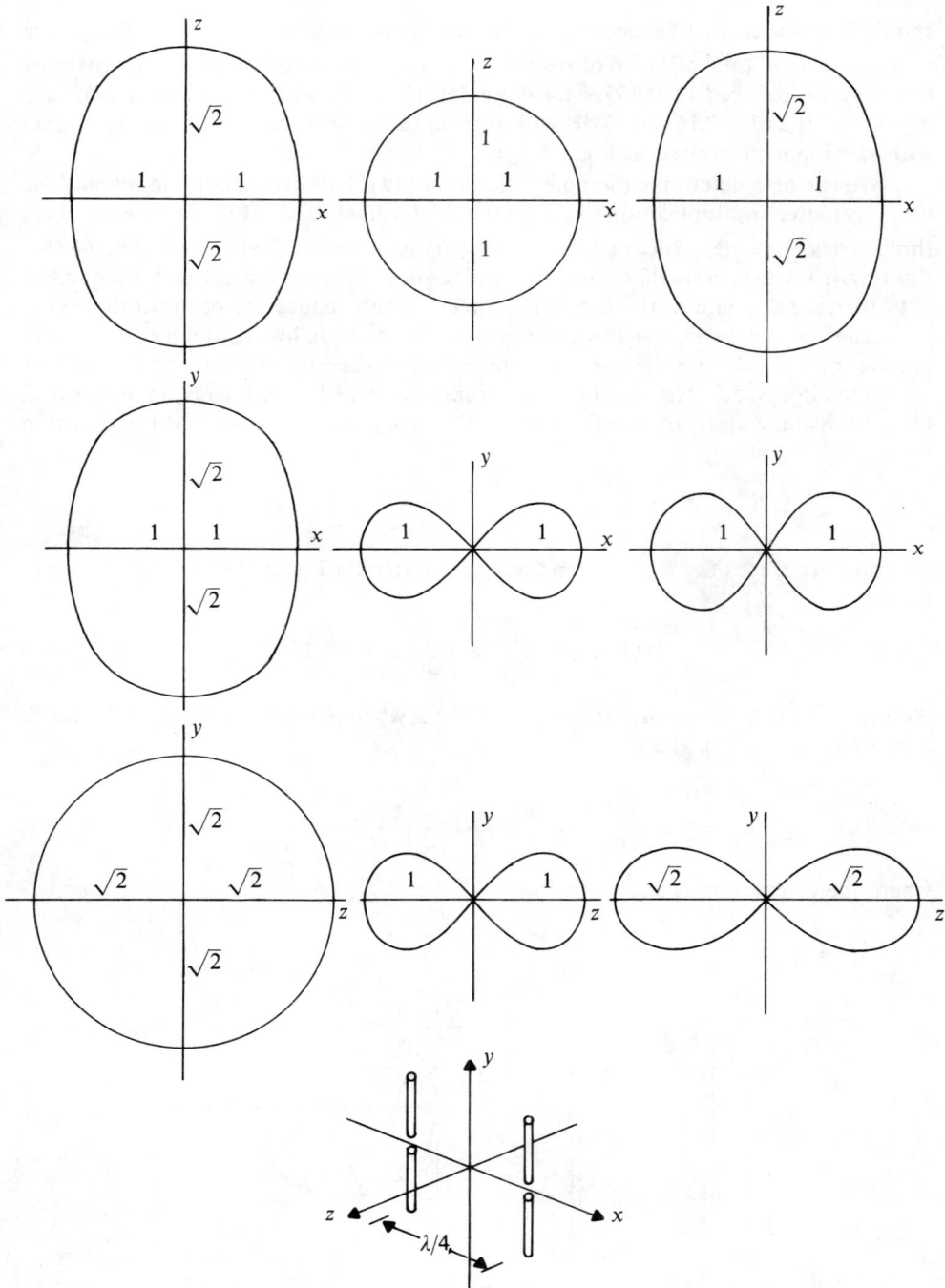

Fig. 5.6 The radiation patterns of two halfwave dipoles in broadside configuration obtained by pattern multiplication in each plane.

should be reduced by this factor from the *ideal* patterns which ignored $\sqrt{R/R_1}$. The maximum of $\sqrt{2}$ (or 3 dB) with respect to the single halfwave dipole ($G = 2.16$ dB) must therefore be reduced to $0.955 \times 1.414 = 1.350$ (2.61 dB) with respect to a halfwave dipole, or to $2.61 + 2.16 = 4.77$ dB with respect to an isotrope. The variation of gain with spacing is illustrated in Fig. 5.5.

We can also determine the polar diagram of two halfwave dipoles in a *broadside* array by pattern multiplication to obtain the results illustrated in Fig. 5.6 for each of the three cartesian planes. Once again it is necessary to scale the ideal patterns by $\sqrt{R/R_1}$, this time using the curves of R_{12} for broadside configuration given in Fig. 5.2. The effect is to increase the gain in the broadside direction and reduce the beamwidth.

One way of achieving *endfire* radiation is to have two halfwave dipoles displaced in parallel by $d = \lambda/4$, with the current in antenna 2 leading that in antenna 1 by 90°, as illustrated in Fig. 5.7. The resultant is the sum of E_1 and E_2, both equal in magnitude, but with E_2 lagging E_1 by $\phi = \beta d \cos \theta - \pi/2$. Firstly, we can obtain the array pattern as:

$$E_\theta = 2E_1 \cos\left(\frac{\beta d \cos \theta - \pi/2}{2}\right) = 2E_1 \cos\left(\frac{\pi}{4}\cos \theta - \frac{\pi}{4}\right)$$

ignoring the effect of $\sqrt{R/R_1}$. Introducing the latter via Eq. (5.3) we obtain the array factor:

$$\text{AF} = 2\sqrt{\frac{R}{2R_1}}\cos\left(\frac{\pi}{4}\cos \theta - \frac{\pi}{4}\right) \tag{5.6}$$

As before, the field due to the array $E_a = \text{AF} \times E_d$, where AF is given by Eq. (5.6) and E_d is given by Eq. (5.1b). Hence:

$$E_a = \sqrt{2}\sqrt{\frac{R}{R_1}}\cos\left(\frac{\pi}{4}\cos \theta - \frac{\pi}{4}\right)\frac{60I_0}{r}\frac{\cos\left(\frac{\pi}{2}\cos \alpha\right)}{\sin \alpha} \tag{5.7}$$

where the orientation is shown in Fig. 5.7.

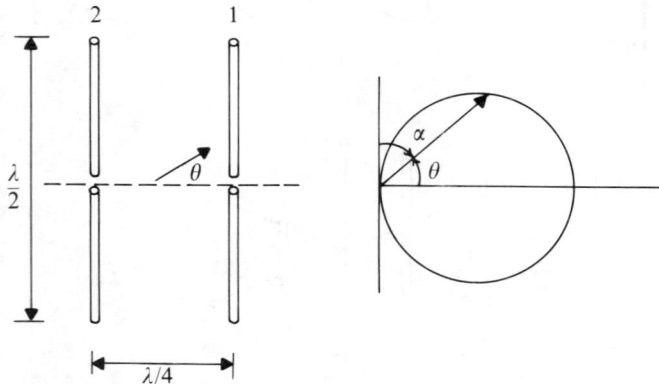

Fig. 5.7 Endfire configuration of two halfwave dipoles. The current in antenna 2 leads that in antenna 1 by 90°. Note the directions indicated by angles α and θ.

The value of R_1 may be obtained from:

$$V_1 = Z_{11}I_1 + Z_{12}I_2 = (R_{11} + jX_{11})(I_1) + (R_{12} + jX_{12})(jI_1)$$

as I_2 leads I_1 by 90° but is of equal magnitude. Figure 5.2 gives the approximate network parameters as $R_{12} = 41\,\Omega$ and $X_{12} = -29\,\Omega$. Thus:

$$V_1 = I_1(73 + jX_{11} + j41 + 29) = I_1[102 + j(X_{11} + 41)]$$

If we can resonate the dipole by adjusting its length very slightly to eliminate the reactance without affecting the other parameters too greatly, $R_1 = V_1/I_1 = 102\,\Omega$. The scaling factor $\sqrt{R/R_1} = \sqrt{73/102}$ is 0.846 (or $-1.45\,\text{dB}$) and the maximum gain is reduced from 3 dB with respect to a halfwave dipole to about 1.55 dB, which is 3.7 dB with respect to an isotrope.

5.3.3 Antenna and Reflector Array

If we place a halfwave dipole antenna in the vicinity of a conducting plane, the *image* of the radiating element can be considered as a second element of a two-element antenna array. The array may be collinear or endfire if the dipole is correctly aligned and spaced, but note in particular the phasing of the antenna and image currents as shown in Fig. 5.8. In addition, the total power P is supplied only to the driven element and no electromagnetic energy is required for the half-space behind the reflecting plane. Mutual coupling effects must also be taken into consideration.

To achieve the endfire effect, the halfwave dipole antenna is aligned parallel to the reflecting plane at a distance initially chosen as one-quarter of a wavelength. The image is then at a distance of half a wavelength from the driven antenna and behaves as a similar dipole but with the image feed current 180° out of phase. The field at some distant point is the phasor sum of $\mathbf{E}_A$ and $\mathbf{E}_B$, both of equal magnitude, but with $\mathbf{E}_B$ lagging $\mathbf{E}_A$ by $\phi = \beta d \cos\theta + \pi$ radians. Figure 5.8 shows that the resultant, after pattern multiplication, is

$$E_\theta = \sqrt{2}E_d \frac{\cos\left(\dfrac{\pi}{2}\cos\alpha\right)}{\sin\alpha} \cos\left(\frac{\pi - \beta d \cos\theta}{2}\right) \tag{5.8}$$

before any correction is made for mutual coupling effects. Using the trigonometrical relationship $\cos(\pi/2 - A) = \sin A$, and noting that $\cos\theta = \sin\alpha$,

$$E_\theta = 2\sqrt{\frac{R}{R_1}}E_d \frac{\cos\left(\dfrac{\pi}{2}\cos\alpha\right)}{\sin\alpha} \sin\left(\frac{\pi d}{\lambda}\sin\alpha\right) \tag{5.9}$$

where $\sqrt{2}$ has become 2 because of the reflected component contribution. The polar pattern for $d = \lambda/2$ is illustrated in Fig. 5.9a. This assumes that the reflector is perfectly conducting and that $R_1 = R_{11} + R_{12}$ is as illustrated in Fig. 5.9b.

If four such dipole and reflector antenna arrays are assembled in the manner

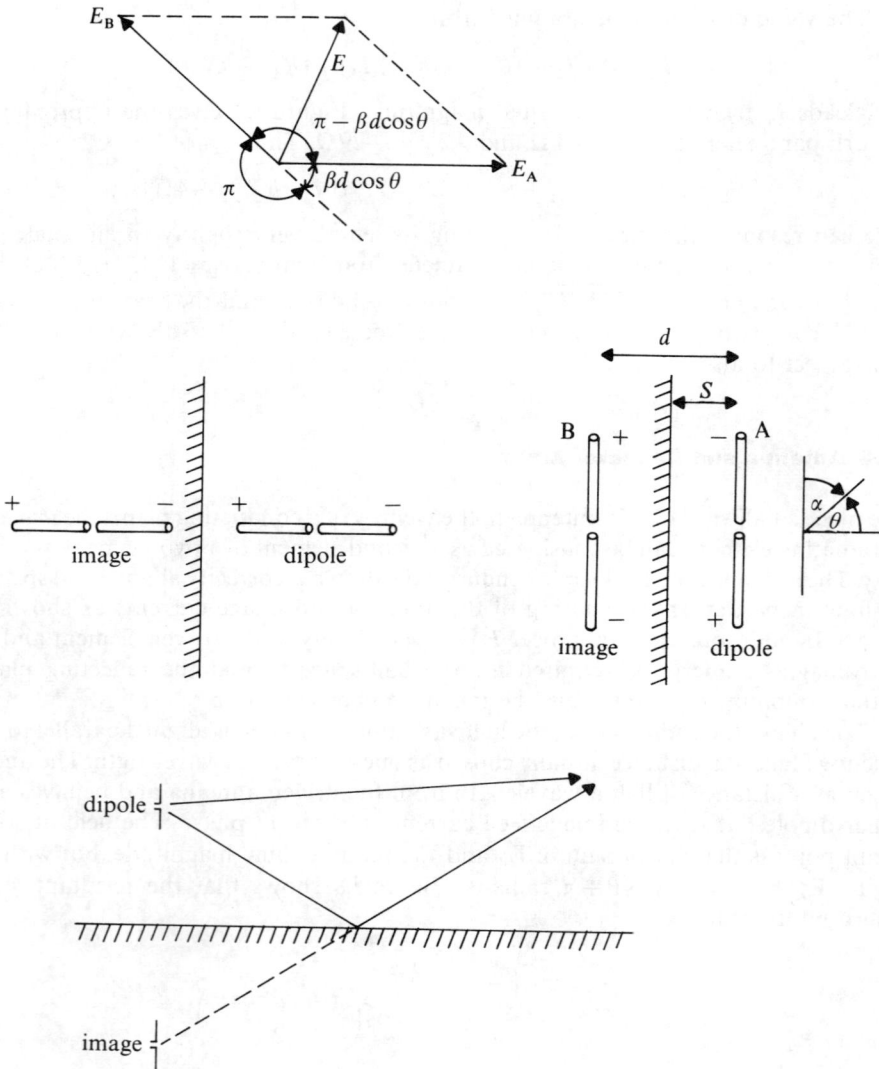

Fig. 5.8 Image elements in a reflector plane. The phasor addition refers to antennas A and B.

illustrated in Fig. 5.10, each quadrant will have a similar overall radiation pattern due to symmetry. The field at some distant point due to each dipole and reflector antenna, from Eq. (5.9), is

$$E_A = E_s \frac{\cos\left(\dfrac{\pi}{2}\cos\alpha_A\right)}{\sin\alpha_A} \sin\left(\dfrac{\pi}{2}\sin\alpha_A\right)$$

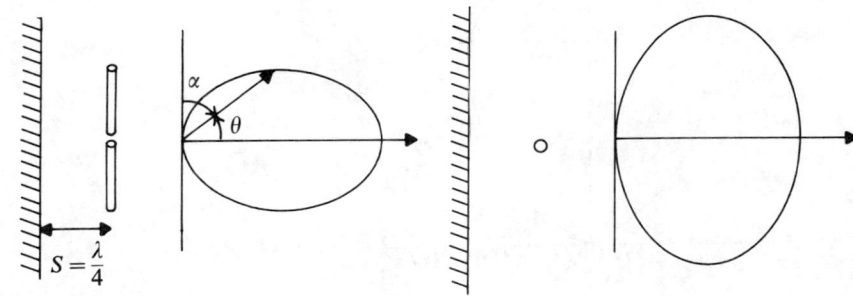

(a)

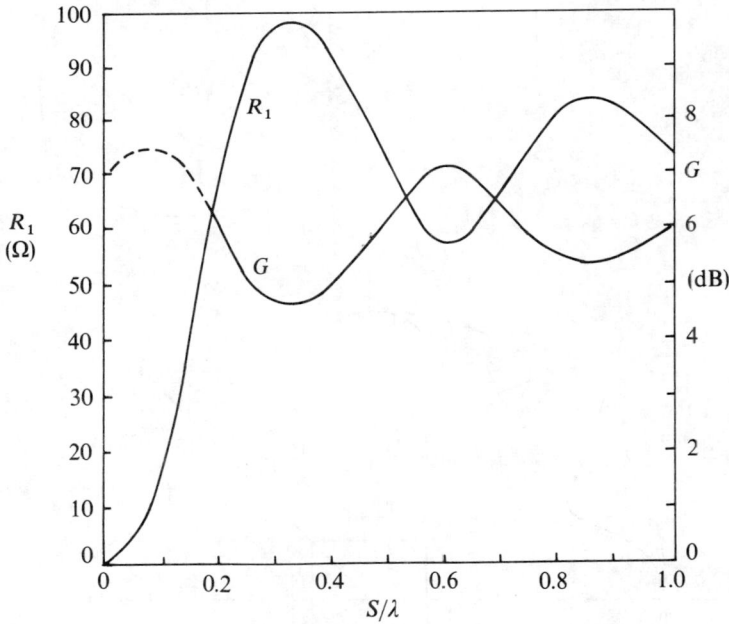

(b)

Fig. 5.9 (a) The radiation patterns of a dipole and reflector endfire array. (b) The input resistance R_1 and gain G (with respect to a dipole) of the endfire array in the presence of a perfectly conducting reflector.

$$E_B = E_s \frac{\cos\left(\dfrac{\pi}{2}\cos\alpha_B\right)}{\sin\alpha_B}\sin\left(\dfrac{\pi}{2}\sin\alpha_B\right)$$

where E_s will be determined later. Note that α_A and α_B are complementary so that $\cos\alpha_A = \sin\alpha_B$, etc., and

$$E_A = E_s \frac{\cos\left(\dfrac{\pi}{2}\sin\alpha_B\right)}{\cos\alpha_B}\sin\left(\dfrac{\pi}{2}\cos\alpha_B\right)$$

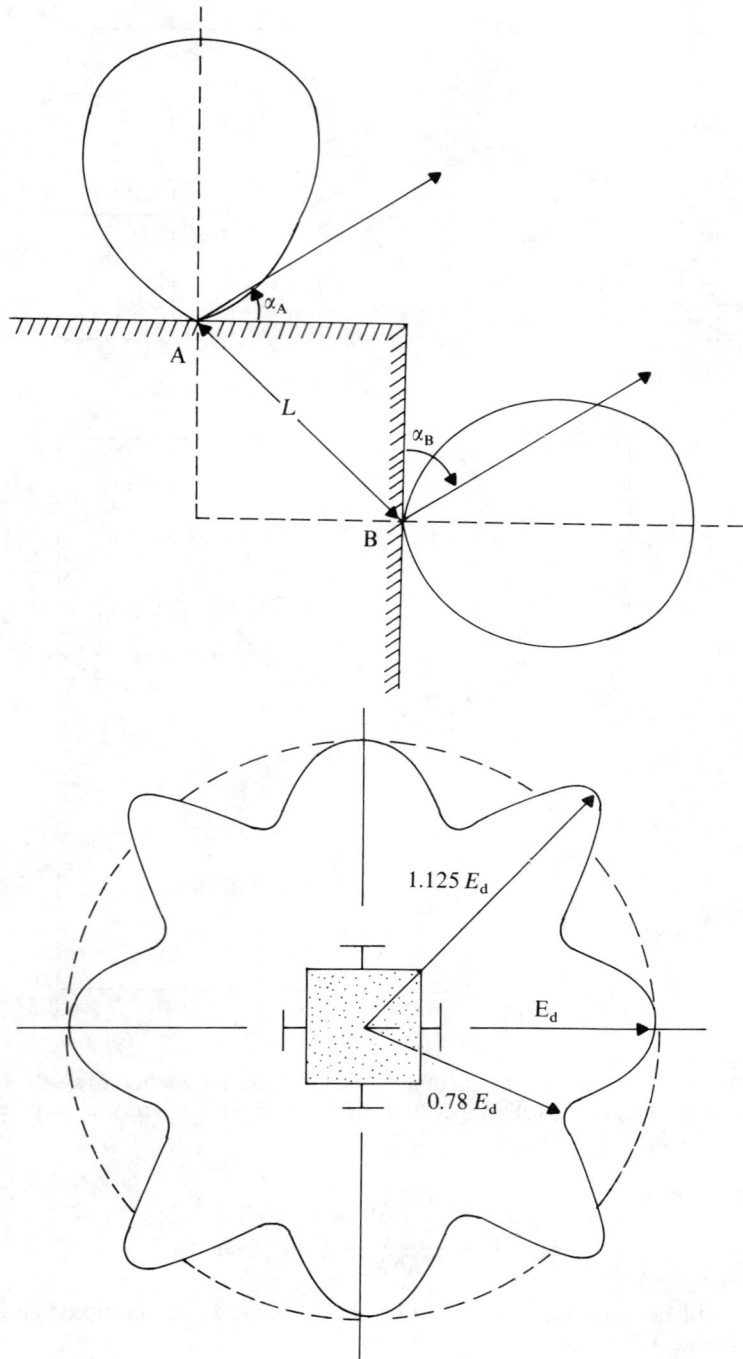

Fig. 5.10 A quadrant array of dipole and reflector type antennas produce an almost omnidirectional radiation pattern in the plane of the array. Factor $\sqrt{R/R_1}$ has been ignored.

The total field at some distant point in direction α_B is thus the phasor sum of $\mathbf{E_A}$ and $\mathbf{E_B}$ with $\mathbf{E_A}$ leading $\mathbf{E_B}$ by $\phi = \beta L \cos(\pi/4 + \alpha_B)$, or by $\phi = 2\pi \cos(\lambda/4 + \alpha_B)$ if we made $L = \lambda$, for example.

The radiation pattern so formed is illustrated in Fig. 5.10. The maximum field strength is about 1 dB greater than E_S when α_B is 45° and the minimum is about 2 dB lower. In the direction $\alpha_B = 90°$ there is no contribution from any antenna except the halfwave dipole B, increased by a factor of almost 2 due to the reflector and decreased by a factor $\sqrt{4}$ because the total power is now shared equally between the four sets of dipole and reflector antenna arrays. Thus the scaling factor E_s is the halfwave dipole gain reduced by $\sqrt{R/R_1} = \sqrt{73/87} = 0.76$ dB due to the mutual coupling of the dipole with the reflector, and ignoring other effects. The pattern is thus almost omnidirectional in the plane of the dipoles.

5.3.4 Parasitic Antenna

If we have two halfwave dipoles in broadside or parallel configuration and we short the input terminals of one of the dipoles, what would be the effect? We can start by using

$$V_1 = Z_{11}I_1 + Z_{12}I_2$$
$$V_2 = Z_{21}I_1 + Z_{22}I_2$$

with $V_2 = 0$ when the second antenna is shorted, and hence

$$I_2 = -I_1(Z_{21}/Z_{22})$$

If we choose (say) $d = 0.2\lambda$ we can obtain $Z_{21} = R_{21} + jX_{21} = 51 - j21\,\Omega$ from Fig. 5.2 and $Z_{22} = 73.2 + j42.5\,\Omega$. By converting to polar coordinates, $Z_{21} = 55.15\,\underline{/-22.38°}$ and $Z_{22} = 84.64\,\underline{/30.14°}$ and hence

$$I_2 = -I_1(Z_{21}/Z_{22}) = 0.65\underline{/180 - 52.5°}I_1 = 0.65\underline{/2.225\,\text{rad}}\,I_1$$

The electric field strength produced by each element is proportional to the currents in each element. Thus E_1 is proportional to I_1 and E_2 is proportional to $I_2 = 0.65\underline{/2.225\,\text{rad}}\,I_1$ with an additional phase lag of $\beta d \cos\theta = 0.4\pi \cos\theta$ because of the displacement $d = 0.2\lambda$ between the elements. The resultant electric field strength is therefore proportional to:

$$E_a = E_1 + E_2 = I_1 + I_1(0.65\underline{/2.225} - 0.4\pi \cos\theta)$$

The polar pattern is shown in Fig. 5.11 with the maximum field strength at $\theta = 0°$ being proportional to $I_1(1 + 0.65\underline{/0.968})$ or $1.47I_1$.

We can find the input impedance of the driven element from:

$$V_1/I_1 = Z_{11} + Z_{12}(I_2/I_1)$$

or

$$Z_1 = 73.2 + j42.5 + 55.15\underline{/-22.38°} \times 0.65\underline{/127.48°}$$

giving

$$Z_1 = 63.8 + j77.2$$

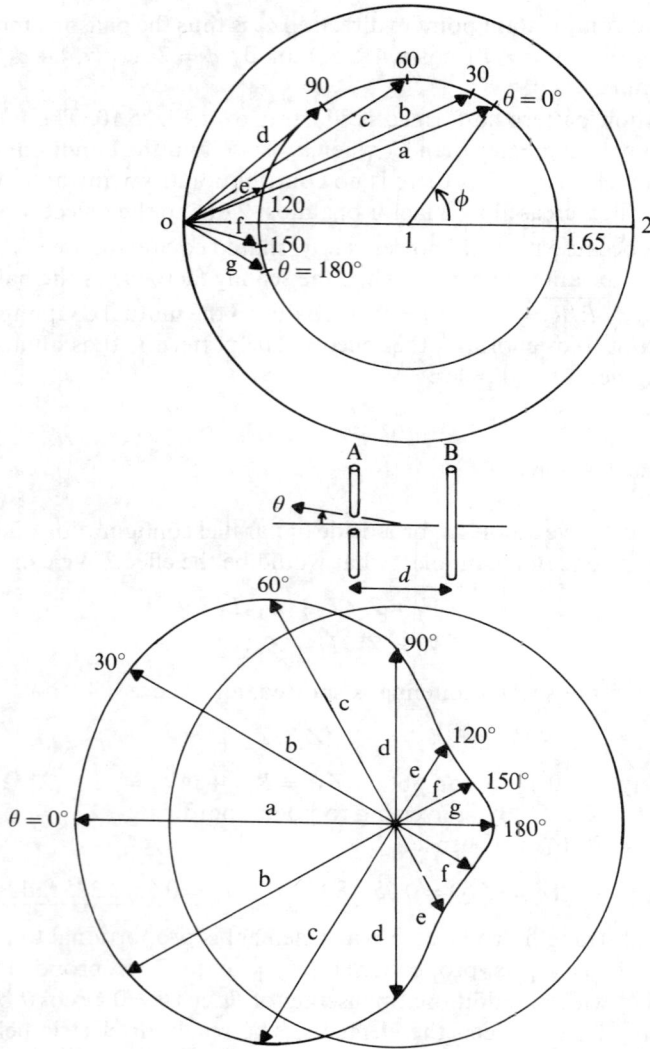

Fig. 5.11 Two-element array with one driven dipole and one parasitic dipole, both half a wavelength long, with $d = 0.2\lambda$.

The power radiated by the two-element parasitic array is $I_1^2 R_1$ whereas the equivalent power radiated by a single halfwave dipole would have been $I^2 \times 73.2$. Thus $I_1^2 \times 63.8 = I^2 \times 73.2$ and $I = 0.93 I_1$. The maximum field of the array, compared with the maximum field of the dipole, is therefore:

$$\frac{E_a}{E_d} = \frac{1.47 I_1}{0.93 I_1} = 1.58 \qquad (\text{or} \quad 4 \, \text{dB})$$

In practice it is customary to arrange the dimensions of the two antennas so that the driven antenna is at resonance and the parasitic antenna is either above or below resonance. The spacing between the elements is also adjusted so that optimum endfire operation can be achieved.

5.4 THREE-ELEMENT ANTENNA ARRAY

The principle of the three-element antenna array is a simple extension of that of the two-antenna array. The mutual impedance concept is

$$V_1 = Z_{11}I_1 + Z_{12}I_2 + Z_{13}I_3$$
$$V_2 = Z_{21}I_1 + Z_{22}I_2 + Z_{23}I_3$$
$$V_3 = Z_{31}I_1 + Z_{32}I_2 + Z_{33}I_3$$

We can choose, for example, three resonant halfwave dipoles whose length-to-diameter ratio is such that

$$Z_{11} = Z_{22} = Z_{33} = 70$$
$$Z_{12} = Z_{21} = Z_{23} = Z_{32} = 40 - j30$$
$$Z_{13} = Z_{31} = -10 - j30$$

when the three are in parallel configuration with $d = \lambda/4$ spacing between each, using Fig. 5.2 as an approximate guide. Then if we choose (say) three currents of identical magnitude but with a progressive phase change of 90° between elements so that $I_1 = 50$ rms mA, $I_2 = -j50$ mA and $I_3 = -50$ mA, we can write:

$$Z_1 = V_1/I_1 = Z_{11} + Z_{12}I_2/I_1 + Z_{13}I_3/I_1 = 50 - j10\,\Omega$$
$$Z_2 = V_2/I_2 = Z_{21}I_1/I_2 + Z_{22} + Z_{23}I_3/I_2 = 60 + j70\,\Omega$$
$$Z_3 = V_3/I_3 = Z_{31}I_1/I_3 + Z_{32}I_2/I_3 + Z_{33} = 110 + j70\,\Omega$$

The power radiated by each element is given by $P_n = \text{Re}(Z_n)|I_n|^2$. This gives $P_1 = 50(50\,\text{mA})^2 = 0.125$ W, $P_2 = 60(50\,\text{mA})^2 = 0.15$ W and $P_3 = 110(50\,\text{mA})^2 = 0.275$ W. The total radiated power is 0.55 W.

The phase change per element is $\phi = \beta d \cos\theta - \pi/2$ with $d = \lambda/4$ giving $\phi = (\pi/2)$ $(\cos\theta - 1)$. In the direction from dipole 1 towards dipole 3, where $\theta = 0°$, the three components are in phase and produce a resultant $E_\theta = 3E$, where E is the field due to a current of 50 mA in each dipole. If the total power of 0.55 W had been supplied by one dipole, the current would have been $I = \sqrt{0.55/73} = 86.8$ mA. Thus the field in direction $\theta = 0°$ is enhanced by a factor $(3 \times 50)/86.5 = 1.73$ with respect to a halfwave dipole. This is equivalent to a gain of 6.9 dBi.

A three-element array can also be made with parasitic elements. One common type of three-element dipole antenna array incorporates a driven dipole with two parasitic elements in parallel, one either side, as shown in Fig. 5.12. The driven element is adjusted to a resonant length, one of the parasitic elements is arranged to be larger

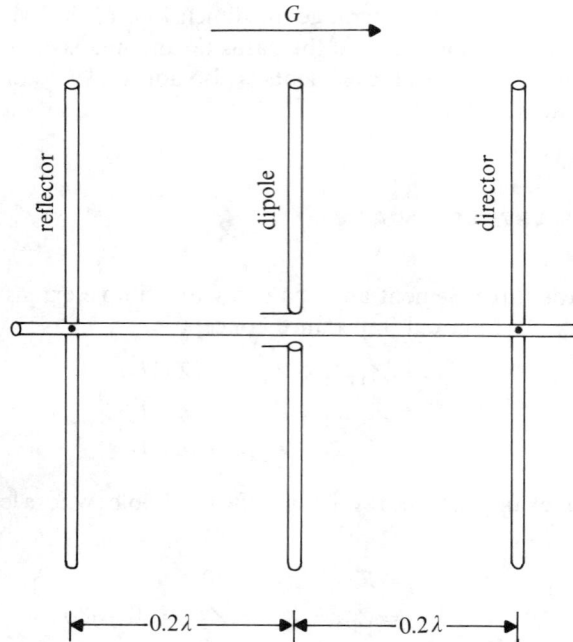

Fig. 5.12 A three-element dipole array with one acting as a reflector and one as a director. Depending on the length-to-diameter ratio, typical lengths may be 0.49λ, 0.47λ and 0.45λ for each of the elements. With displacements around $d = 0.2\lambda$ the overall antenna gain in the direction shown is about 7 dB with respect to a halfwave dipole.

than the resonant length with the consequent inductive reactance, and the other parasitic element is made shorter than the resonant length to produce a capacitive reactance. When combined into a three-element array with appropriate displacements, the overall effect is an endfire antenna in the direction shown. The parasitic elements are commonly called the *director* and the *reflector*.

Analysis of such arrays is complicated. The equations and graphs for R_{12} and X_{12} assume both elements have equal length, and the application to dipoles of different lengths is beyond the scope of this text. There are many practical charts which give some indication of the optimum lengths and spacings and these indicate a maximum gain of about 7 dB over a dipole or 9 dB with respect to an isotrope.

5.5   LINEAR ANTENNA ARRAYS

Arrays can be constructed of more than two or three antennas. If N such antennas are aligned along (say) the x axis and each antenna is displaced distance d from the next, the resultant antenna is called a *linear array*.

5.5.1 Linear Arrays of Isotropes

The linear array of N isotropes illustrated in Fig. 5.13a produces an electric field at some distant point Q which is the vector sum of all the component fields due to each

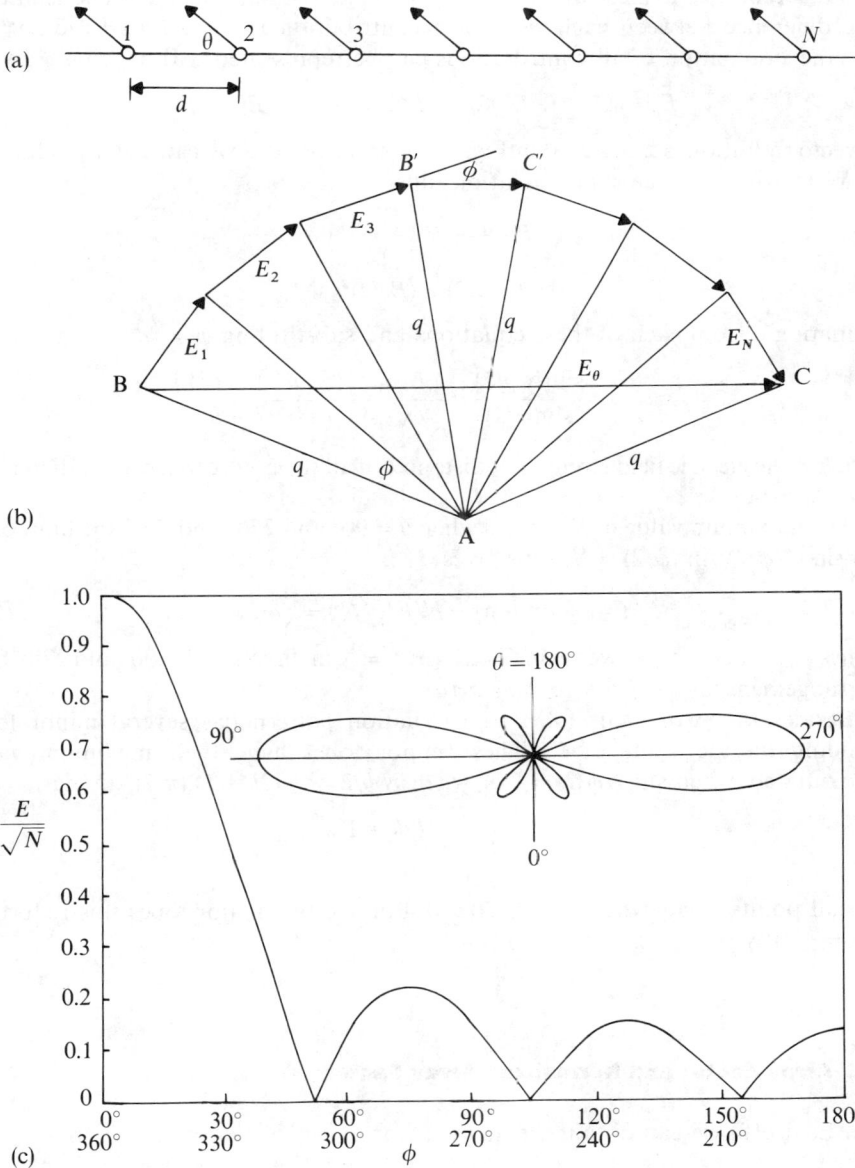

(a)

(b)

(c)

Fig. 5.13 (a) A linear array of $N = 7$ elements, (b) the vector addition for a given value of ϕ, and (c) the radiation patterns in polar and cartesian form.

individual isotrope in the array. When r is large, the magnitude of each contribution is the same, $E^* = E/\sqrt{N}$, because the power is shared equally among the N isotropes and the electric field strength is proportional to $\sqrt{P}$. This assumes that mutual impedance effects are negligible. However, the phases of each of the contributory field components are all different. For a linear antenna array, in which displacement d is a constant, the phase difference between each successive contribution is $\phi = \beta d \cos \theta$ radians and hence the summation of all contributions can be represented with lagging ϕ by

$$E_\theta = E^* \underline{/\phi} + E^* \underline{/2\phi} + E^* \underline{/3\phi} + \cdots + E^* \underline{/N\phi}$$

The vector addition is illustrated in Fig. 5.13b and from this we can use triangles ABC and AB′C′ to obtain the simple relationships

$$\sin(\phi/2) = (E^*/2)/q$$

and

$$\sin(N\phi/2) = (E_\theta/2)/q$$

Eliminating q from each of these equations and substituting $\phi = \beta d \cos \theta$ gives

$$E_\theta = E^* \frac{\sin(N\phi/2)}{\sin(\phi/2)} = \frac{E}{\sqrt{N}} \frac{\sin\{(N\pi d/\lambda)\cos\theta\}}{\sin\{(\pi d/\lambda)\cos\theta\}} \tag{5.10}$$

where E is the electric field strength at distance r if all the power had been radiated by a single isotrope.

The maximum value of E_θ occurs when $\theta = 90°$ and $270°$ and then the limit of the ratio $\sin(N\phi/2)/\sin(\phi/2) = N$. This gives

$$E_\theta(\text{maximum}) = N(E/\sqrt{N}) = \sqrt{N}\,E \tag{5.11}$$

which is equivalent to a power gain $G = (E_\theta/E)^2 = N$ in direction $\theta = 90°$ and $270°$. Such an arrangement is called a *broadside array*.

Besides the two major lobes, the radiation pattern has several minor lobes, depending on the value of N. These minor lobes have their maximum values approximately when $\sin(N\phi/2) = 1$ or when $N\phi/2 = \pm(2k+1)(\pi/2)$, giving

$$\cos\theta = \pm \frac{(2k+1)\lambda}{2Nd} \tag{5.12}$$

The null points occur when $\sin(N\phi/2) = 0$. For the two major lobes this reduces to $\cos\theta = \pm \lambda/(Nd)$.

5.5.2 Array Factor and Normalized Array Factor

From Eq. (5.10) we can obtain the array factor of our N-element array as

$$\text{AF} = \frac{E_\theta}{E} = \frac{1}{\sqrt{N}} \frac{\sin\{(N\pi d/\lambda)\cos\theta\}}{\sin\{(\pi d/\lambda)\cos\theta\}} \tag{5.13}$$

Example 5.2 Seven-element linear array

Draw the E-field polar pattern of a 7-element linear array which incorporates two major lobes broadside and four complete sidelobes.

For a symmetrical polar pattern, each quadrant will contain half of one major lobe, one complete sidelobe and two nulls, one of the latter obviously at $\theta = 0°$ or $180°$. The two nulls in each quadrant occur when

$$\sin(N\phi/2) = 0 \qquad \text{or} \qquad \phi = 2\pi/7 \quad \text{and} \quad \phi = 4\pi/7$$

If $\cos\theta$ is to take all values from 0 to 1 then $\beta d = 4\pi/7$ or $d/\lambda = 2/7$. The nulls when $d = 2\lambda/7$ occur when

$$\frac{N}{2}(\beta d \cos\theta) = \pi, 2\pi, 3\pi, \text{etc.},$$

$$\cos\theta = 0.5, 1.0, 1.5, \text{etc.}$$

Only two values are possible, $\theta = 60°$ and $\theta = 0°$ with corresponding values in each quadrant. Substituting $d/\lambda = 2/7$ into Eq. (5.10) gives the polar plot illustrated in Fig. 5.13c.

---

in which the maximum field occurs when $\theta = 90°$. Sometimes it is useful to have instead a normalized array factor in which the maximum value is unity. Thus

$$\text{NAF} = \frac{E_\theta}{\sqrt{N}E} = \frac{1}{N}\frac{\sin\{(N\pi d/\lambda)\cos\theta\}}{\sin\{(\pi d/\lambda)\cos\theta\}} \tag{5.14}$$

Both of these can be plotted in either polar or cartesian form and either linearly or in decibels, as appropriate or convenient.

5.5.3 Effect of Phase of Feed Current

So far we have assumed that identical currents have been supplied to each antenna in the linear array. If the amplitude of each feed current remains constant but the phase difference between successive feeds is adjusted so that

$$\phi = \beta d \cos\theta - \psi = \beta d(\cos\theta - \cos\theta_0) \tag{5.15}$$

then the resultant field becomes

$$E_\theta = \frac{E}{\sqrt{N}}\frac{\sin\left\{\dfrac{N\pi d}{\lambda}(\cos\theta - \cos\theta_0)\right\}}{\sin\left\{\dfrac{\pi d}{\lambda}(\cos\theta - \cos\theta_0)\right\}} \tag{5.16}$$

and the maximum lobe is aligned in direction $\theta_0 = \cos^{-1}(\psi/\beta d)$. For example, when

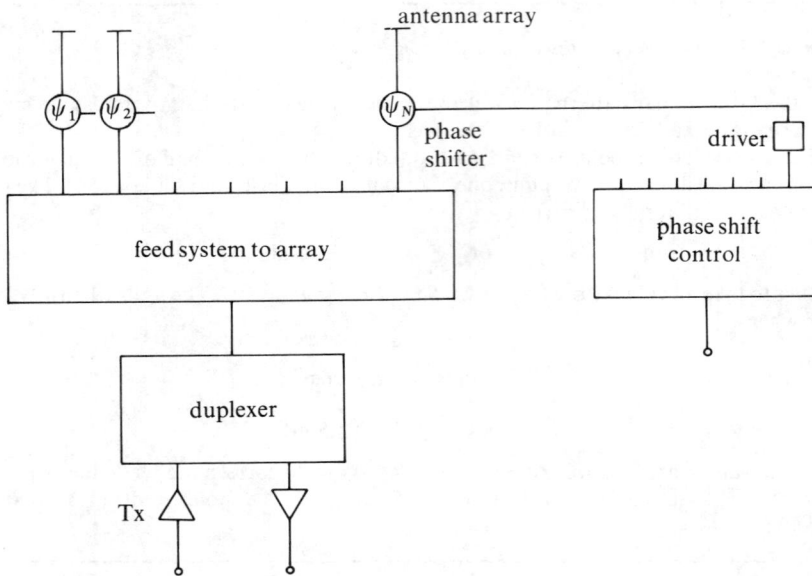

Fig. 5.14 Beam-steering can be achieved with an N-element array if the phases of the currents to each element are controlled (often digitally) via a phase control circuit.

$\psi = 0$, Eq. (5.16) reverts back to Eq. (5.10), but when $\psi = 0.1$ radian, $\theta_0 = 86°$ when $d = \lambda/4$.

One simple arrangement of such an array is shown in Fig. 5.14, where digitally controlled phase shifters are incorporated before each radiating element. As ψ is adjusted between the two available limits, the major lobe can be steered between the corresponding limits of θ_0.

Example 5.3 Beamwidth of N-element broadside array

If the beamwidth of a uniform linear array is defined as the angle between the first nulls either side of the main lobe, derive an expression for beamwidth BW and determine values of BW for various N when $d = \lambda/4$.

Nulls occur when $\sin(N\phi/2) = 0$ and the first nulls either side of the major lobe occur when $(N\phi/2) = \pm \pi$. When $d = \lambda/4$ this leads to $\cos \theta = \pm (\lambda/Nd) = \pm (4/N)$. However, the main lobe is directed $\theta = 90°$ and its half beamwidth $\alpha = 90° - \theta$, or $\theta = 90° - \alpha$, or $\cos \theta = \sin \alpha$. Hence from $\sin \alpha = 4/N$, BW $= 2\alpha = 2 \sin^{-1}(4/N)$.

For $N = 5$ to 10, BW $= 106°, 84°, 70°, 60°, 53°$ and $47°$.

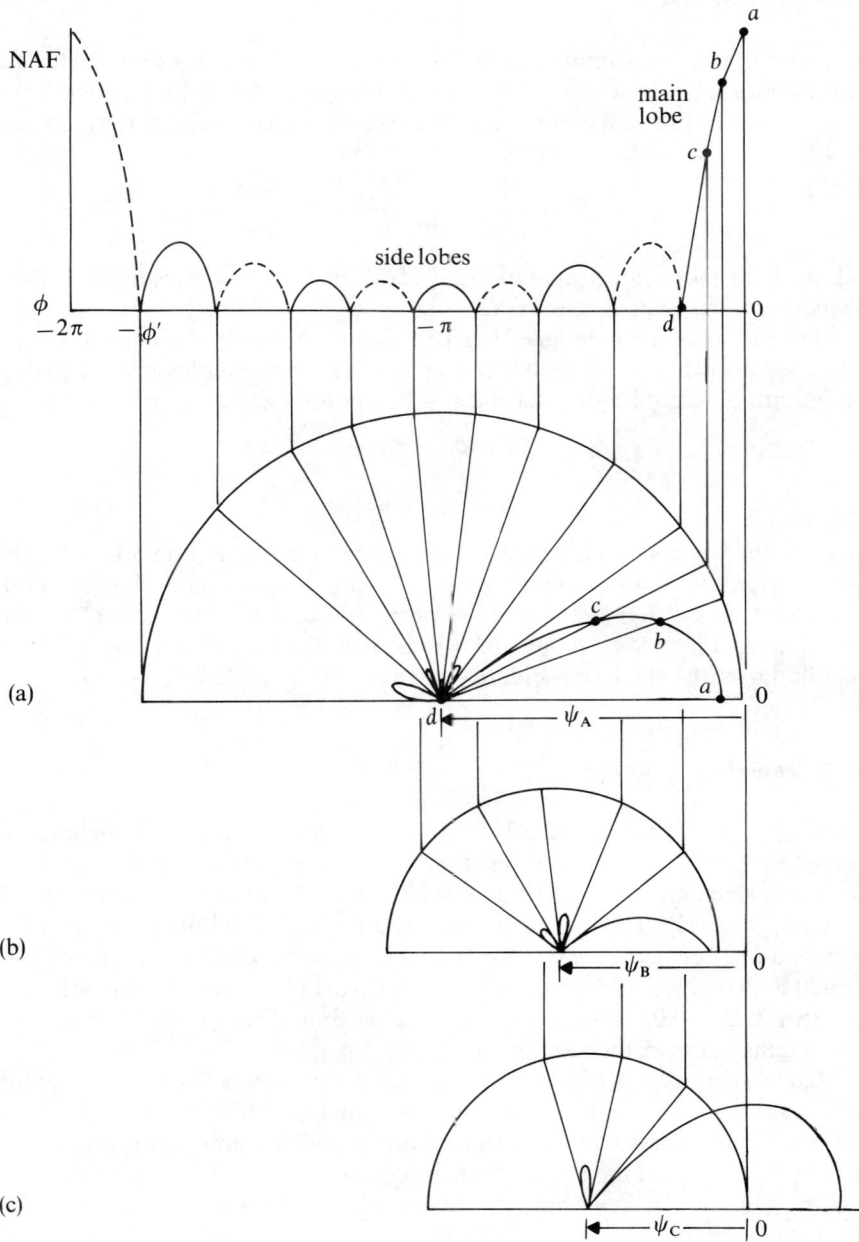

Fig. 5.15 Geometric construction of polar diagrams of a ten-element array (a) for Prob. 5.7, (b) with optimum directivity, and (c) with $d = \lambda/4$ and $\psi = \pi/2$. Only a selection of sidelobes are illustrated.

5.5.4 Endfire Array

If we wish to have maximum gain in the directions $\theta_0 = 0°$, for example, it is necessary for the major lobe to have $\phi = 0$ under this condition. Hence $\phi = \beta d \cos \theta + \psi$ reduces to $\psi = -\beta d$ radians. We can achieve these conditions very simply by arranging $d = \lambda/4$ with $\psi = -\pi/2$ to obtain

$$E_\theta = \frac{E}{\sqrt{N}} \frac{\sin\{N[(\pi \cos \theta - \pi)/4]\}}{\sin\{(\pi \cos \theta - \pi)/4\}} \tag{5.17}$$

and the polar plot is illustrated in Fig. 5.15c. The maximum power gain, ignoring the effects of mutual impedances, is N.

The endfire array so designed has a maximum gain in the direction $\theta = 0°$ but it has a wide beamwidth. Sometimes it is desired to have a narrower beamwidth in the endfire configuration. According to Ref. 5.2 this can best be achieved by using

$$\psi = \beta d + \pi/N \tag{5.18}$$

to give

$$\phi = \beta d(\cos \theta - 1) - \pi/N \tag{5.19}$$

provided there is a suitable limitation on displacement d. Figure 5.15b illustrates the polar diagram for $N = 10$ in comparison with the previous design criterion of Eq. (5.17). Note that the narrower beamwidth between nulls of Fig. 5.15b (about 74°) compared with that of Fig. 5.15c (about 106°) is achieved with an increase in the relative magnitudes of the sidelobes. Figure 5.15a refers to Prob. 5.7.

5.5.5 Yagi–Uda Antenna

The Yagi–Uda antenna (Ref. 5.3) consists of a driven element (sometimes a folded dipole to increase the radiation resistance), a reflector and one or several directors. If the driven element is resonant, the reflector is made a little larger and the directors made progressively smaller, though not necessarily in a uniform manner. Depending on the number of elements, the displacement between reflector and driven element is about 0.18 ± 0.02 wavelengths, with about 0.13 to 0.17 wavelengths to the first director, an extra 0.26 ± 0.05 wavelengths to the second director, and about 0.3 ± 0.03 wavelengths between the remainder.

Exact analysis is obviously difficult, but a rough indication of the antenna gain (with respect to a halfwave dipole) is given in Fig. 5.16 for a specified number of elements. Yagi arrays can be stacked both broadside and collinear, the optimum spacing depending on the required magnitude of the resulting beamwidth. About 3 dB of extra gain can be achieved with two stacked Yagi arrays approximately 1 to 2 wavelengths apart.

There are many applications of the Yagi antenna, particularly in the VHF and UHF bands, for line-of-sight communication links, reception of television, etc. It is essentially a single-frequency antenna but will work reasonably well over a bandwidth of perhaps $\pm 3\%$.

Handwritten annotation (right of graph):

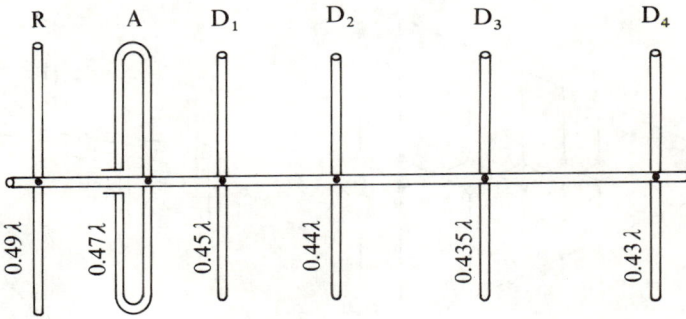

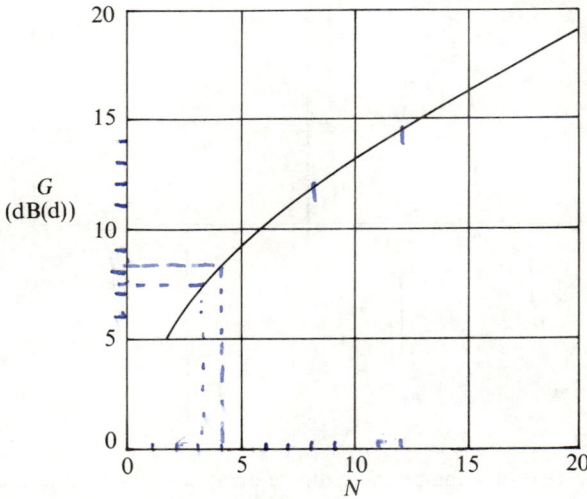

3 ele ~ 7 dBd
4 ele ~ 8 dBd
6 ele ~ 10 dBd
8 ele ~ 12 dBd
12 ele ~ 14 dBd

Fig. 5.16 A typical Yagi–Uda antenna with folded dipole as the driven element. The gain of the N-element array is in decibels with respect to a halfwave dipole and is only an approximation. The length-to-diameter ratio is approximately 200:1.

5.5.6 Log-periodic Dipole Array

The halfwave dipole arrays are frequency-dependent antennas as their self and mutual impedances are related to the antenna lengths and spacing via wavelength λ. One type of dipole array has been designed to operate over a very broad band of frequencies by arranging for the dipoles to have lengths and displacements related in a periodic manner. The particular form described by Isbell (Ref. 5.4) is illustrated in Fig. 5.17. The

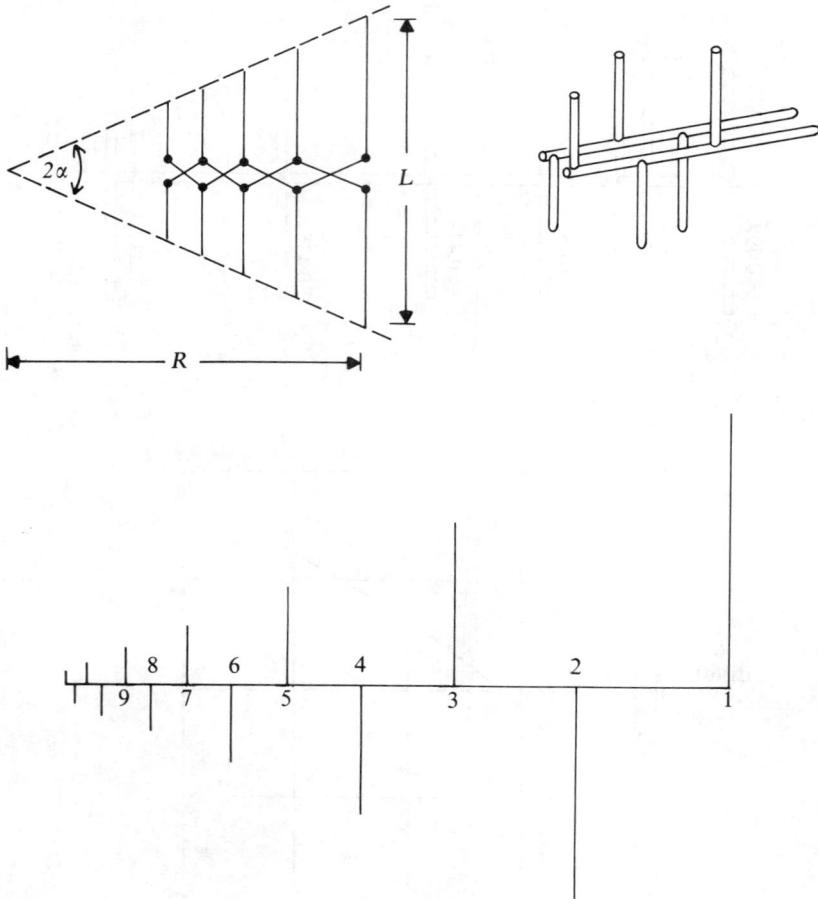

Fig. 5.17 The feed pattern of a log-periodic dipole array and the arrangement of the halfwave dipoles. The lower pattern is drawn to scale from details calculated in Ex. 5.4 and represents one half-section of the antenna.

repetitive pattern is described by a parameter τ such that

$$\frac{1}{\tau} = \frac{f_2}{f_1} = \frac{f_3}{f_2} = \frac{f_4}{f_3} = \cdots = \frac{R_1}{R_2} = \frac{R_2}{R_3} = \frac{R_3}{R_4} = \cdots = \frac{L_1}{L_2} = \frac{L_2}{L_3} = \frac{L_3}{L_4} = \cdots \quad (5.20)$$

and f_1, f_2, f_3, etc., are the frequencies at which the antenna characteristics (such as impedance, radiation resistance or gain) repeat successively. Alternatively, from Eq. (5.20),

$$\log(f_n/f_1) = n \log(1/\tau) \quad (5.21)$$

where $n = 1, 2, 3$, etc. When the array is fed at a specific frequency, the electromagnetic radiation occurs mainly in the vicinity of those dipoles whose overall length is nearest

Table 5.1 Log-periodic array design parameters

G(dBi)	8	9	10	11	12
γ	0.780	0.865	0.920	0.945	0.965
α°	21.9	12.1	6.8	4.6	2.8
BF	2.03	1.75	1.51	1.39	1.29

to half a wavelength at the given frequency. Thus the highest and lowest operating frequencies coincide with the wavelengths which make the smallest and largest dipoles half a wavelength. Such antennas can easily be designed with a 10:1 ratio in the lengths of the extreme dipole elements, and the bandwidths are correspondingly large.

As a rough indication of the optimum values of τ and α that are used in the design of log-periodic dipole antennas, use Table 5.1 to obtain the parameters for a specified overall antenna gain (in dB with respect to an isotrope). Choose the desired bandwidth ratio, f_{max}/f_{min}, and convert to the design bandwidth ratio $B = \text{BF} f_{max}/f_{min}$. The required number of elements can be estimated from

$$N = 1 + \frac{\log B}{\log(1/\tau)} \tag{5.22}$$

Select the length of the largest dipole as being half a wavelength at f_{min} and use the design maximum frequency as $B f_{min}$. Equation (5.20) can be used successively to obtain all remaining values of dipole lengths from $n = 2$ to N. Simple trigonometry indicates that $\tan \alpha = (L_1/2)/R_1$ and thus R_1 is specified. Equation (5.20) is again used for all successive values of R from R_2 to R_N.

Example 5.4 Design of log-periodic antenna

Design a log-periodic dipole antenna capable of operating over the range 100 MHz to 1 GHz with a gain of 8 dBi.

Table 5.1 indicates that $\tau = 0.78$ and $\alpha = 21.9^\circ$. The bandwidth ratio of 10:1 must be increased to the design ratio of 20:1 and the number of elements is $1 + \log 20/\log(1/0.78) = 13$. The lowest frequency gives $L_1 = 1.5$ m as 100 MHz corresponds to a wavelength of 3 m. Hence $L_2 = 1.5\tau = 1.17$ m; $L_n = 1.5\tau^{n-1}$; and $L_{13} = 1.5\tau^{12} = 0.076$ m. The relationship $\tan 21.9^\circ = 0.75/R_1$ gives $R_1 = 1.866$ m. Hence $R_2 = 1.866\tau = 1.455$ m; $R_n = 1.866\tau^{n-1}$; and $R_{13} = 1.866\tau^{12} = 0.095$ m. These are the dimensions of the log-periodic array, a half-section of which is drawn to scale in Fig. 5.17.

5.5.7 Television Transmitting Antennas

An antenna which is used for the transmission of UHF television signals is usually required to have an omnidirectional radiation pattern in the horizontal plane, while the vertical radiation pattern is carefully shaped so that the maximum intensity is directed toward the horizon and reduced levels of intensity toward receiving sites nearer the transmitter. The horizon is about $0.5°$ below the horizontal for an antenna height $h = 200$ m above mean terrain level and about $1°$ below the horizon for $h = 700$ m.

To achieve these goals, one type of television antenna consists of a vertical array of either 16 or 32 sets of the halfwave dipole array described in Sec. 5.3.3. The latter produces almost omnidirectional properties in the horizontal plane (with horizontal polarization) and the array provides the broadside gain in the vertical plane. A little adjustment of the phasing of the currents produces beam shifting so that the major lobe is directed downwards just a little below the horizontal and one of several methods may

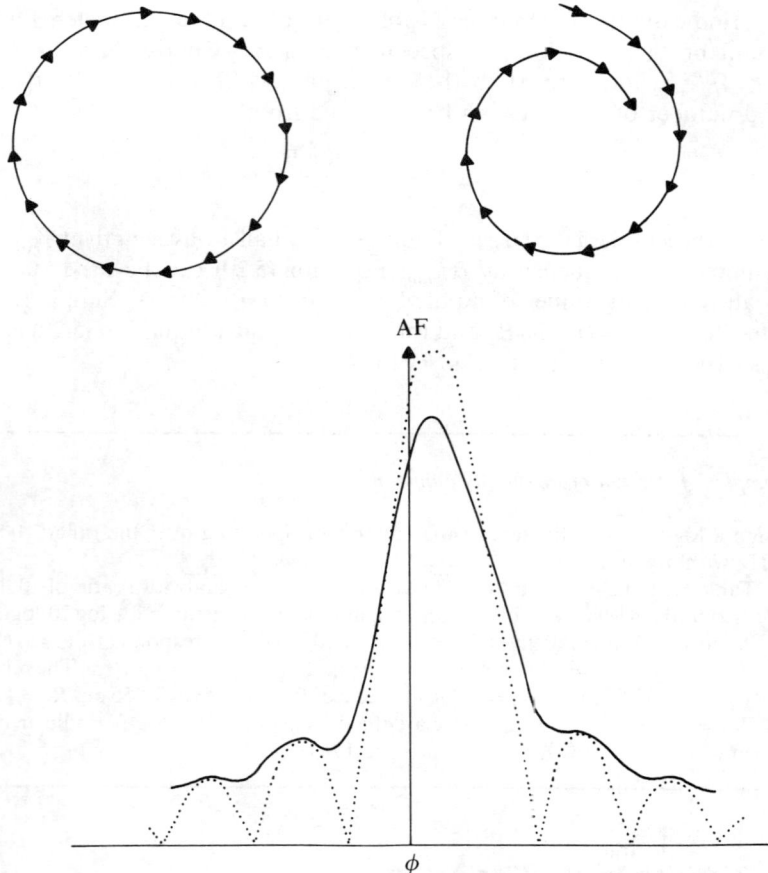

Fig. 5.18 The polygon of vectors illustrating the addition of 16 contributions from a 16-element array when ϕ is such that they result in a null. A slight modification in ϕ causes the vectors to miss producing null conditions. The lower diagram illustrates null-filling at the expense of a reduction in the main lobe gain.

be used to fill in the deep nulls to avoid the consequent deterioration of reception in localized areas nearer the transmitter.

Using four sets of vertical arrays of $N = 32$ dipoles spaced vertically by $d = \lambda/2$, the nominal maximum gain is $G = 32$ or $G = 15$ dB with respect to a halfwave dipole. This is reduced by about 2–3 dB to account for null-filling, the effects of self and mutual impedance, and the reduced size of the reflector, etc. Hence a net gain of about 12–13 dB is realistic. If the maximum effective radiated power (erpd) with respect to a halfwave dipole is nominally 100 kW, and assuming 2–3 dB of transmitter losses, the actual power supplied to the transmitter is of the order of 10 kW.

To achieve broader frequency usage the dipoles are made with a small length-to-diameter ratio, or alternatively there are versions with arrays of bat-wing or superturnstile antennas which have a broader frequency response.

Example 5.5 *Null-filling of array pattern*

Draw the phasor diagram of the sum of all sixteen contributions from a 16-element broadside array in the direction of the first null. If the phase of each current I_n is adjusted so that $\psi_n = n \times \delta\psi$, sketch the phasor addition of all sixteen contributions in the vicinity of the first minimum. Explain how this produces null-filling. Assume the inter-element spacing is $d = \lambda/2$.

For a 16-element array with $d = \lambda/2$, we can write $\phi = \pi \sin \alpha$, where $\alpha = 90° - \theta$ and represents the angle either side of the main lobe. For the first null $(N\phi/2) = \pi$ or $\phi = 22.5°$ or $\alpha = 7.1°$. The 16 contributory phasors $\mathbf{E}_n$ all have amplitude E but different phase lag $n\phi$ with respect to $\mathbf{E}_0$, where $n = 0, 1, 2, \ldots, 15$. The addition is illustrated in Fig. 5.18 as a closed 16-sided polygon, net E-field being zero and a null achieved.

If the phase angle of $\mathbf{E}_n$ is $\phi = n\phi_0 + n\delta\psi$, or $\phi = n(\phi_0 + \delta\psi)$, the phasor summation of 16 contributory phasors no longer completes the 16-sided polygon, but leaves a small net value of E-field. The minimum never quite reaches a full null. Figure 5.18 illustrates the effect on the E–ϕ diagram, showing a small decrease in the maximum gain and the filling of the nulls. This is useful in television broadcasting where deep nulls cause black-spots in the reception.

5.6 POLYNOMIALS IN LINEAR ARRAY THEORY

The electric field strength at distance r from an array of halfwave dipoles is the vector sum of the contributions due to each individual dipole, where

$$E = \frac{60I}{r} \frac{\cos\left(\dfrac{\pi}{2}\cos\theta\right)}{\sin\theta} e^{j\omega t} e^{-j\beta r} e^{j\psi}$$

The first exponential indicates a sinusoidal variation of current, the second indicates the phase retardation over distance r, and the third indicates the phase of the current I relative to some reference value.

At a given point in space, with $\exp(j\omega t)$ understood, r, θ and βr are constants and

$$E = \text{constant} \times I e^{j\psi}$$

or the electric field strength, normalized to this constant, can be represented by $E = I \exp(j\psi)$. This is the basis of the polynomial approach to array theory, due to Schelkunoff (Ref. 5.5), which assumes that each array element can be provided with specified currents of known amplitude and phase.

5.6.1 The *E*-field Equation

The electric field at distance r from a uniform N-element linear array with inter-element displacement d and ψ initially equal to zero can be written as

$$E = I_0 \underline{/0^\circ} + I_1 \underline{/\phi} + I_2 \underline{/2\phi} + \cdots I_{N-2} \underline{/(N-2)\phi} + I_{N-1} \underline{/(N-1)\phi} \qquad (5.23)$$

where the angles θ and ϕ take the special references indicated in Fig. 5.19 so as to make ϕ a leading angle for $\theta < 90^\circ$. It also makes the transfer into polynomial terminology easier if the first array is labeled number 0 and the last is labeled number $N-1$, and the use of $n = N - 1$ will simplify the look of the mathematics. Thus

$$E = \sum_{m=0}^{n} I_m \underline{/m\phi} = \sum_{m=0}^{n} I_m \exp(jm\beta d \cos \theta) = \sum_{m=0}^{n} I_m [\exp(j\beta d \cos \theta)]^m$$

A convenient shorthand approach is to let

$$z = \exp(j\beta d \cos \theta) \qquad (5.24)$$

so that the summation reduces to

$$E = \sum_{m=0}^{n} I_m z^m = I_0 z^0 + I_1 z^1 + I_2 z^2 + \cdots + I_{n-1} z^{n-1} + I_n z^n$$

or

$$E = I_n [a_0 + a_1 z + a_2 z^2 + a_3 z^3 + \cdots + a_{n-1} z^{n-1} + z^n] \qquad (5.25)$$

as $a_n = I_n / I_n = 1$. Thus the electric field strength is represented by a scaling factor I_n and a polynomial function of z which has $n = N - 1$ roots given via

$$E = a_0 + a_1 z + \cdots + a_{n-1} z^{n-1} + z^n = (z - z_1)(z - z_2) \cdots (z - z_n) \qquad (5.26)$$

where z_1, z_2, etc., are the roots of the polynomial.

5.6.2 The Unit Circle

To understand the application of polynomial theory to the analysis of antenna array patterns, we can first consider the most elementary of examples, a two-element linear array with identical currents. The equation for E in polynomial form is

$$E = 1 + z = 1 + 1 \underline{/\phi} \qquad (5.27)$$

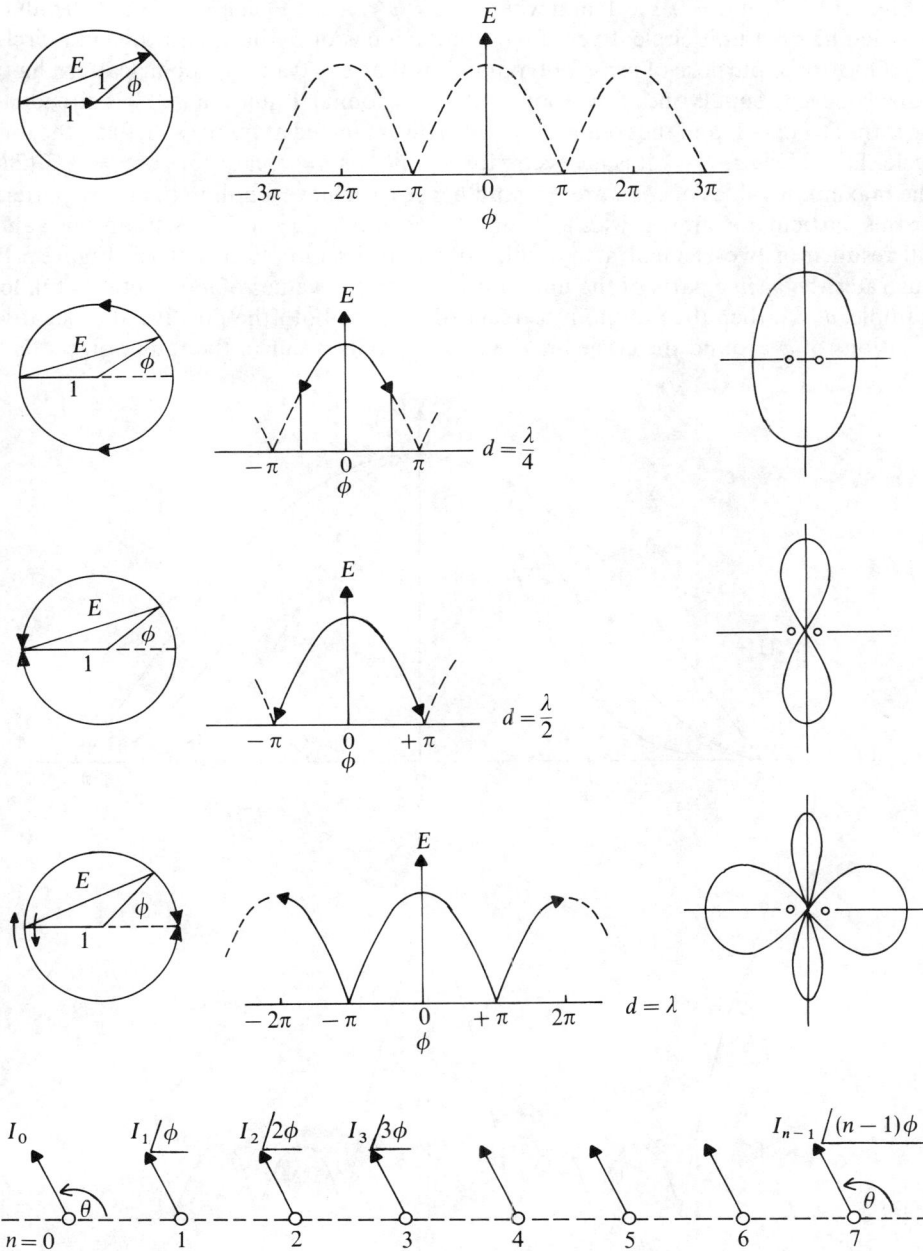

Fig. 5.19 Linear array of N isotropes (note direction θ) with the unit circles for $N = 2$ and $d/\lambda = 0.25, 0.5$ and 1.0. The corresponding polar plots are shown alongside.

The locus of $z = \exp(j\beta d \cos\theta) = \exp(j\phi)$ is obviously a circle of unit radius, as shown in Fig. 5.19. When $\phi = 0$, $z = 1$; and when $-\pi < \phi < +\pi$, $z = \cos\phi + j\sin\phi$, all values of which lie on a unit circle. Even if $|\phi| > \pi$ the locus of z still lies on this unit circle.

One useful purpose of using polynomials is that $E = 0$ and the polar pattern has a null whenever z equals one of the roots of the polynomial. Equation (5.27) is already in the form $E = (z + 1)$ and the root is $z_1 = -1$. This is located at point $(-1, 0)$ on the unit circle. To achieve $z = -1$ it is necessary for $\phi = \beta d \cos\theta = \pm\pi$ or for $\cos\theta = \pm\lambda/(2d)$. The maximum values of $\cos\theta$ are ± 1 and hence to achieve a null in the array pattern for this particular example, $\lambda/(2d) = 1$ or $d = \lambda/2$. Any value of d/λ less than this value will result in a two-element array with no null in its radiation pattern. Figure 5.19 illustrates the *visible* parts of the unit circle for various values of d/λ. Note that if, for example, $d = \lambda$, then the null point is reached twice in both the positive and negative directions of ϕ around the circle and hence four nulls occur in the radiation pattern.

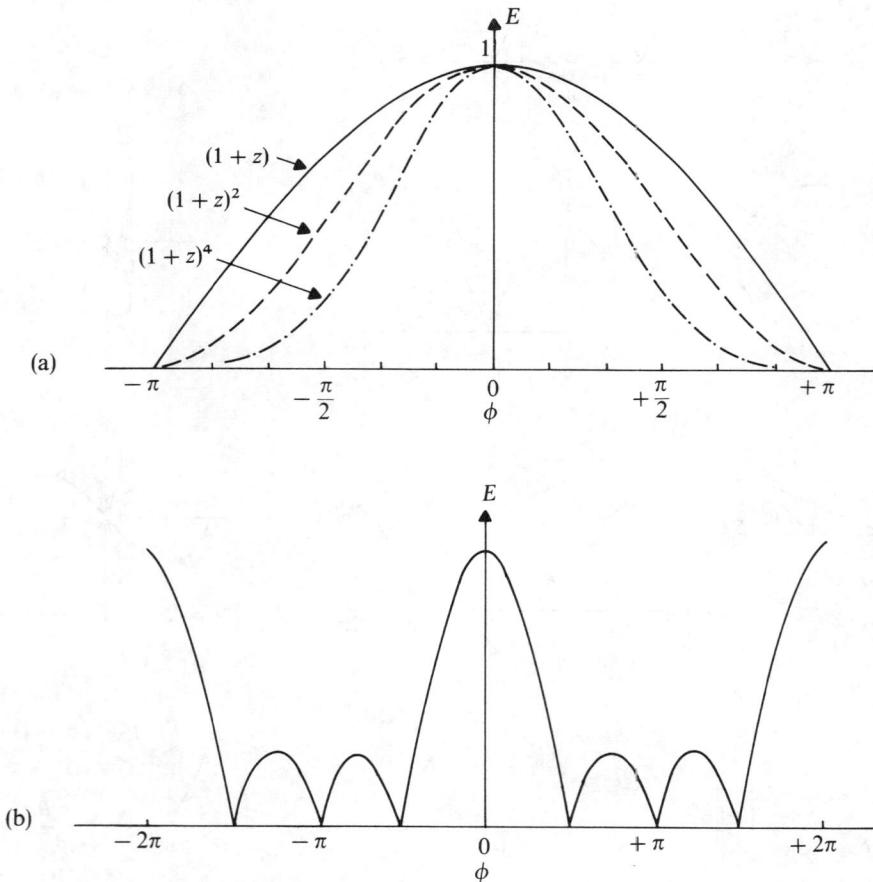

Fig. 5.20 The normalized array factor for (a) a binomial antenna array and (b) a four-element constant-current linear array.

5.6.3 The Binomial Antenna Array

We can develop the ideas outlined in Sec. 5.6.2 a little further by choosing a polynomial function with

$$E = (1 + z)^n \qquad (5.28)$$

For polynomials in general, such as $(z - a)(z - b)(z - c)\ldots$, the zero values occur when $z = a, z = b, z = c, \ldots$, where a, b, c, etc., are complex. For our earlier example, the null occurred when $z = -1$ provided this value was on the visible portion of the unit circle. For the new expression, all n null points occur at the location corresponding to $(-1, 0)$ on the unit circle because all the roots of the polynomial are $z = -1$. However, if we restrict the visible region to one complete circle ($\pm\pi$) by putting $d = \lambda/2$, the polar pattern has only two nulls, one each at $\phi = \pi$ and $\phi = -\pi$, or $\theta = 0°$ and $180°$, and the two major lobes at $\theta = 90°$ and $270°$. The beamwidth, however, and the maximum gain are changed. Figure 5.20a illustrates the cartesian plot of E against ϕ for various values of n in which it is clear that the beamwidth of the array is reduced as n increases.

The gain of the antenna array can be determined in the same manner as that described for two- and three-element arrays, noting that $(1 + z)^n$ has the binomial coefficients (with $N = n + 1$):

$$
\begin{array}{ll}
N = 2 & E = 1 + z \\
N = 3 & E = 1 + 2z + z^2 \\
N = 4 & E = 1 + 3z + 3z^2 + z^3
\end{array}
$$

and that the feed current in each element is $a_n I_n$. Mutual impedance effects must be taken into account. Thus the *binomial antenna array* is a broadside antenna with a single pair of major lobes (or one major lobe if used with a reflector) and no minor lobes. The gain increases with the number of elements and the beamwidth decreases.

5.6.4 Constant Current Array

The earlier approach to linear antenna arrays in Sec. 5.5 assumed that all the currents were of uniform amplitude. How can this be represented using the polynomial approach? It is easily confirmed for a four-element array with equal currents in each element, that

$$E = 1 + z + z^2 + z^3 = (z - j)(z + j)(z + 1) \qquad (5.29)$$

by simply expanding the right-hand section of the expression. All the roots are located on the unit circle and there are therefore three nulls on the unit circle (provided they are within the visible region) at $(0, j)$, $(0, -j)$ and $(-1, 0)$.

With the aid of a little mathematics, we see that

$$(z - j)(z + j)(z + 1) \times \frac{(z - 1)}{(z - 1)} = \frac{z^4 - 1}{z - 1} = \frac{z^N - 1}{z - 1}$$

or

$$f(z) = \frac{\exp(jN\phi) - 1}{\exp(j\phi) - 1} = \frac{\exp(jN\phi/2)}{\exp(j\phi/2)} \frac{\exp(jN\phi/2) - \exp(-jN\phi/2)}{\exp(j\phi/2) - \exp(-j\phi/2)}$$

$$f(z) = \frac{\sin(N\phi/2)}{\sin(\phi/2)} \exp[j(N-1)\phi/2] \tag{5.30}$$

The first term is the normalized form of the magnitude of E as given in Eq. (5.10), with a maximum value of $N = 4$ for a 4-element array, and the second term is the phase of the normalized resultant.

The cartesian plot of the magnitude of E against ϕ is illustrated in Fig. 5.20b. The visible region is $\phi = \pm \pi$ radians for $d = \lambda/2$ and a graphical estimate of the minor lobe maximum gives $\phi = 132°$ (say). For the 4-element array, the sidelobe amplitude is therefore $\{\sin(4 \times 132°/2\}/\{\sin(132°/2)\} = 1.09$ compared with a magnitude of 4 for the major lobe. The sidelobe is therefore 11.3 dB less than the major lobe.

The form of factorization we have just used for a 4-element constant-current array will work for any value of N. If we write:

$$f(z) = 1 + z + z^2 + \cdots + z^{N-1}$$

then

$$z \times f(z) = z + z^2 + z^3 + \cdots + z^N$$

Subtracting the second expression from the first we obtain:

$$(1 - z)f(z) = 1 - z^N$$

or

$$f(z) = \frac{1 - z^N}{1 - z} = \frac{z^N - 1}{z - 1}$$

The zeros of the expression on the right-hand side of the equation are those of $z^N - 1$, excluding the one root $z = 1$. Thus the zeros are the N values of $\sqrt[N]{1}$ (excluding 1 itself), which are $\exp(2\pi jm/N)$ with $m = 1, 2, \ldots, N - 1$. The polynomial must clearly factorize as:

$$(z - \exp(2\pi j/N))(z - \exp(4\pi j/N))(z - \exp(6\pi j/N)) \cdots (z - \exp(2(N-1)\pi j/N)) = 0$$

The roots are uniformly displaced around the unit circle.

5.6.5 Gaussian Array

If it is required to reduce the sidelobe level of the constant-current array even further without additional sidelobes, the solution is to increase N but retain the nulls in the same position. For instance, let

$$f(z) = \{(z - j)(z + j)(z + 1)\}^2 \tag{5.31}$$

as we did for the binomial array with the same consequent reduction in beamwidth. Now the sidelobe level when $\phi = 131°$ is still 1.09 but with respect to the major lobe level of $N = 7$, a difference of 16.2 dB. With

$$f(z) = \{(z - j)(z + j)(z + 1)\}^3 \tag{5.32}$$

the difference when $\phi = 127°$ becomes 19.0 dB for $N = 10$ elements. One step further and the full expansion for $N = 13$ elements is:

$$f(z) = 1 + 4z + 10z^2 + 20z^3 + 31z^4 + 40z^5 + 44z^6$$
$$+ 40z^7 + 31z^8 + 20z^9 + 10z^{10} + 4z^{11} + z^{12} \tag{5.33}$$

The current amplitudes, when plotted as a histogram, tend to the Gaussian distribution as N becomes large. Such an array has a large gain with narrow beamwidth and small sidelobe levels.

We have discussed here the raising of a constant-current pattern function of the form $(1 + z + z^2 + \cdots + z^n)$ to the rth power, leading to $N = 1 + nr$ elements. The binomial array is, of course, the special case where the original constant-current array had only two elements, i.e. the pattern function is $(1 + z)^r$, and the property that the current amplitude histogram tends towards the Gaussian distribution for large N is indeed valid for that case too.

Example 5.6 *The isolation of an interference signal*

It is required to design a 4-element array along the z-axis such that it can receive a maximum signal in the direction $\theta = 90°$ but reject any interference coming in the direction $\theta = 45°$.

We can produce the same solution by two approaches. Firstly, to ensure a null at $\theta = 45°$ and a maximum at $\theta = 90°$ it is useful to create a symmetric null at $\theta = 135°$. The corresponding values of $\phi = \beta d \cos \theta$ require a knowledge of d/λ. Figure 5.21 shows two of many possible solutions, with nulls at $\theta = 45°$ and $\theta = 135°$ within the visible region of the unit circle and a third null at $\phi = 180°$ outside the visible region. This is convenient to provide a solution to the polynomial

$$f(z) = (z - z_1)(z - z_2)(z - z_3)$$

which is symmetrical, though other locations within or without the visible region could be used provided there is no null at or near $0 = 90°$.

The second example in Fig. 5.21 is easier to manipulate as the polynomial solution is simply $z_1 = j$, $z_2 = -j$ and $z_3 = 1$, and

$$f(z) = (z + j)(z - j)(z + 1)$$

or

$$f(z) = 1 + z^2 + z^3 + z^4$$

To achieve $\phi = 90°$ when $\theta = 45°$ it is necessary for d/λ to equal 0.35. Thus a 4-element linear array with uniform current amplitudes and $d = 0.35\lambda$ will achieve the specification.

Secondly, having realized this kind of solution, what we have done is to adjust the beamwidth of a constant-amplitude current linear array so that the first null coincided with $\theta = 45°$. We could have obtained this solution directly from Ex. 5.3 with $\sin \alpha = \lambda/(Nd) = 0.707$. When $N = 4$, $d/\lambda = 0.35$.

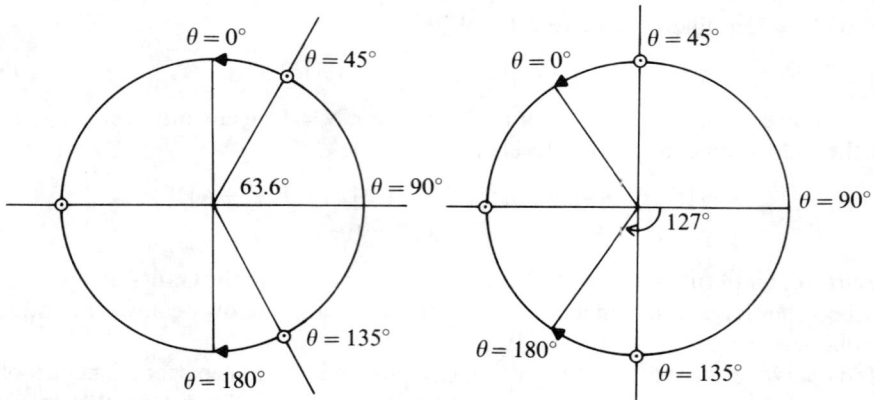

Fig. 5.21 The unit circles for Ex. 5.6. Both solutions will provide the desired result but the second is mathematically simpler to manipulate and easier to control practically.

PROBLEMS

5.1 What are the input impedances Z_1 and Z_2 of a two-element array of halfwave dipoles in broadside configuration if the dipoles are displaced by $d = 0.3\lambda$ and $V_1 = V_2$? If the terminals of the second dipole are shorted so that $V_2 = 0$, what then is Z_1?

5.2 Two halfwave dipoles in broadside array, separated by $d = 0.2\lambda$, are each supplied with 5 W when $V_1 = V_2$. What current is supplied to each dipole? What is the field strength at distance $r = 1$ km in direction $\theta = 45°$ from the array, where angle θ is defined in Fig. 5.7?

5.3 Draw the graphs of $\sin(N\phi/2)$ and $\sin(\phi/2)$ and use these to observe the locations of the maximum and minimum values of the ratio of $\sin(N\phi/2)$ to $\sin(\phi/2)$.

5.4 A broadside array of three halfwave dipoles has displacements $d_{12} = d_{23} = \lambda/2$. Determine the input impedance to each dipole and the maximum gain of the antenna if I_1 and I_3 are in phase and I_2 is in anti-phase. Assume equal amplitude currents.

5.5 Show that a linear array with $Nd/\lambda = 1$ and $\psi = 0$ has only two major lobes and no sidelobes. What is the half-power beamwidth of the major lobe when $N = 4$ and when $N = 8$?

5.6 Design a 7-element broadside array which has one major lobe and two side lobes on each side of the array.

5.7 Draw the endfire radiation diagram for a 10-element linear array in which the visible region extends from $\phi = 0$ to $\phi = -\phi'$, where $-\phi'$ is illustrated in Fig. 5.15.

5.8 Design a Yagi–Uda antenna which has a gain of 11 dB at 50 MHz with respect to a halfwave dipole.

5.9 Design a log-periodic dipole array with a gain of 9 dBi covering the frequency range 140–450 MHz.

5.10 What feed currents are required by each antenna in a 3-element array with $d = \lambda/4$ displacement if a null is to appear in the direction $\theta = 30°$ to the alignment of the array?

5.11 Determine the polar radiation pattern of a 7-element array with a 'triangular' current distribution given via the polynomial $f(x) = 1 + 2z + 3z^2 + 4z^3 + 3z^4 + 2z^5 + z^6$ with a single major lobe on each side of the array.

REFERENCES

5.1 Carter, P. S., 'Circuit relations in radiating systems and applications to antenna problems,' *Proc. IRE*, 20 (June 1932), 1004–1041

5.2 Hansen, W. W., and J. R. Woodyard, 'A new principle in directional antenna design,' *Proc. IRE*, 26 (March 1938), 333–5.

5.3 Yagi, H., 'Beam transmission of ultra-short waves,' *Proc. IRE*, 16 (June 1928), 715–40.

5.4 Isbell, D. E., 'Log-periodic dipole arrays,' *IRE Transactions on Antennas and Propagation*, AP-8 (May 1960), 260–267.

5.5 Schelkunoff, S. A., 'A mathematical theory of linear arrays,' *Bell Syst. Tech. Journ.*, 22 (January 1943), 80–107.

5.6 Stone, J. S., United States Patents No. 1,643,323 and No. 1,715,433.

5.7 Carrel, R. L., 'Analysis and design of the log-periodic dipole antenna,' University of Illinois Antenna Laboratory *Tech. Rept.* 52, Contract AF 33(616) − 6079.

5.8 Weekes, W. L., *Antenna Engineering*. New York: McGraw-Hill Book Company, 1968.

6

Statistical Distributions and Diversity Principles

The study of radiowave propagation frequently makes use of several common types of statistical probability distributions. In certain radio and radar systems which involve scattering processes and electrical noise, the received signal-to-noise ratio can be improved by diversity systems provided the statistics of the propagation parameters and the noise can be determined. It is therefore useful to have at least a brief understanding of the interrelationships between the various distributions which arise in radiowave propagation, without getting too involved in the fundamental theory behind the statistics.

6.1 USEFUL PARAMETERS

A signal (such as $x = E$ mV/m) which is received after radio wave propagation through a varying medium is itself varying, and a plot of the amplitude variation of E or x with respect to time may, for example, look something like that of Fig. 6.1.

To enable such variations to be represented mathematically, the signal amplitude is sampled at numerous frequent intervals of time and the results of the N samples are collated and plotted as a *histogram* similar to that illustrated in Fig. 6.1.

We can divide the horizontal axis of the histogram into a number of *class intervals* separated by *class boundaries* in steps of (say) 10 mV/m from 0 mV/m to 1000 mV/m. This would give exactly 100 class intervals, but it is important that our choice of class interval is appropriate. Had we chosen the ridiculous limits 0–500 mV/m and 500–1000 mV/m, then it would have been impossible to distinguish the true shape of the variation. On the other hand, if we decided to have class intervals of only 1 μV/m steps, then we would need very many millions of sampled measurements to fill the histogram adequately in order to see the shape of the final distribution. Somewhere in between is an optimum state depending on the sample size, and it has been suggested by Sturges (Ref. 6.1) that such an optimum number of class intervals N_c is related to sample size N via

$$N_c = 1 + 3.3 \log_{10} N \tag{6.1}$$

Thus for $N = 100$, 1000 and 10,000, $N_c = 8, 11$ and 15.

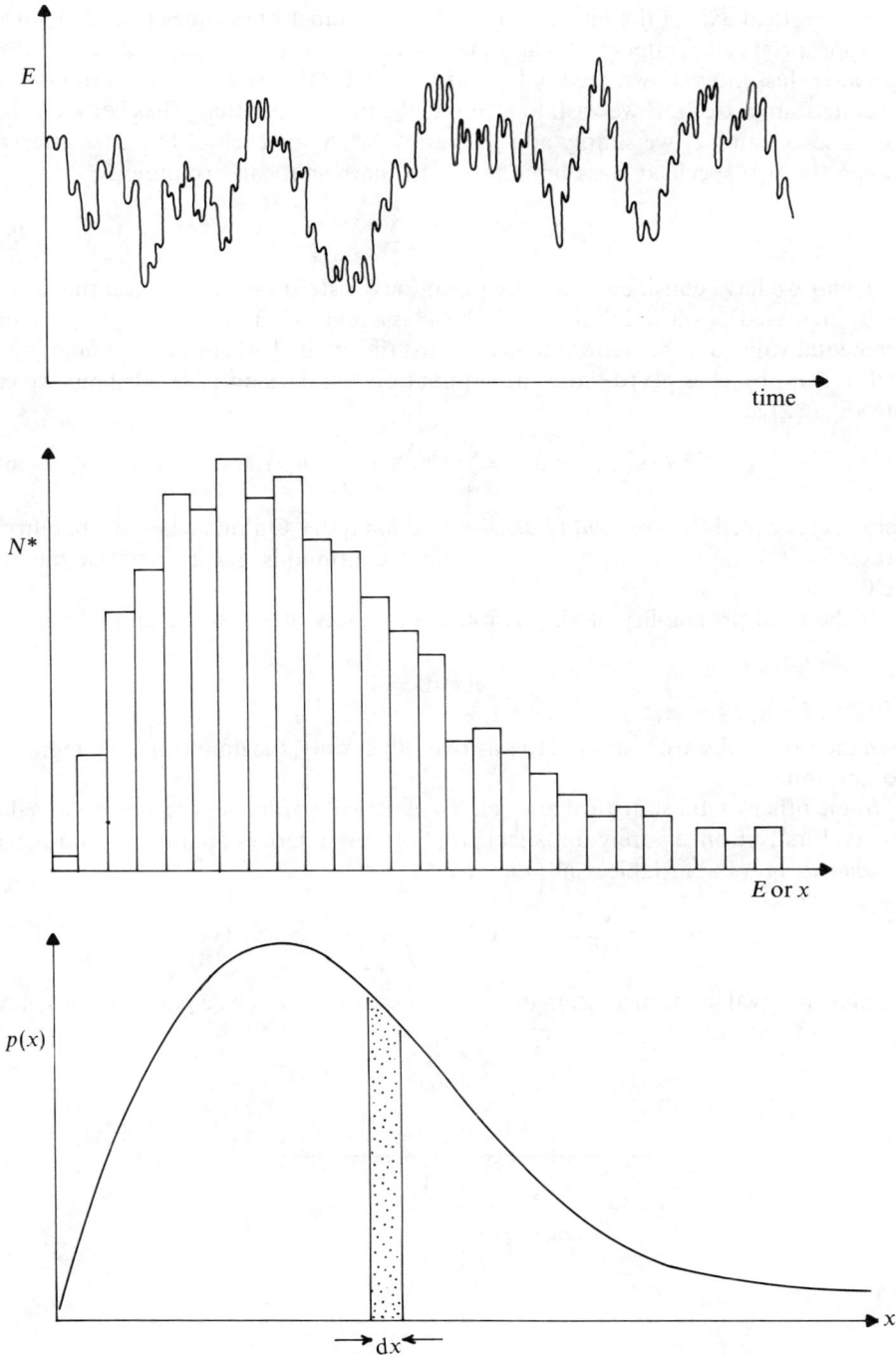

Fig. 6.1 Illustrations of field strength, the histogram of N samples of E, and the equivalent continuous function.

The vertical axis of the histogram is N^*, the number of samples recorded within each appropriate class interval. To find the *probability* $P(x)$ of any given x being within a specified class interval, we need only to divide N^* for the particular class interval by the total sample size N. If we wish to estimate the probability that x lies between class boundaries x_1 and x_2 we simply add the ratios N^*/N for each of the class intervals between the two specified class boundaries. In mathematical terminology,

$$P(x_1 < x < x_2) = \sum_{x_1}^{x_2} N^*/N \tag{6.2}$$

So far we have considered the 'discontinuous' histogram. However, if the sample size is increased toward infinity and the class interval decreased to some small incremental value dx, the pattern tends toward the equivalent continuous function in which N^* approaches $p(x)\,dx$ and the summation becomes an integral. Thus we can write Eq. (6.2) as

$$P(x_1 < x < x_2) = \sum_{x_1}^{x_2} N^*/N = \int_{x_1}^{x_2} p(x)\,dx \tag{6.3}$$

where $p(x)$ is called the *probability density function* (pdf). On this basis, probability is represented by the area under the equivalent continuous graph between the two specified boundaries.

If the total probability of all possible occurrences of x is to be unity, then

$$\int_{-\infty}^{\infty} p(x)\,dx = 1 \tag{6.4}$$

When the values of x are restricted to positive values only, the limits on the integral are zero and infinity.

Some other useful statistical and related electrical parameters can be obtained in this way. Firstly, from a purely statistical point of view, a series of parameters known as the *moments* m_j of a variable x are defined:

$$m_j = \frac{\sum N^* x^j}{N} = \sum x^j \left(\frac{N^*}{N} \right) = \int_{-\infty}^{+\infty} x^j p(x)\,dx$$

Each class interval in the histogram in Fig. 6.1 is characterized by a pair of values of N^*

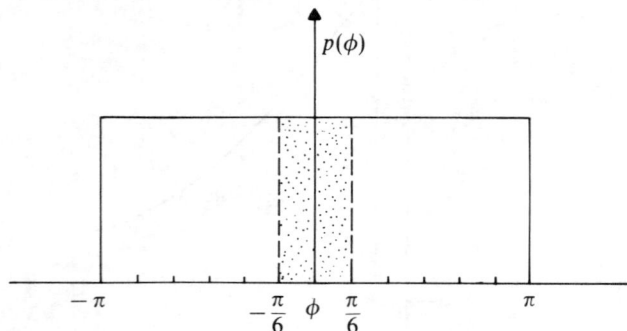

Fig. 6.2 The probability density function and distribution of ϕ for Ex. 6.1.

Example 6.1 Probability calculations

The phases ϕ of signals received via a purely scattering radio wave propagation medium are totally random and there is equal likelihood of ϕ taking any value between $-\pi$ and $+\pi$ radians. Determine $p(\phi)$ and $P(-\pi/6 < \phi < +\pi/6)$.

The distribution is illustrated in Fig. 6.2. If the total probability is to equal unity, we can write

$$\int_{-\pi}^{+\pi} p(\phi)\,\mathrm{d}\phi = 1$$

or

$$p(\phi)[\phi]^{\pi}_{-\pi} = 1$$

$$p(\phi) = \frac{1}{2\pi}$$

Because of the simplicity of the distribution, illustrated in Fig. 6.2, we can deduce that $P(-\pi/6 < x < +\pi/6) = 0.167$. More formally, however,

$$P(-\pi/6 < x < +\pi/6) = \int_{-\pi/6}^{+\pi/6} \frac{1}{2\pi}\,\mathrm{d}\phi = \frac{1}{2\pi}[\phi]^{+\pi/6}_{-\pi/6} = 0.167$$

---

and x, and m_j is therefore the average value of the product of N^* and x^j over the entire histogram.

For radio wave propagation, our interest lies mainly in the first and second moments, m_1 and m_2, given by:

$$m_1 = \frac{\sum N^* x}{N} = \sum x\left(\frac{N^*}{N}\right) = \int_{-\infty}^{+\infty} x p(x)\,\mathrm{d}x = \bar{x} \tag{6.5}$$

and

$$m_2 = \frac{\sum N^* x^2}{N} = \sum x^2\left(\frac{N^*}{N}\right) = \int_{-\infty}^{+\infty} x^2 p(x)\,\mathrm{d}x = \overline{x^2} = r^2 \tag{6.6}$$

In other words, m_1 is the *mean* and m_2 is the *mean square* of variable x, and $r = $ root mean square.

Some variables in radio wave propagation, such as the instantaneous electric field strength at the receiver, have a zero *mean* but a non-zero mean square. Other propagation parameters, such as path loss between transmitter and receiver, have both an average value and a mean square.

A measure of the fluctuation of parameter x can be obtained by noting its variation with respect to $\bar{x}$. Obviously we cannot use:

$$f(x) = \frac{\sum N(x - \bar{x})}{N} = \frac{\sum Nx}{N} - \frac{\sum N\bar{x}}{N} = \bar{x} - \bar{x} = 0$$

and so a better measure is to use:

$$\mathrm{var}(x) = \frac{\sum N(x - \bar{x})^2}{N} = \frac{\sum Nx^2}{N} - 2\bar{x}\frac{\sum N\bar{x}}{N} + \frac{\sum N(\bar{x})^2}{N} \tag{6.7}$$

$$\mathrm{var}(x) = r^2 - (\bar{x})^2 = m_2 - (m_1)^2 = \sigma^2 \qquad (6.8)$$

where σ is known as the *standard deviation* and var (x) as the *variance*.

A parameter which is also of great use in the analysis of statistical distributions is the *median* X_0. This is defined via:

$$\int_{-\infty}^{X_0} p(x)\,\mathrm{d}x = 0.5 \qquad (6.9)$$

It implies that 50% of the sample measurements will have $x < X_0$ and 50% will have $x > X_0$.

6.2 SHAPES OF DISTRIBUTIONS

The several continuous distributions that occur most commonly in radio wave propagation resemble the shapes of the three curves illustrated in Fig. 6.3. When appropriately scaled, single examples of each might be the *exponential*, *Rayleigh* and *normal* distributions.

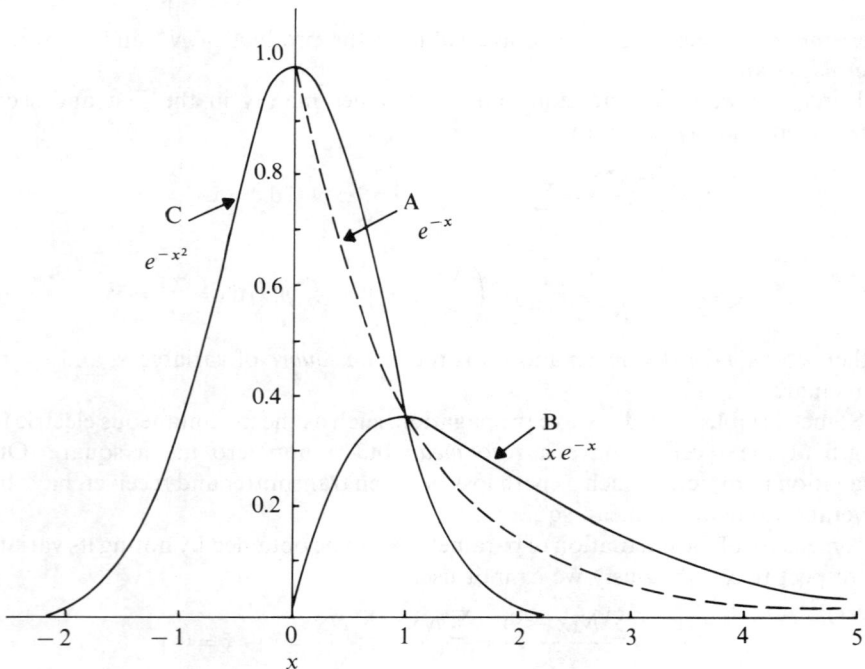

Fig. 6.3 Typical generalized shapes of statistical distributions observed in radio wave propagation.

6.2.1 The Basic Equations

The common feature of these particular curves is that they each involve the negative exponential $\exp(-x)$. For instance, curve A represents

$$p(x) = \exp(-x) \tag{6.10}$$

or some appropriate modification if the scaling is changed, while curve B makes use of

$$p(x) = x \exp(-x) \tag{6.11}$$

so that the increasing effect of the first term predominates at low values of x, when $\exp(-x)$ is approximately unity, and the decreasing effects of the second term predominates at larger values of x. Finally, because of the squaring of the x term, curve C is symmetrical about $x = 0$.

Such simple yet very precise equations, however, give us no freedom to shape and position the curves to fit some observed set of measurements. To obtain such freedom in both x and $p(x)$ directions, one or two *descriptive parameters*, such as n and/or m, may be introduced so that

$$p(x) = x^n \exp(-x^m) \tag{6.12}$$

for example. A whole range of distributions now becomes available for various combinations of n and/or m, and several of these alternatives find use in radio wave propagation.

6.2.2 Normalized Equations

Two further problems need to be cleared up before we use the intuitive expression for $p(x)$ given in Eq. (6.12) in our statistical analyses.

Firstly, the signal being sampled will commonly have a dimension (such as $x = E\,\mathrm{mV/m}$), which is totally inappropriate for the mathematics as there is no meaning to $\exp(-E\,\mathrm{mV/m})$. The solution is to normalize the signal measurement with respect to some known reference which has appropriate dimensions, as we do when we use decibels, for example. We could select something as simple as $E_0 = 1\,\mathrm{mV/m}$, but often the statistics work better with a reference of $b\,\mathrm{mV/m}$, where b may be the mean, median or root mean square of the distribution. Each case has to be tested individually for this purpose. Thus, for this reason,

$$p(x) = \left(\frac{x}{b}\right)^n \exp(-x^m/b^m) = u^n \exp(-u^m) \tag{6.13}$$

where $u = x/b$ is occasionally used as the normalized value of x. In some cases, such as in the analysis of signal-to-noise ratio, the normalization might not seem necessary, but often the statistics are simplified when the SNR is referenced to some statistical parameter, such as the median.

Secondly, the total 'probability' integral of the pdf expression may not equal unity.

This can be corrected by including a coefficient A such that

$$\int_0^\infty p(x)\,dx = \int_0^\infty A^*(x/b)^n \exp(-x^m/b^m)\,dx$$

or

$$\int_0^\infty p(x)\,dx = \int_0^\infty Ax^n \exp(-x^m/b^m)\,dx \qquad (6.14)$$

If $A = A^*/b^n$. Note, however, that if we use the normalized parameter $u = x/b$, we must replace dx by $b\,du$. This gives

$$\int_0^\infty p(u)\,du = \int_0^\infty A^*u^n \exp(-u^m)b\,du = \int_0^\infty A'u^n \exp(-u^m)\,du \qquad (6.15)$$

with $A' = bA^*$. The generalized probability density function which is given by any of the expressions in Eqs. (6.14) and (6.15) is now in a form suitable for use in statistics and is applicable to those distributions in which $x > 0$. It can be adapted for $-\infty < x < +\infty$, but is not required in that form for our purposes.

6.3 GAMMA FUNCTIONS

To help us with our studies of statistical distributions, we can take advantage of the very simple integral known to mathematicians as the *gamma function*. This is defined as

$$\Gamma(k) = \int_0^\infty z^{k-1} \exp(-z)\,dz \qquad (6.16)$$

and values of $\Gamma(k)$ have been determined and tabulated, just like logarithmic, trigonometrical and exponential tables. The implication which we can draw from this definition is that

$$\int_0^\infty \frac{z^{k-1} \exp(-z)\,dz}{\Gamma(k)} = 1 \qquad (6.17)$$

and this simple integral relationship is sufficient for us to *avoid* doing any integration in

Table 6.1   Some useful gamma functions

k	$\Gamma(k)$	k	$\Gamma(k)$	k	$\Gamma(k)$
0.25	3.6256	1.25	0.9064	2.50	1.3293
0.50	1.7725	1.50	0.8862	3.00	2.0000
0.75	1.2254	1.75	0.9191	3.50	3.3234
1.00	1.0000	2.00	1.0000	4.00	6.0000

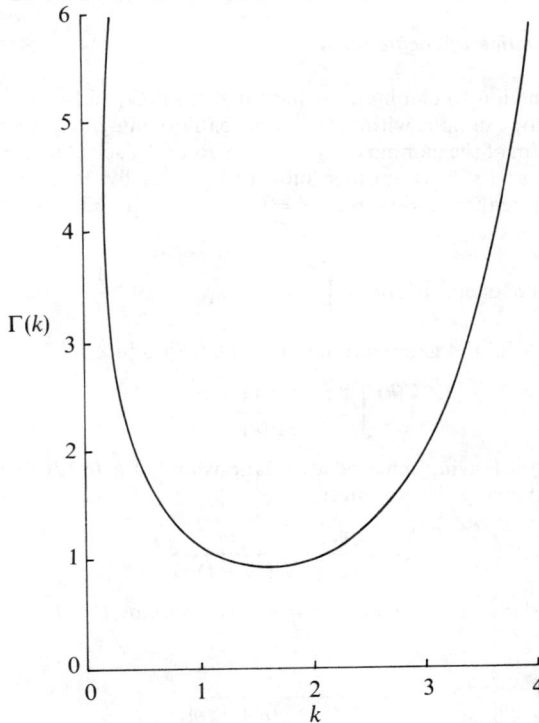

Fig. 6.4 The gamma function for $k < 4$.

our analyses. For this purpose, some useful values of $\Gamma(k)$ are given in Table 6.1 and are illustrated in Fig. 6.4. Note that $\Gamma(k) = (k-1)!$ for integer $k \geqslant 1$.

6.4 SPECIFIC TYPES OF DISTRIBUTIONS

The generalized probability density function derived in Eqs. (6.14) and (6.15) can be made specific by giving suitable values to the two descriptive parameters n and m. Their values must be such that the auxiliary parameter $k = (n+1)/m$ is greater than 0 in order to obtain a realistic $\Gamma(k)$, while $m = 0$ is also an unsuitable choice for obvious reasons. Some of the simpler combinations of n and m are listed in Table 6.2 together with the names that most commonly describe the distributions.

6.4.1 Exponential Distribution

Curve A in Fig. 6.3 resembles the shape of the exponential distribution which is defined from our generalized pdf with $n = 0$ and $m = 1$. With these values of descriptive

Example 6.2 Elimination of coefficient A

Use the gamma function to eliminate coefficient A from Eq. (6.14).

We can start, for example, with the total probability integral given in Eq. (6.15) and 'work' it into the form of the gamma integral of Eq. (6.17) by selecting $z = u^m$. Noting that $u = z^{1/m}$ we obtain $u^n = z^{n/m}$ as another substitution. Finally, to change du into some function of dz, we note that $dz/du = mu^{m-1} = mz^{(m-1)/m}$. With all these substitutions into Eq. (6.15) we observe that

$$\int_0^\infty A'u^n \exp(-u^m)\,du = \int_0^\infty A'z^{n/m}\exp(-z)\frac{1}{mz^{(m-1)/m}}\,dz = 1$$

We must now 'work in' the gamma function, $\Gamma(k)$, to produce

$$\frac{A'\Gamma(k)}{m}\int_0^\infty \frac{z^{k-1}\exp(-z)}{\Gamma(k)}\,dz = 1$$

If we use $k = (n+1)/m$. Having achieved the relationship of Eq. (6.17) we can eliminate the integral to obtain the very simple solution

$$A' = \frac{m}{\Gamma(k)} = \frac{m}{\Gamma\{(n+1)/m\}} \tag{6.18}$$

Using the relationships given in Eq. (6.14) we can obtain $A^* = bA'$ and $A = A^*/b^n = A'/b^{n+1}$. Hence

$$A = \frac{m}{b^{n+1}\Gamma\{(n+1)/m\}} \tag{6.19}$$

Table 6.2   Components of generalized distribution

Type	Distribution name	Conditions
$m = 1$	gamma	$n = n$
$m = 1$	Erlang	$n + 1 = $ PI*
$m = 1$	chi-squared	$b = 2$, $2(n+1) = $ PI*
$m = 2$	Nakagami-m	$n = n$
$m = m$	Stacey	$n = n$
$m = m$	Weibull	$n = m - 1$
$n = 0$	exponential	$m = 1$
$n = 0$	one-sided normal	$m = 2$
$n = 0$	generalized exponential	$m = m$
$n = 1$	Rayleigh	$m = 2$
$n = 1$	generalized Rayleigh	$m = m$

PI* means positive integer

parameters, Eq. (6.15) reduces to

$$p(u) = A' \exp(-u) = \exp(-u) \tag{6.20}$$

as Eq. (6.18) gives $A' = 1$ (Fig. 6.5).

For some particular applications in radio wave propagation it will be useful to let variable y represent the instantaneous *power* level of a signal so that $u = y/b$ and $du = dy/b$ in such a case *if* the observed variation in instantaneous power follows an

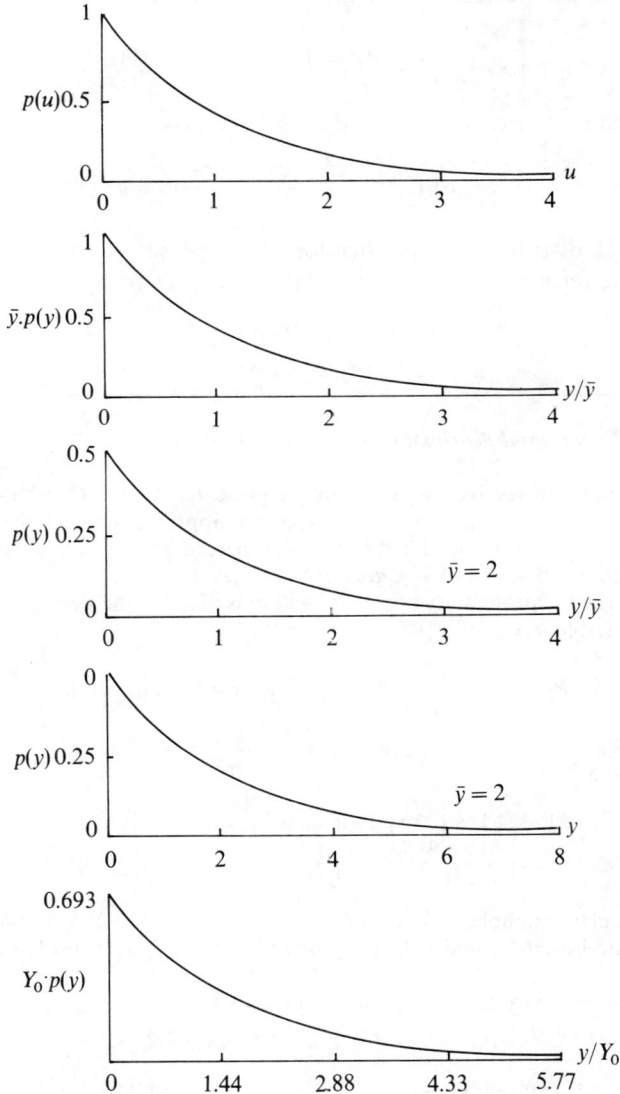

Fig. 6.5 Various ways of scaling the exponential distribution.

exponential distribution. Using

$$\int_0^\infty p(u)\,du = \int_0^\infty \exp(-u)\,du = \int_0^\infty \frac{1}{b}\exp\left(-\frac{y}{b}\right)dy = \int_0^\infty p(y)\,dy = 1$$

we can obtain

$$p(y) = (1/b)\exp(-y/b) \tag{6.21}$$

This distribution has a mean value $\bar{y}$ given via Eq. (6.5) as

$$\bar{y} = \int_0^\infty yp(y)\,dy = \int_0^\infty y\frac{1}{b}\exp(-y/b)\,dy$$

As before, we can replace $z = y/b$ and $dy = b\,dz$ to give

$$\bar{y} = b\Gamma(2)\int_0^\infty \frac{z^{2-1}\exp(-z)}{\Gamma(2)}\,dz = b \tag{6.22}$$

The exponential distribution can therefore be represented by a very simple pdf expression if the reference b is the mean value of the distribution:

$$p(y) = (1/\bar{y})\exp(-y/\bar{y}) \tag{6.23}$$

Example 6.3 *Exponential distribution with known mean value*

Numerous samples of received signal power, represented by $y = P_r$, in a certain radio communication link indicate that the variation approximates to the exponential distribution with a mean value of $2\,\mu\text{W}$. For what percentages of time would you expect the signal to be less than $0.1\,\mu\text{W}$ or greater than $6\,\mu\text{W}$?

We can write the probability expression in terms of either the signal variable y or the normalized variable $u = y/\bar{y}$:

$$P(y < Y) = \int_0^Y (1/\bar{y})\exp(-y/\bar{y})\,dy = [-\exp(y/\bar{y})]_0^Y$$

$$= 1 - \exp(-Y/\bar{y}) \tag{6.24}$$

or

$$P(u < U) = \int_0^U \exp(-u)\,du = [-\exp(-u)]_0^U$$

$$= 1 - \exp(-U) \tag{6.25}$$

In this particular example, $Y = 0.1\,\mu\text{W}$, $\bar{y} = 2\,\mu\text{W}$ and $U = Y/\bar{y} = 0.05$. Hence, by substitution into Eqs. (6.24) and (6.25) the probability of the signal being less that $0.1\,\mu\text{W}$ is 4.88%.

For the second part of the example we note that

$$P(u > U) = 1 - P(u < U) = \exp(-U) \tag{6.26}$$

If $U = 3$ when $Y = 6\,\mu\text{W}$ and $\bar{y} = 2\,\mu\text{W}$, then the probability of the signal exceeding $6\,\mu\text{W}$ is 4.98%.

This is an important distribution in the reception of signals which have passed through a purely scattering radio wave propagation medium.

It turns out that $b = \bar{y}$ is a useful and simple way of normalizing y in the exponential distribution when y represents the instantaneous power level of the received signal. However, if the sampling provides a *median* value Y_0 instead of the *mean* $\bar{y}$, some modification is necessary if Y_0 is to be used as a normalizing reference. From

$$P(u > U_0) = \exp(-U_0) = 0.5$$
$$U_0 = -\log_e 0.5 = 0.693\,147 = Y_0/\bar{y}$$

and

$$\bar{y} = Y_0/0.693 \tag{6.27}$$

This can be used in Eq. (6.23) to produce the alternative form of the probability density function (Fig. 6.5)

$$p(y) = (1/\bar{y})\exp(-y/\bar{y}) = \frac{0.693}{Y_0}\exp\left(\frac{-0.693\,y}{Y_0}\right) \tag{6.28}$$

6.4.2 Rayleigh Distribution

Curve B in Fig. 6.3 resembles the shape of the Rayleigh distribution which is defined via the generalized pdf with $n = 1$ and $m = 2$, or with auxiliary parameter $k = (n + 1)/m = 1$. With these particular values of descriptive parameters we can use Eq. (6.15) to obtain

$$p(u) = 2u\exp(-u^2) \tag{6.29}$$

as Eq. (6.18) gives $A' = m/\Gamma(k) = 2$ (Fig. 6.6).

In some particular example of this type of statistical distribution in radio wave propagation we might find it useful to let variable x represent the received signal in terms of its electric field strength so that $u = x/b$ and $du = dx/b$ in this case. From

$$\int_0^\infty p(u)\,du = \int_0^\infty 2u\exp(-u^2)\,du = \int_0^\infty 2\frac{x}{b^2}\exp(-x^2/b^2)\,dx$$

we can determine that

$$p(x) = 2\frac{x}{b^2}\exp(-x^2/b^2) \tag{6.30}$$

To determine an appropriate value for the normalizing factor b we can derive expressions for the mean, median and mean square values of the distribution and see which will give the simpler or more convenient expression. However, in order to interrelate field-strength variations with the corresponding power variations, the link might appear to be via the mean square value of the field strength. Hence from Eq. (6.6)

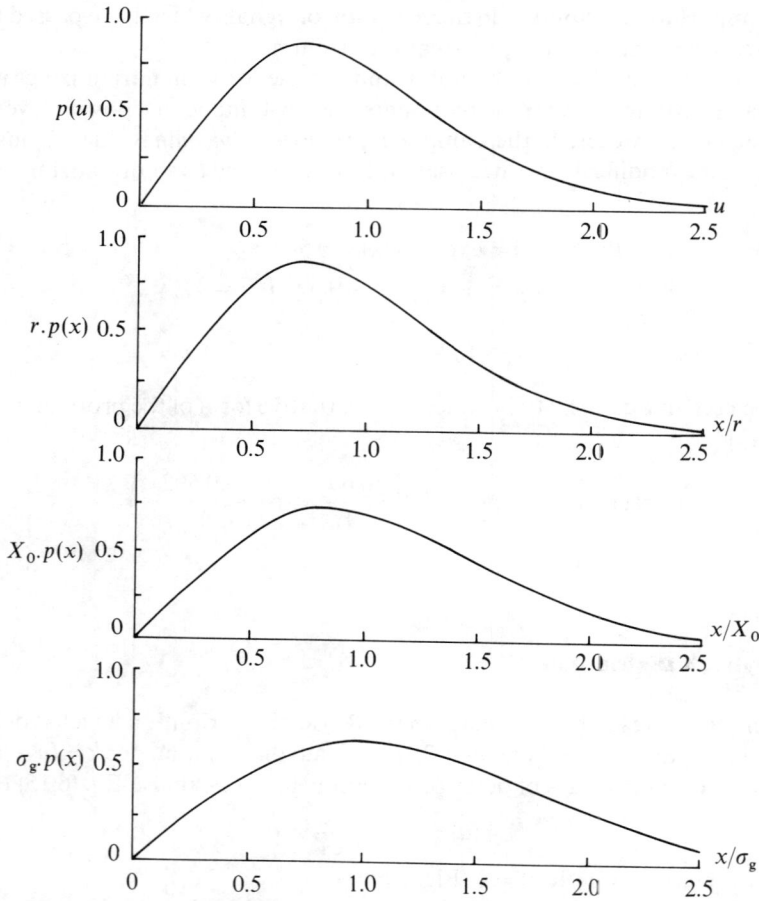

Fig. 6.6 Various ways of scaling the Rayleigh distribution. The parameter σ_g is the standard deviation of the quadrature normal distributions used in the original derivation.

we can write

$$m_2 = r^2 = \int_0^\infty x^2 p(x)\,dx = \int_0^\infty x^2 (2x/b^2) \exp(-x^2/b^2)\,dx$$

To produce the gamma integral of Eq. (6.17) we again use the change of variable technique with $z = x^2/b^2$ and $dx = (b^2/2x)\,dz$. Substitution into the previous expression results in

$$r^2 = b^2 \Gamma(2) \int_0^\infty \frac{z^{2-1} \exp(-z)}{\Gamma(2)}\,dz = b^2 \tag{6.31}$$

Thus the use of the root mean square of the field strength as the normalizing factor produces one version of the pdf of the Rayleigh distribution, namely

$$p(x) = (2x/r^2)\exp(-x^2/r^2) \tag{6.32}$$

If the electric field strength (x) representation of a received signal is observed to follow the Rayleigh distribution, the corresponding power (y) representation of the received signal can be obtained by the substitutions $y = K'x^2$ and $\bar{y} = K'r^2$, where $dy = 2K'dx$ and K' is some scaling factor relating signal power to signal field strength. Hence from

$$\int_0^\infty (2x/r^2)\exp(-x^2/r^2)\,dx = \int_0^\infty (1/\bar{y})\exp(-y/\bar{y})\,dy \tag{6.33}$$

it is seen that the corresponding signal power level y follows the exponential distribution.

In the analysis of scattering mechanisms in radio wave propagation, the description that the signal is Rayleigh distributed implies that the electric field strength (in units such as V/m, and not dBμ) is observed to follow such a distribution whereas the corresponding power levels of the received signal (in units such as W, and not dBW) follow the exponential distribution. Some confusion may arise because the measurements are commonly made or quoted in dBμ and dBW.

6.4.3 Rayleigh Probability Paper

With the aid of calculators or computers, numerical analysis of data is achieved quite easily. However, a graphical method of checking whether an observed set of measurements 'fits' the Rayleigh distribution is still very useful as it gives some visual indication of the pattern of variation.

Such a technique makes use of *Rayleigh probability paper*, one of a series of several different types of probability paper. This is a plot of a certain function of the probability that signal x exceeds some specified value X, against X, in some normalized form (commonly X/X_0 or its decibel equivalent).

To obtain such a plot we must first note that for a variable x with a Rayleigh distribution, the probability P of it exceeding some value X is given by

$$P = P(x > X) = \int_X^\infty (2x/r^2)\exp(-x^2/r^2)\,dx = \exp(-X^2/r^2) \tag{6.34}$$

and hence the median value X_0 can be derived in terms of the rms value via

$$P(x > X_0) = 0.5 = \exp(-X_0^2/r^2)$$

or

$$r^2 = X_0^2/\log_e 2 \tag{6.35}$$

In logarithmic form, to remove the exponential function, we can replace Eq. (6.34) by either

$$\log_{10} P = 0.4343(-X^2/r^2)$$

or

$$\log_{10}(1/P) = 0.4343 \log_e(2)(X^2/X_0^2)$$

i.e.

$$\log_{10}(1/P) = \log_{10}(2)(X/X_0)^2 \tag{6.36}$$

A plot of this relationship between probability P and X/X_0 should therefore produce a straight line with slope 0.3010. Alternatively, if the range of measurements has been made on a decibel scale, Eq. (6.36) may be modified further to

$$10 \log_{10} \log_{10}(1/P) = 10 \log_{10} \log_{10}(2) + 20 \log_{10}(X/X_0)$$

or

$$10 \log_{10} \log_{10}(1/P) = -5.2139 + X(\text{dB}) - X_0(\text{dB}) \tag{6.37}$$

This, too, is a straight-line relationship between a particular function of probability P and the specified field strength measured in decibels with respect to the median value X_0. A plot of the actual measurements may easily be compared with the theoretical line and for this purpose pads of 'Rayleigh paper' are available commercially. The vertical axis is simply the log-log scaling of $1/P$ and the horizontal axis is the linear relationship between X and X_0 in decibels.

Example 6.4 Rayleigh probability paper

One thousand samples of field strength of a radio signal are summarized below, the data being recorded in nine class intervals with the class boundaries at 5 dB intervals. 0 dB refers to the overall median value. Check whether the measurements appear to fit a Rayleigh distribution.

15	10	5	0	−5	−10	−15	−20	−25	−30 dB
1	88	411	303	130	45	15	5	2	N^*

In order that we can plot a graph on the Rayleigh probability paper we must first determine the cumulative frequency CF and percentage cumulative frequency CF% at each of the class boundaries CB:

CB	15	10	5	0	−5	−10	−15	−20	−25	−30
CF	0	1	89	500	803	933	978	993	998	1000
CF%	0%	0.1%	8.9%	50.0%	80.3%	93.3%	97.8%	99.3%	99.8%	100%

The percentage cumulative frequencies must be plotted against class boundary on Rayleigh probability paper to confirm visually the general closeness of the fit. This is shown in Fig. 6.7.

This technique gives only some indication of the possibility that a distribution might conform to the Rayleigh equations. The most sensitive indicator, near the tail of the distribution, is also the most likely to produce an error because the random occurrences and omissions can distort the curve substantially in this region.

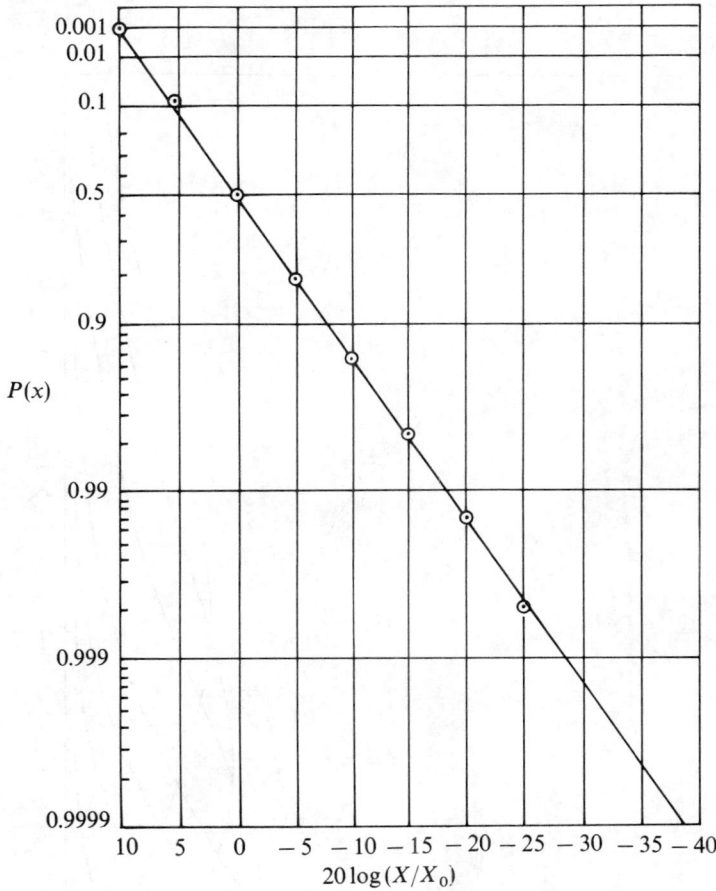

Fig. 6.7 Rayleigh probability paper. This gives the probability that x exceeds some value X (measured with respect to the median). The points refer to Ex. 6.4.

6.4.4 Gamma Distributions

Another family of statistical distributions which occur commonly in radio wave propagation are called *gamma distributions*. These are defined with descriptive parameter $m = 1$ and with auxiliary parameter $k = (n + 1)/m = n + 1$ taking any suitable value for which $\Gamma(k)$ is realistic. Using Eqs. (6.14) and (6.15) with $m = 1$ we obtain either

$$p(y) = \frac{y^{k-1} \exp(-y/b)}{b^k \Gamma(k)} \tag{6.38}$$

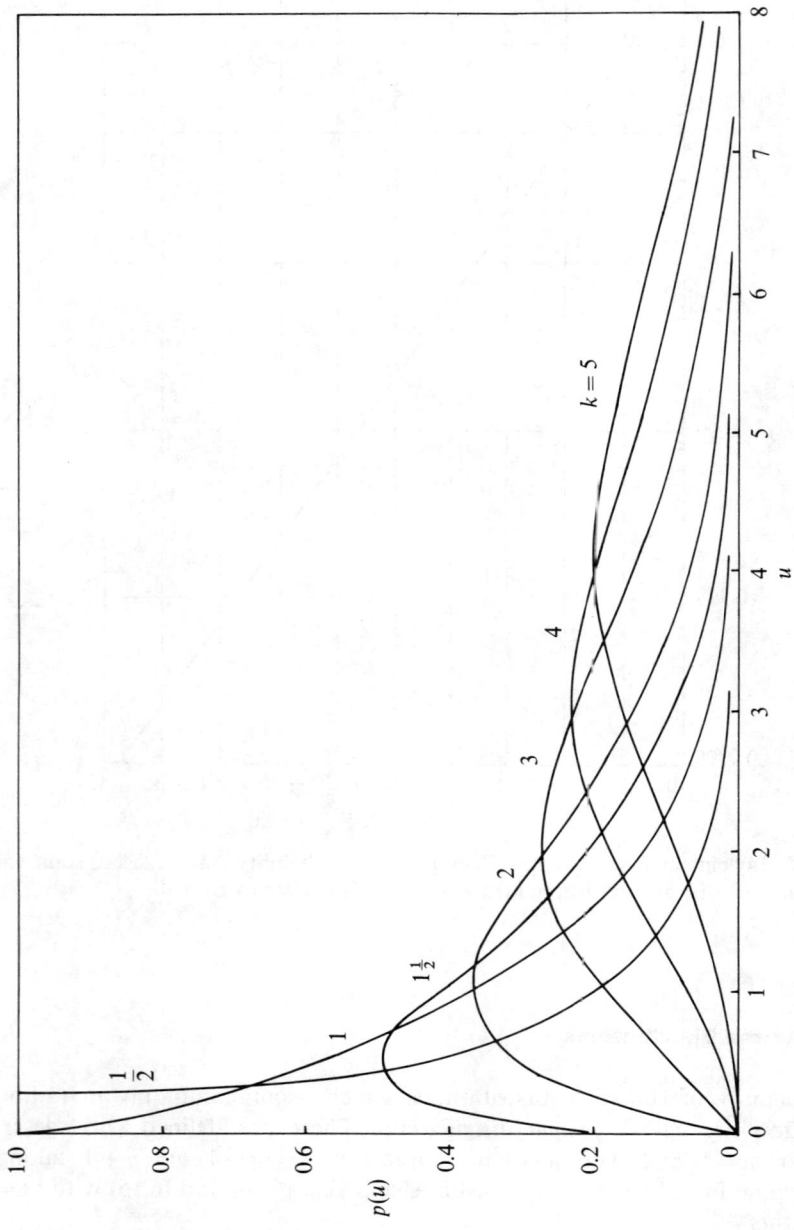

Fig. 6.8 The normalized gamma distributions for various values of *k*.

or

$$p(u) = \frac{u^{k-1}\exp(-u)}{\Gamma(k)} \qquad (6.39)$$

and these are the pdf's which represent the family of gamma distributions. The latter expression, compared with Eq. (6.17), indicates quite clearly why it is called the gamma distribution.

It is a simple matter with an electronic calculator to plot graphs of Eq. (6.39) with $u = 0$ to 8 and $k = 1$ to 4 as

$$p(u) = \exp(-u) \qquad \text{when } k = 1$$
$$p(u) = u\exp(-u) \qquad \text{when } k = 2$$
$$p(u) = \frac{u^2}{2}\exp(-u) \qquad \text{when } k = 3$$

and

$$p(u) = \frac{u^3}{6}\exp(-u) \qquad \text{when } k = 4$$

These are illustrated, with other values of k, in Fig. 6.8. The expressions can be converted into terms of y/Y_{01} with the aid of Eqs. (6.22) and (6.27), which indicate that:

$$u = 0.693(y/Y_{01})$$

where Y_{01} is the median of the exponential (or gamma, $k = 1$) distribution.

The principal application of the various gamma distributions in radio wave propagation lies in the field of diversity, described in more detail later in this chapter. If the radio wave propagation is via a perfect scattering medium, the received signal has an exponential or gamma $(k = 1)$ distribution provided the received signal is measured in terms of its power level or signal-to-noise ratio. If two, three or four independent sets of the signal power are received by some means, such as via separate antennas suitably displaced, and the sets are combined into one single resultant summation, the distribution of this single resultant is the gamma $(k = N)$ distribution, where N is the number of independent and uncorrelated sets of received signals, provided that $b = \bar{y}$ is the same for each set of signals. The proof may be found in mathematical texts on gamma functions. As a result it can be shown that there is a considerable improvement in the overall signal reliability and graphs showing the magnitude of such improvements are given later in the chapter.

However, the improvement can be seen visually in Fig. 6.8. Suppose that we consider that the signal must exceed $u = 1$ to be satisfactory. The probability that it is less than 1 when $k = 1$ is given by the area under the $k = 1$ curve between 0 and 1. When there are two signals to sum, the probability of the resultant being less than $u = 1$ is the area under the $k = 2$ curve between 0 and 1. This is clearly much less than the previous case. By the time $k = 4$ the area under the $k = 4$ curve between 0 and 1 is almost negligible. Hence in the latter case $u > 1$ for almost 100% of time. This is the principle of diversity. As k becomes larger, the area below the curve between 0 and 1 becomes smaller, but eventually the gain no longer warrants the problems and expense of increasing k.

Example 6.5 Cumulative probability curves for gamma distribution

Derive the cumulative probability curves of the gamma distributions in which $k = 1$ and $k = 4$.

The probability density function of the gamma distribution is given by Eq. (6.38) as

$$p(y) = \frac{y^{k-1} \exp(-y/b)}{b^k \Gamma(k)}$$

When $k = 1$ we have shown via Eq. (6.22) that $b = \bar{y}$ of the exponential (or gamma, $k = 1$) distribution, and then in Eq. (6.27) we have further shown that $\bar{y} = Y_0/0.693$, where $\bar{y}$ is the mean and Y_0 is the median of the gamma ($k = 1$) distribution. Thus $b = Y_0/0.693$ can be used as a reference level for the various gamma distributions.

The probability that y *exceeds* some specified value Y is given by

$$P(y > Y) = \int_Y^\infty \frac{y^{k-1} \exp(-y/b)}{b^k \Gamma(k)} \, dy$$

Taking the long approach, we can use initially the transformation $z = y/b$ so that $Z = Y/b$ and $dz = dy/b$. Then the equation becomes

$$P(y > Y) = \int_Z^\infty \frac{(bz)^{k-1} \exp(-z)}{b^k \Gamma(k)} b \, dz = 2^k \int_Z^\infty \frac{z^{k-1} \exp(-z)}{2^k \Gamma(k)} \, dz$$

Now we can replace z by $w/2$ so that $W = 2Z$ and $dz = dw/2$. The equation becomes

$$P(y > Y) = 2^k \int_W^\infty \frac{(w/2)^{k-1} \exp(-w/2)}{2^k \Gamma(k)} (dw/2) = \int_W^\infty \frac{w^{k-1} \exp(-w/2)}{2^k \Gamma(k)} \, dw$$

This integral will be recognized as the commonly tabulated chi-square variate if $W = \chi^2$, $k = N/2$ and $W = 2Z = 2Y/b = 1.386\,29\, Y/Y_0$. Thus the probability that y exceeds Y can be obtained from chi-squared tables with $N = 2k$ degrees of freedom.

For our particular example, with $k = 1$ and $k = 4$, we must use the tabulated values of $N = 2$ and $N = 8$. For the given values of probability the corresponding values of $W = \chi^2$ are available:

$P =$	0.995	0.990	0.975	0.950	0.900	0.750	0.500	0.100	0.010	0.001
$N = 2$	0.0100	0.0201	0.0506	0.1026	0.2107	0.5754	1.3863	4.6052	9.2103	13.816
$N = 8$	1.3444	1.6465	2.1797	2.7326	3.4895	5.0706	7.3441	13.361	20.090	26.125

From this we can calculate for each value $10\log_{10}(Y/Y_0)$ where, as we have just shown, $Y/Y_0 = W^2/1.386\,29 = \chi^2/1.386\,29$.

$P =$	0.995	0.990	0.975	0.950	0.900	0.750	0.500	0.100	0.010	0.001
$k = 1$	-21.4	-18.4	-14.4	-11.3	-8.18	-3.82	0.000	$+5.21$	$+8.22$	$+9.99$
$k = 4$	-0.13	$+0.75$	$+1.97$	$+2.95$	$+4.01$	$+5.63$	$+7.24$	$+9.84$	$+11.6$	$+12.8$

The two graphs for $k = 1$ and $k = 4$ are reproduced in Fig. 6.9, together with those for $k = 2$ and $k = 3$ which have been obtained in the same way. These will prove useful in one of the diversity systems described later in the chapter.

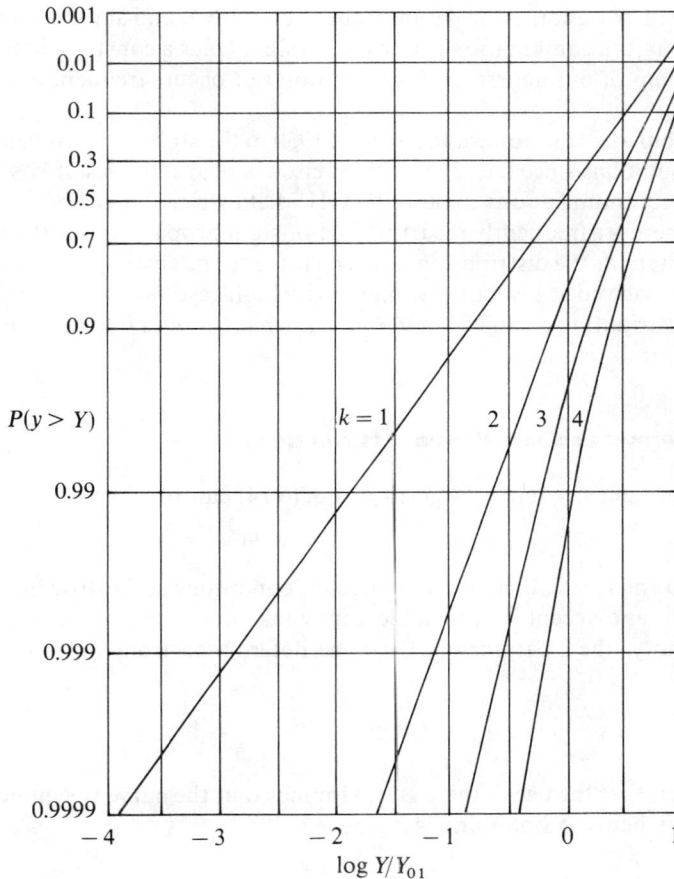

Fig. 6.9 The cumulative probability curves for four of the gamma distributions.

6.4.5 Other Related Distributions

So far we have discussed the exponential, gamma and Rayleigh distributions which are applicable to radio wave propagation via a perfect scattering medium. Several other types of statistical distributions are of interest in radio systems but are outside the scope of this text.

The Nakagami-*m* distribution (where his *m* corresponds to our *k*) was derived in 1943 from large-scale experiments in rapid fading in high-frequency long-distance propagation, but later found to apply equally well at frequencies in the VHF and UHF ranges. It is useful where there is a combination of a scattering mechanism and some additional direct but not constant propagation. In our generalized distribution it is defined as a family of distributions with our $m = 2$ and our $k = n + 1 = 1, 2, 3$, etc. It obviously includes the Rayleigh distribution when $m = 2$ and $n = 1$.

A similar distribution, named jointly after both Nakagami and Rice who produced it independently, is a combination of a scattered signal plus a constant direct signal. It is *not* a component of our generalized distribution but occurs frequently in radio wave propagation.

The Weibull distribution was derived in 1939 in the study of the breaking strength of materials but it has since had applications elsewhere in statistics. It has been used in mobile radio communications to describe UHF field strength distributions over open areas where there is a frequently recurring line-of-sight propagation path in a scattering mechanism. In radar the distribution has been fitted to the envelope of the return signal when the sea is viewed at low grazing angles with high-resolution radar. It is one of our generalized distributions defined via $n = m - 1$, which implies $k = 1$ for each appropriate combination of n and m.

6.4.6 The Normal and Log-Normal Distributions

The probability density function given in Eq. (6.14) can be written as

$$p(x) = A \exp(-x^2/b^2)$$

if we put $n = 0$ and $m = 2$. If, now, we change the conditions of the distributions given in earlier sections and accept that variable x may take any value between minus infinity and plus infinity, then parameter A can be determined from the total probability integral

$$\int_{-\infty}^{+\infty} A \exp(-x^2/b^2)\,dx = 1$$

Note, however, that the use of the x^2 term implies that the curve is symmetrical about $x = 0$, and this helps in obtaining A because

$$\int_{-\infty}^{+\infty} A \exp(-x^2/b^2)\,dx = 2 \int_{0}^{\infty} A \exp(-x^2/b^2)\,dx = 1$$

When compared with Eq. (6.19) it is seen that the magnitude of A obtained there must be halved to account for the factor 2 above, and hence we can write from Eq. (6.19) that

$$A = \frac{1}{2} \frac{m}{b^{n+1}\Gamma\{(n+1)/m\}} = \frac{1}{b\Gamma(\frac{1}{2})} = \frac{1}{\sqrt{b^2\pi}}$$

where $\Gamma(\frac{1}{2}) = \sqrt{\pi}$ from Table 6.1.

The symmetrical nature of the curve means that $m_1 = 0$, but the second moment m_2 may be determined in the usual manner:

$$m_2 = \int_{-\infty}^{+\infty} x^2 \frac{1}{\sqrt{b^2\pi}} \exp(-x^2/b^2)\,dx$$

If we put $z = x^2/b^2$ to force the integral into the form of the gamma function, we obtain

$$m_2 = \frac{b^2}{\sqrt{\pi}}\Gamma(k) \int_{0}^{\infty} \frac{z^{k-1}\exp(-z)}{\Gamma(k)}\,dz$$

with $k = 1\frac{1}{2}$. Removing the gamma function integral, we finish with

$$m_2 = \frac{b^2}{\sqrt{\pi}} \frac{\sqrt{\pi}}{2} = \frac{b^2}{2} = r^2$$

The standard deviation σ can be obtained via the relationship given in Eq. (6.8):

$$\sigma^2 = m_2 - (m_1)^2 = \frac{b^2}{2} = r^2$$

We can therefore replace b^2 by $2\sigma^2$ and A by $1/\sqrt{2\pi\sigma^2}$ in the probability density function to obtain

$$p(x) = \frac{1}{\sqrt{2\pi}\,\sigma} \exp(-x^2/2\sigma^2)$$

This is the probability density function of the *normal* or *Gaussian distribution* with zero mean.

Statisticians sometimes normalize the distribution to reduce the number of variables by putting $t = (x - \bar{x})/\sigma = x/\sigma$ in this case. Then

$$p(t) = \frac{1}{\sqrt{2\pi}} \exp(-t^2/2)$$

and the probability that t is less than some value T is given by

$$P(t < T) = \frac{1}{\sqrt{2\pi}} \int_{-\infty}^{T} \exp(-t^2/2)\,\mathrm{d}t$$

This integral requires the use of tabulated *error functions*, erf. As a result it, too, appears in tabulated form in textbooks of statistics. For $T = 1, 2$ and 3, the probabilities are 84.13%, 97.72% and 99.865% respectively.

The normal distribution occurs in radio wave propagation, but a more frequent version is the *log-normal distribution*. In this form it is the variable $\log x$ (or its decibel equivalent) which varies in accordance with the normal distribution, and the standard deviation is then also in decibels.

Example 6.6 Normal distribution

Some measurements made during the course of a year record the 8760 hourly median values of the path loss over a certain tropospheric scatter radio link. These indicate that the mean path loss was 160 dB and that the standard deviation was 8 dB. For what proportion of a year are the hourly median path losses within one, two and three standard deviations of the annual mean? Assume a normal distribution in decibels.

The data quoted in the preceding text indicates that 15.87%, 2.28% and 0.135% of a normal population lie outside one, two and three standard deviations from the mean *in each tail* of the distribution. The figures for both tails are double these values, namely 31.74%, 4.56% and 0.27%. What remains is within these limits: 68.26% within one standard deviation of the mean, 95.44% within two standard deviations and 99.73% within three standard deviations.

6.5   DIVERSITY

When a signal is received after passing through certain types of radio wave propagation media, the envelope of the signal is observed to vary with time, sometimes only slightly and slowly, sometimes rapidly and deeply. The phenomenon is known as *fading*. If the propagation parameters associated with a given link are sufficiently different, such as by using two or more independent paths at the same frequency or by using two or more different frequencies or polarizations over the same path, then it is observed that the randomness of the variations is such that the probability of simultaneous deep fading on all paths is much less than for a single path. This reduction in the effect of fading makes use of *diversity combining techniques* which result in an improvement of the signal-to-noise ratio at the receiver.

6.5.1   Basic Concepts and Definitions

The mathematical analysis of a diversity system is based on a reference signal $m = m(t)$ which is defined as a continuous tone, at the appropriate frequency, of such amplitude that its mean square value over any period of time $T \gg 1/f$ (i.e. over many cycles) is

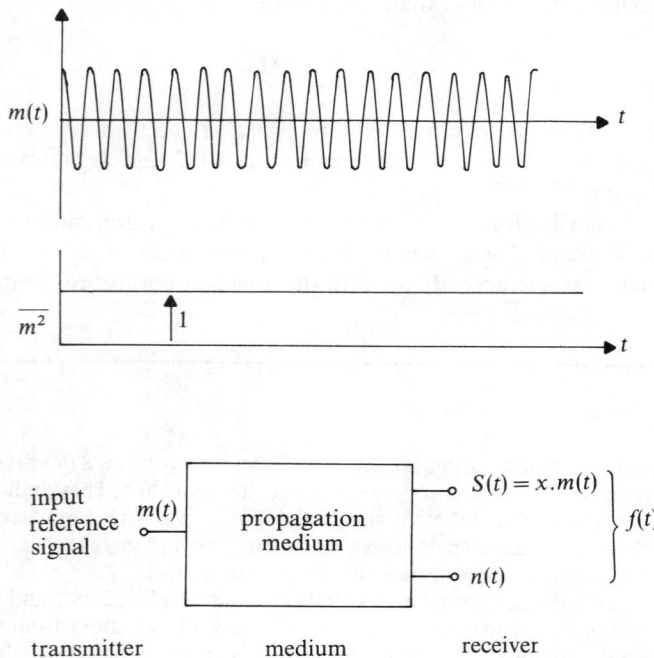

Fig. 6.10 The reference input is a tone with mean square value equal to unity. The output is the combination of attenuated input plus noise.

unity. We can write this condition as

$$\overline{m^2} = 1 \tag{6.40}$$

and it is illustrated in Fig. 6.10. This reference signal travels through the propagation medium between transmitter and receiver and, in doing so, will be modified by the medium: the signal amplitude will be attenuated, noise will be added, and there is a propagation delay. Each of these effects varies with time in some complicated manner and has to be represented in statistical form to permit mathematical analysis.

The output signal $f(t)$ is a combination of the received signal $s(t)$ and the noise $n(t)$ and it is assumed in the mathematical analysis that the process is additive. We can therefore write

$$f(t) = s(t) + n(t) \tag{6.41}$$

If we ignore the added noise, the factor relating the received signal $s(t)$ to the transmitted signal $m(t)$ is called the *fading attenuation* of the propagation medium, which we shall designate as x, our statistical variable. Thus

$$s(t) = xm(t) \tag{6.42}$$

where x is a real number. The model of fading attenuation which we shall follow makes several assumptions based on three definitions of sampling intervals, T(short), T(medium) and T(long). These are illustrated in Fig. 6.11.

Firstly, T(short) $= T_s$ may be of the order of microseconds. It is defined with $T_s \gg 1/f$ and lasts as long as x and $\overline{x^2}$ remain virtually constant.

Secondly, T(medium) $= T_m$ may be of the order of minutes, with $T_m \gg T_s$. In this time interval, $x = x(t)$ varies with time in accordance with some statistical distribution (often, for example, the Rayleigh distribution), but the parameters of the distribution

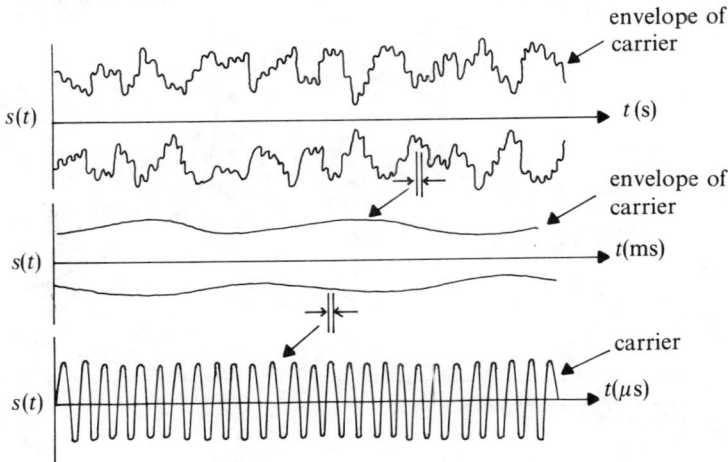

Fig. 6.11 During a short period of time the output signal is virtually constant because the fading attenuation is almost constant.

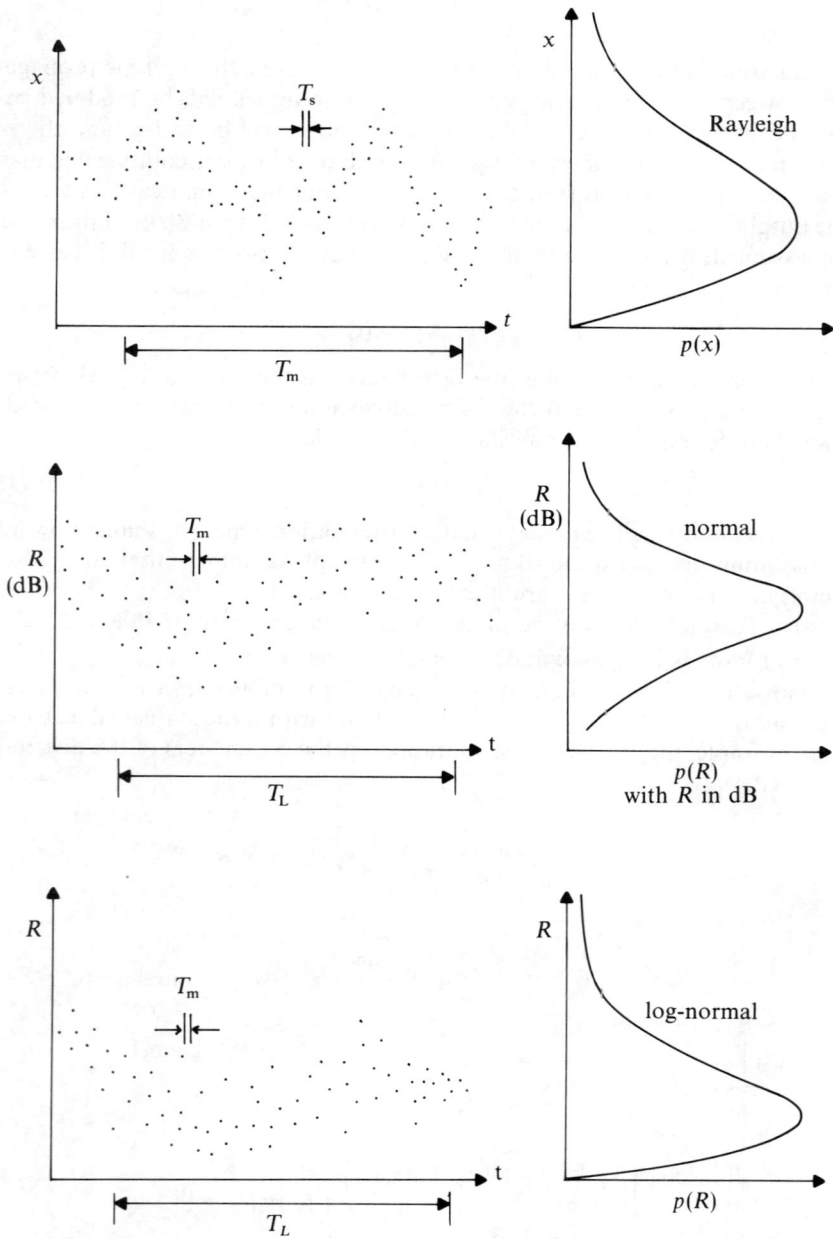

Fig. 6.12 During a short period of time x is approximately constant. Over a medium period of time it varies with a Rayleigh distribution having a mean square R. Over a long period of time R has a log-normal distribution.

(for example, the mean square $R = \overline{x^2}$ in a Rayleigh distribution) remain constant for the duration of T_m. Thus, for the example quoted, we can write

$$p(x) = 2(x/R)\exp(-x^2/R) \qquad (6.43)$$

with $R = \overline{x^2} = $ constant over the time period T_m (Fig. 6.12).

Thirdly, $T(\text{long}) = T_L$ may be of the order of a month or a year, with $T_L \gg T_m$. Our variable x continues to vary with time but the statistical distribution parameter $(R = \overline{x^2})$ associated with each T_m time period varies slowly with time. In some practical cases the variation of R with time follows another known statistical distribution, the log-normal (Fig. 6.13).

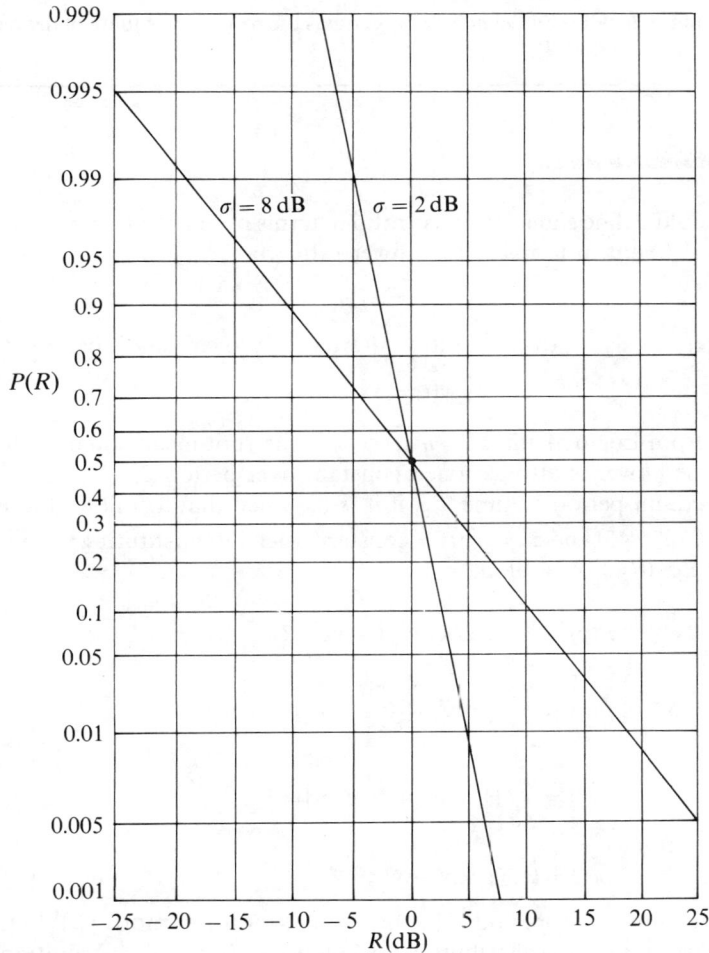

Fig. 6.13 Normal or log-normal probability paper with two lines representing $\sigma = 2\,\text{dB}$ and $8\,\text{dB}$ as examples.

Example 6.7 Fading attenuation and received signal

Show that $x(t)$ is related numerically to $s(t)$ over the period T_s.
  The mean square value of $s(t)$ can be obtained from

$$\overline{s^2} = \frac{1}{T}\int_{t-T}^{t} s(\tau)^2 \, d\tau = \frac{1}{T}\int_{t-T}^{t} \{xm(\tau)\}^2 \, d\tau$$

using Eq. (6.42). Over period T_s, $x = $ constant and hence

$$\overline{s^2} = x^2 \frac{1}{T}\int_{t-T}^{t} m(\tau)^2 \, d\tau = x^2\overline{m^2} = x^2$$

as $\overline{m^2} = 1$ from Eq. (6.40). Thus, *numerically*, x equals the root mean square of the received signal over a short period T_s.

6.5.2 Signal-to-noise Ratio

Although we could define signal-to-noise ratio in terms of signal $s(t)$ and noise $n(t)$, it is more usual to define it in terms of the power ratio via

$$y(t) = \overline{s^2(t)}/\overline{n^2(t)} \tag{6.44}$$

where $y(t)$ represents signal-to-noise ratio. Alternatively, over a short period of time T_s,

$$y(t) = x^2/z^2 \tag{6.45}$$

where $\overline{s^2} = x^2$ from Eq. (6.5) and $z^2 = \overline{n^2(t)}$ or z is the root mean square value of the noise if the noise power is also assumed constant over period T_s.

  Over a medium period of time T_m, if it is assumed that x follows the Rayleigh distribution of Eq. (6.43) and $z^2 = \overline{n^2(t)} = $ constant, then by substituting $y = x^2/z^2$ from Eq. (6.45) into Eq. (6.43) we obtain

$$P(y) = \int_0^Y 2(x/R)\exp(-yz^2/R)(z^2/2x) \, dy$$

$$P(y) = \int_0^Y (z^2/R)\exp(-yz^2/R) \, dy$$

$$P(y) = \int_0^Y (1/\bar{y})\exp(-y/\bar{y}) \, dy$$

or

$$p(y) = (1/\bar{y})\exp(-y/\bar{y}) \tag{6.46}$$

Thus, over medium time period T_m, the signal-to-noise ratio $y(t)$ follows the exponential or gamma ($k = 1$) distribution if x follows the Rayleigh distribution and the mean noise power remains constant. This resembles the conclusions derived in Eq. (6.33).

6.6　MULTIPLE PROPAGATION PATHS

A diversity system is based on the availability of N independent propagation paths, as illustrated in Fig. 6.14. Certain assumptions are required in this model to permit mathematical analysis:

1. x_j and x_k are independent over the short time period T_s.
2. x_j and x_k form independent distributions over medium time period T_m.
3. x_j and x_k are real numbers, though in many practical applications this is not immediately the case and some extra phase-shifting circuitry is necessary to ensure that the signals are coherent.
4. n_j and s_j are independent but additive.
5. n_j and n_k are independent and incoherent.
6. The mean values of n_j and n_k are zero.
7. The mathematical relationship $\overline{n_j n_k} = 0$;
8. The mean square value of n_j and n_k varies slowly during a long time period T_L but is constant during short and medium time periods T_s and T_m.

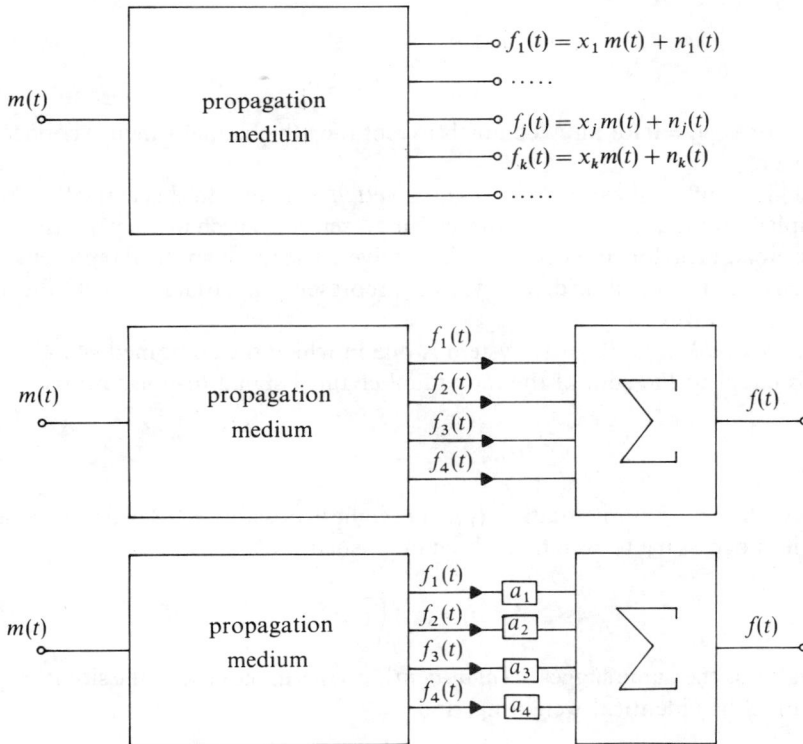

$$f_1(t) = x_1 m(t) + n_1(t)$$
$$f_j(t) = x_j m(t) + n_j(t)$$
$$f_k(t) = x_k m(t) + n_k(t)$$

Fig. 6.14 Models representing propagation through a fading medium and two types of diversity combination.

6.6.1 Some Diversity Combinations

Two types of diversity combination systems are illustrated in Fig. 6.14. In some systems the outputs from the N independent paths are simply summed to provide

$$f(t) = \sum f_j(t) \tag{6.47}$$

In others, the individual components are weighted before being summed so that

$$f(t) = \sum a_j f_j(t) \tag{6.48}$$

One simple example of the latter is called *scanning diversity*. With the aid of some electronic circuitry each of the N outputs is scanned in sequence and the signal level compared with some predetermined threshold. It the signal exceeds the threshold, the output is connected to the next stage of the receiver and will remain connected until the signal level drops below the threshold. Then the scanning sequence continues to the next of the N outputs, and so on. If the signal level of an output, when sampled, is below threshold, the scanning continues until it finds an output above threshold. The mathematics of this process can be written

$$a_j = 0 \qquad \text{for } j \neq k$$
$$a_j = 1 \qquad \text{for } j = k$$

In other words, $a_j = 0$ for all N channels except the kth channel which is connected to the receiver.

A slightly different example is called *selection diversity*. In this case all N outputs are sampled simultaneously and the circuitry selects the channel with the highest signal-to-noise ratio for connection to the receiver. The mathematical representation is the same as that for scanning diversity, with k representing the channel with the highest signal-to-noise ratio.

The *maximal-ratio diversity* system is one in which the combined signal-to-noise ratio y is equal to the sum of the individual channel signal-to-noise ratios, or

$$y = \sum y_j$$

To achieve this optimum situation (i.e. the highest possible SNR from all available signals) it is necessary to weight each channel such that

$$a_j = (x_j)/(\overline{n^2})$$

Finally, as the name suggests, *equal-gain diversity* makes use of the situation where each channel has identical weighting, and

$$a_j = 1$$

describes the mathematics.

6.7 SELECTION DIVERSITY

The analysis of selection diversity makes use of the earlier assumptions that: (1) the mean square value of the noise remains constant over the medium time period T_m; (2) the fading attenuation factor $x(t)$ varies in accordance with the Rayleigh distribution over time period T_m; and (3) that the signal-to-noise ratio $y(t)$ consequently follows the exponential distribution over the same time period.

From Eq. (6.46) the probability that the signal-to-noise ratio y is *less* than some specified value Y is given by

$$P(y < Y) = \int_0^Y (1/\bar{y}) \exp(-y/\bar{y}) \, dy = 1 - \exp(Y/\bar{y}) \tag{6.49}$$

To simplify the mathematics very slightly, we convert to normalized signal-to-noise ratio $w = y/\bar{y}$ with $W = Y/\bar{y}$ so that

$$P(w < W) = 1 - \exp(-W)$$

This distribution applies to each individual channel if it is assumed that they all have the same $\bar{y}$. The probability that $w < W$ for all N channels simultaneously is therefore

$$P(w < W) = [1 - \exp(-W)]^N$$

This is the probability expression associated with probability density function

$$p(W) = N[1 - \exp(-W)]^{N-1} \exp(-W)$$

where $p(W)$ is the pdf of the signal-to-noise ratio achieved by selection. The mean value of this combined distribution is given via

$$\bar{W} = \int_0^\infty W p(W) \, dW$$

$$\bar{W} = \int_0^\infty W N[1 - \exp(-W)]^{N-1} \exp(-W) \, dW$$

This rather long integral has a solution (see Ref. 6.2) which is simply

$$\bar{W} = \sum_{k=1}^N (1/k) \tag{6.50}$$

Thus, if the number of independent channels is $N = 1, 2, 3$ or 4, the mean value of the combined distribution using the principle of selection diversity is $\bar{W} = 1, 3/2, 11/6$ or $25/12$, respectively, where W is the normalized signal-to-noise ratio with respect to the mean value for any given channel.

6.7.1 Graphical Approach

The analysis of selection diversity can be achieved graphically with the aid of exponential probability paper, which is the same as the Rayleigh probability paper

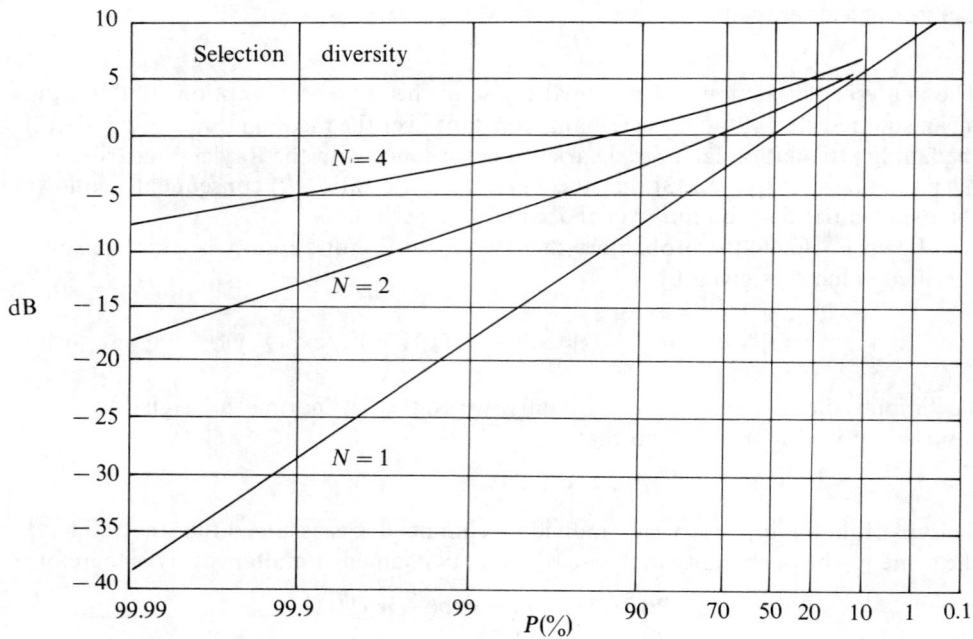

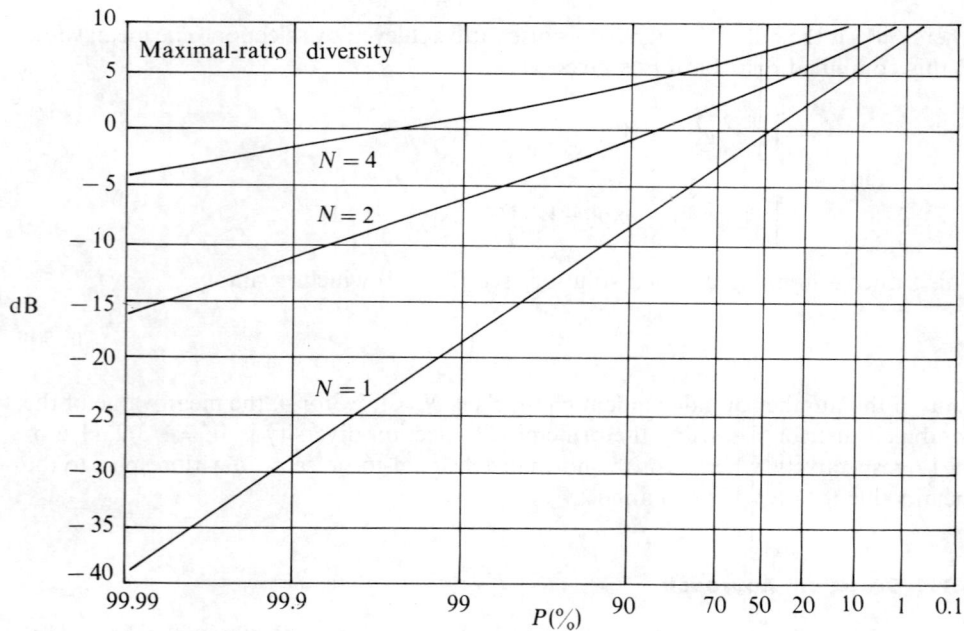

Fig. 6.15 Charts of selection diversity and maximal-ratio diversity.

illustrated in Fig. 6.7 because of the decibel scaling. The graph is constructed with $U = 10 \log(Y/Y_0)$ rather than the $W = y/\bar{y}$ relationship given previously, where Y_0 is the median signal-to-noise ratio of a single channel. The decibel notation is used to produce a straight line on this type of probability paper for the Rayleigh distribution of x or X and the corresponding exponential distribution of y or Y.

To produce the curves for $N = 2, 3$ or 4, for example, from the $N = 1$ straight line, choose any point on the $N = 1$ line. $P = 90\%$ corresponds with $P(u < U) = 10\% = 0.1$ and $U = 8\,\text{dB}$ *below* the single-channel median U_0. In other words, U will be less than $U_0 - 8\,\text{dB}$ for 10% of time. For two channels, the probability that both channels will produce U less than $U_0 - 8\,\text{dB}$ simultaneously is therefore $P(u < U) = 0.1 \times 0.1$ or $P(u < U) = 1\%$. Thus the corresponding point on the $N = 2$ curve is at $P = 99\%$ as $P = 100 - P(u < U)$.

This procedure can be repeated all along the $N = 1$ line to produce the corresponding $N = 2$ curve. A similar procedure produces the $N = 3$ and $N = 4$ curves. The curves of dual and quadruple selection diversity are shown in Fig. 6.15.

Example 6.8 Selection diversity

A certain radio link incorporates quadruple diversity in order to reduce the transmitter power level for 99.9% reliability. It is found that $P_t = 10\,\text{kW}$ is necessary to achieve this reliability over a single link. What transmitted power is necessary with the quadruple selection diversity system in operation?

The graphs are illustrated in Fig. 6.15. These show that at the 99.9% level the single-channel signal may reach $U_0 - 28\,\text{dB}$, and for this reason $P_t = 10\,\text{kW}$ is required. With $N = 4$ selection diversity the corresponding level of received signal is only $U_0 - 5\,\text{dB}$, which is 23 dB higher than that of a single channel. For the same reliability the transmitter power can be reduced by a factor of almost 200, that is from 10 kW to 50 W. This is a considerable saving in power and energy, and it also means that there is less 'pollution' of the radio spectrum due to unnecessary radiation at the particular operating frequency and any consequent interference is similarly reduced.

6.8 MAXIMAL RATIO DIVERSITY

We can see from Fig. 6.14 and Sec. 6.6.1 that the output from a maximal-ratio diversity combiner can be expressed mathematically as

$$f(t) = \sum a_j s_j(t) + \sum a_j n_j(t)$$

with the summation taken over all the contributory channels. The signal-to-noise ratio $y(t)$ over the short time period T_s is defined as the ratio of the mean square signal to the mean square noise:

$$y(t) = \overline{s^2}/\overline{n^2}$$

where

$$\overline{s^2} = \overline{[\sum a_j x_j m(t)]^2} = [\sum a_j x_j]^2 \tag{6.51}$$

because $\overline{m^2} = 1$ and $x_j = $ constant over short time period T_s.

The noise component is a little more complicated. Starting with

$$n^2 = [\sum a_j n_j]^2 = (a_1 n_1 + a_2 n_2 + a_3 n_3 + a_4 n_4)^2$$

for a four-fold diversity system, for example, we see that the expansion will contain terms $(a_j n_j)^2$ and terms $(a_j a_k n_j n_k)$. Noting that

$$\overline{\sum a_j a_k n_j n_k} = \sum a_j a_k \overline{n_j n_k} = 0$$

with assumption 7 in Sec. 6.6, the relationship eventually reduces to

$$\overline{n^2} = \overline{\sum a_j^2 n_j^2} = \sum a_j^2 \overline{v_j^2} \tag{6.52}$$

Combining the two expressions for signal and noise given in Eqs. (6.51) and (6.52), we obtain the signal-to-noise ratio as:

$$y = \frac{[\sum a_j x_j]^2}{a_j^2 \overline{n_j^2}} = \frac{[\sum a_j x_j]^2}{z^2 \sum a_j^2} \tag{6.53}$$

In the special case in which a_j is given the weighting $a_j = x_j / z^2$, where z is the root mean square of the noise normalized as in Eq. (6.45), the signal-to-noise ratio reduces to:

$$y = \overline{s^2}/\overline{n^2} = \sum (x_j^2 / z^2) = \sum y_j \tag{6.54}$$

Thus the signal-to-noise ratio of the N combined channels equals the sum of the individual signal-to-noise ratios of each channel, provided each channel is weighted in the manner described.

As mentioned earlier, one example of the use of the exponential and gamma distributions in radio wave propagation is that of maximal-ratio diversity combining. If a signal is received via Rayleigh-type scatter propagation, the signal-to-noise ratio (defined in Eq. (6.54)) has the exponential or gamma ($k = 1$) distribution. Ref. 6.3 explains that, with the aid of the mathematical theorem that the sum of N independent gamma variates with parameters $k_1, \ldots, k_N$ is a gamma variate with parameter $k = k_1 + \cdots + k_N$, and hence the distribution of the maximal-ratio diversity combiner output signal must be a gamma distribution with $k = N$. This requires that each Rayleigh distribution has to have the same mean square parameter R over the medium time period T_m, which is realistic in practice.

The diversity curves for maximal-ratio combining are the gamma curves of Fig. 6.9 suitably modified into decibel form and reproduced as Fig. 6.15. The reference level is the median of the $N = 1$ line.

Example 6.9 Maximal-ratio diversity

Repeat Ex. 6.8 but with the modification that the radio link makes use of maximal-ratio diversity combining.

The graphs in Fig. 6.15 show that for a single channel the 99.9% level is $U_0 - 28$ dB as before, but with four-fold maximal-ratio diversity the 99.9% level is $U_0 - 2$ dB, an effective gain of 26 dB. The transmitted power can therefore be reduced from 10 kW by 26 dB to 25 W. This is a saving of 3 dB over the selection diversity system.

PROBLEMS

6.1 A small sample of field-strength measurements give the following results in dBμ:

> 31 43 65 81 41 71 54 49 33 51 79 61
>
> 53 38 21 54 41 69 48 56 53 65 63 51

Determine sample size N, mean $\bar{x}$, standard deviation σ, median X_0 and root mean square r, recalling the electrical engineering equivalents.

6.2 During the course of one full year, the 8760 hourly median observations of received power level $P(r)$ dBW were recorded for a tropospheric scatter radio link. The results are displayed in Fig. 6.16. It is usually assumed that such measurements conform to the normal distribution of $P(r)$ dBW or to the log-normal distribution of P_r W. Do these results confirm the assumption? What are the numerical values of the annual mean and the standard deviation? Use probability paper.

6.3 The generalized distribution may be written, from Eqs. (6.14) and (6.19), as:

$$p(x) = \frac{m}{b^{n+1}\Gamma((n+1)/m)} x^n \exp(-x^m/b^m)\, \mathrm{d}x$$

Determine the first moment m_1 and the second moment m_2 of this expression. In each case show how normalizing parameter b may be replaced by either m_1 or m_2.

6.4 A series of measurements on a certain radio link confirmed that the electric field strength followed a Rayleigh distribution. It was found that the median value of $P(r)$ was -80 dBW. What is the probability that the received power level exceeds the -100 dBW limit for satisfactory reception?

6.5 A radio link has a received signal whose variation follows the Rayleigh distribution. If the median field strength is 40 dBμ, what is the probability of the signal exceeding 30 dBμ?

6.6 Two signals are received from a transmitter via a perfect scatter propagation medium. The antennas are sufficiently displaced to make the resulting signal the sum of two independent fading signals, the medians of each being -60 dBW. When ideally combined, as in Sec. 6.4.4, what is the probability of the signal exceeding -50 dBW?

6.7 The annual median path loss on a certain tropospheric scatter radio link is 120 dB. What is the likely order of magnitude of the hourly median path loss during the worst hour of the

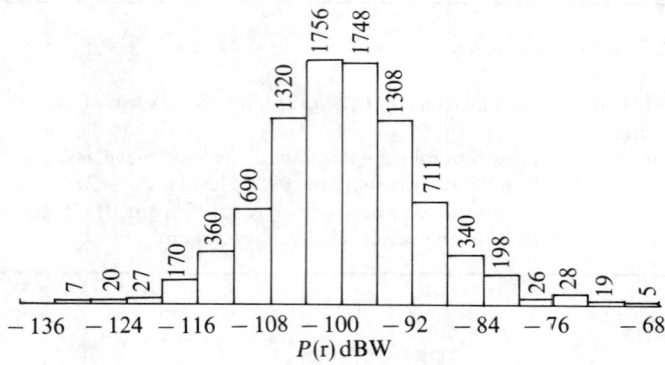

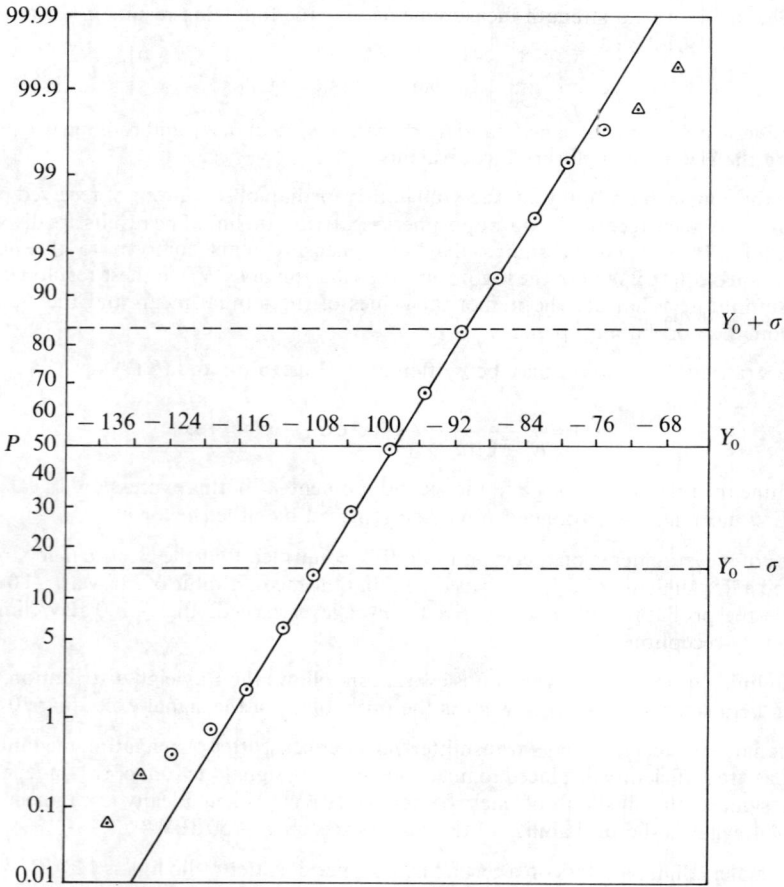

Fig. 6.16 Illustrations associated with Prob. 6.2.

year if path loss follows a log-normal distribution or path loss in decibels follows a normal distribution with 7 dB standard deviation?

6.8 A certain radio link via a Rayleigh type fading medium incorporates four-fold selection diversity and receives satisfactory signals for 99.9% of time when the transmitted power is 100 W. (a) What transmitted power level would have been required for the same reliability if a single non-diversity link had been used instead? (b) If 100 W had been used on a single radio link, over what proportion of time would the signal exceed the same threshold?

6.9 A radio link operates satisfactorily for only 70% of time. What improvement in reliability can be achieved by: (a) doubling the transmitter power, or (b) using two-fold maximal-ratio diversity? Assume a Rayleigh-type fading medium.

6.10 Plot graphs of the Weibull distribution in the form of $p(y)$ against y where $y = x/b$.

REFERENCES

6.1 Sturges, H. A., 'The choice of class interval,' *Journ. Am. Statist. Assoc.*, 21 (1926), 65–66.

6.2 Griffiths, J., and J. P. McGeehan, 'Interrelationship between some statistical distributions used in radio-wave propagation,' *Proc. IEE*, 129, Pt. F, No. 6 (December 1982), 411–17.

6.3 Staras, H., 'The statistics of combiner diversity,' *Proc. IRE*, 44 (1956), 1057–8.

7

Tropospheric Scatter Radio Links

The *scattering* of radio waves is, in several respects, analogous to the scattering of light by dust particles in a smoke-filled cinema. The eye can see the beam from the projector and, because of the limited response of the eye, could receive a kind of 'Morse code' type message if the projector light was appropriately modulated at a slow enough rate. The scattering effect is caused by the reflection or scattering of light or radio waves by a very large number of small particles in the scattering medium and the observation is the resultant summation of all such contributions in both magnitude and phase.

There are several theories for tropospheric scattering. One suggests that it arises because the turbulent atmosphere within the scattering volume produces *blobs* with different refractive indices from that of the surrounding atmosphere and that these blobs scatter the incident energy in all directions. Another theory considers the troposphere to be stratified into many homogeneous layers with different heights and refractive indices and suggests that propagation through such a medium is affected by both refraction and partial reflection at each layer. Whatever the mechanism, a simplified model of a tropospheric scatter radio link assumes that the resultant signal variation is the sum of two principal components: a slow or *long-term variation* and a fast or *short-term variation*.

To achieve low propagation losses via the scattering mechanisms the transmitter launch angle has to be small, perhaps less than 4°. With the *scattering volume* at a height of just a few kilometers above ground level, the tropospheric scatter link ground range is then usually of the order of several hundreds of kilometers. It is therefore particularly useful as a means of broadband radio wave transmission over paths which are beyond the normal range of line-of-sight space wave propagation, say 70–700 km. However, some care is necessary to distinguish whether the received signal is due principally to diffraction around the curved surface of the earth or to scattering by the troposphere. Typical operating frequencies may be in the range 400 MHz to 7 GHz.

7.1 SIMPLE MODEL OF TROPOSCATTER LINK

A simple model of a *tropospheric scatter radio link* starts with an expression for the annual (or sometimes worst-month) median values of

$$\overline{P(r)} = P(t) - L(t) \tag{7.1}$$

where $\overline{P(\mathrm{r})}$ and $P(\mathrm{t})$ are the received and transmitted power levels in dBW and $L(\mathrm{t})$ is called the *transmission loss*. The transmission loss is the net sum of the various losses and gains along the radio link between the transmitter and receiver. For instance, if the transmitting and receiving antennas have gains $G(\mathrm{t})$ dB and $G(\mathrm{r})$ dB in the direction of the link and the combined feeder losses total $L(\mathrm{f})$ dB, then

$$L(\mathrm{t}) = \overline{L(\mathrm{p})} + L(\mathrm{f}) - G(\mathrm{t}) - G(\mathrm{r}) \qquad (7.2)$$

with gains considered as negative losses. In this expression $\overline{L(\mathrm{p})}$ is called the annual median *path loss* between the two antennas.

One of the several components of the path loss is called the *spatial loss* $L(\mathrm{s})$, given in Eq. (1.15c) as

$$L(\mathrm{s}) = 20 \log f(\mathrm{MHz}) + 20 \log d(\mathrm{km}) + 32.45 \qquad (7.3)$$

where the great circle ground range d km is adequate for most practical purposes, as the excess distance of the actual path via the scattering volume is only very slightly greater.

There is an additional loss as a result of the scattering process and this is known as the *scatter loss* $L(\mathrm{sc})$. Its value is found to depend on the magnitude of the *scatter angle* θ, illustrated in Fig. 7.1. Several methods have been proposed for estimating $L(\mathrm{sc})$ from a knowledge of θ but we will consider only one proposal later.

Because the scattered signals arriving at the large receiving antenna do not come from a point source but rather from an extended volume of an infinite number of scatters, the consequent phase incoherence between the many components of the field on the wavefront is responsible for an apparent loss in gain when large-aperture antennas are used. This is known as *antenna-to-medium coupling loss* $L(\mathrm{c})$, for which several empirical expressions are available. Again, we shall use only one of these expressions.

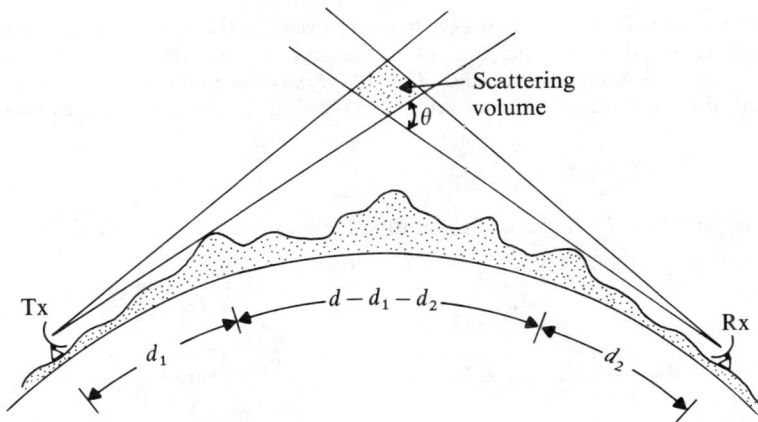

Fig. 7.1 The scatter angle θ of a tropospheric scatter radio link across a mountainous terrain.

By combining these contributions, the annual median path loss can be written as

$$\overline{L(p)} = L(s) + \overline{L(sc)} + L(c) + L(misc) \tag{7.4}$$

where $L(misc)$ accounts for other *miscellaneous losses* in the radio link.

7.2 PATH ANALYSIS

The simplified model of the tropospheric scatter radio link described in Sec. 7.1 concerns only the long-term or slow variation in the received signal. We shall consider the effect of the fast variation later.

The power level $P(g)$ generated by the transmitter is often either 30 dBW (1 kW) or 40 dBW (10 kW), though larger and smaller powers are not uncommon. Not all of the power reaches the transmitting antenna, and thus

$$P(t) = P(g) - L(ft) \tag{7.5}$$

Example 7.1 Scatter angle

Determine both an expression for, and a numerical value of, the scatter angle θ illustrated in Fig. 7.1 if the transmitting and receiving antenna heights are $h_t = 100$ m and $h_r = 50$ m; the height and distance of the first obstacle are $h'_t = 200$ m and $d_1 = 50$ km; the height and distance of the last obstacle are $h'_r = 80$ m and $d_2 = 25$ km; and the great circle distance between transmitter and receiver is $d = 250$ km. Assume an effective radius of earth $R = 8490$ km.

The method is outlined in Fig. 7.2 together with the definitions of heights and distances. It is easy to see that

$$\theta = \theta_0 + \theta_1 + \theta_2 + \theta_3 + \theta_4 - \pi \tag{7.7}$$

where triangles OAB and OAC are constructed as isoceles triangles. When the distances are much larger than the heights, as is the case in practice, we can make the approximations $\theta_1 = (h'_t - h_t)/d_1$ and $\theta_2 = (h'_r - h_r)/d_2$ in milliradians if heights are measured in meters and distances in kilometers. Using the isoceles triangles we obtain

$$\theta_3 = \frac{\pi}{2} - \frac{d_1}{2R} \quad \text{and} \quad \theta_4 = \frac{\pi}{2} - \frac{d_2}{2R}$$

in radians, and $\theta_0 = d/R$ radians. Substituting these expressions into Eq. (7.7),

$$\theta = \frac{1000d}{R} + \frac{h'_t - h_t}{d_1} + \frac{h'_r - h_r}{d_2} - \frac{1000d_1}{2R} - \frac{1000d_2}{2R} \quad \text{mrad}$$

or

$$\theta = \frac{1000}{2Ka}(2d - d_1 - d_2) + \frac{h'_t - h_t}{d_1} + \frac{h'_r - h_r}{d_2} \quad \text{mrad}$$

with heights in meters, distances in kilometers and θ in milliradians. K is the refractivity factor and a is the radius of earth. Substitution of the numerical data gives $\theta = 28.23$ mrad.

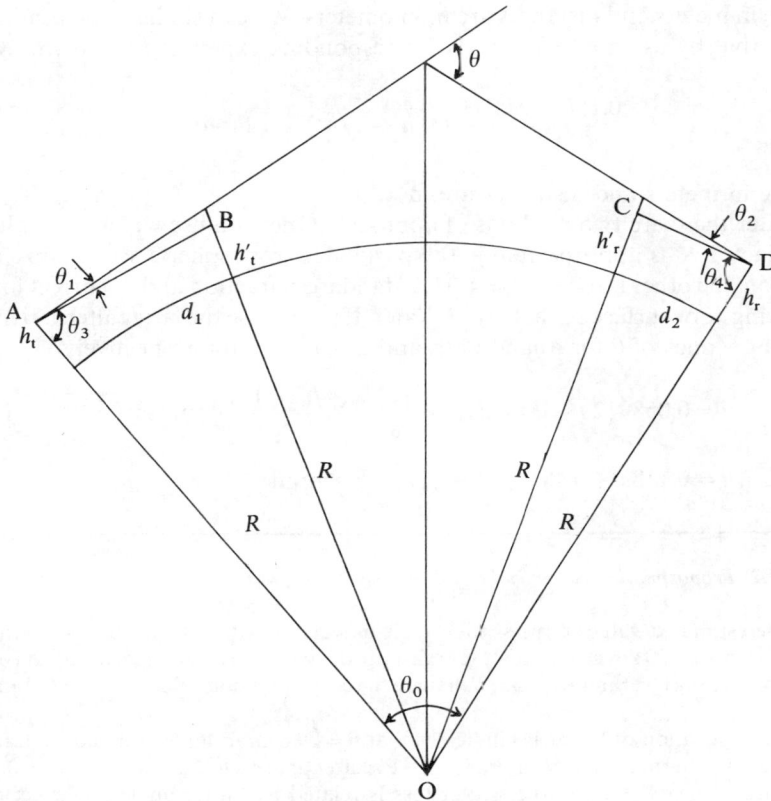

Fig. 7.2 The trigonometry of the scattering angle θ.

where $L(\text{ft})$ is that portion of the feeder losses $L(\text{f})$ associated with the transmitting antenna, and may be of the order of 1 or 2 dB.

The antennas are often of the conventional paraboloidal reflector types with gains given via Eq. (1.32) as

$$G(\text{t}) = G(\text{r}) = 20 \log f(\text{MHz}) + 20 \log D(\text{m}) - 42.3 \qquad (7.6)$$

for paraboloids with diameter D and illumination efficiency $\eta = 0.54$. Other versions of reflector antennas are also used, such as the billboard antenna, and in these cases the antenna gains $G(\text{t})$ and $G(\text{r})$ must be measured.

The scatter angle θ needs to be calculated from the path profile. Example 7.1 indicates how we obtain θ for a tropospheric radio link across a mountainous land path with

$$\theta = \frac{1000}{2Ka}\{2d - d_1 - d_2\} + \frac{h'_t - h_t}{d_1} + \frac{h'_r - h_r}{d_2} \qquad \text{mrad} \qquad (7.8)$$

if heights are in meters and distances are in kilometers. We can similarly determine, as in Prob. 7.1, that by using $h'_t = h'_r = 0$ the corresponding expression for a wholly sea path is

$$\theta = \frac{0.157}{K}d - \frac{0.560}{\sqrt{K}}(\sqrt{h_t} + \sqrt{h_r}) \qquad \text{mrad} \tag{7.9}$$

if heights are in meters and distances are in kilometers.

Thereafter there are two methods of approach. One is to use whichever value of refractivity factor K is appropriate for the particular tropospheric scatter link on a given occasion; the other is to use $K = 4/3$ for standard refraction and to correct for the effect of varying K or surface refractivity N_s later. If we choose the latter alternative, the corresponding values of θ for a land path and a sea path are, respectively:

$$\theta = 0.0589\{2d - d_1 - d_2\} + \frac{h'_t - h_t}{d_1} + \frac{h'_r - h_r}{d_2} \qquad \text{mrad} \tag{7.10}$$

$$\theta = 0.118d - 0.485(\sqrt{h_t} + \sqrt{h_r}) \qquad \text{mrad} \tag{7.11}$$

Example 7.2 Tropospheric scatter radio link across a sea path

What is the estimated value of the scatter angle associated with a tropospheric scatter radio link across a 350 km sea path if the transmitting and receiving antennas are 200 m and 100 m above sea level and $K = 4/3$? What are the corresponding values for $K = 0.7$ and $K = 2.0$?

Direct application of Eq. (7.11) indicates that $\theta = 29.6$ mrad under standard refraction conditions. Alternatively, using Eq. (7.9), the scatter angle is 62.3 mrad when $K = 0.7$, and 17.9 mrad when $K = 2$. As the scatter loss is related to scatter angle, the effect of varying K is quite considerable.

Having calculated the scatter angle θ under standard refraction conditions with $K = 4/3$, there are various empirical expressions for scatter loss, including one quoted by Yeh (Ref. 7.1). With some modifications, this gives

$$\overline{L(\text{sc})} = 21 + 0.57\theta(\text{mrad}) + 10\log f(\text{MHz}) - 0.2(N_s - 310) \qquad \text{dB} \tag{7.12}$$

Example 7.3 Scatter loss of a sea path radio link

Estimate the scatter loss associated with the sea path tropospheric scatter radio link described in Ex. 7.2 if the operating frequency is 2 GHz and the annual median surface refractivity is $N_s = 320$.

We obtained $\theta = 29.6$ mrad as the scatter angle in Ex. 7.2. Using Eq. (7.13) the scatter loss annual median value is $L(s) = 68.9$ dB. Had the scatter angles been 62.3 and 17.9 mrad, the corresponding scatter loss would have been 87.5 and 62.2 dB.

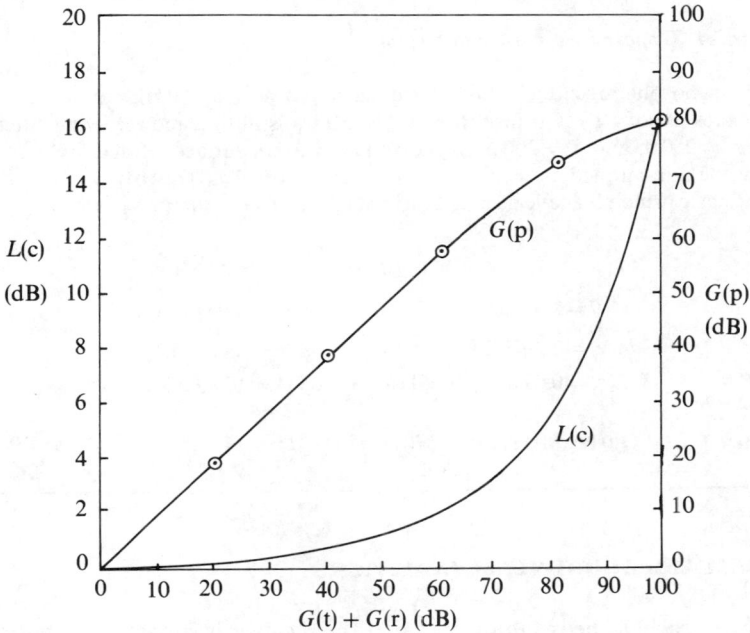

Fig. 7.3 One empirical relationship for the antenna-to-medium coupling loss $L(c)$ in terms of the gains of the transmitting and receiving antennas. The net antenna gain $G(p) = G(t) + G(r) - L(c)$ is most noticeably affected when the gains are large.

where f is the operating frequency and N_s is the annual median value of the surface refractivity. This can obviously be reduced to

$$\overline{L(sc)} = 83 + 0.57\theta + 10\log f - 0.2N_s \qquad \text{dB} \tag{7.13}$$

with the variables in the specified units. A slightly different version is given in Prob. 7.2.

Similarly, there are several alternative expressions for the antenna-to-medium coupling loss $L(c)$. One simple version, given by the CCIR (Ref. 7.2), is

$$L(c) = 0.07 \exp\{0.055[G(t) + G(r)]\} \qquad \text{dB} \tag{7.14}$$

This means that if the total effective gain of the antenna system is $G(p)$ then

$$G(p) = G(t) + G(r) - L(c) \qquad \text{dB} \tag{7.15}$$

and this is illustrated in Fig. 7.3 using Eq. (7.14). In some cases the CCIR estimate is different from that observed in practice, and there is considerable variability between the several suggested expressions for $L(c)$.

Example 7.4 Tropospheric scatter link to oil rig

A 2 GHz tropospheric scatter radio link from a land base to an oil rig situated 250 km out at sea makes use of a 1 kW transmitter and 50 dB transmitting and receiving antennas at heights $h_t = 250$ m and $h_r = 200$ m. If the annual median surface refractivity is $N_s = 320$, estimate the annual median level of the received power if $L(f) = 4$ dB.

The basic principles have been established in the earlier examples in which we showed that

$$L(s) = 20 \log f + 20 \log d + 32.45 = 146.43 \text{ dB}$$

$$\theta = 0.118d - 0.485(\sqrt{h_t} + \sqrt{h_r}) = 14.97 \text{ mrad}$$

$$\overline{L(sc)} = 83 + 0.57\theta + 10 \log f - 0.2N_s = 60.54 \text{ dB}$$

$$L(c) = 0.07 \exp\{0.055[G(t) + G(r)]\} = 17.13 \text{ dB}$$

and

$$\overline{P(r)} = P(t) - L(f) + G(t) + G(r) - L(s) - \overline{L(sc)} - L(c) = -98 \text{ dBW}$$

7.3 LONG-TERM AND SHORT-TERM VARIATIONS

As mentioned earlier, the resultant signal at the receiver is subject to variations in the propagation medium which are considered to be represented by a long-term or slow variation plus a short-term or fast variation.

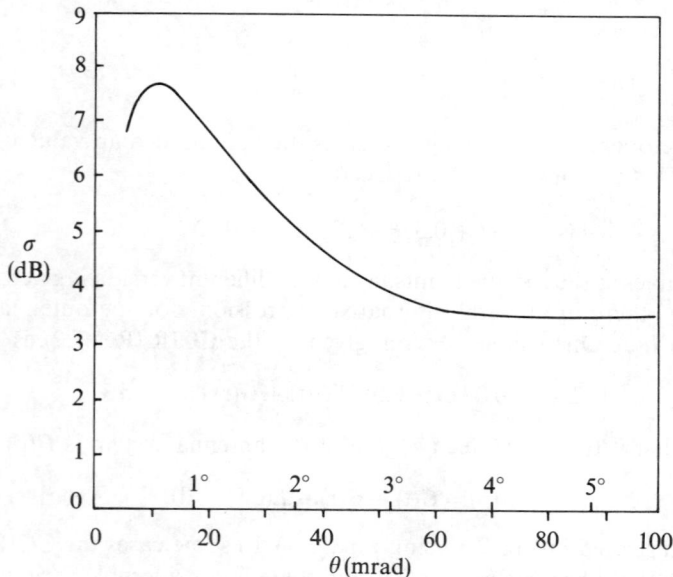

Fig. 7.4 An empirical relationship between the standard deviation of the normal variation in $L(sc)$ and the scatter angle θ.

7.3.1 Slow Variation

The values of $\overline{P(r)}$ and $\overline{L(p)}$ determined in Sec. 7.2 represent the annual median levels, though sometimes a similar analysis is made in terms of the median of the worst month. On a shorter time scale, the median values of $P(r)$ and $L(p)$ could be measured on an hourly basis, with $365 \times 24 = 8760$ samples of hourly medians occurring each year.

If these 8760 samples are collected as a histogram (in dBW or other logarithmic units), it is found that the variation tends to approximate to the normal distribution. This means that P_r (in W) and L_p (linear) approximate the log-normal distribution.

Two of the parameters of this normal distribution are its median (which is the annual median level estimated in Sec. 7.2) and its standard deviation σ, which gives some numerical indication of the dispersion of $P(r)$ or $L(p)$ about the annual median. It

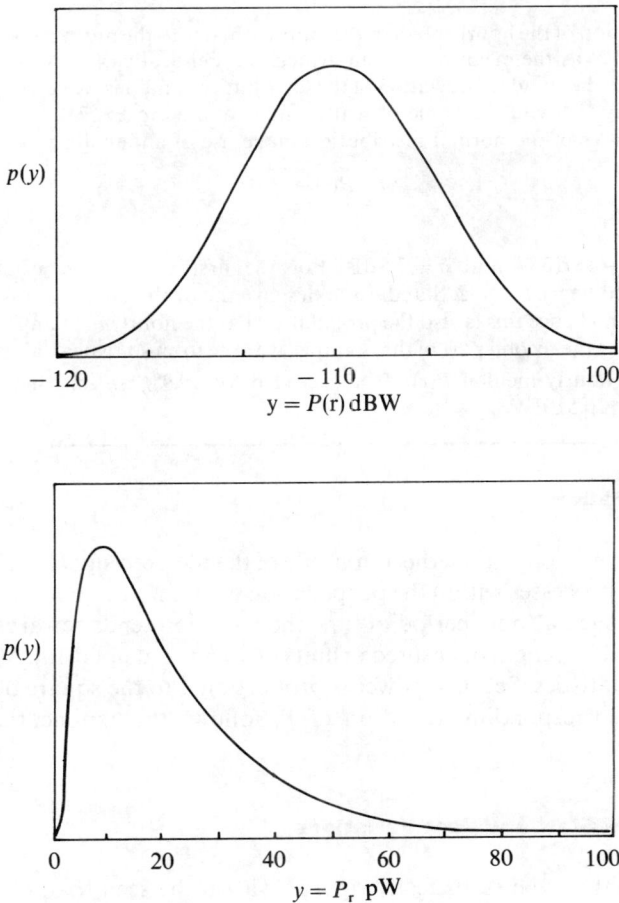

Fig. 7.5 The normal variation of $P(r)$ dBW and the corresponding log-normal variation of P_r pW.

is found empirically that σ depends on the scatter angle θ (calculated with $K = 4/3$), and is commonly between about 3 dB and 8 dB (Fig. 7.4).

Such are the values of σ that the $\pm 3\sigma$ ranges may be of the order of 20 dB to 50 dB wide. In linear form, these imply power ratios from $10^2:1$ to $10^5:1$, and it is not easy to reproduce linear variations of P_r over such extensive ranges. Figure 7.5 has been chosen deliberately with a large scatter angle so that $\sigma = 3.333$ and $\pm 3\sigma = 20$ dB. The corresponding log-normal distribution is then just about realistic.

Example 7.5 Reliability of slow variation

Using the data and solution to Ex. 7.4, estimate: (a) the probability that the hourly median $P(r)$ exceeds -113 dBW; and (b) the level of the hourly median $P(r)$ which is exceeded for 99.9% of time.

The variation of the hourly median $P(r)$ approximates to the normal distribution with $P(r) = -98$ dBW as the mean or median, which are identical with this distribution. We need to know σ, the standard deviation of the distribution, and this we can obtain as about 7.5 dB from Fig. 7.4 with the value of scatter angle quoted in Ex. 7.4.

The statistics of the normal distribution make use of an auxiliary parameter

$$t = \frac{P(r) - \overline{P(r)}}{\sigma} \qquad (7.16)$$

where $\overline{P(r)} = -98$ dBW and $\sigma = 7.5$ dB. For the first part of the example, $P(r) = -113$ dBW and hence $t = -2$. Standard tables give the probability that t is greater than -2 as 97.7% and hence this is also the probability that the hourly median $P(r)$ will exceed -113 dBW. For the second part of the example, we note from statistical tables that t needs to be -3 for hourly median $P(r)$ to be exceeded for 99.9% of time and hence $P(r) = \overline{P(r)} + t\sigma = -120.5$ dBW.

7.3.2 Fast Variation

Over a very short period of time the amplitude of the received signal varies rapidly due to the scattering processes within the propagation medium. It is found that if the signal is sampled over such a short period of time the variation tends towards the Rayleigh distribution of field strength (measured in units of μV/m, and not dBμ). As mentioned in the chapter on statistics, Sec. 6.4, power is proportional to the square of field strength and hence the corresponding variation of P_r follows the exponential distribution (Fig. 7.6).

7.3.3 Combined Slow and Fast Variations

Although it might not be statistically rigorous, it can be seen that:

1. From Ex. 7.4, the annual median level of received power could easily be of the order of -100 dBW.

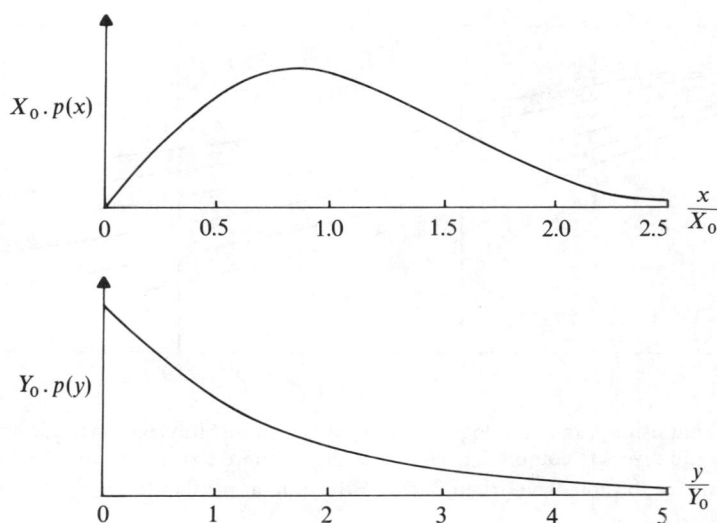

Fig. 7.6 If the distribution of x = electric field strength obeys the Rayleigh equation, then the corresponding distribution of y = power obeys the exponential equation.

Example 7.6 Fast fading

During a short period of time the median value of received power via a troposcatter radio link is $-80\,\text{dBW}$. What is the level of received power which is exceeded for 99.9% of time?

We have already shown in Eqs. (6.34) and (6.35) that the probability of an *electric field strength* x exceeding the median value X_0 in a fast fading situation is given by

$$P(x > X_0) = \exp(-0.693x^2/X_0^2) \tag{7.17}$$

Putting $P = 0.999$ and taking natural logarithms of both sides gives

$$1.4434 \times 10^{-3} = (x/X_0)^2$$

or

$$10\log_{10}(1.4434 \times 10^{-3}) = x(\text{dB}\mu) - X_0(\text{dB}\mu) = -28.4\,\text{dB}$$

The *power level*, also, is 28.4 dB below the median at the 99.9% probability level, namely $-108.4\,\text{dBW}$. Alternatively, this can be read directly from Rayleigh probability paper.

---

2. From Ex. 7.5, some of the hourly medians could be about 20 dB below $-100\,\text{dBW}$.
3. From Ex. 7.6, the fast fading could be up to 30 dB lower still, giving some signal levels down to almost $-150\,\text{dBW}$. This is an extremely low level of received power, bearing in mind that thermal noise kTB is about $-170\,\text{dBW}$ for a 3.4 kHz bandwidth if the receiver noise factor, other forms of noise and the threshold limits are ignored.

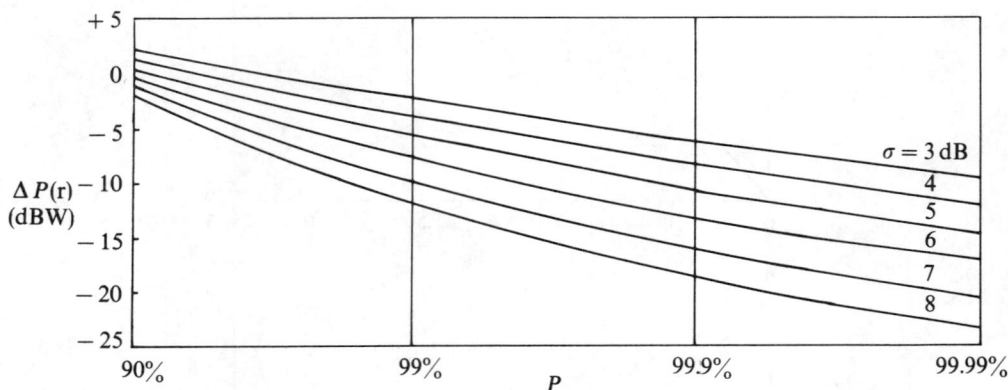

Fig. 7.7 Composite graphs for log-normal slow variation, Rayleigh fast variation and maximal-ratio diversity combining, where σ is the standard deviation associated with the log-normal distribution. The origin refers to the annual median level of $\overline{P(r)}$.

Even accepting that there is a slightly different answer if the problem is analyzed correctly, it is still necessary to increase the transmitter power or apply diversity techniques such as those outlined briefly in Chap. 6. If we applied 4-fold diversity of the type known as maximal-ratio combining, there is a considerable reduction in the effect of fading. For example, at the 99.9% level the Rayleigh fade of a single path is 28 dB below the median; with 4-fold maximal-ratio diversity the corresponding level (from Fig. 6.15) is only 3 dB below the single-path median, an effective gain of 25 dB without any increase in transmitter power.

The difficulty, however, is that a tropospheric scatter radio link with diversity encompasses simultaneously three sets of statistics: the log-normal slow variation of P_r, the exponential fast variation of P_r, and the gamma ($k = 4$) distribution of the diversity

---

Example 7.7 Tropospheric scatter link with diversity

Referring again to the radio link described in Ex. 7.4, assume that the receiver threshold level is -130 dBW and that four-fold maximal-ratio diversity is used. Is the system likely to operate satisfactorily?

    The solution to Ex. 7.4 gives us $\theta = 15$ mrad, and from Fig. 7.4 we can see that the standard deviation of the annual median is of the order 7.5 dB. Using this value of σ in Fig. 7.7 we can see that $\Delta P(r)$ is of the order -2 dB, -11 dB, -17 dB and -22 dB at the 90%, 99%, 99.9% and 99.99% reliability levels, giving corresponding received power levels of -100 dBW, -109 dBW, -115 dBW and -120 dBW with $\overline{P(r)} = -98$ dBW. These are all in excess of the receiver threshold and the system should operate very reliably from this point of view. For example, if 99.9% reliability is specified, there is a margin of about 15 dB to allow for unpredictable losses, variations in transmitter and receiver performances, etc., as well as carrier-to-noise ratio requirements.

combining system. Such combined statistics are beyond the scope of this text, but a graphical approximation of the solution is given in Fig. 7.7 for a 4-fold maximal-ratio diversity combining system. The radio-link path analysis gives the annual median $\overline{P(r)}$ dBW of the given troposcatter system and the standard deviation σ dB of the associated normal distribution. The graphs in Fig. 7.7 give an indication of the levels of $P(r)$ exceeded for a given percentage of time for the given σ for the overall system, with 0 dB referring to the annual median $\overline{P(r)}$ dBW.

7.3.4 Effective Scattering Cross-section

In some respects a tropospheric scatter communication system resembles the type of radar system in which the transmitter and receiver are in different locations. Although the principles of radar are discussed in a later chapter, it is easy to see that the power flux at the common volume *target* is given by

$$P_a = \text{EIRP}/(4\pi r_1^2) = P_t G_t/(4\pi r_1^2)$$

If we consider that some of this incident power is re-radiated as if isotropically from the target, then the re-radiated EIRP' is

$$\text{EIRP}' = \sigma_s P_t G_t/(4\pi r_1^2)$$

where the constant σ_s relating EIRP' to P_a is called the *effective scattering cross-section* of the common volume. The power level at the receiver is obtained via $P_r = P_a A_{\text{eff}}$ as

$$P_r = \frac{\text{EIRP}'}{4\pi r_2^2} G_r \frac{\lambda^2}{4\pi}$$

or

$$P_r = \frac{P_t G_t G_r \lambda^2}{64\pi^3} \cdot \frac{\sigma_s}{r_1^2 r_2^2} \tag{7.18}$$

This is a form of the radar equation for the link illustrated in Fig. 7.8.

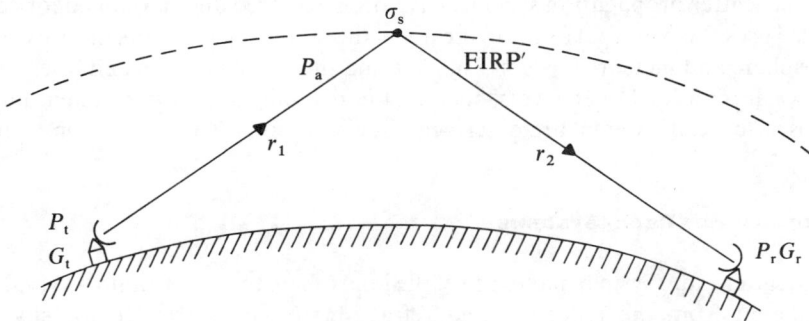

Fig. 7.8 Radio communications via a scattering mechanism resemble bistatic radars with the common volume of the tropospheric scatter system having an effective 'radar' cross-section σ_s.

When applied to tropospheric scatter radio links, the cross-section is sometimes replaced by the *effective scattering cross-section per unit volume*, $\sigma_s(\theta)$, with

$$\frac{\sigma_s}{r_1^2 r_2^2} = \int_v \frac{\sigma_s(\theta)}{r_1^2 r_2^2} \, dv$$

where v is the common volume. Alternatively, if we recall Friis' free-space equation

$$P_r' = \frac{P_t G_t G_r \lambda^2}{(4\pi d)^2}$$

and then define scattering loss L_{sc} via $P_r = P_r'/L_{sc}$ so that it is the excess loss of the tropospheric scatter link over the spatial loss L_s, then by approximating $r_1 = r_2 = d/2$, Eq. (7.18) becomes

$$P_r = \frac{P_t G_t G_r \lambda^2}{(4\pi d)^2} \cdot \frac{4\sigma_s}{\pi d^2} \tag{7.19}$$

From this relationship we can see that the scatter loss is simply

$$L_{sc} = \frac{\pi d^2}{4\sigma_s}$$

or

$$L(\text{sc}) = 59 + 20 \log d(\text{km}) - 10 \log \sigma_s(\text{m}^2) \tag{7.20}$$

In Ex. 7.3 we found that the scatter loss across a 350 km sea path was 68.9 dB. With the aid of Eq. (7.20) we can now determine that its effective scattering cross-section is 12,500 m^2.

7.4   OTHER SCATTER PROPAGATION SYSTEMS

Although the tropospheric scatter radio wave propagation system is the most common form, other scatter propagation systems have been explored and developed for certain specialized uses. Two natural forms make use of the scattering mechanisms provided by the ionosphere and meteor showers (Fig. 7.9), and these will be discussed briefly in the following paragraphs. Other forms, such as the field-aligned scatter system and the forward-scatter chaff system, are described elsewhere (Refs. 7.3 and 7.4, for example).

7.4.1 Ionoscatter Radio Systems

The ionoscatter radio system makes use of the ionosphere to *scatter* radio waves. This is quite different from the reflection and refraction mechanisms used for sky wave propagation at frequencies mainly in the MF and HF bands, and most of the ionoscatter radio links operate in the lower VHF band around 30 to 60 MHz, which is above the MUF of the sky wave propagation.

The idea of ionoscatter was considered as early as 1913 (Ref. 7.5), while in 1932

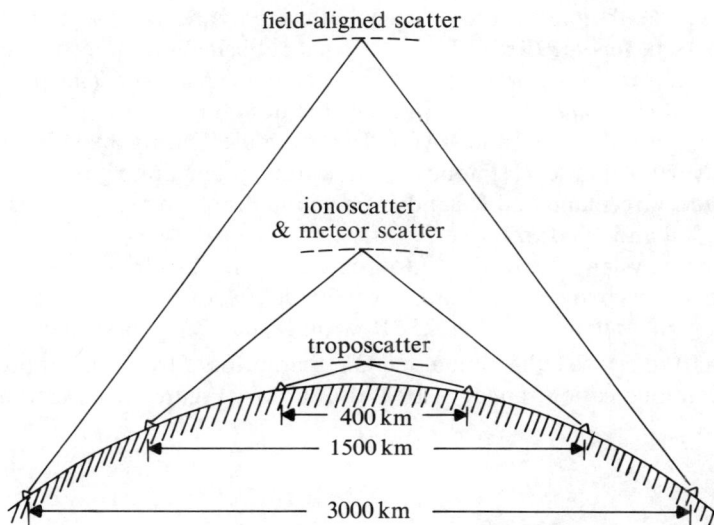

Fig. 7.9 Three of the several forms of radio communication systems which operate via a scattering mechanism.

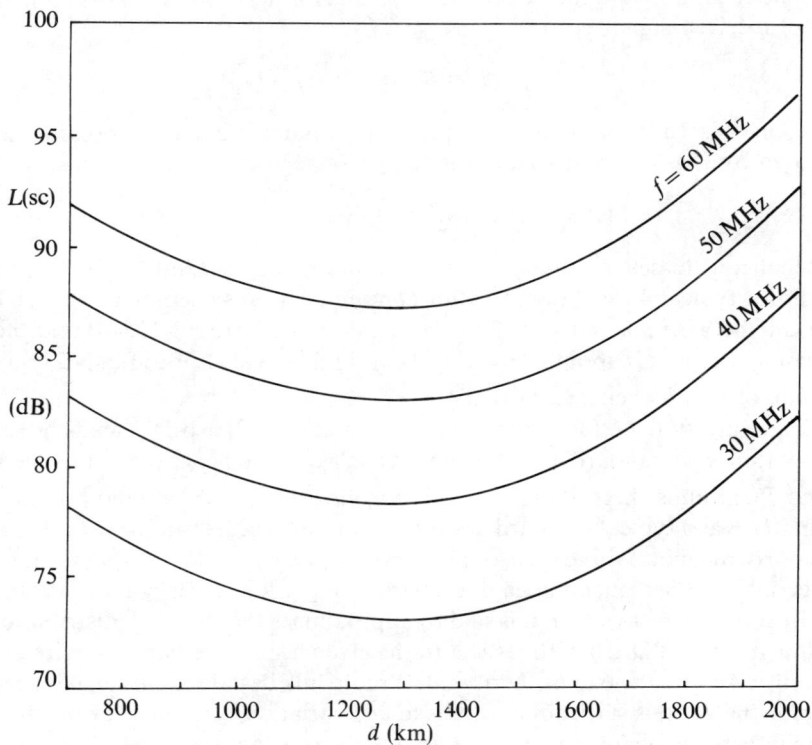

Fig. 7.10 Typical annual median values of ionoscatter loss $L(\text{sc})$.

Eckersley (Ref. 7.6) observed and discussed scattering by the ionosphere at heights of about 70 to 90 km. In 1955 (Ref. 7.7) a paper was published on radio transmission at VHF by scattering and other processes in the lower ionosphere, and *The Proceedings of the Institute of Radio Engineers* contain special issues on radio wave propagation by scattering in volumes 43 (1955) and 48 (1960), for example. During the 1950s the US Air Force had a North Atlantic VHF ionospheric scatter system linking the USA with the UK via Canada, Greenland and Iceland, each link in the chain being between 800 and 1200 miles (1300 and 1900 km).

Several such systems have come into use, each with 12 to 16 telegraphy channels and good reliability (errors better than 1 in 1000 for 99% of the hours in a year). They require high-gain antennas (20 to 25 dB with corner reflectors) and high-power transmitters (10 to 50 kW), the annual median scatter loss $\overline{L(sc)}$ being approximately 90 dB over communication ranges of 1000 to 2000 km. Figure 7.10 illustrates typical values.

7.4.2 Propagation Equations

The radio wave propagation equations are essentially modifications of the standard free-space equation and the noise equation. The former starts with the annual median level of received signal power

$$\overline{P(r)} = P(t) + G(t) + G(r) - \overline{L(p)} \tag{7.21}$$

where, in this case, path loss comprises of annual median ionoscatter loss $\overline{L(sc)}$ indicated by Fig. 7.10 and spatial loss $L(s)$ given by

$$L(s) = 20 \log f(\text{MHz}) + 20 \log d(\text{km}) + 32.45 \tag{7.22}$$

Miscellaneous losses and margins can be represented by $L(m)$.

For a typical 40 MHz ionoscatter communication system with $P_t = 50$ kW, $d = 1000$ km and antenna gains of 22 dB, the spatial loss is about 125 dB and the annual median scatter loss is about 80 dB. Substituting these values into Eq. (7.21) gives us an estimate of $\overline{P(r)}$ as being -114 dBW.

The variation of the hourly median values of $P(r)$ dBW tends to a normal distribution with standard deviation about 6 to 8 dB, and hence the hourly medians will exceed $\overline{P(r)}$ minus three standard deviations for 99.9% of the 8760 hours in a year. Taking 7 dB as a typical standard deviation, our numerical example would suggest that the hourly medians will exceed -135 dBW for 99.9% of the year, ignoring sporadic effects due to other phenomena and overlooking a few other complications.

In addition, fast variation is said to approximate the Rayleigh distribution, where a fading margin of 28 dB with respect to the given hourly median is required for 99.9% reliability. As with the tropospheric scatter radio link described earlier, it is necessary to combine the various statistics of slow and fast variation with those of any appropriate diversity arrangements, but such an analysis is outside the scope of this text.

Example 7.8 Simplified model of ionoscatter system

Assuming a simplified model of an ionoscatter radio link in which the hourly median $P(r)$ does not have any slow variation, while the fast variation follows the Rayleigh pattern, estimate the transmitter power needed for a 2000 bits per second link at 40 MHz over 1000 km with 22 dB antennas if the E_b/N_0 ratio needs to be 16.7 dB for a bit error rate of 10^{-4} and 99.9% reliability is required.

If we rewrite Friis' free-space equation as

$$P_r = \frac{P_t G_t G_r}{L_s L_{sc}} = A_p P_t$$

then the carrier-to-noise spectral density ratio becomes

$$CN_0R = \frac{P_r}{N_0} = \frac{A_p P_t}{N_0}$$

If the transmission consists of binary digits at the rate of R_b per second, the energy per bit is $E_b = P_r/R_b$ joules, and the data transmission rate is

$$R_b = \frac{P_r}{E_b} = \frac{P_r/N_0}{E_b/N_0} = \frac{CN_0R}{E_b/N_0}$$

In logarithmic form this reduces to

$$CN_0R(r) = 10\log(E_b/N_0) + 10\log R_b \qquad (7.23)$$

where, strictly speaking, we are referring to the numerical magnitude of R_b.

For the problem in hand we can determine $L(s) = 125$ dB and note that $L(sc) = 80$ dB from Fig. 7.10. Substitution in the free-space equation produces $P(r) = P(t) - 161$ or the 99.9% minimum $P(r) = P(t) - 189$ dBW. The noise spectral density can be obtained from Fig. 7.11 as $N_0(r) = -188.8$ dBW and hence $CN_0R(r) = P(t) - 0.2$ dB. Substitution into Eq. (7.23) with $E_b/N_0 = 16.7$ dB and $R_b = 2000$ gives $P(t) = 49.91$ dBW.

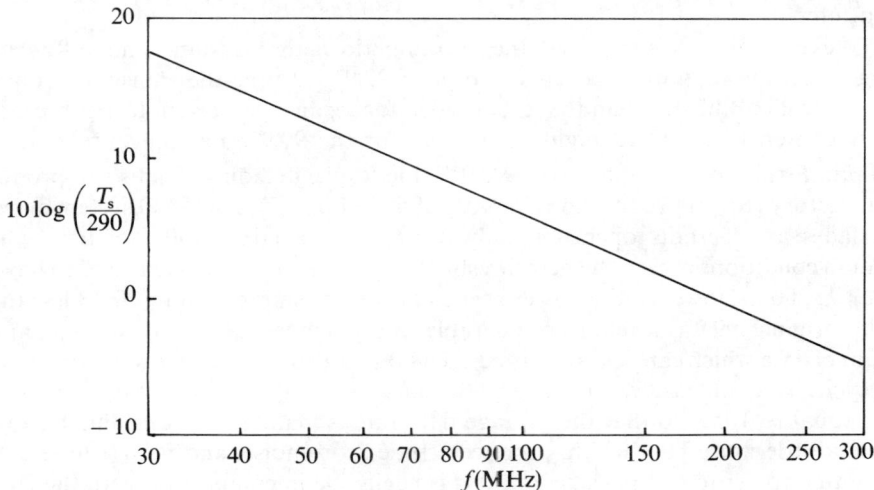

Fig. 7.11 Typical median estimates of cosmic noise in the VHF band with $10\log(T_s\,290)$ $= 52 - 23\log f(\text{MHz})$ corresponding to $N_0(r) = -152 - 23\log f(\text{MHz})$ dBW.

Clearly a transmitter power requirement of 50 dBW, or about 100 kW, is not very practical and some form of diversity or other way of improving the performance is required. Double or quadruple diversity with maximal-ratio combining (Fig. 6.15) will result in 16 dB or 26 dB improvement, respectively, at the 99.9% level. With such assistance the minimum value of transmitter power reduces to 34 dBW (2.5 kW) or 24 dBW (250 W), though again we have ignored the slow variation, etc. Nevertheless, the simplified model indicates the need for diversity.

7.4.3 Automatic Repeat Request

An alternative approach, which can be used with telegraphy, involves what is known as an *automatic repeat request* (ARQ) system. A simplified version is illustrated in Fig. 7.12. The two transmitters operate continuously but the telegraphic message is stored prior to transmission and is transmitted only during those periods when the propagation conditions are adequate for threshold reception. The system waits until the receiver detects the carrier wave from the distant transmitter and indicates that the received signal level exceeds the selected threshold level. The store gate is opened and the telegraphic message is modulated onto the continuous carrier. This arrangement continues until the received signal level next falls below the threshold, when the gate is closed and the system awaits the following interval which is suitable for transmission. The received information is also stored until it is appropriate to output it to a printer.

To explain how this improves the overall performance, assume simple Rayleigh fading. For transmission to be satisfactory for 99.9% of time the transmitter power needs to be 28 dB higher than the level needed for median reception. If the threshold level is chosen to be 20 dB higher than that for the 99.9% reliability, so that the minimum $P(r)$ is now equal to $\overline{P(r)} - 8$ dBW, the Rayleigh fading signals will produce unsatisfactory reception conditions for 10% of time (Fig. 6.7). The ARQ system detects these fades and permits operation only during the remaining 90% of time when reception conditions exceed the new threshold requirements: the ARQ is said to have a *duty cycle* of 0.9. However, the power needed at the transmitter is now 20 dB less than for the original 99.9% level, a considerable improvement, although the maximum amount of data which can be transmitted each hour is also reduced in proportion to the duty cycle. To compensate, if necessary, the data rate can be increased by a speed-up factor equal to 1/0.9, so that the average data rate is maintained constant. This will require a little more bandwidth, produce a little more noise, and need a little extra transmitter power to compensate, but this is negligible in comparison with the 20 dB gain.

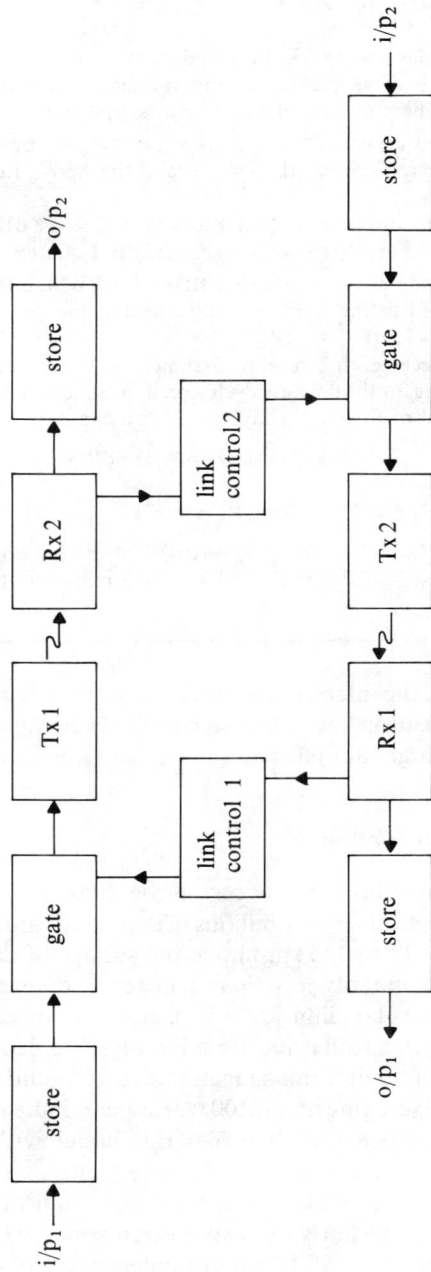

Fig. 7.12 A simplified diagram of an ionoscatter communication link incorporating automatic repeat request (ARQ).

Example 7.9 Ionoscatter system with ARQ

An ionoscatter radio link incorporating ARQ with a duty cycle of 0.7 is required to have an average data transmission rate of 500 bits per second. It operates at 50 MHz over 1200 km with two 20 dB antennas. The statistics of the overall signal pattern indicate that the instantaneous values of $P(r)$ exceed $\overline{P(r)} - 15$ dBW for 99% of time. Determine the minimum transmitter power required at the 99% level if the E_b/N_0 ratio is taken to be 16 dB.

We can use Eq. (7.22) to obtain the spatial loss as $L(s) = 128$ dB and Fig. 7.10 to estimate the annual median ionoscatter loss as $L(sc) = 82$ dB. Equation (7.21) gives us the annual median level of received power as $\overline{P(r)} = P(t) - 170$ dBW and the statistics of the overall signal pattern, quoted in the question, indicate that the 99% minimum power received is $P(th) = \overline{P(r)} - 15 = P(t) - 185$ dBW.

To match this with the receiver threshold we first increase the instantaneous data rate to $R_b = 500/0.7$ to compensate for the 0.7 duty cycle and to produce an *average* data rate of 500 bits/s. Then with the aid of Fig. 7.11 and Eq. 7.23, we can write

$$N_0(r) = -152 - 23 \log f(\text{MHz}) = -191 \text{ dBW}$$

and

$$CN_0R(r) = (E_b/N_0)\text{dB} + 10 \log R_b = 45 \text{ dB}$$

to obtain the threshold $P(th) = 45 - 191 = -146$ dBW. A simple comparison of $P(t) - 185 = -146$ gives us $P(t) = 39$ dBW or $P_t = 8$ kW as the minimum transmitter power for the given specification.

More recently, however, the interest in ionoscatter systems has decreased, mainly because of satellite communication links but also because radio links involving meteor burst scattering operate with greater efficiency.

7.4.4 Meteor Burst Communication

Another form of scattering of radio waves occurs at the ionization layers in the upper atmosphere caused by meteor trails. Many billions of these trails are produced each day worldwide at approximately 80 to 115 km above the surface of the earth, each trail being about 20 km long and initially less than a meter in diameter. Then the trail expands due to normal molecular diffusion and usually disappears in a matter of seconds. The reception via such a trail is illustrated in Fig. 7.13. However, there are so many meteor trails that the radio link can be maintained at a sufficiently reliable level to provide a telegraphic link averaging about 100 words per minute throughout the day at an operating frequency mainly about 40 to 50 MHz, though with possibilities from 30 to 100 MHz.

The scattering height involved with meteor burst communication systems means that the propagation range can be fairly extensive, up to about 2000 km or so, with a meteor-scatter loss of the order 55 to 65 dB, almost independent of range as illustrated in Fig. 7.14. This is roughly 20 dB less than the corresponding value for ionoscatter loss and hence lower transmitter powers are possible.

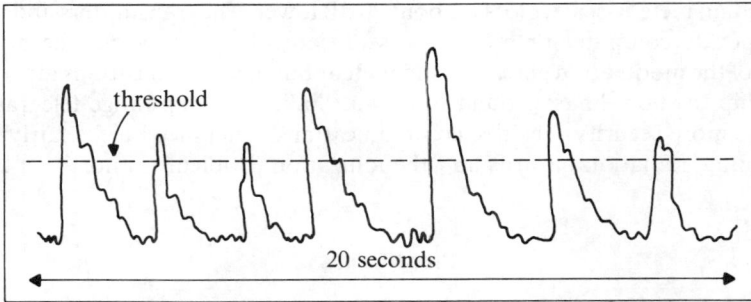

Fig. 7.13 A sketch of the pattern of signal received via a meteor burst communication system. The telegraph signal is transmitted during the brief periods when the signal exceeds a specified threshold, typical duty cycles being between 1 and 10%.

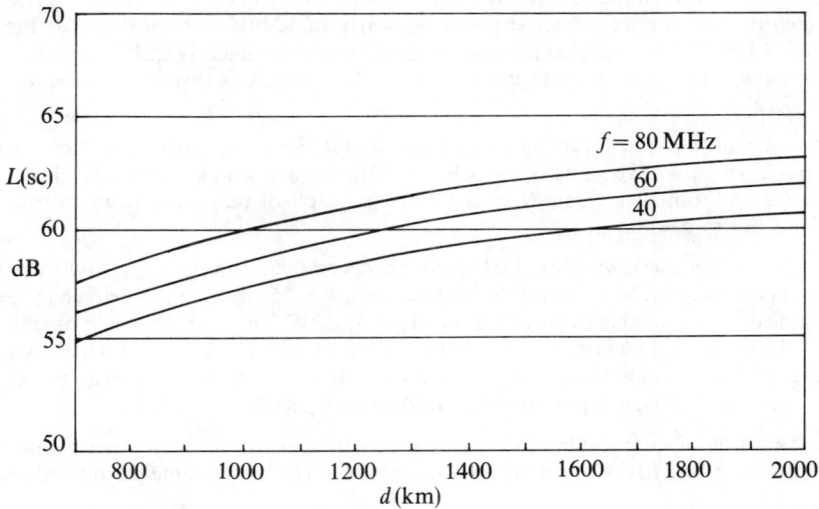

Fig. 7.14 Typical annual median values of the scatter loss associated with meteor burst communication systems.

The meteor burst communication system also uses diversity and automatic repeat request systems but the character of the meteor trails is such that duty cycles are commonly of the order of only 1 to 10%. The instantaneous transmission rates must therefore be moderately high in order to achieve reasonable average transmission rates. A system which operates at 2000 baud with a 5% duty cycle averages only 100 baud over a period of time, yet with sufficiently high transmitter power it is possible to achieve instantaneous data rates approaching 250 kbits per second, equivalent to about 16 kbits per second when averaged over one hour at typical duty cycles.

The analysis is similar to that described for ionoscatter radio links except that the annual median meteor-scatter loss is about 20 dB lower. The transmitters and receivers can incorporate computer or microprocessor controls to carry out the high-speed switching of the modulation gate, etc., and meteor burst communications are useful not only for ship-to-shore links, ground-to-aircraft links and long-range telegraphy, but also for the more security-sensitive communications which need to be fairly immune from jamming, radiation hazards and the ionization problems of northern latitudes.

PROBLEMS

7.1 A certain tropospheric scatter radio link is across a sea path and hence has no obstacles with heights h'_t and h'_r. Derive an expression for the scatter angle θ in terms of distance d, antenna heights h_t and h_r, and the refractivity coefficient K. This is Eq. (7.9).

7.2 Yeh (Ref. 7.1) observed that the annual median scatter loss in a tropospheric scatter radio link was 57 dB when the scatter angle was $\theta = 1°$, the frequency was $f = 400$ MHz, and the annual median surface refractivity was $N_s = 310$. In addition, the scatter loss thereafter increased by 10 dB per degree increase in scatter angle and linearly with frequency, while it was reduced by 0.2 dB per unit increase in N_s above 310. Use this information to explain Eq. (7.12).

7.3 The transmitting and receiving antennas of a 900 MHz tropospheric scatter radio link across a sea path have gains $G = 45$ dB and both are at 100 m above sea-level. Assuming $K = 4/3$ and annual median $N_s = 310$, draw a graph of path loss against distance for $d = 100$ km to $d = 400$ km.

7.4 A 2 GHz tropospheric scatter radio link connects a mainland base to a large oil-rig platform 250 km out at sea. The base station operates with $h_t = 250$ m, antenna diameter $D_t = 18$ m and maximum available radiated power $P(t) = 30$ dBW. The oil-rig receiver operates with $h_r = 200$ m, antenna diameter $D_r = 18$ m and threshold $P(\text{th}) = -130$ dBW. Assuming $L(\text{m}) = 5$ dB, annual median $N_s = 340$ and four-fold diversity with maximal-ratio combining, estimate the overall performance you would expect from the link.

7.5 The variation of surface refractivity N_s over a particular tropospheric scatter radio link is illustrated in Fig. 7.15. Determine the variation in $P(\text{r})$ over the same period and compare

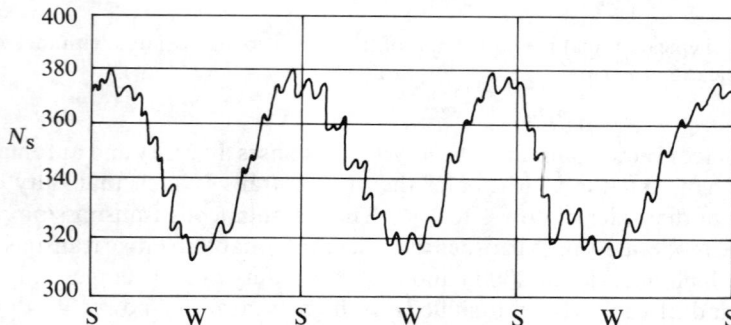

Fig. 7.15 Sketch of the hypothetical variation of N_s over a three-year period (S = summer, W = winter) for Prob. 7.5.

its value with the 99.9% reliability level. Assume $P(t) = 30\,dBW$, $G(t) = G(r) = 45\,dB$, $f = 900\,MHz$, $d = 300\,km$, $\theta = 37.4\,mrad$ and $L(m) = 10\,dB$, while the quadruple diversity arrangement with $\theta = 37.4\,mrad$ requires that the minimum $P(r)$ is 10 dB below the annual median. Explain why the transmitter power is excessive for most of the year. If the receiver can somehow inform the transmitter of excess signal levels, what approximate order of power reduction is possible in mid-summer?

7.6 A short-range tropospheric scatter system links two sides of a mountainous island. The transmitting station has $h_t = 180\,m$ and $h'_t = 500\,m$, $f = 4.7\,GHz$, $P(t) = 30\,dBW$, $G(t) = 40\,dB$ and $d_1 = 20\,km$. The receiving station at $d = 90\,km$ has $h_r = 60\,m$, $h'_r = 500\,m$, $P(th) = -125\,dBW$, $G(r) = 40\,dB$ and $d_2 = 40\,km$. Miscellaneous losses are 5 dB and the annual median $N_s = 340$. Determine the annual median $P(r)$ and estimate the reliability with the four-fold arrangement illustrated in Fig. 7.7.

7.7 A low-capacity tropospheric scatter communication system linking island to island operates at 2 GHz with 50 W transmitter and 2 m diameter antennas. Assuming miscellaneous losses are 5 dB, annual median $N_s = 340$, $P(th) = -130\,dBW$ and the diversity arrangement requires a minimum $P(r)$ which is 5 dB below the annual median at the 99% reliability level, determine the range of the link if it happens to be exactly twice the standard atmosphere line-of-sight distance.

7.8 Draw a graph of the effective scattering cross-section of the common volume of the tropospheric scatter radio link described in Prob. 7.3 for distances up to $d = 400\,km$.

7.9 It is required to establish an ionoscatter radio link over a range $d = 1000\,km$ with scattering from the ionosphere at a height of 80 km. Estimate the launch angle and the scatter angle.

7.10 The ionoscatter radio link described in Prob. 7.9 operates at 45 MHz with 22 dB antennas, 50 kW transmitters and 10 dB miscellaneous losses. The diversity arrangement is such that the minimum level of $P(r)$ required for 99% reliability is 5 dB below the annual median $\overline{P(r)}$. Determine the carrier-to-noise spectral density ratio at the receiver and the maximum data transmission rate R_b if the system requires $10 \log E_b/N_0 = 16.5\,dB$. E_b is the energy per bit in joules.

REFERENCES

7.1 Yeh, L. P., 'Simple methods for designing troposcatter circuits,' *Trans. IEEE in Communication Systems*, CS-8 (September 1960), 193–8.

7.2 CCIR, 'Propagation data required for trans-horizon radio-relay systems,' *Report* 238-2 (1974), XIIIth Plenary Assembly, vol. V, Geneva 1974, pp. 209–229.

7.3 Utlaut, W. F., et al., 'Ionospheric modifications,' *Radio Science*, 9, (November 1974); and *Proc. IEEE*, 63 (July 1975), 1022–43.

7.4 Katz, L. 'Forward scatter chaff system for air-ground long-haul communications,' *National Telecommunications Conference Record*, 72, pp. 10C-1 to 10C-7.

7.5 Kennelly, A. E., 'The daylight effect in radio telegraphy,' *Proc. IRE*, 1 (July 1913), 39–62.

7.6 Eckersley, T. L., 'Studies in radio transmission,' *Journ. IEE*, 71 (1932), 405–54.

7.7 Bailey, D. K., et al., 'Radio transmission at VHF by scattering and other processes in the lower ionosphere,' *Proc. IRE*, 43 (October 1955), 1181–230.

8

Microwave Radio Relay Systems

Microwave communication systems may be classified into several different types: the microwave link which operates on a line-of-sight (LOS) basis and covers large distances with the aid of a number of repeater stations; the tropospheric scatter radio link which relies on the scattering mechanism of the troposphere to achieve a beyond-the-horizon communication system with one single 'hop' of several hundreds of kilometers; and satellite communication links which operate primarily via geostationary satellites orbiting high above the equator in synchronism with the rotation of the earth. There are separate chapters for tropospheric scatter systems (Chap. 7) and for satellite communication systems (Chap. 9), and here we shall give a brief account of the line-of-sight microwave link only, restricting our attention mainly to the factors which affect radio wave propagation.

8.1 ROUTE PLANNING

An early item in the establishment of a line-of-sight microwave communication system is called *route planning*. In a country with full survey maps available, a preliminary route and several possible alternative routes can first be established with the aid of adequately scaled, contoured and detailed maps. Some of the principal criteria to note at this stage are the Fresnel-zone clearances for repeater towers with realistic heights, both physically and economically; the accessibility and size of the site with the availability of a power supply; a lack of disturbance from other microwave links, radars and broadcasting systems, etc.; or such features as airfields, large obstacles or growing vegetation. It is also necessary to anticipate *overshoot* under very good propagation conditions which might interfere with later sections of the link; for this reason a zig-zag path is sometimes chosen.

Once the path has been selected and consideration given to the civil engineering, economic and legal problems associated with the transmitter, repeater and receiver sites, a detailed study is made of the propagation path profile, frequency planning (to avoid interference) and propagation characteristics (reflecting surfaces, meteorological conditions, vegetation growth). Sometimes a radio path propagation test is carried out using mobile or transportable towers, either along each repeater section or perhaps only along the most problematic sections of the route such as an estuary crossing.

Although costly, such tests may reveal radio-frequency interferences or unexpected reflection points which are not immediately obvious from a map or from visual route inspection. The tests can give practical details of the long-term and short-term statistical patterns of path loss, including the effects of multipath fading, rain attenuation or possibly diffraction losses in mountainous terrain. In addition, the tests will indicate optimum heights for the transmitting and receiving antennas along each hop for adequate clearance of obstacles, and on particularly difficult routes some frequency or spatial separation will assist in providing data for diversity possibilities.

Within certain very densely populated areas of the world, the large usage of microwave links may result in frequency congestion and the installation of new links may need very careful preliminary study and investigation to predict the level of interference between systems which may be carrying quite different types of signals – telephony, data or television, for example – and even between sections within the overall length of its own circuit.

One factor in the interference levels between microwave circuits is the transmitter power at each repeater link. A good carrier-to-noise ratio needs adequate transmitter power, but this also produces a high interference signal elsewhere. The design of the circuit must therefore take into account a recommended maximum effective radiated power level and, in the case of the 7.1 to 7.9 GHz private-user frequency band for example, the power may be restricted to provide 10 kW erp. With high antenna gains the actual transmitter powers are of the order of watts. The antenna itself may also be required to have a high front-to-back ratio, which implies a well-designed antenna.

Path clearance causes some problems. It is common practice to express the amount of clearance over obstructions or other reflecting points in terms of *Fresnel zones*, as illustrated in Fig. 8.1. A Fresnel zone is bounded by the locus of points at which the sum of the distances to transmitter and receiver is equal to the direct path length plus an integral number of half-wavelengths. The mathematics is given in Ex. 8.1, which shows that in the 7 GHz private-users' band a radio relay link of about 60 km will require a mid-point clearance of about 25 m to clear the first Fresnel zone

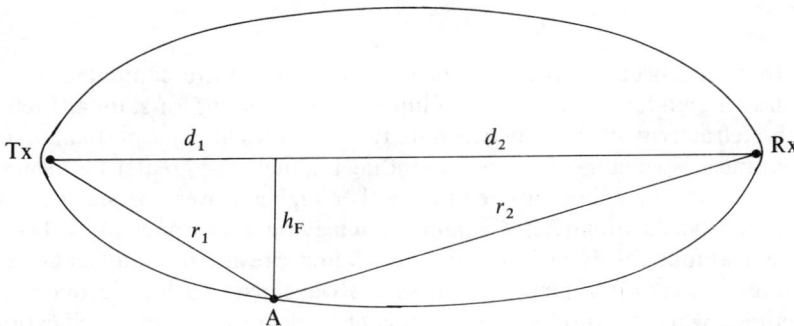

Fig. 8.1 A Fresnel zone is defined as the three-dimensional locus of point A such that the difference in path lengths between the reflected wave $(r_1 + r_2)$ and the direct wave $(d_1 + d_2)$ is an integer number of half-wavelengths.

Example 8.1 Fresnel zone clearance

A microwave radio relay link operating at 7 GHz covers a distance of 60 km. Determine the mid-path clearance needed to coincide with the first Fresnel zone.

The definition of Fresnel zones is that the path difference, shown in Fig. 8.1, between the reflected and direct paths between transmitting antenna and receiving antenna is equal to an integer number of half-wavelengths. Thus we can write

$$r_1 + r_2 - d_1 - d_2 = n\lambda/2$$

Using Pythagoras' theorem and the binomial expansion we can replace

$$r_1 = d_1 \left[1 + \frac{1}{2}\left\{\frac{h_F}{d_1}\right\}^2 \right]$$

and

$$r_2 = d_2 \left[1 + \frac{1}{2}\left\{\frac{h_F}{d_2}\right\}^2 \right]$$

to give

$$\frac{n\lambda}{2} = \frac{h_F^2}{2}\left[\frac{1}{d_1} + \frac{1}{d_2} \right]$$

Transformation of this expression produces (in SI units):

$$h_{Fn} = \left[\frac{n\lambda(d_1 d_2)}{d_1 + d_2} \right]^{1/2} \tag{8.1}$$

while the first Fresnel zone radius, with distances in kilometers and wavelength in meters, is

$$h_F = 31.61 \left[\frac{\lambda d_1 d_2}{d_1 + d_2} \right]^{1/2} \tag{8.2}$$

In our particular example, $d_1 = d_2 = 30$ km and $\lambda = 0.043$ m, and simple substitution gives $h_F = 25.4$ m.

radius. In many instances the clearances so determined are compared with a path profile drawn on a four-thirds earth radius scale, to account for standard refractivity. When the refractivity changes between its two extremes along a particular route, the path clearance also changes in a corresponding manner. The greater the clearance, the more reliable the link, but this requires either higher towers or shorter relay link lengths. One specification (Ref. 8.1) aims at achieving a clearance of 0.6 first Fresnel zone radius at four-thirds earth radius and 0.3 first Fresnel zone radius at two-thirds earth radius, though the latter condition is not always easy to achieve economically and is sometimes waived. Another description of a clearance level specification, on a Canadian microwave link (Ref. 8.2), is to aim for 0.7 first Fresnel zone clearances on a path profile drawn on true radius earth curvature. Pearson (Ref. 8.2) states that antenna heights must be such that the clearance is at least half the first Fresnel zone

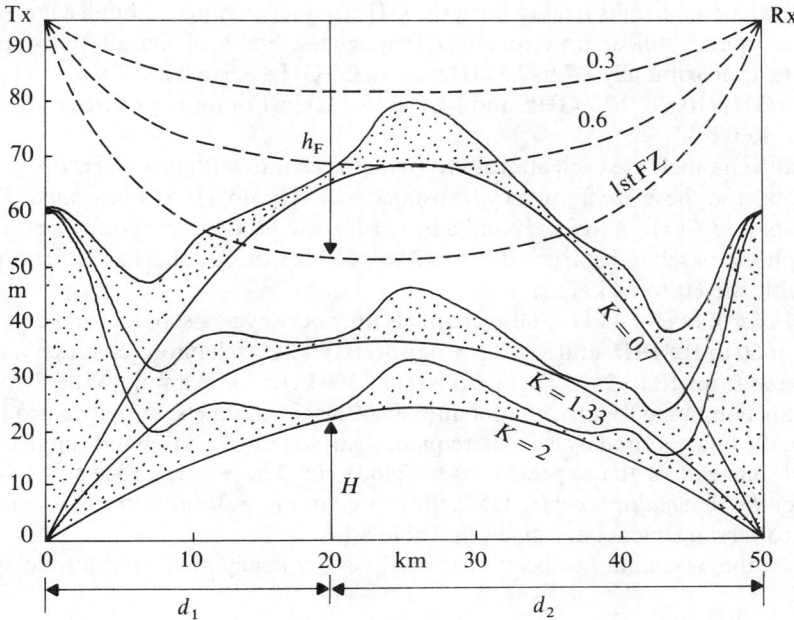

Fig. 8.2 The effect of the refractivity coefficient K upon the path clearance between two antennas. For standard refraction conditions, $K = 1.33$, the clearance exceeds the first Fresnel zone. For sub-standard refraction conditions, $K = 0.7$, the clearance is at the 0.3 first Fresnel zone radius level.

radius when $K = 4/3$ but several authorities use 0.7 first Fresnel zone clearance to provide a clearance which is exceeded for all but a fraction of 1% of time. Figure 8.2 illustrates the difficulty which might be encountered with obtaining clearance at the 0.3 first Fresnel zone level when $K = 0.7$ without the aid of high towers.

8.2 THE FDM/FM RADIO RELAY SYSTEM

The microwave radio relay system was originally a means of transmitting a large number of telephone channels along a major communication trunk route, but gradually it incorporated the transmission of television channels and then high data-rate digital systems. Although microwave links had been established earlier, the first line-of-sight radio relay systems involving many tandem repeaters to cover longer route distances, and operating above 1 GHz, were established in the United States about 1947 by the Bell System Company. In the United Kingdom, similar systems were established by the Post Office.

The large bandwidth needed for hundreds of audio channels or for television channels or high-rate digital systems means that the carrier frequencies of high-capacity microwave links tend to be in the GHz frequency range, though a few smaller-capacity systems still operate at lower frequencies. Some of the allotted frequency bands are approximately 1.7 to 2.3 GHz, 2.5 to 2.7 GHz, 3.5 to 4.2 GHz, 5.9 to 6.4 GHz, 7.3 to 8.4 GHz, 10.5 to 12.7 GHz, and 14.4 to 15.2 GHz. For more precise details see, for instance, Ref. 8.4.

Audio channels are each allocated a frequency bandwidth of 3.1 kHz (from 300 Hz to 3.4 kHz) and these are frequency translated in 4 kHz slots into a *baseband*. The first three slots, 0 to 4 kHz, 4 to 8 kHz and 8 to 12 kHz, are used for line control, etc., and so the telephony baseband starts with the 12 to 16 kHz slot. An alternative arrangement starts with the 60 to 64 kHz slot.

A *group* of twelve 4 kHz audio channels may occupy a baseband from either 12 to 60 kHz or 60 to 108 kHz, and a *super-group* of sixty 4 kHz telephone channels may have a baseband from 12 to 252 kHz or from 60 to 300 kHz, for example. When ten super-groups are combined to form a baseband of 600 audio channels, the super-groups may individually be separated by a small frequency gap so that the baseband runs from 60 to 2540 kHz instead of the expected 60 to 2460 kHz. The whole process is known as *frequency division multiplexing* (FDM), illustrated in Fig. 8.3, and some baseband limits for typical arrangements are given in Table 8.1.

Once the baseband has been processed via frequency division multiplexing, it is

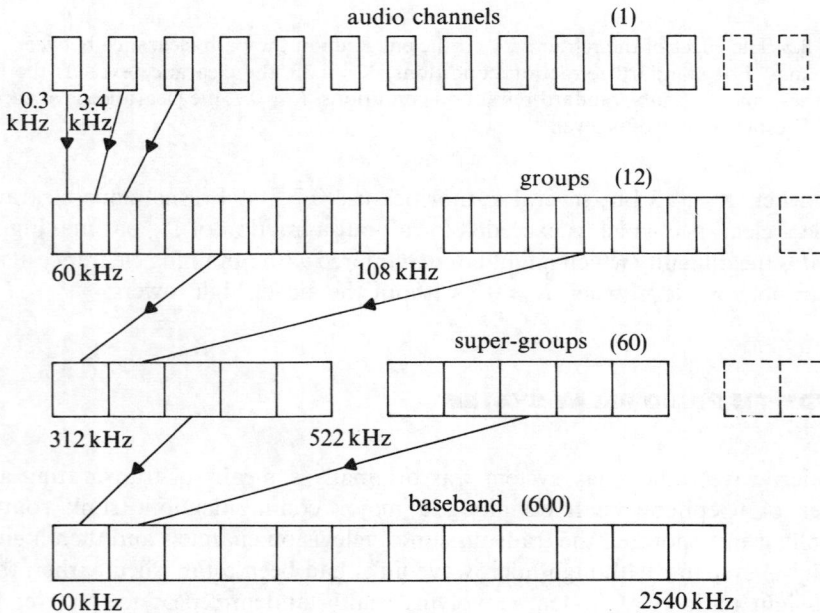

Fig. 8.3 Frequency division multiplexing may be used to translate 12 audio channels (3.1 kHz) into a group (60 to 108 kHz). Five such groups may be combined into a super-group (312 to 522 kHz), and ten such super-groups may be translated into a baseband (60 to 2540 kHz) containing 600 telephone channels suitable for frequency modulating a suitable carrier.

Table 8.1   Some recommended values of f_d and baseband frequencies

No. of channels	$f_d(kHz)$	Baseband limits (kHz)
12	35	12 to 60; 60 to 108
24	35	12 to 108
60	50, 100, 200	12 to 252; 60 to 300
120	50, 100, 200	12 to 552; 60 to 552
300	200	60 to 1300; 64 to 1296
600	200	60 to 2540; 64 to 2660
900	200	316 to 4188
1800	140	316 to 8204

frequency modulated (FM) onto the appropriate carrier frequency and transmitted along the first relay link on the line-of-sight microwave system to the first of a series of *repeater* stations. These are essentially amplifiers and frequency changers, with amplification usually at RF or IF, but some are designated *modulation sections* and are used to demodulate the signal to baseband and then remodulate to carrier after some interchanging of the channels, groups or super-groups in the baseband to reduce the overall intermodulation noise level. There are several or many such relay links in the microwave system, each with a line-of-sight range of roughly 20 to 70 km in length,

---

Example 8.2 Free-space calculations

Determine the minimum transmitter power needed on a 50 km microwave radio relay hop which incorporates 3 m diameter paraboloidal reflector antennas at either end to provide a receiver power level of 1 μW. Assume free-space conditions and use $f = 2, 4, 6$ and 11 GHz for purposes of comparison. $L(\text{misc})$ can be neglected.

This is the basic step in calculations involving microwave links, and the three appropriate equations are:

$$P(r) = P(t) + G(t) + G(r) - L(s) - L(\text{misc}) \tag{8.3}$$

$$G(\text{dB}) = 20 \log D + 20 \log f - 42.3 \tag{8.4}$$

$$L(s) = 20 \log d + 20 \log f + 32.45 \tag{8.5}$$

where frequency f is in MHz, distance d is in km and antenna diameter D is in meters. $P(r)$ and $P(t)$ must be in consistent units such as dBW or dBm, the latter being preferred by some users as the reference signal power per channel is 1 mW. To illustrate the effect of each variable, we can combine the equations to give:

$$P(t) = P(r) - 40 \log D - 20 \log f + 20 \log d + 117.05$$

or

$$P(t) = 71.95 - 20 \log f$$

for this particular example. Direct substitution of the quoted frequencies gives us $P(t) =$ 3.9 W, 0.98 W, 0.44 W and 0.13 W at $f = 2, 4, 6$ and 11 GHz.

depending on the individual path profiles and other factors, and the total length may reach several thousands of kilometers. For each individual hop along the microwave system the essential radio wave propagation calculation is that of free-space propagation, later to be modified to account for fading and miscellaneous losses. Example 8.2 illustrates the method of calculation.

8.3 CIRCUIT SPECIFICATIONS

The International Telecommunications Union (ITU) prepares recommendations for international standards to be used in line-of-sight microwave radio relay systems (as well as in other fields of telecommunications). It does this through two committees, the CCIR (International Radio Consultative Committee) and the CCITT (International Telegraph and Telephone Consultative Committee). The first is responsible for the microwave radio link specifications and the second is concerned with general line telecommunications networks. Hence, specifications concerning both the radio link and the line network must obviously be compatible. Details of both committees can be found in Ref. 8.5.

The CCIR recommendations (Ref. 8.6) for line-of-sight microwave FDM/FM radio relay circuits are incorporated in a fictitious or *hypothetical reference circuit* (HRC) of length 2500 km with 54 hops, each spaced about 46 km (29 miles). Each one-ninth of the circuit (6 hops) is known as a *modulation section*.

8.3.1 Specification of Hourly Mean

In order to analyze the radio wave propagation characteristics of each hop of the circuit it is necessary to have some indication of the noise power level present. This is allocated on a channel basis in the case of a telephony system. The telephone message is assumed to be equivalent to a standard test-tone of 1 mW power at the zero relative level and an overall signal-to-noise ratio of 50 dB will be achieved if the total noise power contribution is 10,000 pW per channel at the same zero relative level.

In CCIR terms, the recommendation is that the total noise power over the entire circuit must not exceed 10,000 pW for more than 1% of the busiest hour. The noise contribution of the frequency translation in the multiplexing equipment is allocated 2500 pW and is assumed to remain constant, and the remaining 7500 pW which is allocated to the radio system may be divided (often equally) between intermodulation noise (3750 pW) and thermal noise (3750 pW), though other proportions are used when appropriate. On this basis the thermal noise allocated to each of the nine modulation sections is about 400 pW.

This CCIR requirement also gives us a reference of 3750 pW of *thermal* noise power per channel, psophometrically weighted, and this means that a signal to *thermal* noise ratio of 1 mW/3750 pW (or 54.3 dB) will produce an overall signal-to-noise ratio of 1 mW/10,000 pW (or 50 dB). The term *psophometrically weighted* means that the voltage is measured with an instrument known as a *psophometer*, a kind of voltmeter with a frequency-dependent network representing the relative interference effects of noise on speech. The voltage readings are then converted into psophometric power. It can be shown that the psophometric weighting factor $P(s)$ is 2.5 dB and this means that the signal-to-thermal noise ratio of 54.3 dB *weighted* is equivalent to a signal-to-thermal noise ratio of $54.3 - 2.5 = 51.8$ dB *unweighted*.

The 3750 pW of thermal noise power is allocated to the worst channel (the one at the highest frequency in the baseband) in the hypothetical reference circuit. Under mean conditions, the equivalent thermal noise power per channel per hop is 3750 pW/54 or 69.4 pW. This is as if we established an equivalent signal-to-thermal noise ratio of 1 mW/69.4 pW or 71.6 dB (weighted) or 69.1 dB (unweighted) along each hop, though in reality such channels are not available at each repeater.

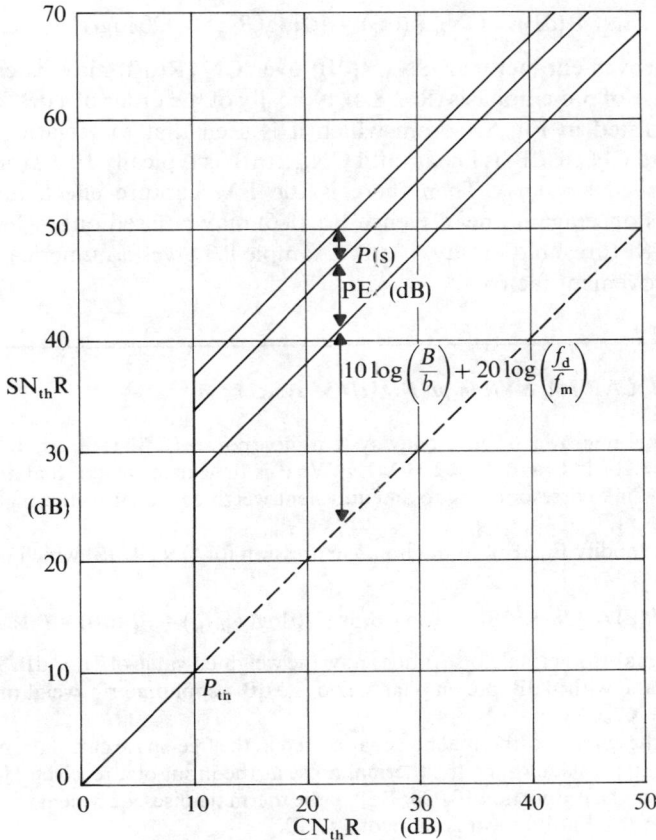

Fig. 8.4 In an FM receiver the relationship between the input carrier-to-thermal noise ratio and output signal-to-noise ratio depends on the receiver threshold level, the FM improvement factor and pre-emphasis. If $SN_{th}R(dB)$ includes psophometric weighting, this is taken into account via the term $P(s)$, normally equal to about 2.5 dB.

The FM receivers used in such circuits produce an output signal-to-thermal noise ratio in the highest telephony channel in the baseband which is related to the carrier-to-thermal noise ratio at the input (e.g. in Ref. 8.4) by the expression

$$SN_{th}R(dB) = CN_{th}R(dB) + 10\log(B/b) + 20\log(f_d/f_m) \qquad (8.6)$$

where $SN_{th}R(dB)$ is the unweighted value, B is the IF bandwidth determined via signal processing techniques, b is the channel bandwidth of 3.1 kHz, f_d is the *rms* deviation of the voice channel for a standard test tone (commonly 100 kHz or 200 kHz by CCIR recommendation; Table 8.1), and f_m is the mid-band frequency of the highest telephone channel in the baseband (commonly equated to the highest frequency in the baseband).

Note that some references to this expression use F_d as the peak deviation, corresponding to $F_d = \sqrt{2}f_d$, and consequently Eq. (8.6) sometimes appears a little differently as

$$SN_{th}R(dB) = CN_{th}R(dB) + 10\log(B/2b) + 20\log(F_d/f_m) \qquad (8.7)$$

The FM improvement factor of $SN_{th}R(dB)$ over $CN_{th}R(dB)$ may be enhanced even more by the use of pre-emphasis (Ref. 8.4), typically of the order of 4 dB. The combined effect is illustrated in Fig. 8.4, from which it is seen that the relationship between $SN_{th}R(dB)$ and $CN_{th}R(dB)$ is linear until $CN_{th}R(dB)$ is typically 10 dB (or a little less in certain improved systems). Then there is the FM capture effect, and finally the relationship is once again linear. Hence Eq. (8.6) may be used only when $CN_{th}R(dB)$ exceeds the FM threshold of, say, 10 dB. Example 8.3 gives a numerical indication of the FM improvement factor.

Example 8.3 CNR and SNR in an FDM/FM system

A 600-channel microwave radio relay system incorporates FM receivers in which $B = 30$ MHz, $f_d = 200$ kHz and $f_m = 2.54$ MHz. What is the carrier-to-thermal noise ratio in a radio relay link corresponding to an equivalent weighted signal-to-thermal noise ratio of 71.58 dB?

We can modify Eq. (8.6) to obtain an expression for $CN_{th}R(dB)$ which incorporates pre-emphasis:

$$CN_{th}R(dB) = SN_{th}R(dB) - 10\log(B/b) - 20\log(f_d/f_m) - PE(dB) - P(s) \qquad (8.8)$$

where the signal-to-thermal noise ratio is now the weighted value of 71.58 dB. Substituting the given data with 4 dB pre-emphasis and 2.5 dB psophometric weighting gives us $CN_{th}R(dB) = 47.3$ dB.

This is the carrier-to-thermal noise ratio used in the free-space calculations because it is an easy matter to determine the thermal noise at the input of a receiver. However, the *total* noise is in the proportion 10,000/3750 to the thermal noise, equivalent to 4.26 dB, and the carrier-to-total noise ratio is therefore 43 dB.

It has been stated that the *worst* audio channel is the highest in the baseband. This is because of the parameter f_m in the previous equation. If, for example, we selected the lowest audio channel in the baseband, with $f_m = 60$ kHz, the minimum $CN_{th}R(dB)$ required is only 14.8 dB.

By the method just described, illustrated by Ex. 8.3, we can estimate the carrier-to-thermal noise ratio at the input to the receiver and if we note that the thermal noise power level at the receiver input is given via $N_{th} = kT_rB = kT_0F_rB$ as

$$N(\text{th}) = -204 + 10\log B + F(\text{r}) \tag{8.9}$$

then the carrier power required over the hop is easily obtained from

$$P(\text{r}) = \text{CN}_{th}\text{R}(\text{dB}) + N(\text{th}) \tag{8.10}$$

If the noise figure is 10 dB, with $B = 30$ MHz given earlier, a $\text{CN}_{th}\text{R}(\text{dB})$ of 47.3 dB corresponds to $P(\text{r}) = 47.3 - 119.2 = -71.9$ dBW. With the data used in Ex. 8.2, the minimum transmitter power, excluding miscellaneous losses and fading, will be of the order of $P(\text{t}) = -12$ dBW or 0.06 W at a frequency of 4 GHz. This may seem an extremely small power level but there has been no allowance for miscellaneous losses (10 dB or more) or fading (perhaps 30 or 40 dB).

8.3.2 Fading

As shown in the previous paragraph, the amount of transmitter power needed in a microwave radio relay link operating under ideal free-space conditions is very small. A problem arises when the received signal or the path attenuation are subjected to *fading*, sometimes up to 30 or 40 dB within a single hop, as in Fig. 8.5, or compositely perhaps 10 to 20 dB more when several hops fade simultaneously in tandem. This fading of the signal can arise in many ways, the more common forms being described briefly in the following paragraphs.

One of the more obvious causes of fading, even in a well-designed path, is

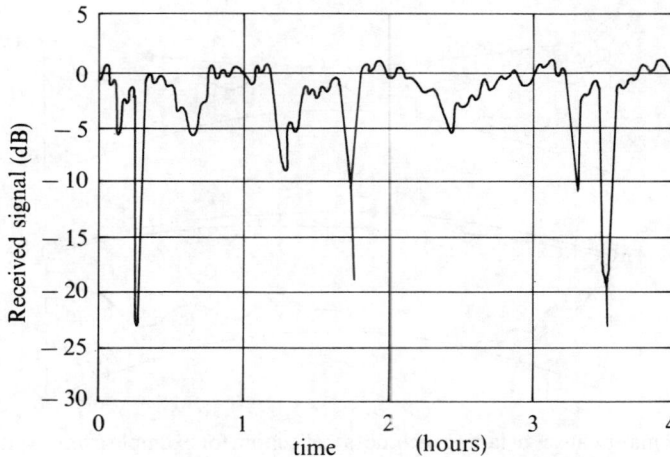

Fig. 8.5 During periods of fading the signal strength at the receiver input may drop for brief intervals by up to 30 or 40 dB.

reflection, and the resulting summation of the direct and reflected signals produces a net signal at the receiver which may be several decibels below the free-space value, depending on the magnitude and phase of the reflection coefficient. Total phase cancellation is rare because reflection coefficients do not commonly approach the − 1 theoretical limit and Bullington (Ref. 8.7), for example, indicates that the average value of reflection coefficients measured over nearly fifty paths operating in the 4 GHz band was about 0.3. However, larger values may be noted in flat areas of extensive plains or deserts.

Reflection from the sea (Fig. 8.6a), even from a rough sea, causes problems under appropriate circumstances, particularly at the higher frequencies. As the simultaneous effect of atmospheric refraction is to change the effective radius of the earth, the nulls of the resultant field at the receiver vary diurnally as well as with weather conditions.

Besides reflections from the ground or sea, reflections from an elevated inversion

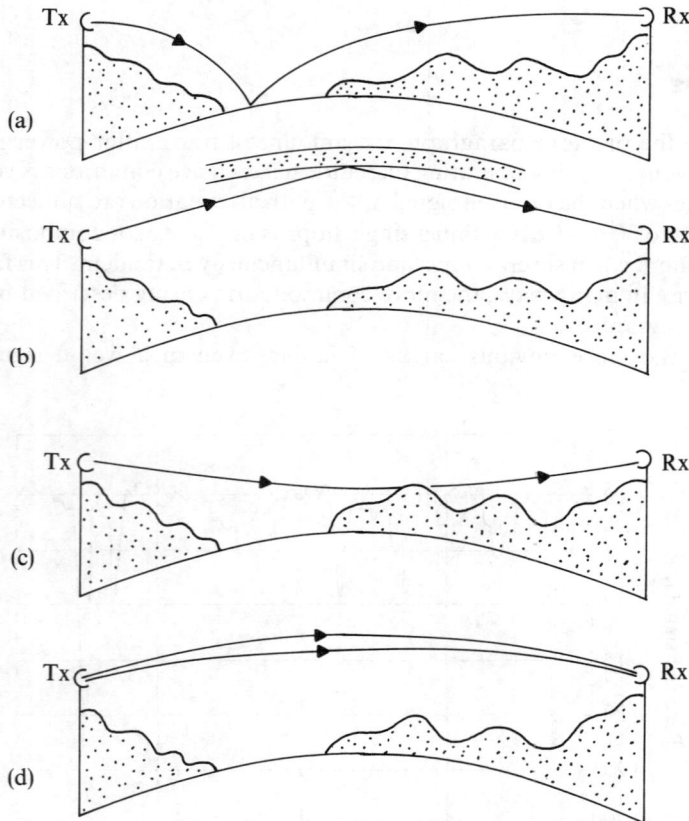

Fig. 8.6 The many causes of fading include (a) reflection, for example from a water surface, (b) reflection from an elevated layer, (c) sub-refraction propagation conditions, and (d) multipath propagation conditions. Attenuation by atmospheric gases (mostly above 10 GHz) as well as by rain, fog and snow also contribute to fading.

layer are possible (Fig. 8.6b). These occur when the meteorological conditions provide a sharp discontinuity in the vertical refractive index gradient of the atmosphere. Small shifts in height can cause large changes in the phase difference between the direct and reflected signals, somewhat analogous to tidal variation in the height of the surface of the sea in estuaries. If the propagation path actually passes through such an elevated inversion layer, the received signal is much attenuated and the particular hop is not amenable to the enhancement processes of diversity.

Fading due to sub-refraction may occur in certain paths if the reduction in the K-factor and effective radius of earth appear to raise the height of a terrain obstacle into the direct line-of-sight path between transmitter and receiver (Fig. 8.6c). There may be some diffraction over the obstacle but the effect is still to produce a large fade in the level of the received signal. A well-designed path should ensure that this does not occur.

Other forms of multipath effects are caused by irregularities in the atmospheric refractive index (Fig. 8.6d) so that, once again, the received signal is the sum of two or more contributions with differing amplitudes and, in particular, differing phases. De Lange (Ref. 8.8) transmitted short pulses of 4 GHz over a 35 km path. The pulses, about 3 ns long, were received at the end of the hop after traveling along the multipaths. A single received pulse would suggest the path differences were less than about 1 m, but de Lange found that under fading conditions the path lengths could differ by up to 2 m. This is several times the wavelengths of the 4 GHz carrier and hence the fading effects.

8.3.3 Short-term Specifications for the HRC

The recommendations of the CCIR (Ref. 8.6) also include permissible degradation of the signal-to-noise ratio in each channel for short-term periods of time. For the hypothetical reference circuit the recommendations are that:

1. 7500 pW of psophometrically weighted mean noise power shall not be exceeded in any *hour*.
2. 7500 pW of psophometrically weighted *one-minute* mean noise power shall not be exceeded for more than 20% of any month.
3. 47,500 pW of psophometrically weighted *one-minute* mean noise power shall not be exceeded for more than 0.1% of any month.
4. 1,000,000 pW of *unweighted* noise power shall not be exceeded for more than 0.01% of any month (measured with a 5 ms time-constant instrument).

The variation in noise power per channel with time is due mainly to fading, and the CCIR recommend that if n hops are fading in a microwave radio relay circuit containing a total of q hops, the circuit should be designed on a free-space basis with N' pW of weighted thermal noise power per hop so that the total hourly mean thermal noise $\bar{N}$ is given by:

$$\bar{N} = 10nN' + 1.44(q - n)N' \tag{8.11}$$

In other words, if we follow the CCIR suggestion that 20% of the hops may be subjected to fading in tandem, and take the mean thermal noise in the worst hour to be 3750 pW

Table 8.2 Multi-hop fading factor ΔP

Prob (%) \ n	1	2	5	10	15	20	30	50	100
0.01	39	41	46	48	49	51	52	54	58
0.1	29	31	36	38	39	41	42	44	48
1.0	19	21	26	29	30	32	33	34	38
20	5	9	13	17	19	21	22	24	28
50	0	4	10	14	16	18	20	21	25

(half of the allocated 7500 pW), then:

$$3750 = 10(0.2 \times 54)N' + 1.44(0.8 \times 54)N'$$

or $N' = 22$ pW weighted. If we assume that one-third of the hops are likely to fade simultaneously in tandem, then $N' = 16$ pW weighted.

For the short-term recommendations, the composite effects of tandem fading has been determined analytically by Dukta (Ref. 8.9). On this basis it is possible to construct Table 8.2 which relates the number of fading hops n, the effective reduction factor per hop ΔP, and the probabilities at which the noise levels are exceeded. Depending on global location and the local meteorological and environmental conditions, it appears that between one-fifth and one-third of hops in a long circuit are simultaneously affected by fading. Factor n can be determined either on this basis or from local knowledge and experience.

To explain the application of Table 8.2 to the analysis of fading in the hypothetical reference circuit, consider the short-term conditions given in the CCIR specifications. The first requires that 7500 pW of one-minute weighted noise power must not be exceeded for more than 20% of a month. Taking the usual 3750 pW as referring to thermal noise, the number of simultaneous fading hops in the 54-hop HRC is likely to be between $n = 54/5$ and $n = 54/3$. If we select $n = 15$, ΔP can be obtained from Table 8.2, column $n = 15$ and row $P = 20\%$, as 19 dB (79.4 linear). This means that if each hop is allocated $3750/79.4 = 47$ pW of thermal noise power (or less), the CCIR specification will be satisfied.

The second of the short-term specifications requires that 47,500 pW (i.e. 43,750 pW thermal) of one-minute weighted noise power must not be exceeded for more than 0.1% of any month. Using column $n = 15$ and row $P = 0.1\%$, we obtain $\Delta P = 39$ dB (7943 linear). This means that if each hop is allocated $43,750/7943 = 5.51$ pW or less, the CCIR specification will be satisfied.

In the latter case it implies that if the circuit is designed so that the signal-to-thermal noise ratio for each hop is effectively at least 1 mW/5.51 pW (82.6 dB) when there is no fading, the overall performance of the 15 hops in simultaneous fades is identical to each of the hops fading simultaneously and identically to produce an $SN_{th}R(dB)$ of 55.4 dB. In other words, 39 hops with 5.51 pW noise power leaves 43,535 pW for the 15 fading hops, or 2902 pW per hop. This is equivalent to an effective signal-to-thermal noise ratio per hop of 1 mW/2902 pW = 55.4 dB.

Table 8.3   Specifications for a real radio relay system

Total circuit length	(1)	(2)	(3)	
50 km $< L <$ 280 km	$3L + 200$	$3L + 200$	47,500	$(280/2500) \times 0.1\%$
280 km $< L <$ 840 km	$3L + 200$	$3L + 200$	47,500	$(L/2500) \times 0.1\%$
840 km $< L <$ 1670 km	$3L + 400$	$3L + 400$	47,500	$(L/2500) \times 0.1\%$
1670 km $< L <$ 2500 km	$3L + 600$	$3L + 600$	47,500	$(L/2500) \times 0.1\%$

8.3.4 Specifications for a Real Radio Relay System

The CCIR specifications have so far concerned the hypothetical reference circuit with 54 equal hops over a total distance of 2500 km. For real radio relay systems in which the total length L is between 280 km and 2500 km, and whose composition does not differ appreciably from the HRC, the psophometrically weighted noise power at a point of zero relative level in the audio channel should not exceed:

1. $3L$ pW mean value in any hour;
2. $3L$ pW one-minute mean power for more than 20% of any month;
3. 47,500 pW one-minute mean power for more than $(L/2500) \times 0.1\%$ of any month.

Otherwise, for a dissimilar circuit, the corresponding values are given in Table 8.3, where the number in each column refers to the specifications 1, 2 and 3 above. Example 8.4 illustrates numerically the application of these specifications.

Example 8.4 Minimum transmitter power requirement

A microwave radio relay FDM/FM system, operating at 6 GHz, carries 1800 audio channels across a total length of 1000 km with 24 hops. Estimate the minimum transmitter power required to satisfy the CCIR specifications. Assume that the antennas are 2.5 m in diameter and that L(misc) = 8 dB.
    The hourly mean thermal noise power per hop N' can be obtained from Eq. (8.11) with $n = 5$, if we assume that typically 20% of the hops may fade in tandem, and $q = 24$. To find $\bar{N}$, we note that the HRC contains nine 6-hop modulation sections and that each modulation section is allocated 3750/9 pW of noise power. The 24-hop radio relay is therefore likely to have four modulation sections requiring about 1667 pW total allocation of noise power. Thus out of the specification of $3L + 400$ (3400) pW, 1733 pW is available for $\bar{N}$ in Eq. (8.11). Direct substitution into

$$\bar{N} = 10nN' + 1.44(q - n)N'$$

gives $N' = 22.40$ pW and $SN_{th}R = 1$ mW/22.40 pW (which is either 76.5 dB weighted or 74.0 dB unweighted).
    The carrier-to-thermal noise ratio can be determined from Eq. (8.6) with $f_d = 140$ kHz and $f_m = 8204$ kHz for a 1800 audio channel system (Table 8.1), and PE(dB) = 4 dB. Direct substitution into

$$SN_{th}R(dB) = CN_{th}R(dB) + 10\log(B/b) + 20\log(f_d/f_m) + PE(dB)$$

gives us $CN_{th}R(dB) + 10 \log B = 140.27$. For a receiver with a noise figure of 10 dB we can write

$$P(r) = CN_{th}R(dB) + N(th) = CN_{th}R(dB) - 204 + 10 \log B + F(r)$$

and, on substitution of the data, we obtain $P(r) = -53.73$ dBW. Note that the $10 \log B$ terms cancel.

The propagation equations described in Ex. 8.2 may be used to estimate the antenna gains as 41.22 dB and the path loss as 140.41 dB, so that

$$P(r) = P(t) + G(t) + G(r) - L(s) - L(\text{misc})$$

gives us $P(t) = 12.24$ dBW (16.75 W).

The second specification, with the aid of Table 8.2, indicates that the starting point should be $N' = 1733$ pW reduced by ΔP, where ΔP is approximately 13 dB. This gives $N' = 86.86$ pW, approximately 5.88 dB higher than the $N' = 22.40$ pW required for the first specification, and this requires $P(t)$ to be 5.88 dB lower than the 12.24 dBW derived earlier. The new requirement for $P(t)$ is therefore 6.36 dBW (4.3 W).

The third specification requires a probability level of only 0.04% or $(L/2500) \times 0.1\%$. Extrapolating Table 8.2 suggests that ΔP is of the order of 41 dB, and N' is barely $43,750/12,589 = 3.475$ pW. This is 8.1 dB below the $N' = 22.4$ pW required for the first specification and the transmitter power needs to be 8.1 dB higher than the earlier $P(t) = 12.24$ dBW to achieve this third specification, viz. $P(t) = 20.33$ dBW (108 W).

It is seen that in this case the most difficult specification to achieve is the third.

8.4 CIRCUIT CALCULATIONS WITH FAST FADING

One of the short-term specifications for the CCIR hypothetical reference circuit is that 47,500 pW of one-minute mean power is not exceeded for 0.1% of any month. If we include the 2500 pW allowance for the FDM multiplexing equipment, the corresponding overall signal-to-noise ratio is 1 mW/50,000 pW or 43 dB. Thus the one-minute mean signal-to-noise ratio must not be less than 43 dB for more than about 44 minutes in the worst month. If, however, it is required to have a higher degree of reliability even in the presence of fast fading of the Rayleigh type, then the circuit must be designed to even higher standards.

8.4.1 The Rayleigh and Inverse Rayleigh Distributions

During a period of stable conditions the received signal over a particular hop in a microwave link remains sensibly constant. The hourly mean is almost equal to the free-space value as calculated by the method illustrated in Ex. 8.2, though in practice possibly 1–2 dB lower.

When the propagation path is subjected to fading which approximates to Rayleigh-type variation, there are rapid fluctuations in the received signal, something like that sketched in Fig. 8.5. If sampled in an appropriate manner, the variation of the electric field strength (in V/m) or the receiver input voltage (in V) can be represented by the Rayleigh probability density function

$$p(v_r) = \frac{2v_r}{r^2} \exp\left(\frac{-v_r^2}{r^2}\right) \tag{8.12}$$

where r^2 is the mean square value of v_r over the sampling period. Alternatively, as shown in Chap. 6, the signal *power* levels under such circumstances follow the exponential distribution and the probability that P_{rf} exceeds any value P is given by

$$\text{Prob}(P_{rf} > P) = \exp(-P/P_0) \tag{8.13}$$

In this expression, P_{rf} is the received power (in W) when the propagation path is subjected to Rayleigh-type fading and P_0 is the *mean* value of the corresponding distribution. As shown by Eq. (6.27) there is a difference of 1.6 dB between the *mean* and the *median*.

We thus have a simple expression for the variation of received power under Rayleigh-type fading conditions. The input noise power is sensibly constant at $N_{th} = kT_rB = kT_0F_rB = Q_1$ (say).

To determine the corresponding variation at the output, we note that the FM receiver incorporates *automatic gain control* (AGC) to ensure that, within certain limits, the output signal level remains almost constant (equivalent to 1 mW at the zero relative level in the audio channel). Any variation of input signal is thus reflected as opposite variation in output noise power per channel (Fig. 8.7). If $S = GP$ is to remain constant (say Q_2), then the receiver gain is controlled by the AGC so that $G = Q_2/P$. This implies that the output noise $N = GN_{th} = GQ_1 = Q_1Q_2/P = \text{constant}/P$.

If the value of N which occurs when $P = P_0$ is designated N_0 (and this must not be confused with the mean value of N because the distribution is different), we can write $N_0 = \text{constant}/P_0$. Combining both relationships we have

$$\frac{P}{P_0} = \frac{N_0}{N} \tag{8.14}$$

and this can be introduced into Eq. (8.13) to produce

$$\text{Prob}(N_{rf} > N) = 1 - \exp(N_0/N) \tag{8.15}$$

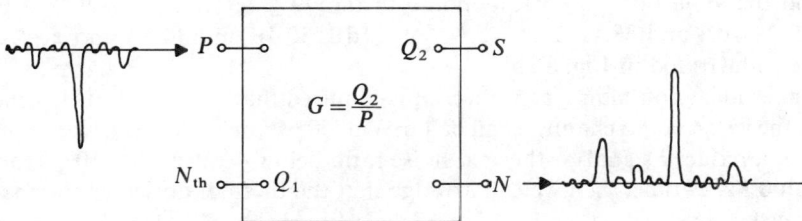

Fig. 8.7 An FM receiver incorporating automatic gain control causes the pattern of fading in the received signal level to be inverted when applied to the pattern of variation of output noise. If the input variation follows the Rayleigh distribution, the output variation follows the inverse Rayleigh distribution.

This is called an *inverse Rayleigh distribution* and involves an expression of the type $\exp(-a/x)$ rather than the more usual $\exp(-ax)$. Because input power P can theoretically drop to zero, N_{rf} can theoretically reach infinity, and hence the mean value of N_{rf} is indeterminate. However, in practice, $P > P_{th}$, where P_{th} is the threshold level of the FM receiver, and if P_0/P_{th} is known, a mean value of the inverse Rayleigh distribution can be derived, as in Refs. 8.3 and 8.4. These show that:

$$\frac{\bar{N}_{rf}}{N_0} = \log_e\left(\frac{P_0}{P_{th}}\right) - 0.577 \tag{8.16}$$

If the ratio P_0/P_{th} is expressed in decibels as $P''(\text{dB})$,

$$\frac{\bar{N}_{rf}}{N_0} = 0.23 P''(\text{dB}) - 0.577 \tag{8.17}$$

For $P''(\text{dB}) = 30$, 40 and 50 dB, $\bar{N}_{rf}/N_0$ will be 6.3 (8.0 dB), 8.6 (9.4 dB) and 10.9 (10.4 dB), respectively.

8.4.2 Signal-to-Thermal Noise Ratio with Rayleigh Fading

We have seen that the thermal noise per channel at the output of the FM receiver has a statistical variation which can be represented by a mean value $\bar{N}_{rf}$ which is roughly 8 to 10 dB higher than N_0 (the value corresponding to $P = P_0$), but to further complicate matters P_0 does not normally coincide with P_r, the free-space value obtained by the normal calculations. There is commonly a *depression* in mean values, described in more detail in Ref. 8.3, which depends on path length and roughness. If, for simplicity, we assume a round-figured difference of exactly 10.0 dB between $\bar{N}_{rf}$ and N_r, then the variation of N_{rf} may resemble something like that sketched in Fig. 8.8a.

In Fig. 8.8b we start with N_r as the value of output noise corresponding to free-space P_r at the input. P_0 is depressed a little lower than P_r and so N_0 is a little higher than N_r, say 1.5 dB in a particular case. $\bar{N}_{rf}$ is somewhat higher than N_0 as given by Eq. (8.14), say 8.5 dB in the same particular case to give the rounded difference of 10 dB between $\bar{N}_{rf}$ and N_r.

The *median* N_m of the inverse Rayleigh distribution is 1.6 dB higher than the mean $\bar{N}_{rf}$, and the values of N_{rf} corresponding to the 99%, 99.9% and 99.99% are about 18.4 dB, 28.4 dB and 38.4 dB above N_m or 20 dB, 30 dB and 40 dB above $\bar{N}_{rf}$. These levels are illustrated in Fig. 8.8b.

Under ideal non-fading conditions it is required that the signal-to-thermal noise ratio in the worst audio channel shall be 1 mW/3750 pW or 54.26 dB. If we are prepared to accept a reduced signal-to-thermal noise ratio below 44.26 dB (i.e. 10 dB lower) for less than 0.1% of time, then we can arrange that the thermal noise per channel at the 99.9% level is 44.26 dB below 1 mW or 37.497 pW. At the 99% level the thermal noise level will be 10 dB less (3750) and the signal-to-thermal noise level will be back to 54.26 dB again. Thus the quality of reception will be below the specified *mean* level for only 1% of the time during which there is severe Rayleigh-type fading (36 seconds per hour) and only 10 dB worse for less than 0.1% of time (3.6 seconds per hour).

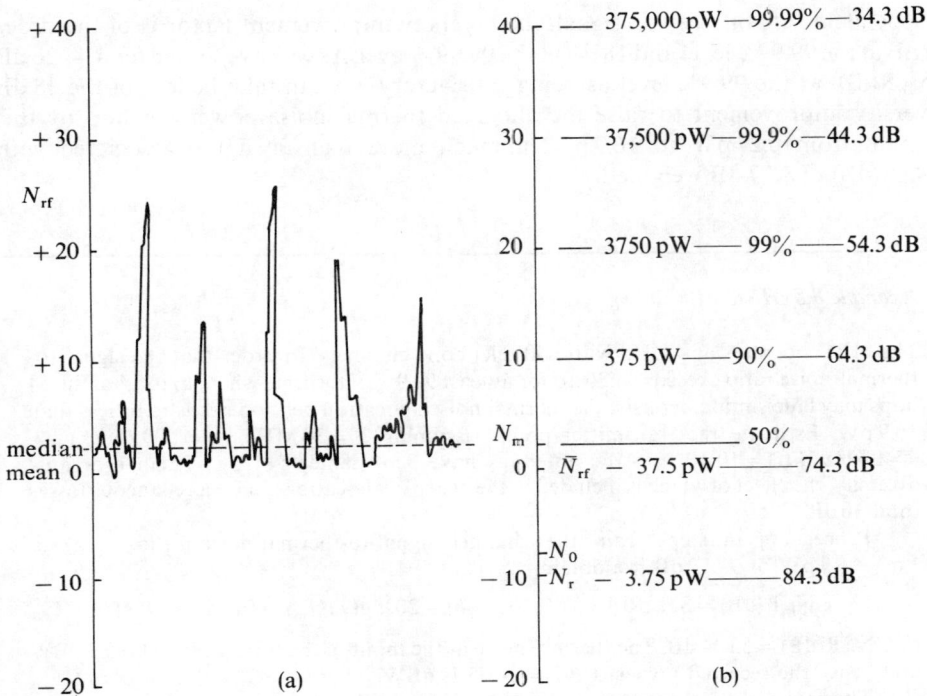

Fig. 8.8 The variation of output noise with fading N_{rf}. It should be noted that the mean $\bar{N}_{rf}$ of the inverse Rayleigh distribution does not coincide with N_0, the value corresponding to the mean of the input Rayleigh distribution P_0.

Continuing the processes downwards in Fig. 8.8b, $\bar{N}_{rf}$ corresponds to 30 dB below 37,497 pW, or 37.5 pW, and N_r corresponds to a further 10 dB reduction to 3.75 pW. This is the value which we must use in the free-space calculation to achieve the required level of reliability on the assumption that one hop is subjected to severe Rayleigh-type fading.

In a real circuit there may be about 25% of the total number of hops fading in tandem, or about 14 of the 54 hops in the HRC. If each hop happened to be at a 30 dB fade at the same instant, the total fade would amount to 420 dB. The statistics are such, however, that the composite fade of n hops in tandem is nowhere as severe. Optimistically, the *tandem fading factor* at the 99.99% level is of the order $10 \log n$, but a slightly higher figure is sometimes used in practice.

Using the optimistic approach, with 14 fading hops the tandem fading factor is $10 \log 14 = 11.5$ dB and the equivalent free-space noise allowance per hop should be reduced by 11.5 dB from the 3.75 pW (when only one hop is fading) to 0.27 pW to account for the tandem fading of 14 hops. We are left with a very small noise allowance, equivalent to a signal-to-thermal noise ratio of 95.7 dB weighted per hop.

To overcome this problem we must resort to diversity improvements of the sort described in Chap. 6. For instance, if we chose to use maximal-ratio combining of two independent sets of received signal, the diversity improvement factor is of the order 23 dB at the 99.99% level and 18 dB at the 99.99% level. As we have settled for a 44.26 dB $SN_{th}R(dB)$ at the 99.9% level as being satisfactory, we can take benefit of the 18 dB diversity improvement to raise the allocated thermal noise power per hop by this amount from 0.27 pW to 16.9 pW, a much more realistic figure associated with $SN_{th}R(dB)$ of 77.7 dB weighted.

Example 8.5 HRC with fading

The HRC operates at 4 GHz with 600 telephone channels. In order that the signal-to-thermal noise ratio exceeds 44.26 dB for at least 99.9% of the time when up to 14 of the 54 hops may fade simultaneously, the thermal noise allocation per hop needs to be less than 16.9 pW. Estimate the transmitter power assuming $B = 30$ MHz, $f_d = 200$ kHz, $f_m = 2540$ kHz, $F(r) = 10$ dB and the antennas have 3 m diameters. The system uses dual diversity, the effect of which is included in the 16.9 pW allocation, and miscellaneous losses total 10 dB.

Proceeding in steps from the channel signal-to-thermal noise ratio equal to 1 mW 16.9 pW or 77.72 dB, we obtain

$$CN_{th}R(dB) = SN_{th}R(dB) - 10\log(B/b) - 20\log(f_d/f_m) - PE(dB) - P(s)$$

or $CN_{th}R(dB) = 53.44$ dB. The thermal noise at the input of the receiver is -119.23 dBW and hence the received power is $P(r) = -65.79$ dBW.

The gain of the 3 m antennas at 4000 MHz is 39.3 dB and the spatial loss over 46.3 km at 4000 MHz is 137.8 dB. Using

$$P(t) = P(r) - G(t) - G(r) + L(s) + L(misc)$$

we obtain $P(t) = 3.41$ dBW or 2.2 W.

8.5   DIGITAL RADIO RELAY SYSTEMS

The majority of the earlier radio relay systems operate at frequencies up to about 10 GHz and involve the analog modulation of up to 2700 telephone channels per carrier. One of the disadvantages of this method is that the noise powers created along each hop and associated repeaters and modulation sections are additive and, if the circuit consists of many hops spaced at short distances, the overall noise power level per channel would eventually exceed the CCIR specifications.

At frequencies up to about 10 GHz the path loss has been considered as consisting mainly of the spatial loss $L(s)$, together with various circuit losses between transmitter and antenna at one end and antenna and receiver at the other end $L(misc)$, and fading. Above this frequency the attenuation due to certain atmospheric gases begins to be of some significance, particularly that due to water vapor and oxygen.

Certain weather conditions also contribute to path attenuation. Small droplets of water in the form of fog or rain, as well as snow and sleet, cause some scattering and

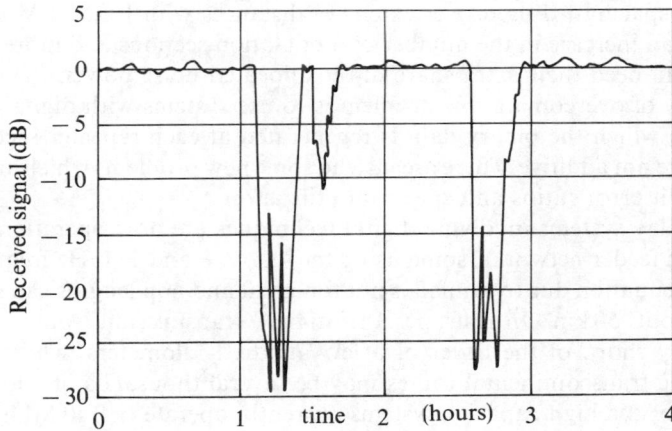

Fig. 8.9 The pattern of fading, usually associated with frequencies above about 10 GHz, when a rain storm causes severe fading along a particular hop in a radio relay system.

absorption of the propagating energy and this is represented by additional path attenuation during periods of rain or snow, something similar to that sketched in Fig. 8.9. To compensate for this increase in path attenuation it is necessary to reduce the hop spacing or increase transmitter power, but the latter option would cause increased interference with other systems and needs to be restricted. The attenuation due to rain is illustrated in Fig. 8.10, the effect increasing with frequency, and consequently some of the communication links in excess of 10 GHz tend to have

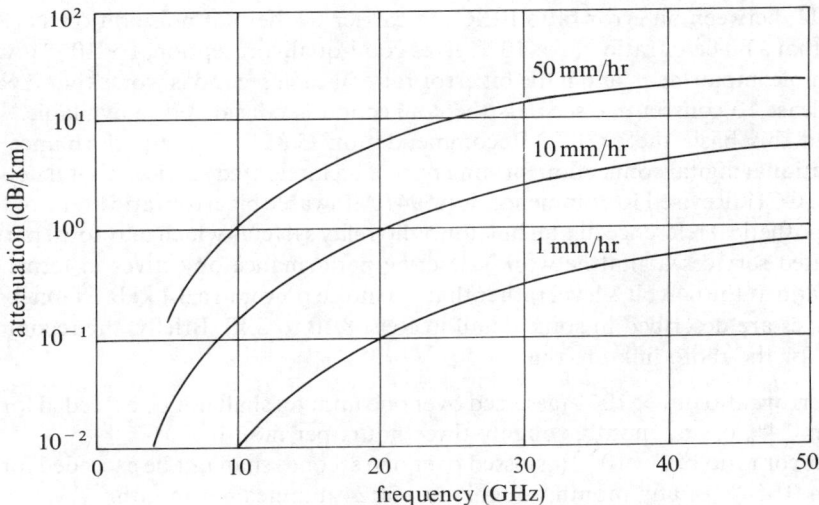

Fig. 8.10 An approximate indication of the variation of path attenuation due to rain. The intensity of the storm is indicated in mm of rainfall per hour.

reduced hop spacings. If used over extensive distances with FDM/FM systems, this would mean an increase in the number of modulation sections and more modulation sections would need their extra share of the allocated noise power.

One way of overcoming this problem is to use systems with digital modulation techniques in which the binary data is regenerated at each repeater section and the noise levels are not additive. There are instead some new problems which are now in the province of bit error ratios and spectrum utilization.

Radio relay systems involving digital techniques are now operational for trunk networks and feeder networks, some using the 4, 6, 7, 8 and 11 GHz frequency bands where the attenuation due to rainfall is not too great and hop lengths can still be of the order of about 50 km. In Europe, the digital transmission routes tend to be comparatively short, of the order of a few hundred kilometers, whereas in North America some transcontinental routes may be several thousands of kilometers long.

Several of the high-capacity systems currently operate at 140 Mbit/s per trunk channel, providing a circuit capacity of 1920 telephone circuits at 64 kbit/s. An 11 GHz digital radio relay system with six both-way trunk channels (five in use and one standby) will have an operational capacity of 9600 telephone circuits at 64 kbit/s. Small and medium-capacity systems operate at lower bit-rates.

At the higher allocated frequencies the attenuation due to rain, fog, and snow have the effect of requiring smaller hop spacings. However, there are 19 GHz digital relays which operate as medium-capacity feeders into a main trunk system (for example, with six both-way channels at 8 Mbit/s providing 720, 64 kbit/s telephone circuits) and also as part of a high-capacity main trunk route (for example, with six both-way channels at 140 Mbit/s providing some 11,520, 64 kbit/s telephone circuits).

The radio wave propagation aspects of digital relay systems are comparatively simple if the digital system analysis provides a relationship, similar to that illustrated in Fig. 8.11, between bit error ratio BER and carrier-to-thermal noise ratio $CN_{th}R$. It is found that a bit error ratio of 1×10^{-6} gives good-quality reception, 1×10^{-5} gives just discernible impairment, and if the bit error ratio in each second is worse than 1×10^{-3} for at least 10 consecutive seconds the connection is considered unavailable.

On this basis the CCITT Recommendation G.821 ('Error performance of an international digital connection forming part of an integrated services digital network') and the CCIR Revised Recommendation 594 ('Allowable bit error ratio at the output of the hypothetical reference digital link for radio relay systems which may form part of an integrated services digital network') describe performance objectives in terms of bit error ratio at the 64 kbit/s level rather than the noise power in a 3.1 kHz channel. These objectives are described in some detail in Refs. 8.10 to 8.13. Briefly, the recommendations for the radio link are that:

1. An error ratio of 1×10^{-6} (assessed over one minute) shall not be exceeded for more than 0.4% of any month, roughly three hours per month.
2. An error ratio of 1×10^{-3} (assessed over one second) shall not be exceeded for more than 0.054% of any month, roughly about 24 minutes per month.
3. The errored seconds shall not be exceeded for more than 0.32% of any month, roughly 28 hours.

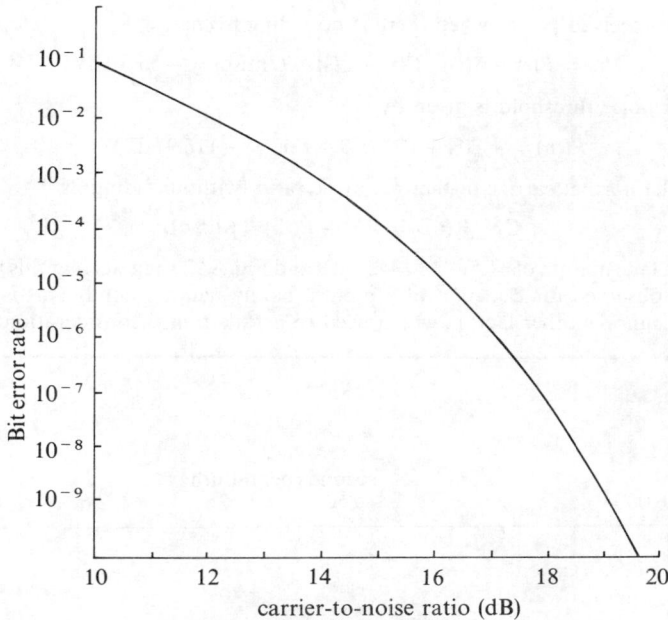

Fig. 8.11 The typical shape of a curve relating bit error ratio (BER) to the carrier-to thermal-noise ratio $CN_{th}R(dB)$ for a particular form of digital processing system.

Unlike the method used for the FDM/FM system, described earlier, the transmitter power required for digital radio relay systems is determined by calculating the carrier-to-noise ratio corresponding to the maximum permitted error ratios and then adding a suitable fading margin. The latter needs to be determined from propagation tests over specified paths and may be similar to the details illustrated in Fig. 8.11.

Example 8.6 Digital radio relay link at 11.2 GHz

A digital radio relay system operates at 11.2 GHz over a propagation path in which $d = 45$ km and atmospheric attenuation is 1.5 dB. The system involves a 10 W transmitter with 9 dB circuit and feeder losses, two 49 dB antennas, an 80 MHz bandwidth receiver with 8 dB noise figure and 8 dB circuit and feeder losses. An analysis of the digital processing system indicates that a minimum carrier-to-thermal noise ratio of 18 dB is necessary if the bit error ratio is not to exceed 1×10^{-6}. What fade margin is available over this link? If the sketch illustrated in Fig. 8.12 is representative of this particular link, over what fraction of a month is the system reliable?

It is necessary to determine the spatial loss over 45 km at 11.2 GHz from:

$$L(s) = 20 \log d + 20 \log f + 32.45 = 146.5 \, dB$$

and then the received power when there is no fading from:

$$P(r) = P(t) + G(t) + G(r) - L(s) - L(\text{misc}) = -55.5 \, \text{dBW}$$

The receiver noise threshold is given by

$$P(\text{th}) = -204 + 10 \log B + F(r) = -116.97 \, \text{dBW}$$

and hence the normal carrier-to-thermal noise ratio (without fading) is

$$\text{CN}_{\text{th}}\text{R(dB)} = P(r) - P(\text{th}) = 61.5 \, \text{dB}$$

This leaves a fade margin of $61.5 - 18 = 43.5 \, \text{dB}$ and Fig. 8.12 suggests that this magnitude of fading is observed for 2×10^{-7} of a month, barely half a second. This is 99.9999% reliable, but ignores other factors which need to be taken into consideration.

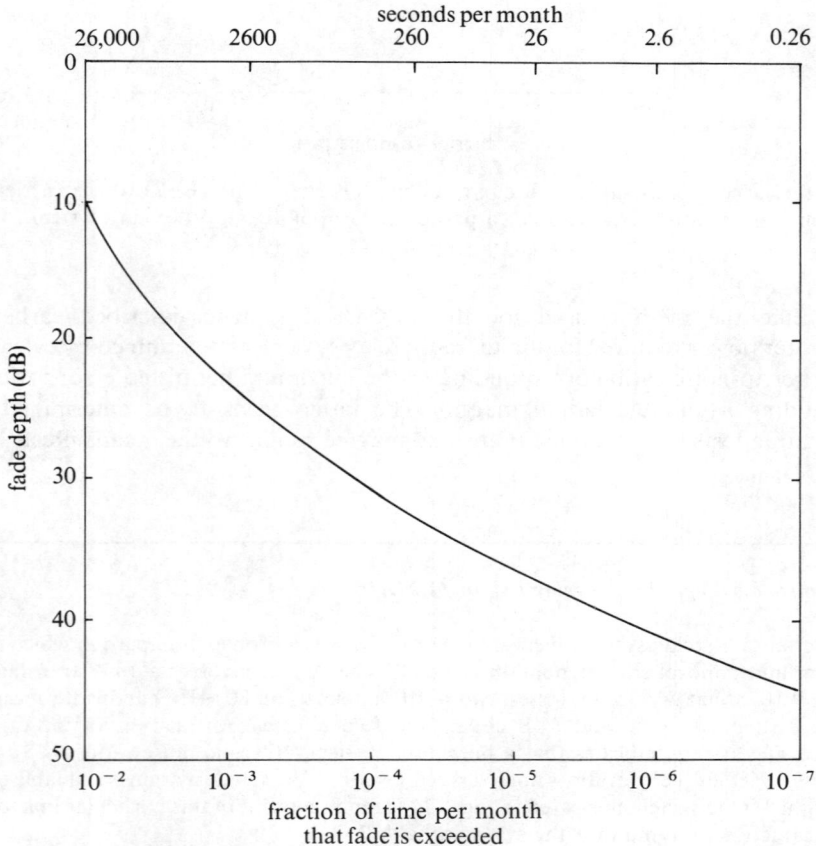

Fig. 8.12 The typical shape of the curve relating fade depths with the fraction of time per month that the fade is exceeded.

PROBLEMS

8.1 A 2 GHz radio relay link covers a distance of 50 km. The transmitting and receiving antennas are respectively at 250 m and 300 m above a reference datum level. It is found that the principal obstacle occurs at a distance of 15 km before the receiver and its height above the reference datum is such that the clearance is at the 0.3 first Fresnel zone radius when the atmospheric refractivity factor $K = 0.7$. What is the height of the obstacle?

8.2 An 11 GHz radio relay hop is required to have a received power level of -60 dBW when the propagation conditions are fade-free and the transmitter power level is 1 W. What paraboloidal reflector antenna diameters are required if miscellaneous losses account for 6 dB at *each* end of the 40 km hop? What is the effective radiated power?

8.3 A microwave radio relay system links two terminals separated by $d = 1500$ km via 35 equi-spaced hops. What are the design levels of channel thermal noise power per hop N' needed to satisfy CCIR requirements if the modulation sections are allocated a total of 2000 pW? Assume a maximum of 20% of the hops fade in tandem.

8.4 An FDM/FM microwave link operates at 6 GHz and in order to achieve the third CCIR specification that 45,500 pW of one-minute mean *thermal* noise power is not exceeded for more than 0.06% of time (for $L = 1500$ km) it is calculated that $N' = 3$ pW of weighted thermal noise power per channel per hop is appropriate. The hop parameters are: distance $d = 43$ km, antenna diameter $D = 2$ m, receiver noise figure $F(r) = 10$ dB, pre-emphasis and psophometric weighting account for 6.5 dB, and dual diversity improvement at the 99.94% level is 18 dB. Estimate the transmitter powers required, with and without diversity, if the link operates with: (a) 1800; (b) 900; or (c) 600 telephone channels in the baseband.

8.5 Consider a single hop of a radio relay system in which $d = 40$ km, $f = 4$ GHz, $D = 3$ m and $P_t = 1$ W. Neglect miscellaneous losses. The receiver has a thermal noise level of -120 dBW and an FM threshold $P(\text{th}) = -110$ dBW. Assume $B = 30$ MHz, $f_d = 200$ kHz, $f_m = 2540$ kHz, PE(dB) = 4 dB, P(s) = 2.5 dB and $b = 3.1$ kHz.

(a) Calculate the power at the receiver input $P(r)$ under free-space conditions.

(b) Assume that under Rayleigh fading conditions the mean of the Rayleigh distribution P_0 is equal to the free-space level P_r. What is the probability of P_{rf} (the power received during fading) exceeding P_r? What value of P_{rf} is exceeded for 50% of time? What is the relationship between P_0 and median P_m?

(c) What value of P_{rf} is exceeded for 99.99% of time? How is it related to P_0?

(d) What is the $CN_{th}R(dB)$ under fade-free conditions? What value of $CN_{th}R(dB)$ is exceeded for 99.99% of time under Rayleigh fading conditions?

(e) What is the $SN_{th}R(dB)$ in the highest baseband channel under fade-free conditions? What is N_0 under these conditions?

(f) What is the difference in magnitude between $\bar{N}_{rf}$ and N_0 during periods of Rayleigh fading? What are their magnitudes?

(g) What values of N_{rf} are *exceeded* for 0.1% and 0.01% of time and what are the corresponding $SN_{th}R(dB)$?

8.6 If, under Rayleigh fading conditions, the signal-to-thermal noise ratio in the top baseband channel is 35 dB, 45 dB and 55 dB at the 99.99%, 99.9% and 99% probability levels when the link is operated without diversity, what values of $SN_{th}R(dB)$ would you expect if dual diversity combiners with the equal-gain type of processing are used?

8.7 A microwave radio relay system covers a total distance of 333 km in q equal hops and it is assumed that in the worst case all hops are likely to fade in tandem. Determine the optimum number of hops to satisfy the CCIR mean hour specification if appropriate data are: $G(\text{dB}) = 45\,\text{dB}$, $L(\text{s}) = 140\,\text{dB}$, $P(\text{t}) = 0\,\text{dBW}$, $L(\text{misc}) = 10\,\text{B}$, the receiver $N(\text{th}) = -120\,\text{dBW}$ and the relationship between carrier-to-thermal noise ratio and signal-to-thermal noise ratio is $CN_{\text{th}}R(\text{dB}) = SN_{\text{th}}R(\text{dB}) - 25\,\text{dB}$. Assume equipment noise allocation is $100\,\text{pW}$ for each end termination and $70\,\text{pW}$ for each interim repeater.

8.8 A digital radio relay link operates with the QPSK characteristic illustrated in Fig. 8.11. If $P(\text{t}) = 0\,\text{dBW}$, $L(\text{misc}) = 12\,\text{dB}$, $d = 40\,\text{km}$, $D = 2.5\,\text{m}$, $f = 11\,\text{GHz}$, $B = 40\,\text{MHz}$ and $F(\text{r}) = 8\,\text{dB}$, determine the normal fade-free values of $P(\text{r})$ and $CN_{\text{th}}R(\text{dB})$ and bit error ratio BER when (a) $37\,\text{dB}$, or (b) $40\,\text{dB}$ fades occur as a result of heavy rainfall.

REFERENCES

8.1 Salkeld, B., 'Planning microwave links for television,' *IBA Technical Review*, No. 7 (July 1976), 37–47.

8.2 Pearson, K. W., 'Method for the prediction of the fading performance of a multisection microwave link,' *Proc. IEE*, 112 (July 1965), 1291–1300.

8.3 Sheffield, H. C., 'Microwave relay system between Saint John and Halifax,' *Electrical Communication*, (December 1955), pp. 215–237.

8.4 Panter, P. F., *Communication Systems Design*. New York, McGraw-Hill, 1972.

8.5 Bellchambers, W. H., et al., 'The International Telecommunication Union and development of worldwide telecommunications,' *IEEE Communications Magazine*, 22 (May 1984), 72–82.

8.6 CCIR, *Documents of the XIth Plenary Assembly*, vol. IV, Radio-Relay Systems, Space Systems and Radio Astronomy, International Telecommunications Union, Oslo 1966.

8.7 Bullington, K., 'Reflection coefficients of irregular terrain,' *Proc. IRE*, 42, (August 1954), 1258–1262.

8.8 De Lange, O. E., 'Propagation studies at microwave frequencies by means of very short pulses,' *Bell Syst. Tech. Journ.*, January 1952, pp. 91–103.

8.9 Dukta, S., and Mehlman, S. J., 'Noise power probability distributions for multihop FM radio relay systems,' Globe Commun. Symp., St. Petersburg, Flor., (December 1957).

8.10 Bennet, G. H., 'The evolution of transmission and switching for integrated services digital networks,' *The Radio and Electronic Engineer*, (February 1984), pp. 59–63.

8.11 McLintock, R. W., and B. N. Kearsey, 'Error performance objectives for digital networks,' *The Radio and Electronic Engineer*, (February 1984), pp. 79–85.

8.12 Kearsey, B. N., 'The status of international studies relating to the transmission performance of ISDNs,' *The Radio and Electronic Engineer*, (February 1985), pp. 71–77.

8.13 Hart, G., 'The performance of digital radio relay systems,' *The Radio and Electronic Engineer*, (April 1984), pp. 155–162.

9

Satellite and Space Communication

Radio communication via artificial earth satellites is a comparatively new technique, yet its development has proceeded with considerable rapidity. The first satellite to be launched was Russia's *Sputnik I* on 4 October 1957, followed by *Sputnik II* on 3 November 1957 and America's *Explorer I* on 31 January 1958. A few years later the joint NASA/UK1 satellite was put into orbit on 26 April 1962, with NASA/Canada's *Alouette I* on 28 September 1962 and NASA/Italy's *San Marco I* on 15 December 1964. France's *A-1* (*Asterix*) was launched on 26 November 1965.

During the intervening years the space surrounding earth has seen many satellites of varying shapes and sizes and these have been used for a whole range of communication, navigational, experimental and observational purposes, both military and commercial. In addition, there have been similar radio links between earth and moon and between earth and several observational probes which have been sent on planetary exploration.

9.1 SATELLITE ORBITS

The orbits of these satellites may be circular (with earth at the center) or elliptical (with earth at one of the foci) and, before we consider the radio wave propagation problems associated with satellites or with space probes, a little understanding of the orbital mechanics and geometry will be useful, though we shall restrict ourselves to circular or near-circular orbits only as illustrated in Fig. 9.1.

9.1.1 Orbital Period of Satellite

Following on from Kepler's hypotheses for planetary orbits (1618), Newton described certain laws of motion and gravitation (1686) for objects with mass m. These laws indicated: (1) that the force of attraction between two masses, such as M (earth) and m (satellite), separated by distance r between centers is given by

$$F_1 = G \frac{Mm}{r^2} \qquad \text{(SI units)}$$

where G is called the *universal constant of gravitation*; and (2) that when mass m orbits

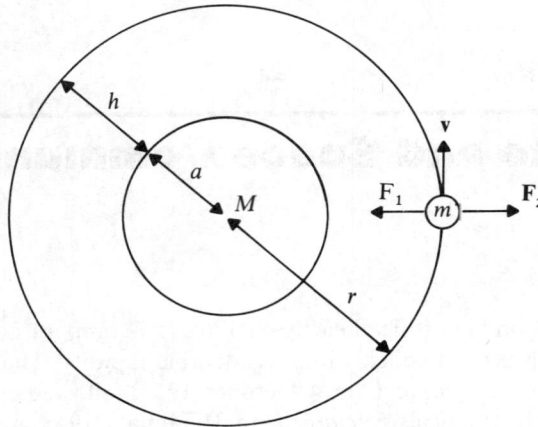

Fig. 9.1 The geometry of a satellite in circular orbit around the earth.

mass M with circumferential velocity v, the centrifugal force is

$$F_2 = \frac{mv^2}{r} \quad \text{(SI units)}$$

When the orbit is circular, the acceleration is constant, and the orbital relationship is

$$G\frac{Mm}{r^2} = \frac{mv^2}{r}$$

or

$$v = (GM/r)^{1/2} \tag{9.1}$$

where v is the satellite orbital velocity in meters per second; M is the mass of earth, about 5.977×10^{24} kg; G is the universal constant of gravitation, about 6.668×10^{-11} m^3/kg s^2 or N m^2/kg; and r is the distance from the satellite to the center of the earth (m).

Example 9.1 Orbital period of satellite

A non-synchronous satellite orbits earth with a mean altitude of 1500 km. Approximately how many times does it orbit the earth in one day?

We can use Eq. (9.3) to obtain

$$\tau = 0.0099527(1500 + 6378)^{3/2} = 6959 \text{ seconds}$$

and hence the number of orbits within a 24 hour period is about 12.4. This is the number of times the satellite orbits the earth; it is not the number of times it passes over any particular location because the earth is itself rotating.

The *orbital period* τ of the satellite is given by $\tau = 2\pi r/v$, and hence from Eq. (9.1) in SI units,

$$\tau = 2\pi(r^3/GM)^{1/2} = 3.1473 \times 10^{-7}r^{3/2} \tag{9.2}$$

or

$$\tau = 0.0099527r^{3/2} \tag{9.3}$$

if τ is in seconds and r is in kilometers.

9.1.2 Satellite Range and Relative Velocity

In order that we can later estimate several radio wave propagation parameters, including the received power level, we need to know the distance between the satellite and the receiver or between the ground-based transmitter and the satellite. If a = radius of earth (6378 km at the equator), h = altitude of the satellite above the *sub-satellite point* on earth, and θ = elevation of the ground antenna located directly on the path of the satellite, then Fig. 9.2 indicates (using the sine rule) that

$$\frac{a}{\sin A} = \frac{d}{\sin B} = \frac{a+h}{\sin(90 + \theta)}$$

where d is the *range* of the satellite. There is sufficient information to obtain angles A and B via

$$\sin A = \frac{a}{a+h}\cos\theta \tag{9.4}$$

and

$$B = 90° - A - \theta \tag{9.5}$$

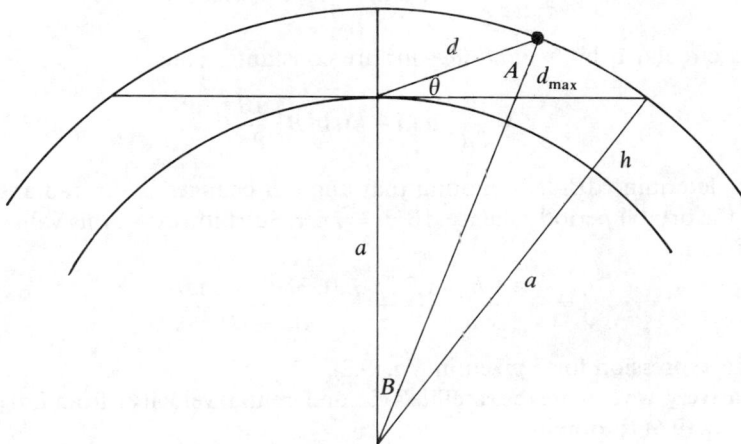

Fig. 9.2 The range of a satellite in non-synchronous orbit.

Two alternative expressions for range d make use of either the sine rule to give

$$d = \frac{a \sin B}{\sin A} \tag{9.6}$$

or the cosine rule to give

$$d = \{a^2 + (a+h)^2 - 2a(a+h)\cos B\}^{1/2} \tag{9.7}$$

Example 9.2 Satellite range

A satellite orbits earth with a mean altitude $h = 1500\,\text{km}$. What is its range from the receiver if the receiving antenna is correctly aligned with the path of the satellite at an angle of elevation of 20°? What is the maximum visible range of the satellite under these circumstances ignoring refraction?

Knowing that $a = 6378\,\text{km}$, we can use Eqs. (9.4) and (9.5) to determine $A = 49.5°$ and $B = 20.5°$. Substitution into either Eq. (9.6) or Eq. (9.7) gives us $d = 2931.6\,\text{km}$.

The maximum visible range under these circumstances, ignoring refraction, occurs when $\theta = 0°$. Using the same method we can obtain $d_{max} = 4624\,\text{km}$, but more simply we can note that

$$d_{max}^2 = (a+h)^2 - a^2$$

and this, too, gives us $d_{max} = 4624\,\text{km}$ for this particular example.

If the satellite is moving with respect to the receiving antenna, the range d varies with time. The relative velocity v_r at which the range changes is called the *range rate* and we can obtain an expression for this parameter for a satellite in circular orbit over the receiving site in two ways. Firstly, by differentiating Eq. (9.7) we obtain

$$v_r' = \frac{1}{2d} \frac{\text{d}}{\text{d}t} \{a^2 + (a+h)^2 - 2a(a+h)\cos B\}$$

where, for a circular orbit, a^2 and $(a+h)^2$ are constants. Thus

$$v_r' = \frac{1}{2d} \{2a(a+h)\sin B\} \frac{\text{d}B}{\text{d}t}$$

and we can determine $\text{d}B/\text{d}t$ by noting that angle B changes by 2π radians when t changes by the orbital period τ; hence $\text{d}B/\text{d}t = 2\pi/\tau$. Substitution of this value into the equation gives us

$$v_r' = \frac{a(a+h)\sin B}{d} \frac{2\pi}{\tau} = \frac{(GM)^{1/2} a \sin B}{d(a+h)^{1/2}} \tag{9.8}$$

if we use the expression for τ given in Eq. (9.2).

Alternatively, we can use the satellite's circumferential velocity v from Eq. (9.1) and $\sin A$ from Eq. (9.6) to obtain

$$v_r' = v \sin A = v \sin\left\{\cos^{-1}\left(\frac{d^2 + r^2 - a^2}{2dr}\right)\right\} \tag{9.9}$$

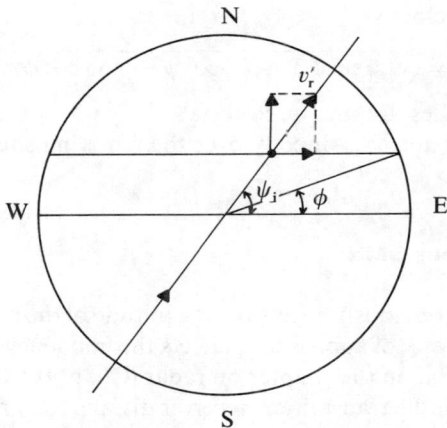

Fig. 9.3 The angle of inclination of the orbit of a non-synchronous satellite.

where the angle A can also be obtained via the cosine rule:

$$a^2 = d^2 + r^2 - 2dr \cos A \tag{9.10}$$

This is, of course, not the true relative velocity or range rate observed by the receiver as the earth also rotates about its axis once every sidereal day of 86164.1 seconds and the satellite orbit has an angle of inclination ψ_i in the direction illustrated in Fig. 9.3.

The surface velocity of earth at latitude ϕ is the ratio of the earth's circumference at that latitude to the rotational period:

$$v_e = \frac{2\pi a \cos \phi}{86,164} = 465 \cos \phi \qquad \mathrm{m\,s}^{-1} \tag{9.11}$$

The satellite's relative velocity v'_r is in direction ψ_i and it has two components, $v'_r \sin \psi_i$ and $v'_r \cos \psi_i$, the latter in the same direction as the earth's surface velocity. For the orientation given in Fig. 9.3 the net relative components of the satellite velocity with respect to a moving receiver are $v'_r \sin \psi_i$ as before and $(v'_r \cos \psi_i - 465 \cos \phi)$, leaving a

---

Example 9.3 Relative velocity of satellite

A satellite orbits the earth with a mean altitude of $h = 1500$ km and an angle of inclination $\psi_i = 50°$. A receiving antenna at latitude $\phi = 40°$N, and in line with the satellite's path, observes the satellite when the angle of elevation $\theta = 20°$. What is the relative velocity or range rate of the satellite at this instant?

We can use Eq. (9.1) to determine the satellite's orbital velocity as $v = 7112.65\,\mathrm{m\,s}^{-1}$, and from Ex. 9.2 we note that $d = 2931.6$ km. The relative velocity v'_r can then most easily be obtained from Eq. (9.8) as $v'_r = 5411\,\mathrm{m\,s}^{-1}$ relative to a stationary earth. To correct for a rotating earth we use Eq. (9.12) to give $v_r = 5189\,\mathrm{m\,s}^{-1}$.

resultant true relative velocity

$$v_r = \sqrt{(v'_r \sin \psi_i)^2 + (v'_r \cos \psi_i - 465 \cos \phi)^2} \tag{9.12}$$

providing all the velocities are measured in $m\,s^{-1}$.

   If the satellite does not pass directly over the receiving site the mathematics is a little more complicated.

9.1.3 Doppler Frequency Shift

The range d of a satellite obviously affects the magnitude of the received signal strength whereas the relative velocity of approach v_r affects the frequency of the received signal. As described in more detail in the chapter on radar (Chap. 10), the frequency f_r of the received signal changes if the transmitter–receiver distance d varies with time, because of the *Doppler effect*. The difference between the transmitted and received frequencies is called the *Doppler frequency shift* f_D and is given by

$$f_D = \frac{v_r f_t}{c} \tag{9.13}$$

where v_r is the relative velocity of approach, $c = 3 \times 10^8\,m\,s^{-1}$ and f_t is the transmitted frequency, all in consistent units. The received frequency is then given by

$$f_r = f_t + f_D \tag{9.14}$$

If the satellite is approaching the receiver, the positive v_r will produce a positive f_D, and f_r is greater than f_t. If the satellite is moving away from the receiver, the negative v_r will produce a negative f_D, and f_r is less than f_t.

Example 9.4 Doppler bandwidth

The satellite described in Ex. 9.3 transmits a test frequency at exactly 435 MHz. What is the minimum bandwidth of the receiver necessary to detect the test signal between the acquisition and loss of the satellite as it passes over the receiver?

   If the satellite passes directly over the receiving site, the maximum range (either side) will occur when $d = d_{max} = 4624\,km$, from Ex. 9.2. From Eqs. (9.9) and (9.12) we obtain $v'_r = 5758\,m\,s^{-1}$ and $v_r = 5536\,m\,s^{-1}$, and from Eq. (9.13) $f_D = 8027\,Hz$. The received frequency may vary between 435 MHz plus 8.027 kHz and 435 MHz minus 8.027 kHz, and the receiver bandwidth needs to be about 16 kHz to accommodate this variation. This example assumes the satellite passes directly over the receiver. If this is not the case then the Doppler frequency shift range will be less than that calculated.

9.1.4 Geostationary Orbit

The idea of geostationary or synchronous satellites for worldwide communication systems is generally attributed to Arthur C. Clarke (Ref. 9.1) in 1945. With the aid of

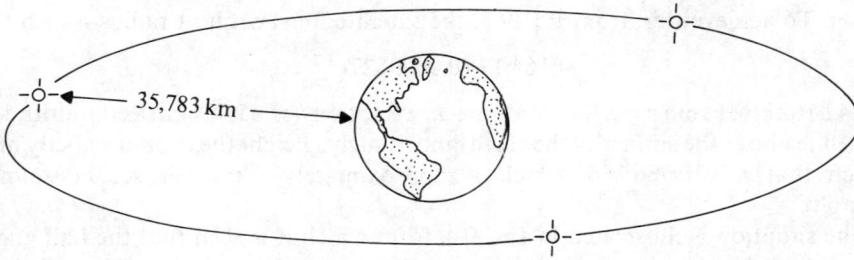

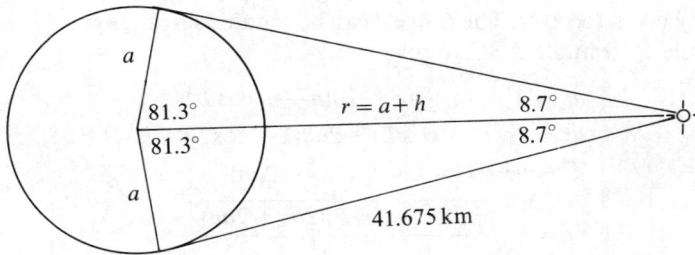

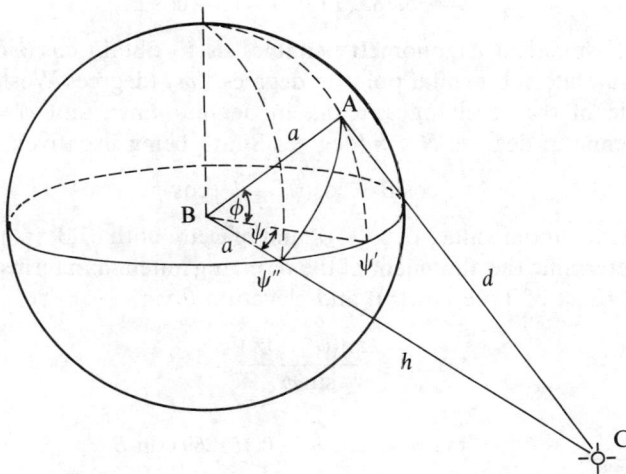

Fig. 9.4 A satellite in geostationary orbit 35,783 km above the equator.

Eq. (9.2) we can see that a satellite in the equatorial plane, having an orbital period τ equal to one sidereal day of 84164.1 seconds, will orbit in synchronism with earth's rotation. To achieve this, from Eq. (9.3) the satellite must orbit at radius r such that

$$86164.1 = 0.0099527 r^{3/2}$$

or $r = 42,162$ km. From $r = a + h = 6378 + h$, we obtain $h = 35,784$ km as the altitude of the satellite above the surface of the earth, and at such a height the orbital velocity must be such that $v = 2\pi r/86,164$, which is approximately 3 km per second (almost 7000 mph).

The situation is illustrated in Fig. 9.4, from which it is seen that the half-angles subtended at the satellite and at the center of the earth are about 8.7° and 81.3°, respectively. This means that the satellite cannot be 'seen' beyond latitudes 81.3° N or 81.3°S, leaving the polar regions outside the range of geostationary satellites.

Two of the parameters needed in the analysis of geostationary satellite communication systems are the range and bearing (azimuth α and elevation θ) of the satellite with respect to any given receiver. The range d can be obtained from Fig. 9.4 if we first apply the cosine rule to triangle ABC to give

$$d^2 = a^2 + (a + h)^2 - 2a(a + h)\cos B$$
$$d^2 = 2a^2(1 - \cos B) + 2ah(1 - \cos B) + h^2$$
$$d^2 = h^2 + (1 - \cos B)(2a^2 + 2ah)$$
$$d = h\sqrt{1 + \frac{(1 - \cos B)(2a^2 + 2ah)}{h^2}}$$

We know that $a = 6378.28$ km and $h = 35,783.91$ km, and simple substitution reduces the expression to

$$d = 35,783\sqrt{1 + 0.42(1 - \cos B)} \tag{9.15}$$

Application of spherical trigonometry enables us to obtain $\cos B$ in terms of $\psi' =$ longitude of satellite sub-orbital point in degrees *East* (degrees West being negative), $\psi'' =$ longitude of the receiving antenna in degrees *East*, and $\phi =$ latitude of the receiving antenna in degrees *North* (degrees South being negative):

$$\cos B = \cos(\psi'' - \psi')\cos\phi \tag{9.16}$$

Note that the maximum values of $\psi'' - \psi'$ and of ϕ are both 81.3° in magnitude. From this we can determine the alignment of the receiving antenna in terms of azimuth α (in degrees ETN (East of True North)) and elevation θ via

$$\tan \alpha' = \frac{\tan(\psi'' - \psi')}{\sin \phi} \tag{9.17a}$$

and

$$\tan \theta = (\cos B - 0.151269)/\sin B \tag{9.17b}$$

If the receiving antenna is in the northern hemisphere, $\alpha = \alpha' + 180°$ ETN, whereas in the southern hemisphere $\alpha = \alpha'$ ETN.

For a geostationary satellite situated above the equator at longitude 0°, with

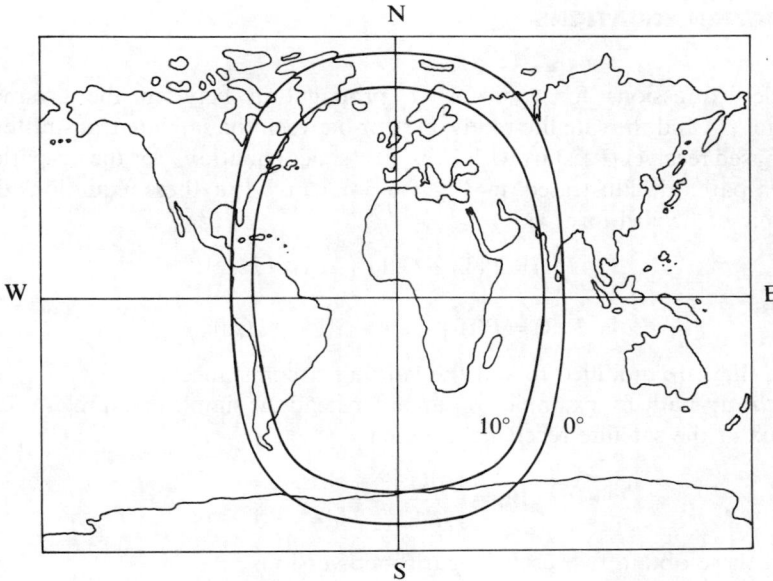

Fig. 9.5 The approximate coverage of a satellite located at longitude 0° and latitude 0° if the angle of elevation of the receiving antenna is 0° and 10°.

$r = 42, 162$ km and $a = 6378$ km, the line-of-sight limits extend as far as 81.3° North, East, South and West along latitude 0° and longitude 0°. In any other direction the longitude and latitude coordinates ψ and ϕ must be related via

$$\cos \psi \cos \phi = \cos 81.3° = 0.1513 \qquad (9.18)$$

so that at any point on the maximum line-of-sight range, $d = d_{max} = 41,677$ km.

For angles of elevation greater than $\theta = 0°$ simple trigonometrical analysis produces either ψ_{max} or ϕ_{max} and again

$$\cos \psi \cos \phi = \cos \psi_{max}$$

The coverage for $\theta = 0°$ and $\theta = 10°$ has been estimated in this manner and is illustrated in Fig. 9.5 for a satellite located at longitude 0°. For satellites at any other longitude the coverage curves are simply moved to the east or west by the appropriate amount.

Example 9.5 Antenna orientation for geostationary satellite

An antenna is located at longitude 2°W and latitude 53°N and it is required to align it with a geostationary satellite located over the equator at longitude 24°W. Determine the appropriate bearing of the antenna and the range of the satellite.

From Eq. (9.16) we obtain $\cos B = \cos 22° \cos 53° = 0.558$ and from Eq. (9.15) the satellite range is approximately 38,963 km. The bearing and elevation are given by Eq. (9.17) as $\alpha = 206.8°$ and $\theta = 26.11°$.

9.2 PROPAGATION EQUATIONS

The basic expressions for radio wave propagation between the ground-based transmitter (T) and the satellite receiver (r), or between the satellite transmitter (t) and ground-based receiver (R), start with Friis' free-space equation. For the up-path (u) and the down-path (d) illustrated in Fig. 9.6 we can write these equations in their logarithmic or decibel form as

$$P(r) = P(T) + G(T) + G(r) - L(su) \tag{9.19}$$

and

$$P(R) = P(t) + G(r) + G(R) - L(sd) \tag{9.20}$$

but often they are modified to suit the varying circumstances.

On the up-path, for example, we can use instead the simple linear relationships for power flux at the satellite receiving antenna:

$$P_{ar} = \frac{\text{EIRP}_T}{4\pi d_u^2} = \frac{P_r}{G_r(\lambda^2/4\pi)}$$

and from these obtain two useful logarithmic versions

$$P_a(r) = \text{EIRP}(T) - 20 \log d_u(\text{km}) - 71 \tag{9.21}$$

and

$$P(r) = P_a(r) + G(r) + 10 \log(\lambda^2/4\pi) \tag{9.22}$$

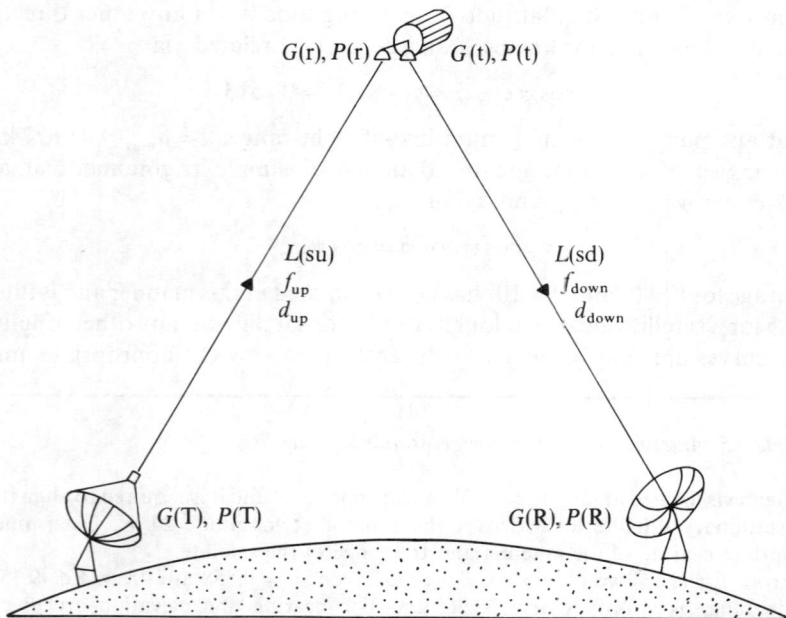

$G(r), P(r)$ $G(t), P(t)$

$L(su)$
f_{up}
d_{up}

$L(sd)$
f_{down}
d_{down}

$G(T), P(T)$ $G(R), P(R)$

Fig. 9.6 The parameters used in the up-path and down-path propagation equations for a satellite link between two ground stations.

The satellite design is such that there is a maximum value of power flux $P_a(r)$ which can be accepted by the satellite before its processing system (amplification and frequency changing) is saturated, and the actual operating level of $P_a(r)$ may be 1–2 dB below this maximum, the difference being called the *input back-off*. Hence Eq. (9.21) is particularly useful in determining the maximum EIRP(T) of the ground station for a given satellite system at a given range d.

9.2.1 Carrier-to-Noise Ratio

Noise is also an important factor in determining the overall quality and performance of the satellite radio link. If we initially consider only the thermal noise, represented by the system noise temperature T_r (satellite) or T_R (ground station), then from the basic linear form of the noise equation $N = kTB$ we can write

$$N(r) = 10 \log k + 10 \log T_r + 10 \log B_r \tag{9.23}$$

and

$$N(R) = 10 \log k + 10 \log T_R + 10 \log B_R \tag{9.24}$$

where k is Boltzmann's constant ($10 \log k = -228.6$) and B refers to the noise bandwidth of the system.

Two forms of *carrier-to-noise ratio* are used in various texts. The basic definition, using the up-path as an example, is

$$\text{CNR}(r) = P(r) - N(r) \qquad \text{dB} \tag{9.25}$$

Example 9.6 Satellite up-link calculations

The 6 GHz up-link of a certain 60 MHz bandwidth satellite communication system involves a satellite designed with $G/T(r) = -3$ dB/K and a saturation power flux $P_{ar} = 0.02 \, \mu\text{W/m}^2$. The link is operated over a range $d_u = 40{,}000$ km with 2 dB input back-off. What are: (a) the maximum carrier-to-noise ratio CNR(r) on the up-link; and (b) the maximum power $P(T)$ radiated from the $D = 22$ m diameter ground station paraboloidal reflector antenna?

This example clearly makes use of Eq. (9.27) for the up-link. From this we see that

$$\text{CNR}(r) = P_a(r) + G/T(r) + 10 \log(\lambda^2/4\pi) - 10 \log k - 10 \log B$$

Direct substitution of data, with $P_a(r) = -77.0$ dBW/m^2 and $\lambda = 0.05$ m, gives us maximum CNR(r) = 33.8 dB. For use with a later example, it is useful to note that 33.8 dB corresponds linearly with a ratio of 2399:1.

For the second part of the example we can modify Eq. (9.21) slightly to obtain

$$\text{EIRP}(T) = P_a(r) - L(\text{ibo}) + 20 \log d_u(\text{km}) + 71 = 84 \text{ dBW}$$

and then equate EIRP(T) = $P(T) + G(T)$, with $G(T) = 60.1$ dB, to obtain $P(T) = 23.9$ dBW or about 246 W. The antenna gain is determined from

$$G(T) = 20 \log f(\text{MHz}) + 20 \log D - 42.3 = 60.1 \text{ dB}$$

but an alternative definition in common use is to use

$$\mathrm{CN_0R(r)} = \mathrm{CNR(r)} + 10\log B_r = P(r) - 10\log k - 10\log T_r \quad (9.26)$$

This is the ratio of carrier-to-noise spectral density kT, and is useful in the comparison of the needs of different types of signal communication techniques such as telegraphy, telephony or television, etc. Substituting $P(r)$ from Eq. (9.22) into Eq. (9.26) we obtain

$$\mathrm{CN_0R(r)} = P_a(r) + 10\log G_r + 10\log(\lambda^2/4\pi) - 10\log k - 10\log T_r$$

or

$$\mathrm{CN_0R(r)} = P_a(r) + 10\log(G_r/T_r) + 10\log(\lambda^2/4\pi) - 10\log k \quad (9.27)$$

From this we can see that the value of $\mathrm{CN_0R(r)}$ has a maximum value dictated by the satellite system parameters: the maximum permitted power flux $P_a(r)$, the almost constant value of the ratio of antenna gain to effective noise temperature, often expressed oddly as dB/K; and the operating frequency via wavelength λ. As mentioned earlier, the maximum is often reduced by 1–2 dB input back-off L(ibo).

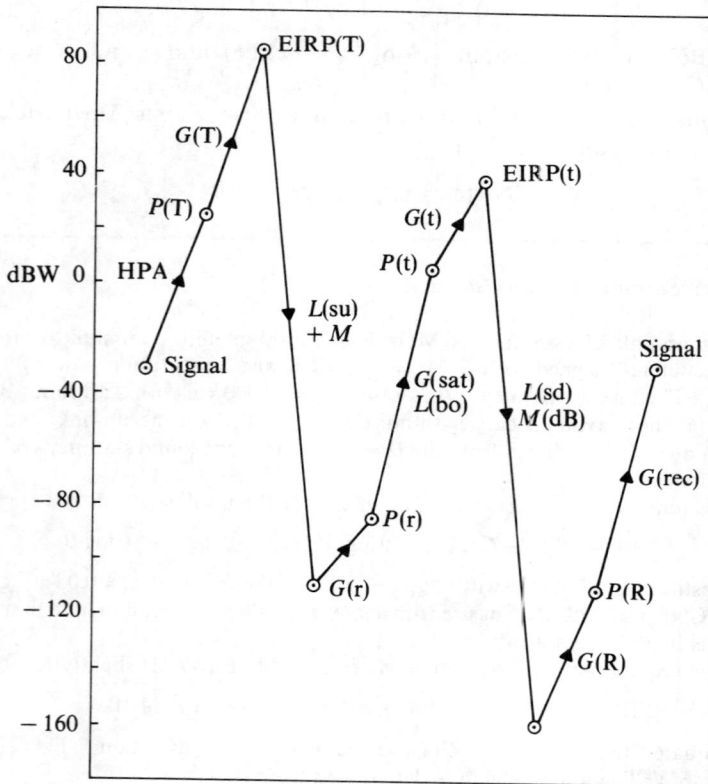

Fig. 9.7 The power profile for a satellite communication link via two ground stations.

On the down-path it is easier to use one of the alternative equations because of the different limitations and restrictions. If we substitute Eq. (9.20) and Eq. (9.24) into Eq. (9.25) we obtain

$$\mathrm{CNR(R)} = P(\mathrm{t}) + G(\mathrm{t}) + G(\mathrm{R}) - L(\mathrm{sd}) - 10\log k - 10\log T_{\mathrm{R}} - 10\log B_{\mathrm{R}}$$

With just a little simplification this reduces to

$$\mathrm{CN_0R(R)} = \mathrm{EIRP(t)} + G/T(\mathrm{R}) - L(\mathrm{sd}) - 10\log k \qquad (9.28)$$

Once again we see that $\mathrm{CN_0R(R)}$ has a maximum value dictated by the maximum available EIRP(t) from the satellite, perhaps reduced half a decibel as *output back-off*; the receiver $G/T(\mathrm{R})$ parameter, again almost constant for the specified operating frequency; and the spatial loss which is almost constant for specified values of frequency and distance.

In certain cases some modification is needed to the down-link equation, as appropriate to the particular application and operation. A small additional loss may be needed to account for atmospheric attenuation and an allowance must be made for the fact that the equation refers to the beam-centre: if the receiver is near the beam-edge, for example, the signal strength is likely to be about 3 dB lower. In addition, a margin $M(\mathrm{dB})$ may be introduced to account for atmospheric loss during heavy rain, variations in path length, transmitter power levels, receiver noise levels, antenna alignments, satellite movements, etc. (Fig. 9.7).

Example 9.7 Satellite down-link calculations

The satellite in the communication system described in Ex. 9.6 operates at $f_{\mathrm{d}} = 4\,\mathrm{GHz}$ with a maximum EIRP(t) = 33 dBW beam-center, equivalent to 30 dBW at the beam-edge, and with an output back-off of 0.5 dB. The earth receiving station operates with $G/T(\mathrm{R}) = 41\,\mathrm{dB/K}$ and bandwidth $B_{\mathrm{d}} = 60\,\mathrm{MHz}$. What is the carrier-to-noise ratio CNR(R) at the receiver if it is assumed $d = 40{,}000\,\mathrm{km}$ near the beam-edge?

This time we can use Eq. (9.28), slightly modified to include both the output back-off and the bandwidth, with EIRP(t) referring to the beam-edge:

$$\mathrm{CNR(R)} = \mathrm{EIRP(t)} - L(\mathrm{obo}) + G/T(\mathrm{R}) - L(\mathrm{sd}) - 10\log k - 10\log B_{\mathrm{d}}$$

where

$$L(\mathrm{sd}) = 20\log d_{\mathrm{d}}(\mathrm{km}) + 20\log f_{\mathrm{d}}(\mathrm{MHz}) + 32.45 = 196.5\,\mathrm{dB}$$

Direct substitution of the data gives us CNR(R) = 24.9 dB. Once again, in anticipation of the next example, 24.9 dB is equivalent to a linear ratio of 309:1.

9.2.2 Carrier-to-Noise-Plus-Interference Ratio

If the unwanted components of a signal are called noise, then noise exists in several forms in a radio communication link. However, for our purposes it is sufficient to group the contributions into just two categories: *noise*, meaning the thermal noise described in the previous section; and *interference*, due to other causes such as intermodulation effects, frequency reuse, etc. We can therefore represent the ratio of carrier-to-thermal noise as CNR(r), CNR(R) or CNR(S), referring to the up-path (r), down-path (R) or the overall system (S). Similarly, it is possible to determine the carrier-to-interference ratio as CIR(r), CIR(R) and CIR(S). When both noise and interference are present the measure of the quality or performance of the link is now represented by the ratio of carrier-to-noise-plus-interference, denoted variously as CNIR(r), CNIR(R) and CNIR(S) or even by $CN_0IR(r)$, $CN_0IR(R)$ and $CN_0IR(S)$, if more appropriate, where for example,

$$CN_0IR(S) = CNIR(S) + 10 \log B \tag{9.29}$$

To combine the various forms of carrier-to-noise or carrier-to-interference ratios it is necessary to represent them in linear form, as they do not combine directly in logarithmic or decibel form. The total average power of noise and interference together is just the sum of their individual powers, and it is then obvious that, for example:

$$\frac{1}{CNIR_r} = \frac{1}{CNR_r} + \frac{1}{CIR_r} \tag{9.30a}$$

$$\frac{1}{CNIR_R} = \frac{1}{CNR_R} + \frac{1}{CIR_R} \tag{9.30b}$$

$$\frac{1}{CNIR_S} = \frac{1}{CNIR_r} + \frac{1}{CNIR_R} \tag{9.30c}$$

When the mathematics has been completed it is then a simple task to reconvert the answer back into decibels as the system CNIR(S) dB.

Example 9.8 Satellite CNIR(S)

The satellite communication link described in Ex. 9.6 and Ex. 9.7 operates under conditions in which the carrier-to-interference ratios CIR(r) and CIR(R) on the up-path and down-path, respectively, are both equal to 22 dB (linear 158 : 1). Determine the carrier-to-noise-plus-interference ratios for the up-path, down-path and the overall system.

Using the linear values $CNR_r = 2399$ from Ex. 9.6 and $CNR_R = 309$ from Ex. 9.7 with $CIR_r = CIR_R = 158$ given in the question, we can use Eq. (9.30) to obtain

$$\frac{1}{CNIR_r} = \frac{1}{2399} + \frac{1}{158} = \frac{1}{148}$$

$$\frac{1}{CNIR_R} = \frac{1}{309} + \frac{1}{158} = \frac{1}{105}$$

$$\frac{1}{CNIR_S} = \frac{1}{148} + \frac{1}{105} = \frac{1}{61}$$

From this we see that CNIR(r) = 21.7 dB, CNIR(R) = 20.2 dB and CNIR(S) = 17.9 dB, within the approximations used in the above mathematics. This is the maximum value of CNIR(S) at the beam-edge. It may be necessary to reduce this by 2–3 dB to allow for miscellaneous losses and margins so that a more realistic value is nearer (say) 15 dB.

Although perhaps outside the scope of this text, it is of useful background information to note that the carrier-to-noise-plus-interference ratio per 1 Hz of bandwidth $CN_0IR(S)$ is of use in digital communication systems. If E_b represents the energy of one bit of information then the dimensionless ratio $E_b/N_0 = E_b/kT$ is the energy of one bit of information in units of kT joules. With the aid of signal processing theory it is possible to relate the bit error rate of a digital system to the ratio E_b/N_0, usually expressed as $10\log(E_b/N_0)$ dB. In our radio wave propagation equations we can determine $CN_0IR(S)$ as in Ex. 9.8, and this can be related to E_b/N_0 via

$$CN_0IR(S) = 10\log(E_b/N_0) + 10\log R_b \qquad (9.31)$$

where R_b is the bit rate of the system. In our Exs. 9.6 to 9.8 the bit rate is 120 Mb/s ($10\log R_b = 80.8$) and the probable CNIR(S) is about 15 dB. We can therefore write, with $10\log B = 77.7$:

$$CNIR(S) + 10\log B = 10\log(E_b/N_0) + 10\log R_b \qquad (9.32)$$

and $10\log(E_b/N_0) = 11.9$ dB.

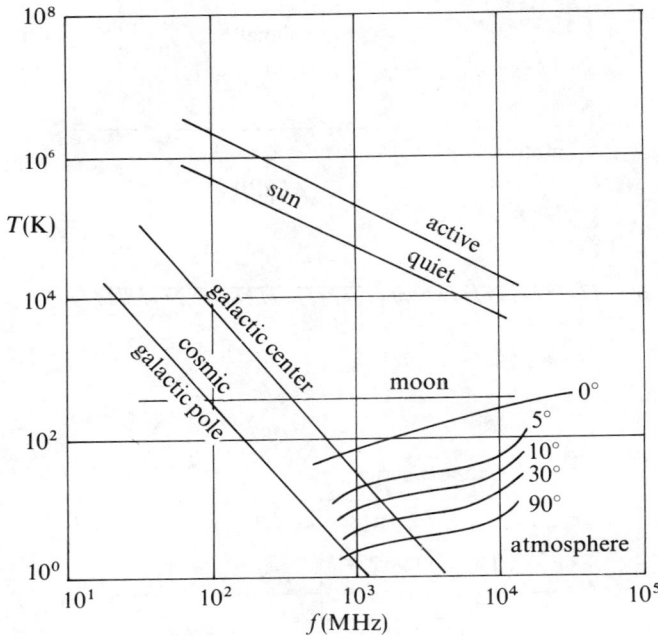

Fig. 9.8 When a receiving antenna is pointed towards the sky it receives noise from cosmic or galactic sources, from the sun or moon if pointed in the right direction, and from the atmosphere surrounding the earth. In the latter case, the angles of elevation of the antenna are indicated.

9.3 NOISE IN SATELLITE COMMUNICATION LINKS

One of the restrictions in satellite communication systems is the presence of radio noise, the r.f. energy within the system bandwidth which is not produced by the signal. It arises in many forms and, in order to use the radio wave propagation equations described earlier, we need to have some numerical idea of each noise contribution.

9.3.1 Cosmic Noise and Sky Noise

Natural background noises are caused by sources within the earth's atmosphere, within the surrounding galaxy and from such objects in space as the sun, moon and radio stars, etc. A rough indication of the magnitude of each of these sources is shown in Fig. 9.8.

If we ignore all other factors, the effective noise temperature T_a of a ground-based receiving antenna would be equal to the cosmic noise temperature T_c. However, as shown in Fig. 9.9, radio signals and noise originating outside the atmosphere are attenuated as they pass through the atmosphere towards the ground receiving antenna. The amount of attenuation depends on several factors, including the elevation angle θ

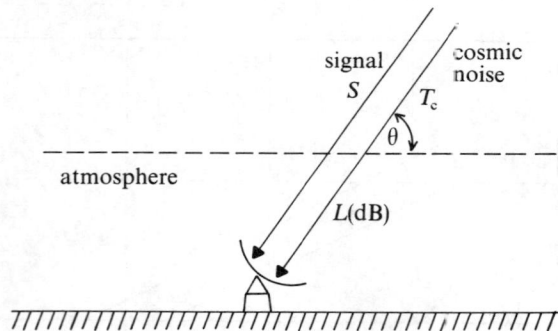

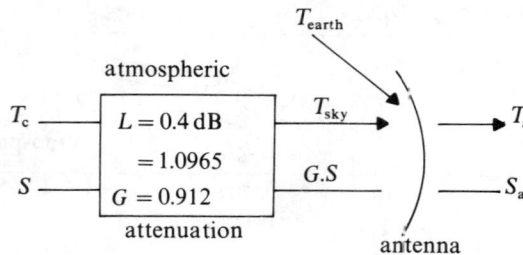

Fig. 9.9 The atmosphere around the earth acts as an attenuator which contributes towards the total sky noise observed by the receiving antenna.

of the antenna. The path length through the atmosphere is greatest when $\theta = 0°$ and least when $\theta = 90°$.

For a given set of conditions there will be an attenuation of L(dB) of both the wanted signal S and the unwanted cosmic noise N_c, represented by the effective noise temperature T_c. If, for example, in the vicinity of $f = 4\,\text{GHz}$ the cosmic noise contributed $T_c = 2.7\,\text{K}$ and the atmospheric attenuation was $0.4\,\text{dB}$ (say), the effect is to produce an effective input noise temperature $T_{ma} = T_0(L-1) = 290(1.0965-1) =$

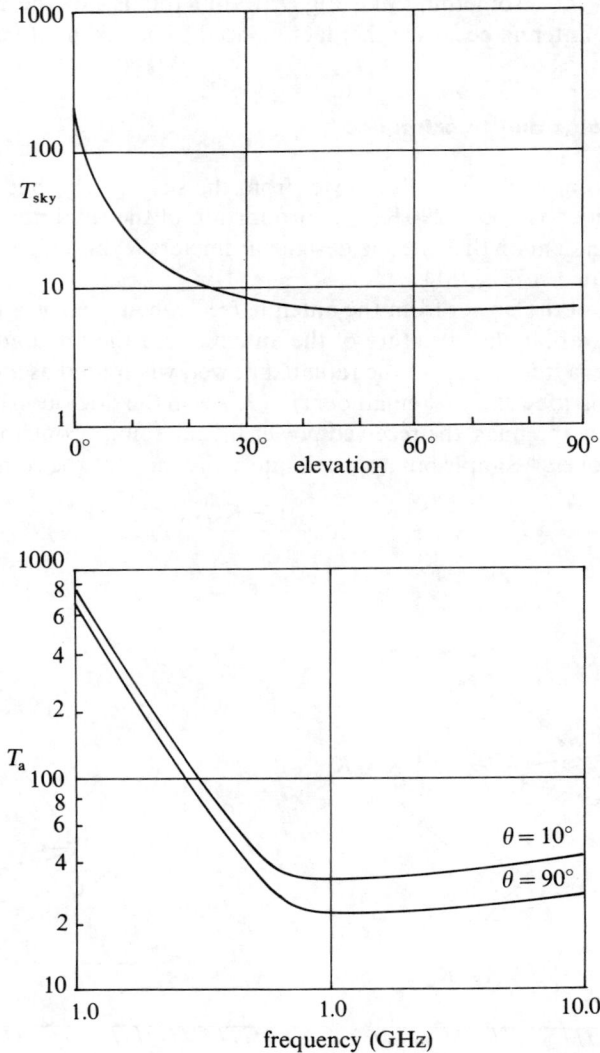

Fig. 9.10 Typical variations of sky noise temperature with elevation, and of antenna noise temperature with frequency.

28 K, where T_{ma} is derived in Appendix 2. At the antenna itself the signal level is $G \times S = S/L = 0.912\,S$ and the noise temperature $T_{sky} = G(T_c + T_{ma}) = 28$ K. This is the effect of atmospheric attenuation. What was originally a carrier-to-noise ratio of S/kT_cB outside the atmosphere becomes a carrier-to-noise ratio of $GS/kT_{sky}B$ or $S/k(T_c + T_{ma})B$, a decrease of $10\log\{(T_c + T_{ma})/T_c\}$ dB. In this example the decrease is 10.6 dB, but note that we have ignored all other factors (Fig. 9.9).

With such an approach it is possible to estimate an overall value of T_{sky} available to the ground-based antenna. Under certain conditions the variation of T_{sky} with angle of elevation θ will look something like Fig. 9.10, but this is not the effective noise temperature of the antenna because other factors need to be taken into consideration.

9.3.2 Antenna Noise and Interference

The receiving antenna picks up noise power from the sky via T_{sky} but also from the earth via T_{earth}, which is about 290 K. The proportion of the total noise due to each contributor is dependent on the antenna design parameters K_a and K_b and the angle of elevation θ illustrated in Fig. 9.11.

A fraction K_a of the power from the antenna feed, when used as a transmitter, is reflected by the paraboloidal structure of the antenna and the remainder $(1 - K_a)$ is spillover. Similarly, a fraction K_b of the radiated power, when used as a transmitter, is confined to the main lobe and the remainder $(1 - K_b)$ is in the sidelobes. In reverse, due to the reciprocity of antennas, the received power includes noise contributions via the spillover and sidelobes. A simple but approximate expression for the combined effect is

$$T_a = \frac{1 + K_a K_b}{2} T_{sky} + \frac{1 - K_a K_b}{2} T_{earth} \tag{9.33}$$

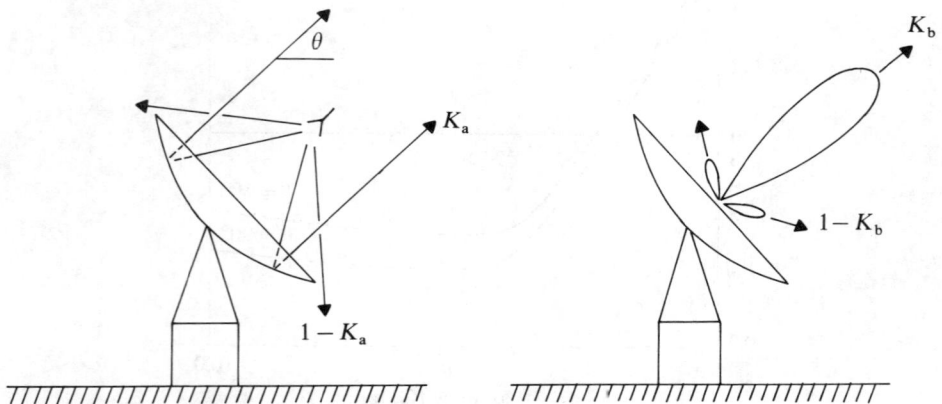

Fig. 9.11 The two antenna design factors, K_a and K_b, which represent spillover and sidelobe levels in a paraboloidal antenna.

Note that T_{sky} varies with angle of elevation and frequency, as described earlier, and that the minimum realistic angle of elevation is about $5°$.

At the minimum angle of elevation, $T_{sky} = 33$ K (say). If a paraboloidal reflector antenna with focal feed has (say) $K_a = K_b = 0.9$, then application of Eq. (9.33) indicates an antenna noise temperature of 57 K. If a Cassegrain antenna with design parameters of (say) $K_a = K_b = 0.95$ is used instead, the antenna noise temperature will be about 45 K. The Cassegrain antenna is illustrated in Fig. 9.12.

Transmitting antennas need also to be considered as sources of radio noise. For instance, geostationary satellites are located in the equatorial plane at a height of 35,783 km above the equator and displaced (say) every $2°$–$4°$ apart. As shown in Fig. 9.13 this means that the ground-based transmitter intended for use with satellite A must have a beamwidth sufficiently narrow to avoid any communication with satellite B. Initially, using the rough approximation

$$BW = 2\theta_b = 70\lambda/D \qquad \text{degrees} \qquad (9.34)$$

with λ about 0.05 m, we have $\theta_b^0 = 1.75/D$(m) and hence a beamwidth of $1°$ $(\theta_b = \frac{1}{2}°)$ can be achieved with a diameter of approximately 3.5 m.

However, as mentioned earlier, antennas have sidelobes. One of the effects of sidelobes is to reduce the antenna illumination efficiency; another is to increase the antenna noise temperature; and the third is to broadcast (and receive) radio waves in a direction other than that desired. To prevent this, the antenna designer is required to restrict the off-axis radiation to within certain specifications.

In Ex. 9.6 we noted that, under the specified conditions, the maximum EIRP(T) required for a given satellite link was 84 dBW. At first it might seem possible to achieve this with any combination of $G(T)$ and $P(T)$, the higher the gain the less power required, though large antennas are not without their problems and costs. Now, however, it is seen that $G(T)$ and $P(T)$ must be chosen with a little more care. Using Fig. 9.13 we can

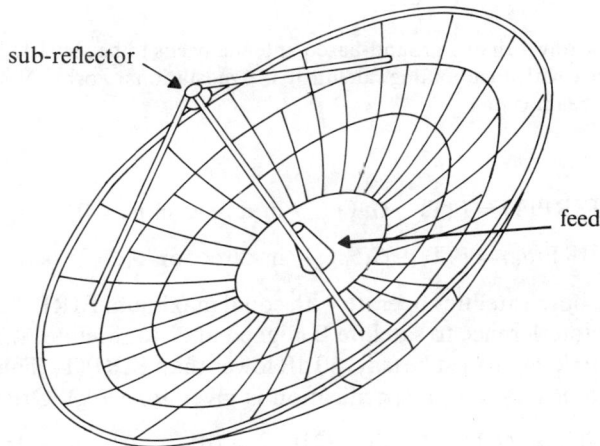

Fig. 9.12 The Cassegrain paraboloidal reflector antenna.

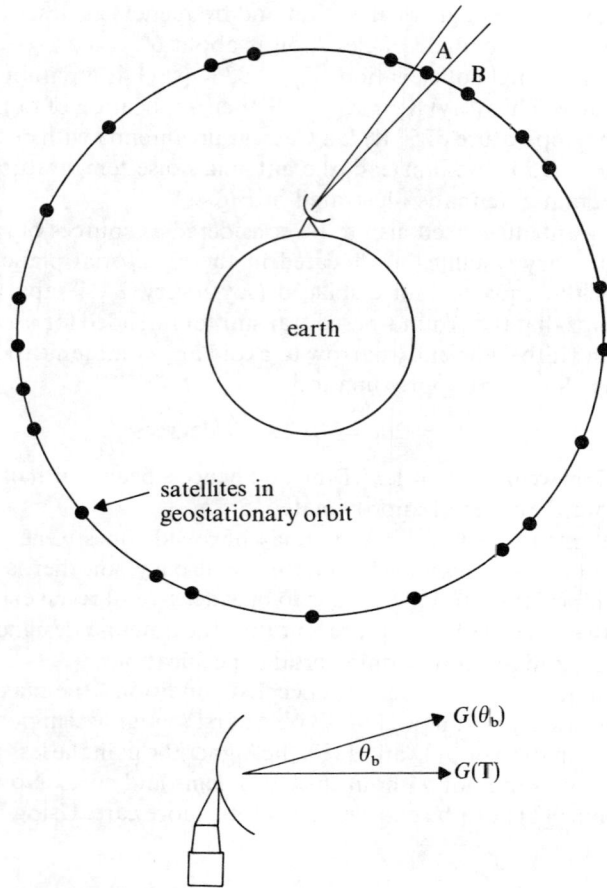

Fig. 9.13 The beamwidth of a ground-based antenna needs to be restricted sufficiently to avoid interference with an adjoining satellite in the geostationary orbit. Note that θ_b is the angle from the beam center.

write:

$$\text{EIRP}(T) = P(T) + G(T) \quad \text{in direction } \theta_b = 0° \qquad (9.35a)$$

$$\text{EIRP}(\theta_b) = P(T) + G(\theta_b) \quad \text{in direction } \theta_b = 2° \text{ (say)} \qquad (9.35b)$$

In the case where both satellites operate with equal maximum EIRP(T), EIRP(θ_b) must be considered as interference to satellite B aligned at 2° to satellite A. To ensure it has negligible effect, EIRP(θ_b) must be (say) 30 dB lower than EIRP(T). This still leaves two variables to choose, so a design specification is given for $G(\theta_b)$. Originally this was:

$$G(\theta_b) = G(T) - \{G(T) - 32\}\theta_b° \quad \text{dB} \qquad 0° \leqslant \theta_b < 1°$$

$$G(\theta_b) = 32 - 25 \log_{10} \theta_b° \quad \text{dB} \qquad 1° \leqslant \theta_b < 48° \qquad (9.36)$$

$$G(\theta_b) = -10 \, \text{dB} \qquad 48° \leqslant \theta_b$$

but more recently it has been felt necessary to tighten this to:

$$G(\theta_b) = 29 - 25 \log_{10} \theta_b^\circ \quad \text{dB} \quad 1^\circ \leqslant \theta_b^\circ < 48^\circ \tag{9.37}$$

for the equation most appropriate to our problem.

Yet another source of interference is called *cross-polarization*. Two signals processed at the same frequency are broadcast with different polarizations, such as vertical and horizontal, or circularly polarized in opposite directions, for example, Cross-polar isolation is a measure of the immunity between the two signals. Typical values are of the order of 30 to 35 dB.

Example 9.9 Interference limitations

Two satellites in geostationary orbits, displaced 2°, each operate with the same maximum value of $\text{EIRP(T)} = 84$ dBW at 6 GHz. What is a suitable value of minimum ground antenna gain $G(\text{T})$ if the carrier-to-interference ratio is to be at least 30 dB?

Using the more recent recommendation of Eq. (9.37) we obtain

$$G(\theta_b = 2^\circ) = 29 - 25 \log 2 = 21.5 \text{ dB}$$

Hence from the specification $\text{EIRP(T)} - \text{EIRP}(\theta_b) = 30$ dB, we note that

$$\{P(\text{T}) + G(\text{T})\} - \{P(\text{T}) + G(\theta_b)\} = 30$$

and hence $G(\text{T}) = 51.5$ dB. This suggests a maximum $P(\text{T}) = 32.5$ dBW and a minimum diameter of 8.1 m for a 'typical' paraboloidal reflector antenna. Improved antenna designs permit antennas of lower dimensions to provide the same standards, some being elliptical rather than circular in cross-section.

9.4 SOME SATELLITE SYSTEMS

There are too many satellites in orbit around the earth to permit any realistic description of all the uses to which they are put. This is the province of the specialist texts. Instead, we shall consider briefly just a few typical examples, concentrating mainly on a simplified insight into the radio wave propagation relationships of each.

9.4.1 Commercial Satellites

One of the most well known family of satellites is called *Intelsat* (International Telecommunications Satellite Consortium, formed in 1964). Since the launch of *Early Bird* or *Intelsat I* in 1965 with a capacity of about 240 transatlantic telephone channels, continued modifications and developments have produced *Intelsat*s Mark II (1967, Atlantic and Pacific Oceans), III (1968, Atlantic, Pacific and Indian Oceans), IV (1971), IVA (1976), V (1980, 12,000 channels), VA (1984, 15,000 channels) and VI (about 1986, about 35,000 channels). Very roughly, each 240–300 telephone channels can be replaced by one television channel.

The satellites are now located over the Atlantic, Pacific and Indian Oceans for virtually worldwide communication and the more recent versions have antenna systems with global, 'hemi', 'zonal' and 'spot' beam coverages, linking two or more groundstations with an up-frequency of 5.925–6.425 GHz and a down-frequency of 3.700–4.200 GHz. Two types of earth station are permitted in this frequency range: standard A earth station with $G/T(R) = 40.7$ dB/K and antennas of about 32 m diameter; and standard B earth station with reduced $G/T(R) = 31.7$ dB/K and smaller antennas of about 11–13 m diameter. The more recent versions of *Intelsat* satellites also use the 14/11 GHz up-down frequency bands, and these require standard C earth stations with $G/T(R) = 39$ dB/K on a clear day (or probably nearer 40–41 dB/K to make an allowance for local weather conditions) with antennas of about 18–19 m in diameter.

Example 9.10 Intelsat down-link

If it is assumed that the global beam of an Intelsat satellite provides EIRP(t) = 23 dBW at the beam edge, estimate the carrier-to-noise ratio at the receiver for a standard A earth station. If one of the spot beams provides EIRP(t) = 32 dBW at its beam edge, estimate the CNR(R) for a standard B earth station. Assume bandwidth $B = 36$ MHz in each case.

The down-path analysis resembles that in Ex. 9.7, where it is shown that

$$\text{CNR(R)} = \text{EIRP(t)} + G/T(R) - L(\text{sd}) - 10 \log k - 10 \log B$$

For the global beam, the maximum range at the beam edge is 41,675 km from Fig. 9.4. However, if $\theta = 5°$ is taken as being the practical limit, then $d = 41,125$ km is more appropriate. Either way, the spatial loss is approximately

$$L(\text{sd}) = 20 \log d + 20 \log f + 32.45 = 197 \text{ dB at } 4.0 \text{ GHz}$$

Substitution of EIRP(t) = 23 dBW, $G/T(R) = 40.7$ dB/K, $L(\text{sd}) = 197$ dB, $10 \log k = -228.6$ and $B = 36$ MHz into this expression gives CNR(R) = 19.7 dB for the global beam. Increasing EIRP(t) by 9 dB to 32 dBW and reducing $G/T(R)$ by a similar difference gives us an identical answer for the CNR(R) of the spot beam, but the spatial loss will obviously be less in this case. The system will operate satisfactorily with a CNR(R) of about 13–14 dB, and hence we have a margin of about 6 dB to allow for other losses ignored in the analysis.

The Intelsat system uses both *frequency division multiple access* (FDMA) and *time division multiple access* (TDMA) as well as other radio systems. In the FDMA system the frequency bandwidth of the satellite transponders are grouped into specified bands or channels and these are allocated as appropriate to specified links and users. In the TDMA system the link is established with the full frequency bandwidth of the satellite but allocated to individual users on a synchronized time-sharing basis. A third arrangement (SCPC) is self-explanatory, a single channel per carrier, though the channel is not necessarily a standard telephony channel.

The many commercial users include, for example, the British Broadcasting Corporation (BBC), which transmits its World Service program from the Madley earth station in England via *Intelsat V* over the Indian Ocean to such places as Cyprus,

Masirah in Oman and Singapore. The signal is received in, say, Masirah via a standard B earth station and then broadcast from the BBC transmitters there in the normal manner. This gives a much better quality 'live' broadcast compared with earlier techniques and also avoids the need to transport magnetic tapes of programs (Fig. 9.14).

The Masirah standard B earth station uses a 10 m Gregorian antenna at 4095 MHz with a system noise temperature of only 76 K. It therefore must achieve the Intelsat specification for $G/T(R) = 31.7\,\text{dB/K}$ with a minimum gain of $G(R) = 31.7 + 10\log T = 50.5\,\text{dB}$. This is equivalent to $G = 112{,}412$ linearly. Using

$$G = \frac{\pi D^2}{4} \times \frac{4\pi}{\lambda^2} \times \eta = 112{,}412$$

we can obtain an estimate of the illumination efficiency η as being 61%. The Gregorian antenna is a variation of the Cassegrain antenna with an elliptic sub-reflector. It achieves both a higher illumination efficiency η (up to 75%) and cross-polarization isolation. It is likely that the Masirah antenna's efficiency is actually greater than 61% and its $G/T(R)$ figure comfortably exceeds the 31.7 dB/K specification.

The link is one of the special SCPC types (single channel per carrier), occupying two normal telephony channels at baseband to produce the high-grade broadcast quality audio channel reception with 6.4 kHz bandwidth.

There are similar domestic commercial satellite systems throughout the world, some in operation, some still in the planning stage. It is estimated (Ref. 9.2) that there

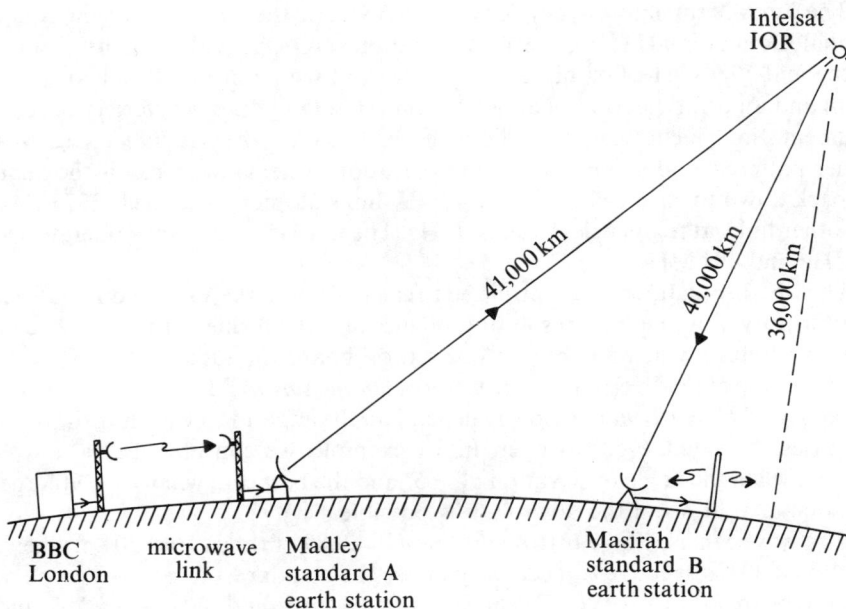

Fig. 9.14 An example of the use of a geostationary satellite to provide 'live' high-quality MF broadcasting at a distant transmitter.

may be some 60 systems by the turn of the century including about 160 geostationary satellites besides the low-orbit non-synchronous types. One of the former includes the geostationary European Communication Satellite (ECS) from the EUTELSAT Organization. This satellite includes four 'beams': a large 'Eurobeam' for reception of all up-path carriers and transmission of down-path television and special services; and three spot beams to conserve power with $3.7°$ beamwidth and known as spot east, spot west and spot Atlantic.

The ECS satellite uses digital transmission systems for telephony circuits using 120 Mb/s TDMA and also provide Eurovision programs with FM video and digital audio. It has twelve transponders, though not all in actual use: some are standby units. Each transponder can deal with about 3200 channels for telephony or two television channels. The links operate in the ranges 14.0–14.5 GHz up, 10.95–11.20 GHz and 11.45–11.70 GHz down, with special services for business systems around 14.0/12.5 GHz. The G/T(R) is required to be 39 dB/K for 90% of time, which means a slightly higher design value to account for local weather variations.

9.4.2 Weather Satellites

Various satellites have been used in recent years to enable meteorologists to observe the patterns of weather throughout the world. There are the usual two types of arrangements: the low orbiting satellites (including the well-known Tiros-N group) and the geostationary satellites.

The Tiros-N satellite was developed by NASA and then operated by the American National Oceanic and Atmospheric Association (NOAA), with launches from 1978 into the mid-1980s. The first has a near polar orbit with a mean altitude of the order 850 km and an orbital period of about 102 minutes. On board is a variety of scientific instrumentation, including the radiometer with which the satellite recognizes the weather patterns, and some radio communication systems to transmit the acquired data back down to earth. These use two VHF links at frequencies near 137 MHz and three other links at frequencies near 1.7 GHz. The up-path radio links operate around 148 MHz and 401 MHz.

The weather pictures are transmitted to earth via both the VHF and 1.7 GHz links, the latter providing the high-resolution picture and the former providing a reduced-resolution picture which has been processed on board the satellite.

In the case of the VHF *automatic picture transmission* (APT) link, the quality of the received signal on earth will obviously depend on the angle of elevation of the receiving antenna as the satellite passes over. If, for example, we consider the worst case as occurring when the angle of elevation $\theta_1 = 5°$ and the best case when $\theta_2 = 90°$, then an approximate analysis of the down-link is possible.

Firstly we can use Eq. (9.6) to determine $d_1 = 2889$ km and $d_2 = 850$ km when $\theta_1 = 5°$ and $\theta_2 = 90°$, respectively. Then we can use the standard radio wave propagation equations with P(t) $= 7$ dBW (5 W linear), G(t)$_1 = 0$ dBi and G(t)$_2 = 3.5$ dBi; and the losses between the transmitter and antenna total L(t) $= 2$ dB, say. At $f = 137.5$ MHz the down-path spatial losses are L(sd)$_1 = 144.5$ dB and L(sd)$_2 = 133.8$ dB, respectively.

If we substitute these data into

$$P(R) = P(t) - L(t) + G(t) - L(sd) + G(R) \tag{9.38}$$

and ignore other factors, the nominal values of received power levels are $P(R)_1 = -139.5 + G(R)$ and $P(R)_2 = -125.3 + G(R)$ dBW, respectively.

The larger value of received power may be used to estimate the maximum permitted antenna gain $G(R)$ to avoid saturating the receiver. If the latter had an upper limit of, say, -60 dBW, then $P(R)_2 + G(R)$ must be less than this value, or $G(R)$ must not exceed 65 dB.

The lower of the two estimated values of received power is used with the minimum acceptable carrier-to-noise ratio $CNR(R) = 12$ dB and a margin $M(dB)$ of, say, 6 dB or more to determine the noise performance, etc. Using:

$$CNR(R) + M(dB) = P(R) - N(R)$$

we obtain

$$12 + 6 = -139.5 + G(R) + 228.6 - 10 \log T - 10 \log B$$

$$G/T(dB) = -71.1 + 10 \log B \tag{9.39}$$

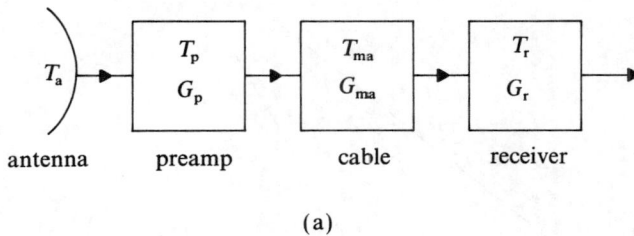

(a)

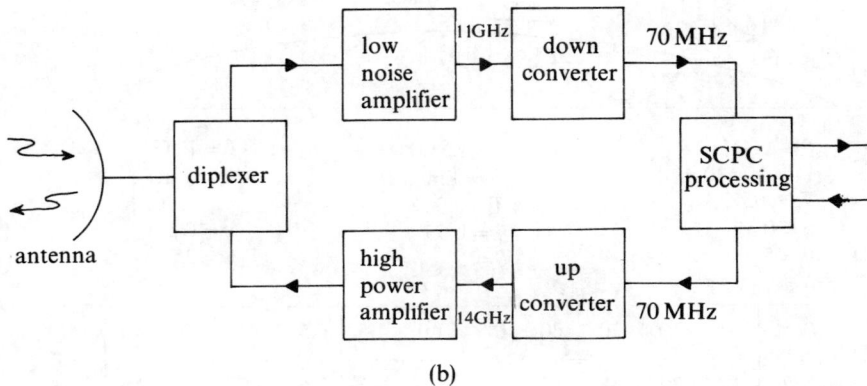

(b)

Fig. 9.15 (a) A simple receiver for the VHF automatic picture transmission link from a weather satellite. (b) A basic transceiver for an SCPC satellite communication link.

The receiver illustrated in Fig. 9.15a will be used as a typical example with $T_a = 1500$ K at 137.5 MHz, $F(p) = 6.5$ dB or $T_p = 1005$ K, $L(\text{dB}) = 2$ dB or $T_{ma} = 170$ K, $F(r) = 6$ dB or $T_r = 864.5$ K, $G(p) = 30$ dB or $G_p = 1000$. Following the procedure outlined in Appendix 2 we can write:

$$T_s = T_a + T_p + \frac{T_{ma}}{G_p} + \frac{T_r}{G_p G_{ma}} = 1500 + 1005 + \frac{170}{1000} + \frac{864.5}{1000 \times 0.63} = 2606 \text{ K}$$

or $10 \log T_s = 34$. Substituting this value into Eq. (9.39) we obtain the minimum gain with this receiver as being $G(R) = -37.1 + 10 \log B$. For receiver bandwidths of 50 kHz or 100 kHz the minimum antenna gains are 9.9 dB and 12.9 dB, respectively. To allow for some variation from a precise analysis, a gain of 15–20 dB would be adequate.

The geostationary weather satellites include the *geostationary operational environmental* satellites (GOES) operated by NOAA for America, the *geostationary meteorological satellite* (GMS) operated by NASDA in Japan, and the European Space Agency's METEOSAT. For good reception of signals the receiving stations require antennas of about 4 m diameter and receiving systems with G/T around 10 dB/K.

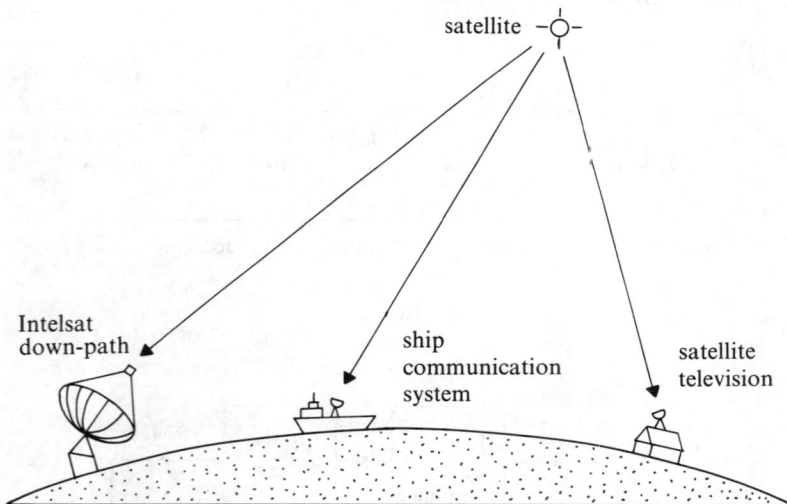

satellite

Intelsat
down-path

ship
communication
system

satellite
television

$f = 4$ GHz
$D = 10$–30 m
$T_s = 65$K
$\eta = 0.55$
$\theta = 5°$

$f = 1.5$ GHz
$D = 1$ m
$T_s = 600$K
$\eta = 0.55$
$\theta = 5°$

$f = 12$ GHz
$D = 1$ m
$T_s = 1200$ K
$\eta = 0.55$
$\theta = 25°$

D	1	10	20	30	0.5	1.0	1.5	0.5	0.7	0.9	1.0
G/T	12	32	38	41	-12	-6	-3	2.6	5.5	7.7	8.6

Fig. 9.16 Some typical parameters for satellite communication systems.

9.4.3 Broadcasting from Satellites

One increasing application for geostationary satellites is for broadcasting, with a number of countries operating or planning for such facilities. The World Administrative Radio Conference (WARC) in 1977 considered arrangements for this particular application of satellites and allotted frequencies at various points in the radio frequency spectrum. In the vicinity of 12 GHz, for example, Region I may use 11.7–12.5 GHz, but in general the limits worldwide are 11.7–12.75 GHz.

In Region I this allocation allows 40 channels for television, each with 27 MHz bandwidth, but spaced at 19.18 MHz intervals. With frequency reuse from four such satellites at 13°W, 19°W, 25°W and 31°W there will be a total of 160 available channels. If each satellite is shared by, say, eight countries then each country can have five such channels suitably displaced throughout the frequency range. Their use need not be for television, provided there are certain interference limits. The UK is allotted five channels on the satellite at 31°W, channels 4, 8, 12, 16 and 20, with circular polarization. Adjacent channels have alternate left-hand and right-hand polarizations to reduce interference.

If we restrict our consideration to the problems of radio wave propagation, the design of the link is such that the television receiving system is required to have a nominal or minimum $G/T(R)$ figure of 6 dB/K (Fig. 9.16) and the antenna to have a maximum beamwidth of 2° (1.8° in Region 2). In addition, the satellite's EIRP(t) per channel is restricted so that the maximum power flux $P_a(R)$ does not exceed -103 dBW/m^2 (50 pW/m^2) at the 3 dB *edge* of the beam. This figure is -105 dBW/m^2 in Region 2.

Example 9.11 Minimum receiver signal level for a good picture

Determine the minimum level of received signal power required for a good television picture from a direct broadcast satellite if the minimum CNR(R) is 12 dB, including 10 dB threshold and 2 dB for ageing, etc., and the receiver noise figure is (a) 7 dB, or (b) 5 dB.

Assuming that the bulk of the system noise temperature is contributed by the receiver, we can write the noise level as

$$N(R) = -228.6 + 10 \log T_0 + 10 \log B + 10 \log F_s$$

where $T_0 = 290$ K, $B = 27$ MHz and $F(s)$ is 7 dB. This gives the noise level as -122.7 dBW. With a required CNR(R) of 12 dB, the minimum level of received power is

$$P(R) = N(R) + CNR(R) = -110.7 \text{ dBW}$$

With the better receiver, $F(s) = 5$ dB and the minimum level of received signal power is $P(R) = -112.7$ dBW.

The receiving antenna must be capable of converting the nominal power flux at earth's surface (-103 dBW/m^2 at the beam edge) into the minimum required level of receiver input signal. If the receiving antenna has an effective aperture A_e, the power received P_R is related to the power flux via

$$P(R) = P_a(R) + 10 \log A_e \tag{9.40}$$

Using the data given in Ex. 9.11, with $F(s) = 7\,dB$ and $P(R) = -110.7\,dBW$, for example, Eq. (9.40) can be used to obtain $A_e = 0.17\,m^2$. If we assume that the illumination efficiency is typically $\eta = 0.55$, the physical area A of the (paraboloidal reflector) antenna can be obtained via $A_e = \eta A$ as $A = 0.31\,m^2$. The corresponding diameter D is 0.63 m and the approximate beamwidth, from

$$BW = 70\lambda/D \tag{9.41}$$

is 2.8°. This exceeds the 2° specification and hence, approaching the problem in reverse, with a bandwidth of 2° and wavelength (at 12 GHz) of 0.025 m, the diameter should be at least $D = 0.875\,m$.

We thus have the two requirements: $D > 0.63\,m$ to achieve the minimum power level and $D > 0.875\,m$ to achieve the maximum beamwidth specification. Obviously the latter value must be used in

$$G(R) = 20 \log f + 20 \log D - 42.3 \tag{9.42}$$

to obtain $G(R) > 38.1\,dB$. To provide $G/T(R) = 6\,dB$ with such an antenna we need $10 \log T_s = 32.1$ and hence $F(s) = 10 \log T_s - 10 \log T_0 = 7.5\,dB$. This exceeds the 7 dB quoted earlier.

Example 9.12 Satellite EIRP(t) for television broadcasting

Estimate the satellite EIRP(t) per channel for the direct broadcasting of a television picture to an earth receiver located $d = 39,600\,km$ from the satellite. Assume a 3 dB margin to account for antenna alignment, ageing, atmospheric attenuation, etc. What satellite antenna gain is necessary if the satellite channel power is restricted to, say, 20 dBW (100 W)?

We can start with the assumption that the power flux is $-100\,dBW/m^2$ at *beam center* and note that

$$P_a(R) = EIRP(t) - 10 \log(4\pi d^2)$$

with d in meters. This gives us $EIRP(t) = 63\,dBW$ plus a 3 dB margin, and hence an antenna with $G(t) = 46\,dB$ and $D = 2.2\,m$.

9.4.4 Marine Communication Satellites

In order to provide ships at sea with additional worldwide communication links to land bases via coastal receiving stations, an international organization INMARSAT was set up with effect from 1982 to organize and operate such a system via geostationary satellites at about 15°W, 176°E and 73°E over the Atlantic, Pacific and Indian Oceans. Each satellite operates with about 8 telephony links and 44 teletext links.

The principal parameters for radio wave propagation include the operating frequency bands: 1·6 GHz up and 1·5 GHz down for the satellite-to-ship link; 6 GHz up and 4 GHz down for the coast station-to-satellite link. The coastal stations themselves are the responsibility of the individual countries and not of INMARSAT. They employ antennas with about 13 m diameter and G/T(R) around 32 dB/K.

The ship has an antenna of about 1.2 m diameter and requires an overall CN_0R(R) of the order 52 dB for a voice channel and nearer 31 dB for teletext.

Example 9.13 Ship receiver G/T(R) ratio

Estimate the G/T(R) ratio of the ship receiving system if the antenna is 1.2 m diameter with maximum noise temperature $T_a = 100$ K at an elevation of $10°$ and the receiver incorporates a low-noise amplifier with F(r) = 2.3 dB. The connection between antenna and amplifier has a loss of 1.5 dB at 1540 MHz.

We can use the standard equation for system noise temperature:

$$T_s = T_a + T_0(L - 1) + \frac{T_0(F_r - 1)}{1/L} = 506.6 \text{ K}$$

where $T_0 = 290$ K, $L = 1.413$ (1.5 dB) and $F_r = 1.70$ (2.3 dB). Similarly, we can use the standard equation for antenna gain assuming 0.55 illumination efficiency:

$$G(\text{R}) = 20 \log f + 20 \log D - 42.3 = 23.03 \text{ dB}$$

if $f = 1540$ MHz and $D = 1.2$ m. Hence the G/T(R) ratio is -4 dB/K.

The down-path link from satellite to ship can be analyzed by using Eq. (9.28) with a margin of 4.2 dB. With a minimum elevation near $10°$ the spatial loss is about 188.1 dB maximum. Hence

$$CN_0R(\text{R}) = EIRP(\text{t}) + G/T(\text{R}) - L(\text{sd}) - 10 \log k - M(\text{dB}) \qquad (9.43)$$

or

$$51 = EIRP(\text{t}) - 4 - 188.1 + 228.6 - 4.2$$

and the satellite EIRP(t) per voice channel is about 18.7 dBW.

For the up-path link from ship to satellite at, say, 1640 MHz, the spatial loss is about 188.7 dB and EIRP(T) about 37 dBW (or 25 W into 23 dB antenna). Using the same equation with a required CN_0R(R) now about 61 dB (with 10 dB allowance for degradation later), it is seen that the satellite G/T(r) ratio is roughly -12 dB/K.

The satellite antenna has to have a global beamwidth, approximately, and we can see from Fig. 9.4 that this is about $17.4°$. For the up-path frequency of, say, 1640 MHz (wavelength 0.183 m) we can use the approximation $BW = 70\lambda/D$ to estimate $D = 0.735$ m and then substitute this value into

$$G(\text{r}) = 20 \log f + 20 \log D - 42.3 \qquad (9.44)$$

to obtain G(r) about 19 dB. For the earlier G/T(r) of about -12 dB/K we would need $10 \log T_s = 27$ or $T_s = 501$ K. This is not easily achieved as the **antenna temperature** with global beamwidth is about 290 K. Using the data in Ex. 9.13 we can roughly estimate that $T_s = 700$ K is more realistic, and hence $G(r)$ needs to be about 16.45 dB. This corresponds to $D = 0.53$ m (Fig. 9.16).

There are corresponding analyses for the up-path and down-path links of the satellite-to-coastal station link.

9.5 DEEP SPACE COMMUNICATION

For the reception of radio signals from space vehicles in deep space the problem is one of distance and noise levels. The basic propagation equation remains unchanged, with

$$P(R) = P(t) + G(t) + G(R) - L(s) - L(\text{misc}) \tag{9.45}$$

where $L(\text{misc})$ includes such losses as transmitter and receiver losses, antenna pointing and polarization losses, and atmospheric attenuation. The noise equation remains

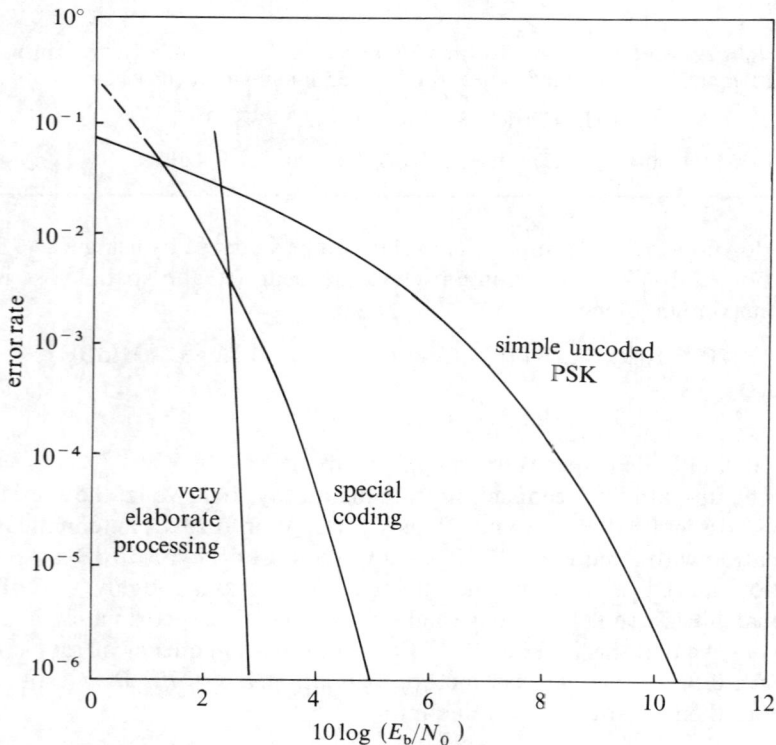

Fig. 9.17 The relationship between bit error rate and the ratio of E_b/N_0 from curves similar to these depending upon the type of signal processing involved in the communication system.

unaltered, with

$$N(R) = 10 \log k + 10 \log T_s + 10 \log B \qquad (9.46)$$

and every effort is needed to reduce T_s, for example by using cryogenically cooled maser amplifiers and by selecting the frequencies at which the atmospheric loss is least.

Following on from earlier work, we can use these equations to obtain either the carrier-to-noise ratio CNR(R) or the carrier-to-noise spectral density ratio $CN_0R(R)$ via

$$CN_0R(R) = P(R) - 10 \log k - 10 \log T_s \qquad (9.47)$$

and hence the bit-energy-to-noise spectral density ratio E_b/N_0 via

$$CN_0R(R) = 10 \log(E_b/N_0) + 10 \log R_b \qquad (9.48)$$

where R_b is the bit-rate. With the aid of signal processing theory (outside the scope of this text), it is possible to construct graphs relating E_b/N_0 dB with bit error rates (Fig. 9.17).

Example 9.14 Signals from deep space

Two space vehicles were launched by NASA in 1977 to study the planets Jupiter (1979), Saturn (1980–81), Uranus (1986) and Neptune (1989). Assume that 13.3 dBW of power is transmitted back to earth at 8.4 GHz via a 48 dB gain antenna when the vehicle is in the neighborhood of Jupiter, $d = 6.8 \times 10^8$ km, and that miscellaneous propagation losses are 1.6 dB. The earth receiving station employs a 64 m diameter antenna with 42% illumination efficiency and a maser-type receiver with overall noise temperature of only 29 K. The signal processing techniques require a minimum $CN_0R(R)$ of 53 dB-Hz (to correspond with a bit error rate of 5×10^{-3} when the data rate is about 115 kb/s). Is reception possible over this great range?

We can start by estimating the spatial loss at 8400 MHz over distance $d = 6.8 \times 10^8$ km:

$$L(s) = 20 \log d + 20 \log f + 32.45 = 287.6 \, dB$$

The gain of the receiving antenna must be determined from first principles as the usual equation assumes an illumination efficiency of 54%. Hence

$$G_R = \frac{\eta \pi^2 D^2}{\lambda^2} = 13{,}311{,}440 \qquad \text{or} \qquad 71.24 \, dB$$

Substituting these figures into the standard propagation equation gives

$$P(R) = P(t) + G(t) + G(R) - L(s) - L(misc) = -156.65 \, dBW$$

Finally, the carrier-to-noise ratio per 1 Hz bandwidth can be determined from

$$CN_0R(R) = P(R) - 10 \log k - 10 \log T_R = 57.33 \, dB\text{-}Hz$$

Hence we can conclude that the received $CN_0R(R)$ of 57 dB-Hz exceeds the minimum required $CN_0R(R)$ by about 4 dB, which permits a small margin for discrepancies.

PROBLEMS

9.1   It is required that a non-synchronous satellite in inclined circular orbit ($\psi_i = 80°$) passes over a receiving site ($\phi = 50°$) exactly every three hours. Determine: (a) the orbital radius; (b) the mean altitude above earth's surface; (c) the maximum visible range of the satellite; (d) the orbital velocity; and (e) the Doppler frequency shift when the satellite is observed at an angle of elevation of 15° approaching the receiver. Assume the satellite transmits a 435 MHz tone.

9.2   Plot a graph of the Doppler frequency shift f_D against angle of elevation ($0° < \theta < 180°$) for the satellite and receiving station described in Prob. 9.1. At what elevation is $f_D = 0$?

9.3   Estimate the maximum power of a ground-based transmitter which is used with a 25 m diameter antenna to transmit a signal to a geostationary satellite at a distance of 40,000 km if the frequency is 6100 MHz and the satellite is designed to operate with a maximum power flux $P_a(r) = -76\,\mathrm{dBW/m^2}$ after allowing for back-off.

9.4   The satellite described in Prob. 9.3 has a maximum $P_a(r) = -76\,\mathrm{dBW/m^2}$ in the vicinity of $f = 6100\,\mathrm{MHz}$, allowing for back-off. If the antenna is a paraboloidal reflector type with 55% illumination efficiency designed for global coverage, estimate the $CN_0R(r)$ at the satellite input. Assume for simplicity that the satellite is an amplifier with $F(r) = 5\,\mathrm{dB}$.

9.5   By making the assumption that the illumination efficiency of a paraboloidal reflector antenna is roughly constant at 55% and that its beamwidth is virtually $BW = 70\lambda/D$ degrees, show that the gain of a satellite transmitting antenna is about 19 to 20 dB if designed for full global coverage and is independent of the operating frequency.

9.6   The down-link of an imaginary SCPC satellite communication system with a global coverage antenna has a range of about 40,000 km. The receiving paraboloidal reflector antenna is to be very small, roughly 0.1 m to 0.2 m, and the associated receiver is to have a low noise figure around 4 dB to 6 dB with 0.5 dB loss between antenna and receiver. Assuming T_a has a maximum value of 100 K and that initially $CN_0R(R) = 52\,\mathrm{dB}$ is required, estimate the satellite EIRP(T) and $P(t)$ per channel, ignoring margins, for all combinations of D and $F(R)$ just given. If a poorer-quality reception is acceptable, with $CN_0R(R) = 44\,\mathrm{dB}$, and an allowance of 4 dB for margin is introduced, determine the satellite power radiated per channel when $D = 0.2\,\mathrm{m}$ and $F(R) = 4\,\mathrm{dB}$.

9.7   A satellite communication system with bandwidth $B = 60\,\mathrm{MHz}$ has an up-path $CN_0IR(r) = 98\,\mathrm{dB}$. The down-path $CN_0R(R) = 99\,\mathrm{dB}$ and the down-path $CIR(R) = 22\,\mathrm{dB}$. Ignoring other factors, determine CNIR(S).

9.8   A signal arriving from beyond the atmosphere (where $T_c = 4\,\mathrm{K}$) is attenuated 0.25 dB as it passes through the atmosphere. It is received via a paraboloidal reflector antenna ($K_a = 0.92$, $K_b = 0.9$) coupled via a 0.5 dB waveguide to a 30 dB amplifier with $F(r) = 8\,\mathrm{dB}$. Determine the SNR at various points along the system and observe the progressive deterioration in SNR, assuming SNR = 50 dB outside the atmosphere.

9.9   Three adjacent satellites A, B and C in geostationary orbit are progressively displaced by 3°. It is required that the minimum CIR(r) of satellites A and C due to the up-link of satellite B is to be 35 dB. What are the appropriate values of gain and diameter of the ground antenna for satellite B assuming $f = 6\,\mathrm{GHz}$ and 75% illumination efficiency? The system follows the specification of Eq. (9.37).

9.10   An Intelsat standard B ground station is located at $d = 39,500\,\mathrm{km}$ from a geostationary satellite. It receives an SCPC link, effectively two normal baseband channels. The antenna is 10 m diameter, 70% illumination efficiency, and the receiver has $T_s = 80\,\mathrm{K}$ at 4100 MHz. Is the G/T ratio satisfactory? What is the approximate magnitude of the power radiated by the satellite per normal channel?

9.11 Prior to man's journey to the moon, NASA sent *Lunar Orbiter* to orbit the moon at an altitude of about 40 km and to transmit data, including video, back to earth. Assuming a 10 W transmitter at 2295 MHz with a 24 dB antenna, a 400,000 km path with 2 dB miscellaneous losses, a 53 dB receiving antenna on earth with $T_s = 165$ K, $B = 3.5$ MHz and 7 dB threshold signal-to-noise ratio, determine whether there is any margin along the link.

9.12 Assuming that a space vehicle travels towards the planets, with typical spatial losses of 211 dB (Moon), 212 dB (Mercury), 255 dB (Venus and Mars), 275 dB (Jupiter), 286 dB (Saturn) and 297 dB (Pluto) at 2.3 GHz, estimate the maximum bandwidth in each case if minimum CNR(R) = 10 dB, P(t) = 20 dBW, G(t) = 20 dB, G(R) = 50 dB and $T_s = 40$ K.

REFERENCES

9.1 Clarke, A. C., 'Extra terrestrial relays,' *Wireless World*, October 1945, pp. 305–7.

9.2 Moralee, D., 'Satellites: their impact on world communication,' *Electronics & Power*, 24 (June 1978), 429–35.

10

Radar

The early ideas of *detection* of targets include some sound-ranging devices which were used during the First World War to locate enemy guns, and some acoustic methods which were developed in the 1920s to locate aircraft. Around the same time it was suggested that it might be possible to direct high-power radio waves at aircraft in order to interfere with the aircraft's electric ignition system.

Then, by the 1930s, it was observed that electromagnetic waves emitted by a radio broadcasting station were reflected by aircraft and produced *echoes* which could be detected by suitable electronic equipment. This idea developed into a series of *radio direction-finding stations* which were set up around Britain by 1937 and operated in the HF band around 20–30 MHz. Later, the acronym RADAR was coined in the USA for *radio detection and ranging*.

The need to have radar beams with higher directivity, to obtain increased EIRP and directional accuracy, required the use of smaller wavelengths. This led to the development of microwave (10 cm and 3 cm) sources and amplifiers, initially for wartime and military uses. Since then other developments of radar have moved into civilian applications such as marine and air-traffic control, Doppler navigation, meteorological observations, speed detection of traffic, etc., and the operating frequencies have extended into the millimetric range for some applications.

The progress of radar has been rapid and extensive and, within the limits of a single chapter, we shall have to restrict ourselves to the fundamental principles of radar and, in particular, to the basic equations associated with radio wave propagation.

10.1 PULSE RADAR

One particular type of radar system, illustrated in Fig. 10.1, sends out short pulses of carrier-wave signals. Each pulse of r.f. energy travels a distance d to the target at velocity c and reacts with the target such that some of the incident energy is re-radiated. The pulse is eventually received back at the radar system, much diminished in magnitude, after a total time delay of t' seconds. We can determine the range of the target via the simple relationship

$$d = \tfrac{1}{2}ct' \tag{10.1}$$

The instantaneous radial velocity of a moving target is not indicated directly by this

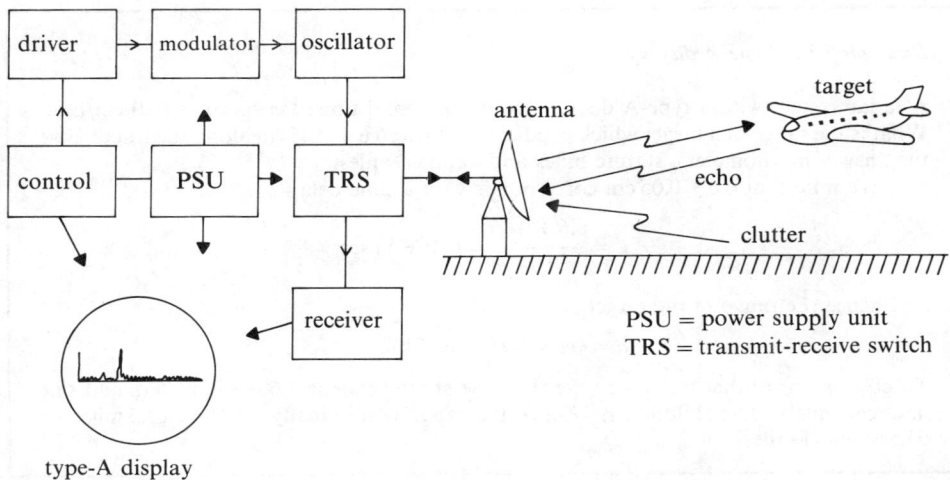

Fig. 10.1 A block diagram of a pulse radar system using a type-A cathode-ray oscilloscope display. A single antenna is used with a transmit–receive switch. The power supply unit is connected to all components.

method though, of course, the change in radial range over a known time period will produce an estimate of the average radial velocity.

A long-established technique for measuring the time delay t' and hence calibrated range d is to use a conventional cathode-ray oscilloscope (CRO). The timebase is triggered to coincide with the instant the pulse is transmitted and the return echo can be observed as a blip on the screen a few centimeters along the trace (Fig. 10.1). If we know the CRO x-deflection sensitivity in cm/μsec, we can calibrate the display directly in terms of range d. This arrangement is called a *type-A display*.

The range may be indicated simply by a coarse set of range markers representing kilometers, nautical miles or other appropriate unit of length. Alternatively, a movable strobe marker may be used, coupled to an accurate range counter. Automatic arrangements are also available which produce gating signals that can be matched into coincidence with the pulses so that a measurement of time delay is achieved electronically with a high degree of precision.

Another type of oscilloscope display used in radar systems is called a *plan position indicator* (PPI). A suitably arranged timebase system produces rotating radial deflections from the center of the screen. The CRO brilliance is adjusted below cut-off and the return pulses from the target are amplified, converted to video signals, and applied to the brightness modulation or z-terminal of the CRO. This causes a brightening-up of the trace for the duration of each pulse, producing spots on the screen at the appropriate ranges and bearings. The latter is achieved via a rotating antenna operating in synchronism with the rotating timebase. The system is illustrated in Fig. 10.2. There are several other types of display.

Two of the parameters of the pulse radar system illustrated in Fig. 10.3 are the

---

Example 10.1 Type-A display

A radar system with a type-A display has a horizontal timebase speed of 0.01 cm/μsec. What is the range of a target which produces a blip at 6.6 ± 0.05 cm along the trace? Give the answer in kilometers, statute miles and nautical miles.

We note that 6.6 ± 0.05 cm corresponds with a time delay

$$t' = \frac{6.6 \pm 0.05}{0.01} = 660 \pm 5 \,\mu\text{sec}$$

and hence the range of the target is

$$d = \tfrac{1}{2}ct' = 99.0 \pm 0.75 \,\text{km}$$

To convert into other units we note that one statute mile is 1.609 kilometers and one nautical mile is 1.852 kilometers. Hence the range is nominally 99.0 km, 61.5 miles or 53.5 nautical miles.

---

pulse width τ and the *pulse repetition frequency* $f_R = 1/T$. The minimum and maximum ranges of a pulse radar system are related to these parameters via

$$d_{min} = \tfrac{1}{2}c\tau \tag{10.2}$$

and

$$d_{max} = \tfrac{1}{2}cT = \frac{1}{2}\frac{c}{f_R} \tag{10.3}$$

In the first case the signal cannot be processed until the full pulse has been transmitted. In the second case there will be an ambiguity in the range if the time delay exceeds $T = 1/f_R$ because the timebase will have completed one trace and recommenced a second trace. Thus a radar system operating with a pulse width of 1 μsec and a pulse repetition frequency of 1000 pps will theoretically be capable of detecting targets unambiguously

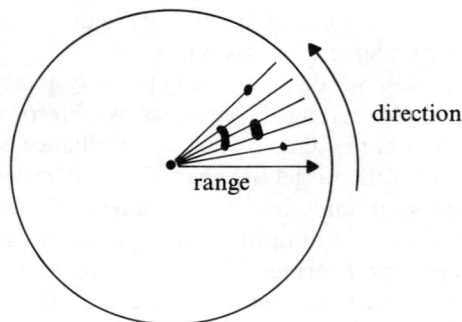

Fig. 10.2 The plan position indicator (PPI) indicates both range and bearing of a target.

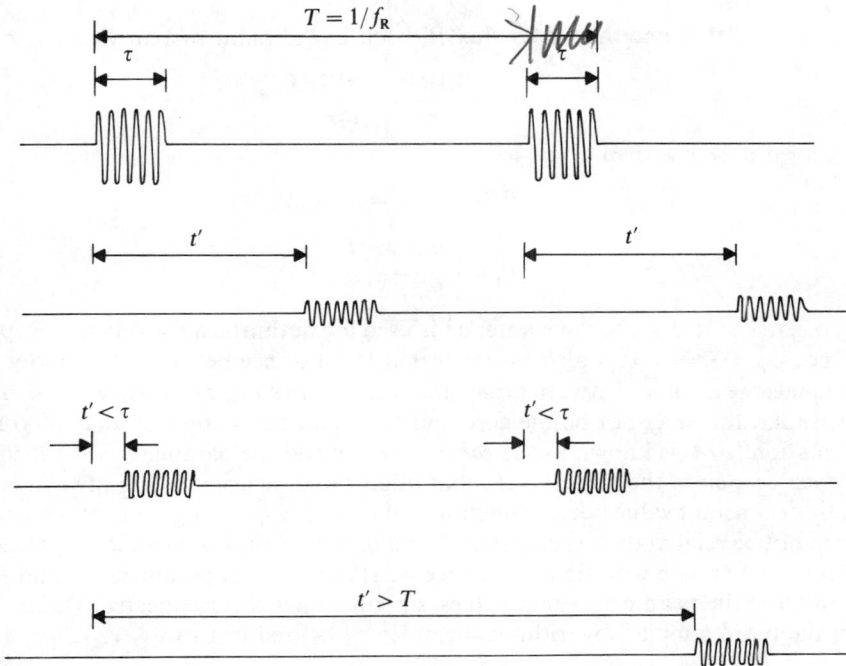

Fig. 10.3 Two parameters used in pulse radar include the pulse width τ and the pulse repetition frequency f_R. For unambiguous reception of a return echo it is necessary for $t' > \tau$ and $t' < T$.

between $d_{min} = 0.15\,\text{km}$ and $d_{max} = 150\,\text{km}$ provided there is adequate power, etc. The ratio of τ to T is called the *duty cycle*.

10.1.1 The Radar Equation

The principal radio wave propagation equation in radar can be derived by assuming that a transmitter with mean pulse power P_t and associated antenna gain G_t combine to produce an effective isotropically radiated power $\text{EIRP}_t = P_t G_t$ in the direction of the target. At some distance d in this direction the mean power flux is

$$P_a = \frac{\text{EIRP}_t}{4\pi d^2}$$

If this power flux impinges upon a target which is able to re-radiate some of the incident power, then the target behaves as a kind of transmitter at the same frequency. It is assumed to have an equivalent EIRP' related to the incident power flux P_a via some parameter σ so that

$$\text{EIRP}' = \sigma P_a = \sigma \frac{\text{EIRP}_t}{4\pi d^2}$$

Hence we can determine the power flux P'_a back at the radar system via

$$P'_a = \frac{\text{EIRP}'}{4\pi d^2} = \frac{\sigma \text{EIRP}_t}{(4\pi d^2)^2}$$

The received power is then given by

$$P_r = P'_a A_{\text{eff}} = P'_a G_r (\lambda^2/4\pi)$$

or

$$P_r = \frac{\sigma P_t G_t G_r \lambda^2}{(4\pi)^3 d^4} \qquad (10.4)$$

where, in some systems, the same antenna is used for both transmission and reception and hence $G_r = G_t = G$. It is also assumed that the distance between transmitter and target equals the distance between target and receiver. In some radar systems, such as a bistatic radar, this may not be the case and the equation is appropriately modified.

Equation (10.4) is known as the *radar equation* and the parameter σ is called the *radar cross-section* of the target. Note that σ has the dimensions of an area. It is not normally of constant value but is a function of the target aspect angle and other factors, and need not be related to the geometrical area of the target. For any moving target, σ will therefore fluctuate with time and hence we should specify parameters σ and P_r as being either their mean or median values, or sometimes their 'expected' values.

In the more familiar logarithmic form, Eq. (10.4) reduces to

$$P(\text{r}) = P(\text{t}) + G(\text{t}) + G(\text{r}) + 10 \log \sigma (\text{m}^2) - 20 \log f(\text{MHz}) - 40 \log d(\text{km}) - 103.4 \quad (10.5)$$

The equation can be a little misleading in that the gains of the antennas are also frequency-dependent and the apparent decrease in $P(\text{r})$ with frequency is countered by the increase in $P(\text{r})$ due to the antenna gains increasing with frequency.

Example 10.2 Radar contact with Venus

The relative orbits of the planets Venus and Earth are such that their nearest approach is about 40 million kilometers. Estimate the received carrier-to-noise spectral density ratio $\text{CN}_0\text{R}(\text{r})$ of a radar contact with Venus if it is assumed that a 12 kW, 2.4 GHz transmitter with a 54 dB antenna is used with a low noise ($T_s = 60$ K) receiver system. Take the diameter of Venus as 12,500 km and assume that during the experiment the distance between Venus and Earth is not more than 50 million kilometers. The radar cross-section of Venus is approximately 10% of its actual cross-section. Ignore other factors.

We can use Eq. (10.5) to obtain

$$P(\text{r}) = P(\text{t}) + G(\text{t}) + G(\text{r}) + 10 \log \sigma - 20 \log f - 40 \log d - 103.4$$

where $\sigma = 0.1 \times \pi r^2$ and $10 \log \sigma = 130.9$. This gives $P(\text{r}) = -199.3$ dBW. The carrier-to-noise spectral density ratio is

$$\text{CN}_0\text{R}(\text{r}) = P(\text{r}) - N_0(\text{r}) = P(\text{r}) + 228.6 - 10 \log T_s = 11.5 \text{ dB-Hz}$$

Various signal processing techniques enable the return signal to be processed with this value of $\text{CN}_0\text{R}(\text{r})$ if the frequency of the system is controlled extremely accurately. Note that the delay time between the transmission and reception of the pulse is about 333 seconds at this range.

10.1.2 Noise

In its simplest form it is usual to consider the *mean* level of the noise power present at the input or output of a receiver as being

$$N_i = kT_sB$$

and

$$N_o = kT_sBG_r$$

T_s is called the *system noise temperature* and includes noise contributions generated in the receiver itself and also arriving via the antenna; 'input noise level' N_i is the equivalent input noise level required to produce the total output noise if the receiver is noise-free; G_r is the gain of the ideal linear amplifier; B is the effective noise bandwidth; and k is Boltzmann's constant.

The instantaneous value of the *noise voltage* v_n, however, varies continuously with time, perhaps something like the pattern shown in Fig. 10.4. The idea that noise is the net sum of innumerable independent chance processes has been used (e.g. see Ref. 10.1) to indicate that if we were to sample v_n over a period of time, the resulting histogram would eventually approach the normal (or Gaussian) distribution for a linear process. This is shown in Fig. 10.4, where the statistical standard deviation σ_n corresponds to the electrical root mean square value of v_n. Hence we can write

$$N_o = \frac{\sigma_n^2}{R} = kT_sBG_r = N_iG_r \tag{10.6}$$

where N_o represents the mean noise output from the receiver amplifier into a matched load R before envelope detection.

After envelope detection, the pulse of r.f. energy becomes a video pulse. For a perfect video pulse, the bandwidth before and after the detector needs to be infinite.

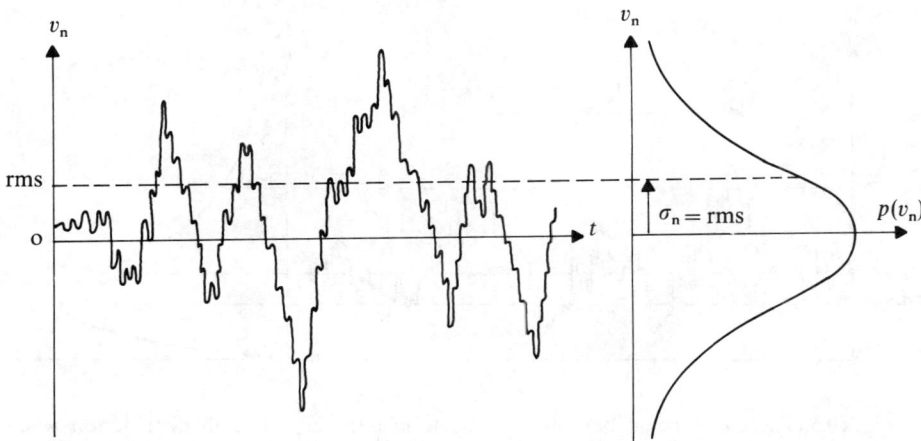

Fig. 10.4 A sketch of the instantaneous noise voltage v_n and its Gaussian distribution when sampled over a suitably long period of time.

However, it is a common practice to use the convenient initial design relationship

$$B\tau = 1 \qquad (10.7)$$

for pre-detector bandwidth B and pulse width τ, though perhaps $B\tau = 1.2$ may give a more optimum performance. Criteria such as these are designed to meet the requirements of optimizing the signal-to-noise ratio while not distorting the pulse waveform excessively. These requirements conflict to some extent. Distortion of the waveform may impair the ability to resolve multiple or complex targets. The so-called 'matched' or 'optimum' filter achieves the highest possible signal-to-noise ratio with moderate 'smearing out' of the pulse waveform shape. The simple basis of a relationship such as $B\tau = 1$ is just that the receiver should have about the same bandwidth as the pulse being received; wider bandwidth merely lets in more noise without useful signal, while narrower bandwidth throws away useful parts of the signal spectrum being received. It can also be shown that an optimized pre-detector filter is superior to an optimized post-detector one. With a well-designed pre-detector filter it is sufficient just to have additionally a simple low-pass filter after the detector, with the cut-off frequency just low enough to remove the carrier or intermediate frequency and accurately reproduce the envelope.

If there is no r.f. pulse, the output of the envelope detector is simply the envelope v_e of the band-limited noise voltage v_n, which might resemble the pattern shown in Fig. 10.5. This time the statistics are such that if we were to sample v_e over a period of time the histogram would eventually resemble the Rayleigh distribution in which

$$p(v_e) = \frac{v_e}{\sigma_n^2} \exp\left[\frac{-v_e^2}{2\sigma_n^2}\right] \qquad (10.8)$$

and the probability that v_e exceeds some specified threshold level v_t is given by

$$P(v_e > v_t) = \exp\left[\frac{-v_t^2}{2\sigma_n^2}\right] = P_{fa} \qquad (10.9)$$

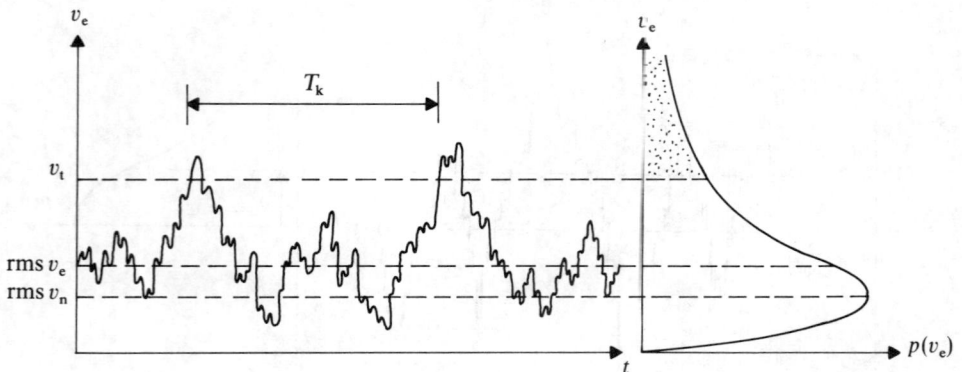

Fig. 10.5 The envelope of band-limited noise v_e and the Rayleigh distribution when sampled over an adequate period of time. The selected threshold voltage level is indicated by v_t.

The receiver is constantly processing some combination of noise and signal. In a pulse radar system the combination is mostly noise with the occasional signal pulse of small amplitude. How can the receiver be adapted to recognize from the combination of signal and noise whether a pulse is present? One of several methods is to select a certain threshold voltage v_t and compare it with v_e: if $v_e < v_t$ it is assumed that only noise is present; if $v_e > v_t$ it is assumed that a pulse is present.

Unfortunately, as shown in Fig. 10.6, there are occasions when the noise envelope v_e exceeds the threshold level even when no pulse is present and these excursions of v_e above v_t produce what are known as *false alarms*. The probability of false alarms is P_{fa}. as given in Eq. (10.9) and it can be shown (e.g. Ref. 10.2) that the time between false alarms has a mean value T_{fa} given by

$$T_{fa} = \frac{1}{P_{fa}B} \qquad (10.10)$$

where B is commonly equal to $1/\tau$ as indicated in Eq. (10.7).

Increasing the threshold level v_t results in a lower value of P_{fa} but it also means a reduction in the *probability of the detection* of a pulse P_d. We have seen how the sampled

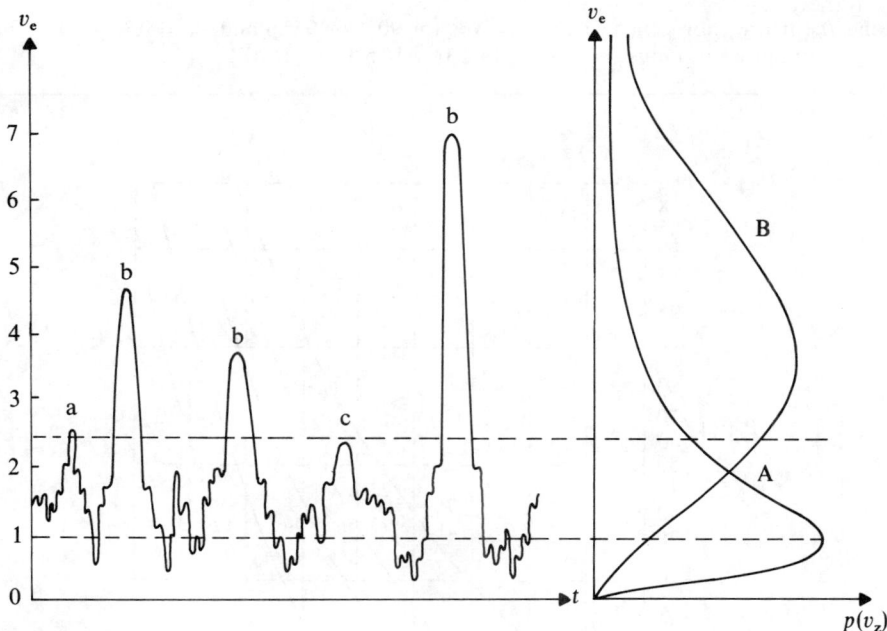

Fig. 10.6 The envelope of band-limited noise plus return echoes. Point a represents the noise envelope exceeding the threshold; points b indicate echoes in excess of the threshold; and point c is an echo which is below the selected threshold level. Point a represents a false alarm and point c represents a failure in detection. Curve A is the distribution of band-limited noise when no pulse is present; curve B indicates the distribution of echo plus noise when a return pulse is present.

envelope of the band-limited noise voltage v_e produces a Rayleigh distribution. When the receiver is processing the combination of noise *and a radar pulse*, the resulting statistical distribution is more complicated. It depends upon the character of the pulse and the target and the analysis is outside the scope of this text. However, Fig. 10.6 can still be used to illustrate qualitatively how P_d is calculated if the precise form of statistical distribution is known.

For the very special case of a steady sinusoidal signal plus noise the resulting

---

Example 10.3 Probability of false alarms

A certain pulse radar system with bandwidth $B = 1\,\text{MHz}$ is required to have no more than one false alarm in 5 minutes and a probability of detection of 80% (or 90% or 95%). What signal-to-noise ratio is needed to achieve this specification if it is assumed that the return pulse is of a constant magnitude in excess of the noise contributions?

The solution can be obtained from Eq. (10.10) and Fig. 10.7. The equation gives us

$$P_{fa} = \frac{1}{T_{fa}B} = 0.333 \times 10^{-8}$$

and the figure shows that when $P_d = 80\%$ (or 90% or 95%) and $P_{fa} = 0.33 \times 10^{-8}$ the signal-to-noise ratio must be $\text{SNR} = 14\,\text{dB}$ (or $14.5\,\text{dB}$ or $15\,\text{dB}$).

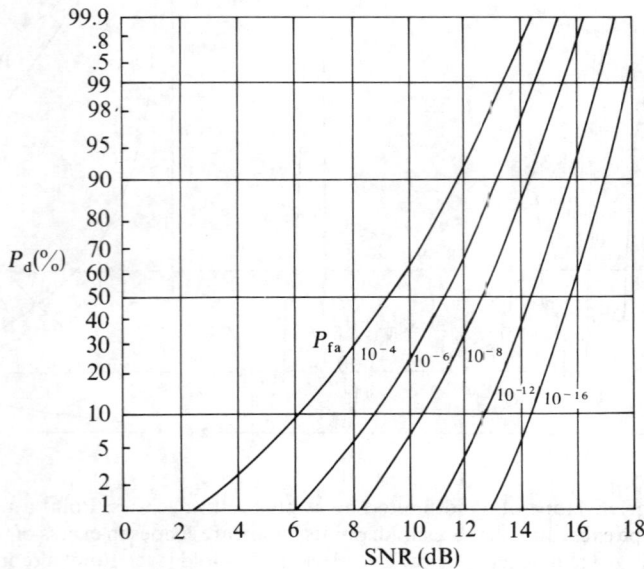

Fig. 10.7 The relationship between probability of detection P_d, probability of false alarms P_{fa} and signal-to-noise ratio SNR.

Rician distribution can be approximated (Ref. 10.2) to give

$$P_d = \frac{1}{2}\left[1 + \text{erf}\left\{ \sqrt{\frac{1}{2} + \frac{S}{N}} - \sqrt{\log\left(\frac{1}{P_{fa}}\right)} \right\} \right] \qquad (10.11)$$

provided the probability of detection P_d exceeds 0.5, where erf is the *error function* available in tabulated form in various mathematical texts and S/N is the signal-to-noise ratio. The graphical representation in Fig. 10.7 is a more accurate relationship.

10.1.3 Signal-to-Noise Ratio

There are several related definitions of signal-to-noise ratio in texts on radar. The ideal radar pulse shown in Fig. 10.8a has the instantaneous power variation shown in Fig. 10.8b. The peak value of the instantaneous power has envelope A (Fig. 10.8c) whereas the mean value of the instantaneous power has envelope B (Fig. 10.8d). These two descriptions of pulse power lead to definitions of signal-to-noise ratio SNR and peak signal-to-noise ratio PSNR, where one is twice the linear magnitude of the other.

Alternatively, if the bandwidth B is related to the pulse width τ via the simple empirical design rule $B = 1/\tau$, then

$$\text{SNR} = \frac{P_r}{kT_sB} = \frac{P_r\tau}{kT_s} = \frac{E_r}{N_o} \qquad (10.12)$$

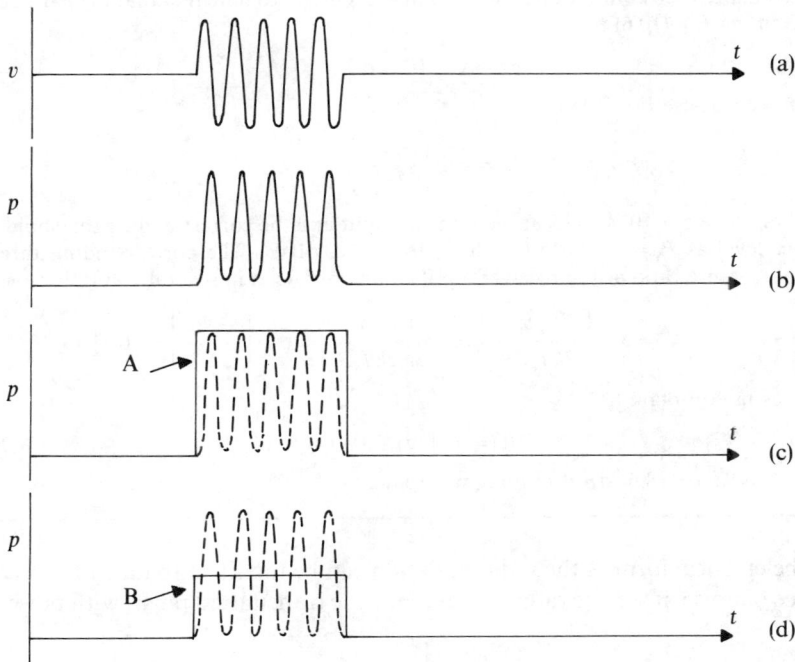

Fig. 10.8 Two methods commonly used to measure the power level and signal-to-noise ratio associated with a pulse.

and

$$PSNR = 2SNR = \frac{E_r}{N_o/2} \tag{10.13}$$

where P_r is the mean pulse power received in watts, E_r is the energy of the received pulse in joules, and N_o is the noise spectral density. This equation is general: it works for any shape of pulse provided the receiver uses a matched filter. Detectability thus depends on the pulse energy, regardless of its shape.

Example 10.4 Threshold SNR of pulse radar receiver

A certain pulse radar system operates with pulse width $\tau = 1$ μsec and the 50 Ω receiver has a system noise temperature $T_s = 1000$ K. It is required that there should be not more than one false alarm in 100 seconds. Determine: (a) the probability of false alarms; (b) the receiver input threshold voltage; (c) the corresponding threshold power input; and (d) the corresponding threshold signal-to-noise ratio.

The probability of false alarms is given by Eq. (10.10) as

$$P_{fa} = \frac{1}{BT_{fa}} = \frac{\tau}{T_{fa}} = 10^{-8}$$

To calculate receiver input threshold voltage it is useful to note first that the parameter σ_n is given via Eq. (10.6) as

$$\sigma_n^2 = kT_sBR = 6.9 \times 10^{-13}$$

Then we can use Eq. (10.9),

$$P_{fa} = \exp\left[\frac{-v_t^2}{2\sigma_n^2}\right]$$

to obtain $v_t = 5 \times 10^{-6}$ volts at the receiver input and the corresponding threshold input power level as $P_r = v_t^2/R = 5.1 \times 10^{-13}$ (or -123 dBW). The corresponding threshold signal-to-noise ratio at the receiver input can be obtained in several ways. For example,

$$P_{fa} = \exp\left[\frac{-v_t^2}{2kT_sBR}\right] = \exp\left[\frac{-P_r}{2kT_sB}\right] = \exp\left[\frac{-E_r}{2N_o}\right] = 10^{-8}$$

enables us to obtain

$$SNR(r) = 10\log(E_r/N_o) = 15.66\,dB$$

or a $PSNR(r) = 18.66$ dB if more appropriate.

The classical form of the radar equation can be modified to take into account the presence of noise in a pulse radar system. For the transmitted pulse, with pulse width τ,

$$E_t = \int_0^\tau p_t \, dt = P_t\tau \tag{10.14}$$

where p_t is the instantaneous power and P_t is the mean power of the pulse. If, for example, $P_t = 1 \, \text{mW}$ and $\tau = 1 \, \mu\text{sec}$, $E_t = 1 \, \text{J}$. Similarly, for the received pulse

$$E_r = \int_{t'}^{t'+\tau} p_r \mathrm{d}t = P_r \tau \tag{10.15}$$

where t' is the delay time between the transmission and reception of the pulse (though it does not affect the integration) and P_r is the mean value of the received power during the pulse period.

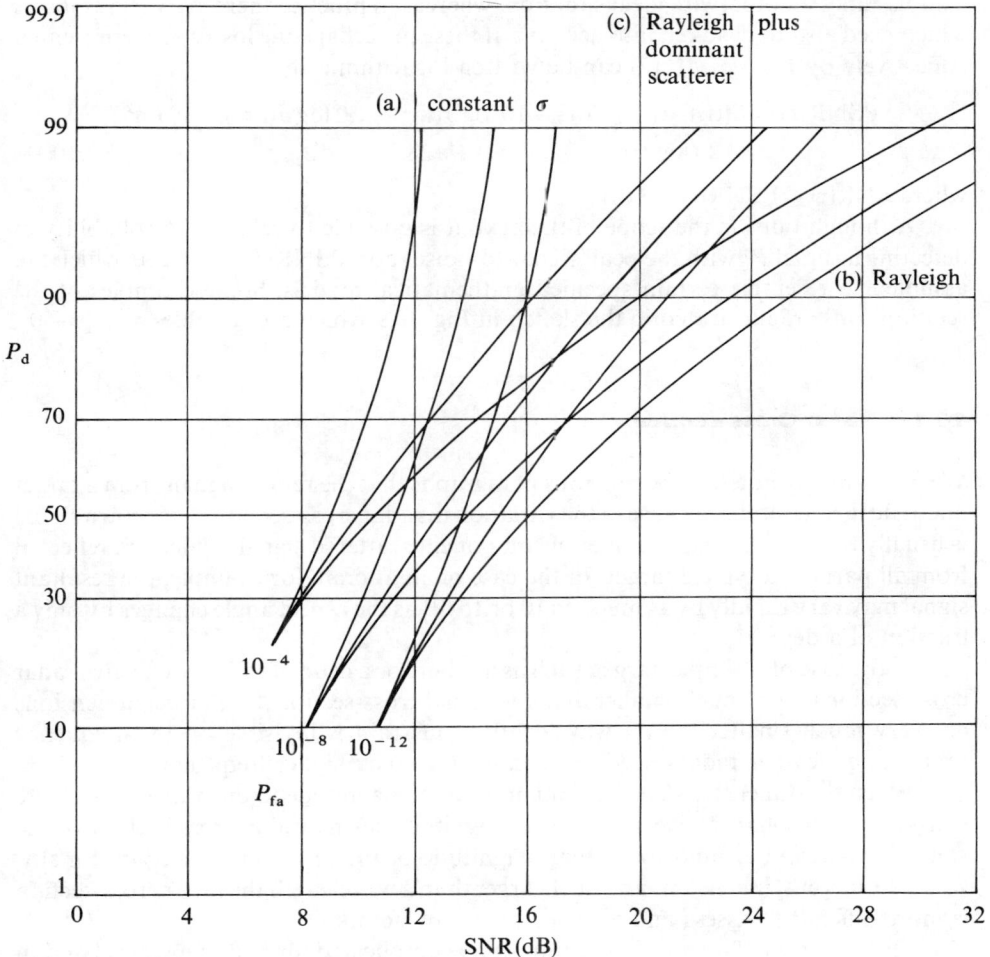

Fig. 10.9 The relationship between probability of detection P_d, probability of false alarm P_{fa} and signal-to-noise ratio SNR(dB) for three models of a radar target: (a) a target with constant cross-section, (b) a target represented by the sum of numerous random scatterers, and (c) a target represented by numerous scatterers, one of which is very much larger than the others.

Combining Eqs. (10.4),(10.14) and (10.15),

$$E_r = E_t \frac{G_t G_r \lambda^2 \sigma}{(4\pi)^3 d^4} \tag{10.16}$$

and from this we can obtain

$$\text{PSNR} = 2\text{SNR} = \frac{2E_r}{N_o} = \frac{2E_t G_t G_r \lambda^2 \sigma}{(4\pi)^3 d^4 k T_s} \tag{10.17}$$

This assumes an otherwise ideal situation, whereas in practice there are various losses which need also to be taken into account. If these miscellaneous losses are represented collectively by L_m, Eq. (10.17) can be written logarithmically as

$$\text{PSNR}(r) = E(t) + G(t) + G(r) + 10\log\sigma(m^2) - 40\log d(km) - 10\log T_s$$
$$- 20\log f(\text{MHz}) - L(m) + 128.2 \qquad \text{dB} \tag{10.18}$$

where $\text{SNR}(r) = \text{PSNR}(r) - 3\,\text{dB}$.

Although outside the scope of this text, it is possible to relate the probability of detecting a signal P_d with the peak signal-to-noise ratio $\text{PSNR}(r)$ for targets which are assumed to resemble certain specified mathematical models. Some examples of the relationships are illustrated in the sketch in Fig. 10.9 which are suitable for $P_d > 50°$.

10.1.4 Radar Cross-section

We have already noted in the previous paragraphs that the return signals from a target fluctuate in magnitude because of the variation in radar cross-section. The return signal is usually the sum of a vast number of component scattered signals which are reflected from all parts of a typical target. In the case of an aircraft, for example, the resultant signal may vary rapidly by as much as 10 or 15 dB as the aspect angle changes by only a fraction of a degree.

In the case of a 'simple' target, such as a sphere or a drop of rain, the effective radar cross-section is very much smaller than the actual cross-section if the target dimensions are very much smaller than a wavelength. Thus rain is invisible on low-frequency radars and weather radars need to operate at a much higher frequency.

When the dimensions are of a similar order of magnitude to one wavelength there is a kind of 'resonance' effect, with the magnitude of the radar cross-section of the sphere fluctuating around the actual magnitude of the target cross-section. Finally, when the target's dimensions are much larger than a wavelength the radar cross-section is more difficult to assess and depends on many factors.

The variation of σ with time is thus quite complicated for a moving target and in order to produce numerical estimates for radio wave propagation analyses it is customary to describe several mathematical or statistical models of the radar cross-section of a target. For instance, if the target consists essentially of a large number of scatters, the largest of which is not significantly larger than many of the others, the variation of σ is found to approximate to the exponential distribution with

$$p(\sigma) = (1\sqrt{\sigma})\exp(-\sigma/\bar{\sigma}) \tag{10.19}$$

where $\bar{\sigma}$ is the average radar cross-section of the target.

In Chap. 6 we noted two points relevant to this topic. Firstly, the exponential distribution is closely linked with the chi-squared distribution with 2 degrees of freedom. Secondly, if the voltage or electric field strength of a signal follows the Rayleigh distribution, the corresponding power level follows the exponential distribution. Consequently, a *Rayleigh target* is one in which the linear magnitude of the electric field strength of the return signal, or associated receiver voltage, follows a Rayleigh distribution, and not the radar cross-section of the target. There are further differences if the field strength, voltage or signal-to-noise ratio are expressed in decibel form. Typical Rayleigh targets, using this definition, are rainstorms, *chaff* and certain types of sea clutter.

If the target has one dominant scatterer plus various smaller scatterers, the distribution of σ follows the chi-squared distribution with 4 degrees of freedom with

$$p(\sigma) = (4\sigma/\bar{\sigma}^2)\exp(-2\sigma/\bar{\sigma}) \tag{10.20}$$

and not the Rician distribution, because the latter specifically assumes the dominant scatterer is non-fluctuating in magnitude. However, other models of radar targets include the Rician distribution as well as the log-normal and Weibull distributions.

Example 10.5 Probability of detection

A 3 GHz pulse radar system emits 1 J per pulse from a 49 dB common transmit–receive antenna. It is intended to detect targets nominally 8 m^2 in effective radar cross-section with a probability of false alarms of 10^{-8} using a receiver with a system noise figure $F(s) = 8$ dB. Assuming 6 dB miscellaneous losses, plot graphs of the probability of detection against range d if the target characteristics correspond with the three types illustrated in Fig. 10.9.

We can initially use Eq. (10.18) to obtain an estimate of the signal-to-noise ratio:

$$SNR(r) = E(t) + G(t) + G(r) + 10\log\sigma - 40\log d - 10\log T_s$$
$$- 20\log f - L(m) + 125.2 \quad \text{dB}$$

Substitution of numerical data, with $10\log T_s = 10\log 290 + F(s)$, gives us

$$SNR(r) = 106 - 40\log d(\text{km}) \quad \text{dB}$$

By selecting a suitable series of range d (for example, $d = 50, 100, 150, 200, 250$ and 300 km) we can obtain the corresponding values of SNR(r) (approximately $38, 26, 19, 14, 10$ and 7 respectively). With the probability of false alarms fixed at 10^{-8} we can note from Fig. 10.9 the corresponding values of probability of detection (about $100, 100, 100, 91, 26$ and 6% for constant-σ target; $100, 99.6, 90, 56, 18$ and 3% for the Rayleigh-plus-constant target; and $100, 95, 75, 44, 16$ and 3% for the Rayleigh target). These approximate data are sketched in Fig. 10.10 and illustrate the variation of P_d with range d.

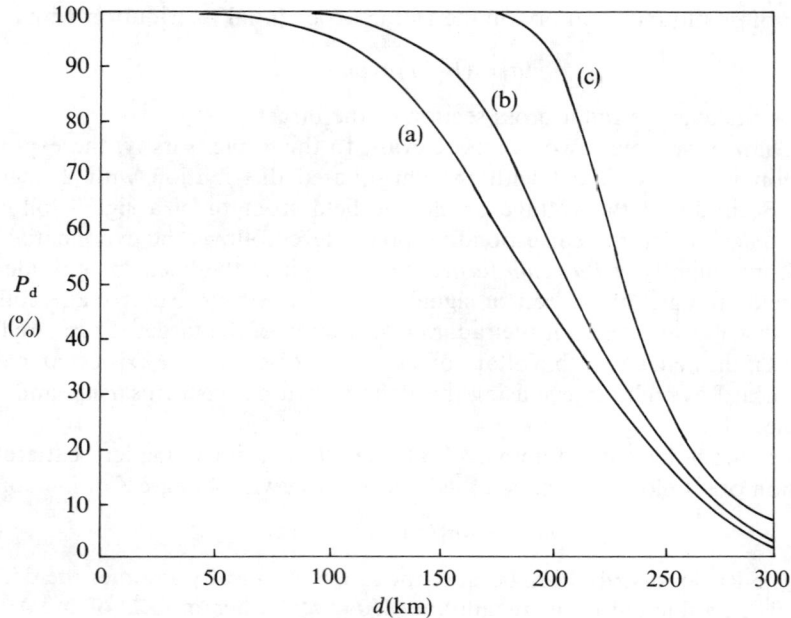

Fig. 10.10 The variation of P_d with target range d for the circumstances described in Ex. 10.5. The targets are (a) Rayleigh, (b) Rayleigh plus dominant and (c) constant-σ. $P_{fa} = 10^{-8}$.

The full analysis of radar targets is outside the scope of this text and the sketches in Fig. 10.9 are intended to indicate an approximate solution for the ideal cases of a constant-σ target, a Rayleigh type of target, and a Rayleigh plus one dominant scatterer type of target. These are suitable for tutorial examples. Detailed analyses of target fluctuation are given in Refs. 10.3 and 10.4.

10.2 SEARCH AND TRACKING RADAR

A *search radar* system is required to cover a specified volume of space and to detect any targets which may occur within that volume. To do this the radar repeatedly scans a given solid angle ψ_s in a particular pattern and makes use of various signal processing techniques to discriminate between targets and noise. The discrimination level needs some specification in terms of probability of false alarms P_{fa}, probability of detection P_d, etc., but for the radar equation it is convenient to combine these requirements into an indication of the minimum signal-to-noise ratio needed to achieve the specification, usually of the order 12 to 16 dB for a single-pulse system as illustrated approximately in Fig. 10.9.

10.2.1 Integration of Pulse Trains

When a pulse radar system detects the target it does not normally receive back just one echo pulse. If it has a pulse repetition frequency f_R of, say, 1000 pps then it will receive about 1000 echoes for each second it maintains contact with the target, depending on the probability of detection, etc. Each pulse contains the same information about range, assuming the range does not alter significantly during the observation time, and thus a considerable improvement in the signal-to-noise ratio can be achieved by summing a succession of the echoes. The process is called *integration of pulse trains* and the resulting signal-to-noise ratio is denoted SNR_i linearly or SNR(i) in decibels.

Initially, let us assume that the radar antenna rotates in the horizontal plane only at R rpm with a beamwidth BW in the same plane and a pulse repetition frequency f_R. For a point target the radar will receive back N sets of echoes in succession per scan of the antenna. The time for one rotation is $1/R$ minutes or $60/R$ seconds and the target is in view for $BW/360$ of one rotation or $BW/(6R)$ seconds. The maximum number of returning echoes is therefore

$$N = \frac{BW f_R}{6R} \tag{10.21}$$

If some signal processing circuitry (a predetection integrator), represented by block diagrams in Fig. 10.11, can integrate the r.f. or i.f. pulses coherently, the signal voltages add in phase but the noise voltages do not. Thus the SNR_i presented to the threshold detector is greater than that due to the individual pulses. Without going into the signal processing analyses, the relationship between SNR_i and SNR is simply

$$SNR_i = N \times SNR$$

or

$$SNR(i) = 10 \log N + SNR(dB) \tag{10.22}$$

For example, if the SNR of a single pulse is of the order of 0 dB, then if we arrange a system which coherently integrates trains of 16 consecutive pulses, the resulting

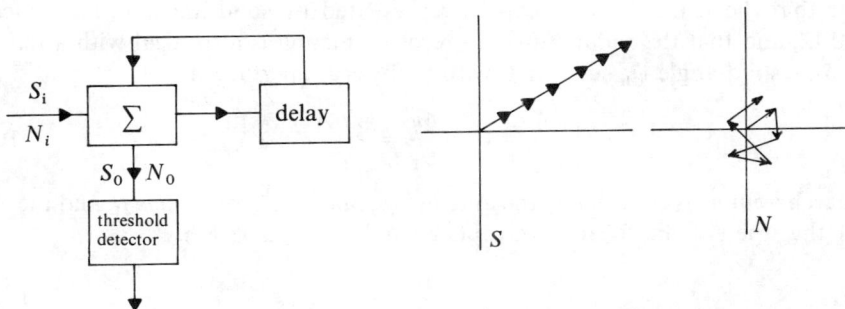

Fig. 10.11 A signal processing circuit which integrates the radar pulses coherently to enhance the overall signal-to-noise ratio.

$SNR(i) = 10 \log 16 + 0 \, dB = 12 \, dB$, and the probability of detection, etc., will be as specified for this larger value.

For *post-detection or video integration*, without using a coherent reference for phase information, the situation is not quite so good:

$$SNR(i) = 10\gamma' \log N + SNR(dB) \tag{10.23}$$

where γ' is called the *integration efficiency* and is approximately 0.75 when $N = 1000$, 0.80 when $N = 30$, and 0.9 when $N = 3$. Using the data of the previous paragraph, if we integrate trains of 32 consecutive echoes, $SNR(i) = 8 \log 32 = 12 \, dB$, as before, but the system requires more pulses to achieve this level.

The terms *integration gain G_i* and *integration loss L_i* are sometimes used, where the gain indicates the improvement of $SNR(i)$ over $SNR(dB)$ and the loss indicates the difference between the non-coherent gain and the theoretical maximum coherent gain. In logarithmic form

$$G(i) = SNR(i) - SNR(dB) \tag{10.24a}$$
$$L(i) = 10 \log N - 10\gamma' \log N = 10(1 - \gamma') \log N \tag{10.24b}$$

Thus a post-detection integrator sampling $N = 100$ echoes in succession will have an integration gain of $7.9 \log 100 = 15.8 \, dB$ (estimating $\gamma' = 7.9$ when $N = 100$) and an integration loss of $10(1 - 0.79) \log 100$ or $4.2 \, dB$ compared with the coherent integrator.

10.2.2 Search Radar Equation

A modification of the radar equation is called the *search radar equation* and it can be derived from Eq. (10.4):

$$P_r = \frac{\sigma P_t G_t G_r \lambda^2}{(4\pi)^3 d^4 L_m}$$

where L_m is added to represent miscellaneous losses compared with the ideal model. We assume that the search volume can be represented by solid angle ψ_s illustrated in Fig. 10.12, and that the radar transmit–receive antenna is also ideal with a uniform beam over solid angle ψ_b such that, with a physical aperture A,

$$G_t = G_r = \frac{4\pi}{\psi_b} = \frac{4\pi\eta A}{\lambda^2} \tag{10.25}$$

The search volume is scanned by the antenna ψ_b once each *frame time* t_f, and the *dwell time* t_d that the antenna beam intercepts a single point target is

$$t_d = \frac{\psi_b}{\psi_s} t_f \tag{10.26}$$

In order to receive N pulses for integration within each dwell time it is necessary for t_d to be greater than N/f_R, where f_R is the pulse repetition frequency. The total number of

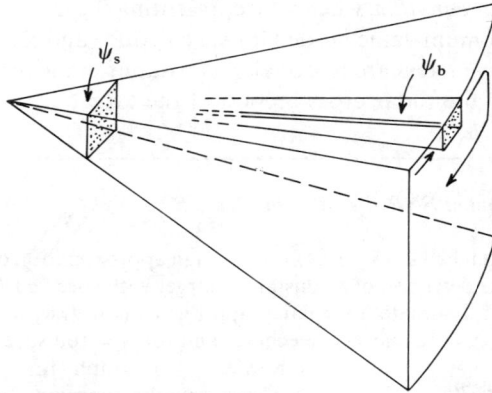

Fig. 10.12 The parameters ψ_s and ψ_b associated with the search volume of a radar system.

pulses per frame is $f_R t_f$ and the total number of pulses per dwell time is

$$N = \frac{t_d}{t_f} f_R t_f = t_d f_R = \frac{\psi_b}{\psi_s} t_f f_R \tag{10.27}$$

Example 10.6 Frame time

A 40 dB radar antenna is used to scan a search volume occupying $\psi_s = 0.1$ steradian with a pulse repetition frequency of 1000 pps. What is the minimum value of frame time that will permit integration of a 100-pulse train of echoes in succession?

We can use Eq. (10.25) to obtain ψ_b and Eq. (10.27) to obtain minimum frame time $t_f = 8$ seconds.

The signal-to-noise ratio with a single pulse is then

$$\text{SNR} = \frac{P_r}{N_r} = \frac{\sigma P_t G_t G_r \lambda^2}{(4\pi)^3 d^4 L_m k T_s B} \tag{10.28}$$

and the signal-to-noise ratio with N integrated echoes is $N \times \text{SNR}$ or

$$\text{SNR}_i = \frac{\psi_b}{\psi_s} t_f f_R \frac{\sigma P_t G_t G_r \lambda^2}{(4\pi)^3 d^4 L_m k T_s B} \tag{10.29}$$

This expression is modified by noting that: (a) the average value of the transmitted power is $P_{av} = P_t \tau f_R = P_t f_R / B$; (b) $G_t = 4\pi/\psi_b$; and (c) $G_r = 4\pi\eta A/\lambda^2$. With these substitutions in Eq. (10.29) we obtain

$$\text{SNR}_i = \frac{\sigma t_f}{\psi_s 4\pi d^4 k T_s (L_m/\eta)} P_{av} A \tag{10.30}$$

This is the *search radar equation,* where d^4 (representing the maximum range) and SNR_i (representing the minimum value needed for specified P_{fa} and P_d) can be interchanged. Most of the other parameters are reasonably constant for a given search radar system except, of course, for the radar cross-section of the target.

Example 10.7 Minimum SNR for video integration

One of the graphs sketched in Fig. 10.9 gives a rough approximation of the signal-to-noise ratio required for the detection of a constant-σ target with specified P_d between 10% and 99% and $P_{fa} = 10^{-8}$. Estimate a similar graph for a radar system incorporating video integration with (a) $N = 10$ successive echoes, and (b) $N = 100$ successive echoes.

We can take some readings for the $N = 1$ graph in Fig. 10.9 with $P_d = $ 99%, 90%, 70%, 50%, 30% and 10%, say, to estimate the approximate SNR required with a single-pulse system as 14.7, 13.9, 12.8, 11.2, 10.2 and 8.2 dB. Using the data in Eq. (10.23) with (a) $N = 10$ and $\gamma' = 0.87$, and (b) $N = 100$ and $\gamma' = 0.79$ gives us two sets of data which we can plot as in Fig. 10.13; (a) SNR = 6.0, 5.2, 4.1, 2.5, 1.5 and -0.5 dB; (b) SNR = -1.1, -1.9, -3.0, -4.6, -5.6 and -7.6 dB.

Thus it is possible to detect weak echoes from targets, with signal-to-noise ratios less than 1, by the processes of integration.

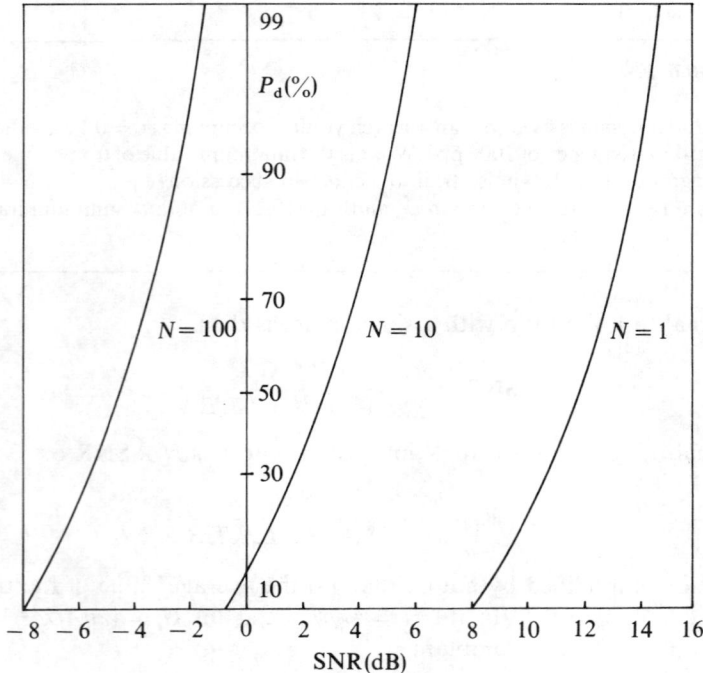

Fig. 10.13 The relationship between probability of detection P_d and signal-to-noise ratio SNR(dB) for the radar system described in Ex. 10.7 employing video integration.

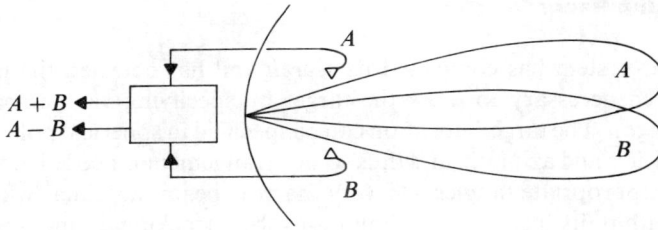

Fig. 10.14 A paraboloidal reflector antenna incorporating two offset feeds for use in tracking radars.

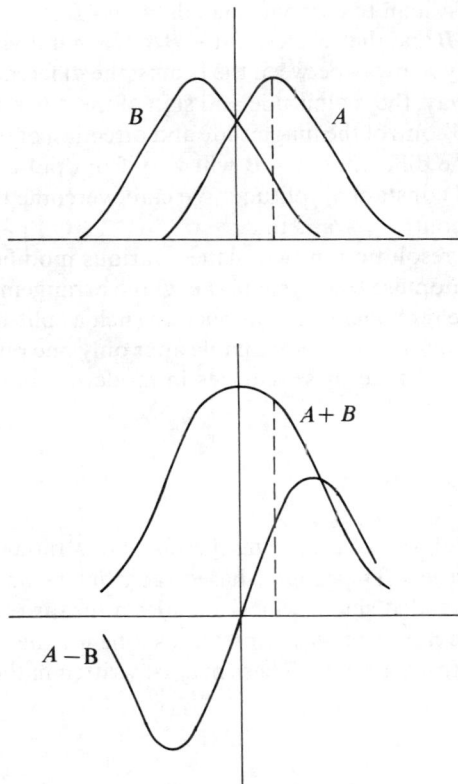

Fig. 10.15 The signals received via feeds A and B for various antenna alignments and the results of addition $A + B$ and subtraction $A - B$. When the antenna is aligned as shown by the broken line, the ratio $(A + B)/(A - B)$ is roughly $+ 2$. As the antenna aligns itself upon the target, the ratio $(A + B)/(A - B)$ becomes very large. To the left of the target the ratio is negative.

10.2.3 Tracking Radar

Once the radar system has completed its *search* and has detected the presence of a target, it is then necessary to *track* the target by specifying various parameters in addition to range d. The target's location can be specified in spherical polar coordinates by range, elevation and azimuth, and thus some arrangement is needed in two planes to measure the appropriate angles. A very narrow-beam antenna will give these parameters within its beamwidth limitations but tracking a moving target via automatic controls needs some further information about the target's motion.

Consider the measurement of angle in only one plane. If we use a special paraboloidal reflector antenna with two displaced or offset feeds, as illustrated in Fig. 10.14, the system behaves as if it had two antennas with their main lobes at slightly different orientations. The beam patterns are illustrated in cartesian coordinates in Fig. 10.15, together with the angle of arrival of an echo from the target. If we include some signal processing system to compare signals A and B from each antenna feed we can obtain the sum $A + B$ and the difference $A - B$ for each individual pulse received. If the target angle is exactly midway between the beams, the difference $A - B$ is zero. If the target angle is not midway, the magnitude and sign of the difference signal gives some feedback information to control the magnitude and direction of the antenna alignment correction. However, the difference $A - B$ will vary from pulse to pulse because the return echoes are not of constant amplitude. We can overcome this problem by using $(A - B)/(A + B)$ as the control parameter.

To obtain angular resolution in two planes, various modifications can be made, such as a four-horn monopulse feed system, the whole arrangement forming part of a *null tracking system*. The term *monopulse* applied to such a split-beam tracking system means that the tracking information is available after only one pulse has been received. Computers are used to enhance these features in modern radar systems.

10.3 BEACON AND BISTATIC RADARS

A *radar beacon* is a useful way of extending the range of a radar system with suitably equipped *cooperative targets*. The ground-based radar *interrogates* the target and the beacon *responds* with a coded signal which includes a means of target identification. The radar equations are now one-way expressions, one for the radar-to-beacon path and one for the beacon-to-radar path. These may be written in the standard free-space form as

$$P_r = \frac{P_t G_t G_r \lambda^2}{(4\pi d)^2 L_m} \tag{10.31}$$

where L_m has been added to the expression to represent miscellaneous system and propagation losses in excess of the ideal model. The signal-to-noise ratio is

$$\text{SNR} = \frac{P_t G_t G_r \lambda^2}{(4\pi d)^2 L_m k T_s B} = \frac{E_t G_t G_r \lambda^2}{(4\pi d)^2 L_m k T_s} = \frac{\text{PSNR}}{2} \tag{10.32}$$

The same equations are used for both paths with appropriate values for each parameter.

Non-cooperative targets include the case in which the beacon is replaced by a *noise-jammer*. This is used in an attempt to confuse the radar system by a target, such as an aircraft, which wishes to avoid detection. The jammer distributes its available transmitter power P_j as uniformly as possible over the frequency band covered by the interrogating radar, and it can therefore be considered by the radar system as additional noise represented by noise spectral density N_j watts per Hz.

For jammer power P_j spread over frequency band Δf the equivalent noise spectral density at the radar receiver is given by the free-space equation with P_r replaced by $N_j \Delta f$:

$$N_j = \frac{P_j}{\Delta f} \times \frac{G_j G_r \lambda^2}{(4\pi d)^2 L_j} = \frac{P_j}{\Delta f} \frac{G_j A_{\text{eff}}}{4\pi d^2 L_j} \tag{10.33}$$

where A_{eff} is the effective aperture of the radar antenna, G_j is the gain of the jammer antenna and L_j is the jammer system and propagation losses in excess of the ideal model.

In most effective applications $N_j \gg kT_s$, as this is the prime purpose of the jammer. Hence the conventional search radar equation given by Eq. (10.28) can be modified by replacing kT_s by N_j in order to determine the signal-to-noise ratio in the presence of

---

Example 10.8 *Effect of jammer on radar performance*

A ground-based radar surveillance system operates with $P_{\text{av}} = 1800\,\text{W}$, $\lambda = 0.23\,\text{m}$, pulse repetition frequency $f_R = 250\,\text{pps}$, $G(t) = 40\,\text{dB}$, $L(m) = 13\,\text{dB}$ and $T_s = 2900\,\text{K}$. Determine the maximum range of the radar system for the reference condition $\text{SNR}(r) = 0\,\text{dB}$ and $\sigma = 1\,\text{m}^2$. If the target has a noise jammer centered on the same wavelength with $P_j = 100\,\text{W}$, $\Delta f = 120\,\text{MHz}$, $G(j) = 3\,\text{dB}$ and $L(j) = 10\,\text{dB}$, determine the maximum range of the radar for the same reference $\text{SNR}(r) = 0\,\text{dB}$ and $\sigma = 1\,\text{m}^2$.

We can use Eq. (10.28) to obtain the maximum range without any jammer to disturb the system, noting that $P_{\text{av}} = P_t \tau f_R$ or $P_t \tau = P_{\text{av}}/f_R$. Hence:

$$\text{SNR}(r) = P(\text{av}) - 10\log f_R + G(t) + G(r) + 20\log \lambda + 10\log \sigma$$
$$- 30\log(4\pi) - 40\log d - L(m) - 10\log k - 10\log T_s$$

Substituting the appropriate data we obtain $40\log d = 223.8$ with the range in meters, giving $d_{\text{max}} = 393.6\,\text{km}$.

Similarly, we can use Eq. (10.34) to obtain the maximum range with the specified noise jammer operating, again noting the P_{av} relationship:

$$\text{SNR}(r) = P(\text{av}) - 10\log f_R + 10\log \Delta f + G(t) - G(j) + L(j)$$
$$- P(j) - 10\log(4\pi) - 20\log d - L(m) + 10\log \sigma$$

This gives $20\log d = 92.37$ with the range in meters, or $d_{\text{max}} = 41.54\,\text{km}$.

The effect of the jammer is thus to reduce the effective reference range of a radar system from about 394 km to about 42 km.

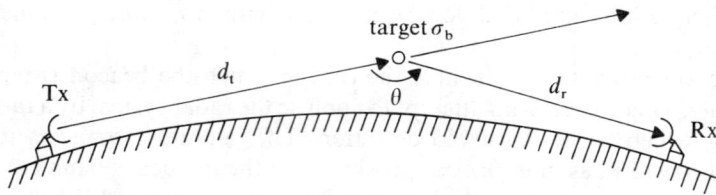

Fig. 10.16 A bistatic radar system. When θ is near $180°$ the forward-scatter cross-section can considerably exceed the head-on cross-section seen by the more normal radars.

jamming:

$$\text{SNR} = \frac{P_t G_t G_r \lambda^2 \sigma}{(4\pi d)^3 d^4 B L_m} \cdot \frac{\Delta f (4\pi d)^2 L_j}{P_j G_j G_r \lambda^2}$$

$$\text{SNR} = \left[\frac{P_t \tau}{P_j / \Delta f} \right] \frac{G_t}{G_j} \frac{\sigma}{4\pi d^2} \frac{L_j}{L_m} \tag{10.34}$$

It is thus possible to assess the effect of jamming by comparing the radar performance with and without the jammer.

In certain applications of radar the transmitter and receiver may be widely separated for various reasons. The radar equation for such a *bistatic* system may be derived in the same way as Eq. (10.28) to produce

$$\text{SNR} = \frac{P_t G_t G_r \lambda^2 \sigma_b}{(4\pi)^3 d_t^2 d_r^2 k T_s B L_t L_r} \tag{10.35}$$

where d^4 is now replaced by $d_t^2 d_r^2$ and L_m by $L_t L_r$. The radar cross-section of the target σ_b is defined in a slightly different way to account for re-radiation in the new direction of the receiver.

For certain types of targets, such as oncoming missiles, the radar cross-section of the nose of the missile as it approaches a monostatic radar may be very small and hence difficult to detect until the range is smaller. However, the effective radar cross-section at other aspect angles is often much larger and a network of monostatic radars can be used in a kind of space diversity arrangement to detect the target, each radar seeing the target at a different aspect. The bistatic radar also makes use of this principle and if the aspect angle is near $180°$ (as shown in Fig. 10.16), the *forward-scatter cross-section* can considerably exceed the head-on cross-section, sometimes giving enhancements of the order of $30\,\text{dB}$ to the system performance.

10.4 DOPPLER RADARS

The Doppler effect takes its name from the Austrian scientist Christian Doppler, who first explained in 1842 why the sound from a moving source increased in frequency as the source approached an observer.

10.4.1 Doppler Principle

Although originally derived for acoustical propagation, the Doppler principle applies equally well to radio wave propagation. If a mobile radio transmitter at frequency f_t approaches a stationary receiver at velocity v, the receiver will observe a frequency f_r which exceeds f_t by the Doppler frequency shift f_D. However, some care is needed in radar as both the transmitter and receiver are stationary with respect to each other and it is the target which travels with a radial component of approaching velocity v that reduces the range by $2v$ metres every second, and not v m s^{-1} as described in the case of a mobile transmitter.

Thus, for radars of this type, the Doppler frequency shift can be obtained as

$$f_D = |f_t - f_r| = 2f_t(v/c) = 2v/\lambda_t \qquad (10.36)$$

where v = radial velocity. Transposing the equation, we can determine the target's

---

Example 10.9 Doppler frequency shift

Show that the radial velocity of a moving target can be determined with a Doppler radar system based on Eq. (10.37).

The solution can be obtained via Fig. 10.17, which indicates how to relate the received frequency f_r to the transmitted frequency f_t for a single cycle of the continuous carrier wave. Starting with

$$\frac{1}{f_r} = t_2 - t_1$$

we note that t_1 is the time delay associated with the starting point of the cycle, $t_1 = 2d/c$. The end of the cycle leaves the transmitter t seconds later. By the time it reaches the *approaching* target the target will be a distance vt' *nearer* the transmitter or $d' = d - vt'$, and t_2 is then given by

$$t_2 = t' + \frac{2d - 2vt}{c} = t' + \frac{2d}{c} - \frac{2vt}{c}$$

Thus we can write

$$\frac{1}{f_r} = t_2 - t_1 = t + \frac{2d}{c} - \frac{2vt}{c} - \frac{2d}{c}$$

or

$$f_r = f_t \left\{ \frac{1}{1 - 2v/c} \right\} = f_t \left\{ \frac{c}{c - 2v} \right\}$$

The Doppler frequency shift can then be derived from

$$|f_r - f_t| = f_t \left\{ \frac{c}{c - 2v} - 1 \right\}$$

or

$$f_D = f_t(2v/c)$$

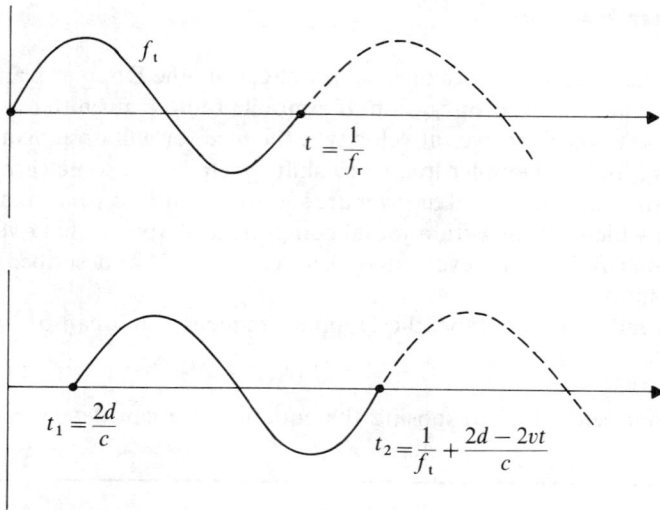

Fig. 10.17 The transmitted and received signals of a Doppler radar system used in Ex. 10.9 to determine the Doppler frequency shift f_D.

radial velocity as

$$v = \tfrac{1}{2} f_D c / f_t \tag{10.37}$$

For a fixed, known transmitter frequency f_t, the radial velocity of the target can be calibrated directly in terms of the much lower Doppler frequency shift f_D, but there is no information concerning the target's range.

10.4.2 Simple Applications

There are various applications of Doppler radar systems. For instance, if an aircraft is flying at constant height above ground- or sea-level and it transmits a narrow beam at a

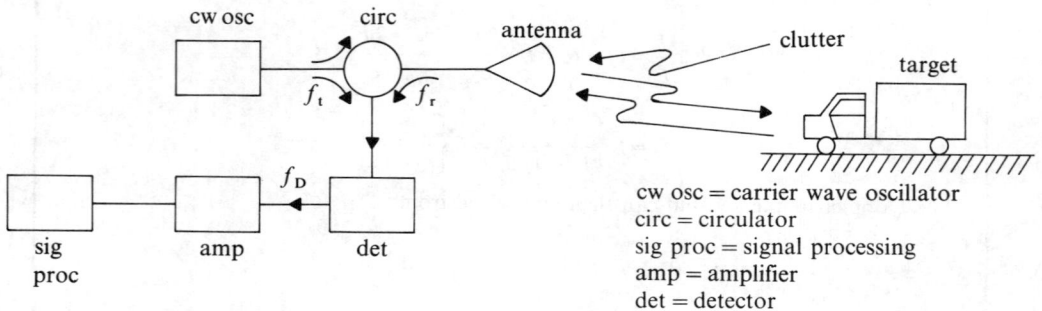

cw osc = carrier wave oscillator
circ = circulator
sig proc = signal processing
amp = amplifier
det = detector

Fig. 10.18 A commonly used application of a Doppler radar system is the measurement of the oncoming velocity of automobiles or other targets.

forward angle θ, the echo from the ground will be frequency-increased by $f_D = (2v \cos \theta)/\lambda_t$. When two such arrangements are attached to the aircraft, one beamed forward at angle θ and one beamed backwards at the same angle, the difference between the two received signals is $(4v \cos \theta)/\lambda_t$, where v is measured in the direction of alignment of the two Doppler systems. Thus, if a further two beams are aligned perpendicular to the first two, the arrangement records the two velocity vectors, longitudinal and transverse, and the resultant *drift* of the aircraft can thereby be observed.

A Doppler radar may also be used to measure and indicate the velocity of an oncoming vehicle on a highway. The very simple arrangement shown in Fig. 10.18 makes use of a three-port circulator in which the limited isolation of the circulator between ports 1 and 3 acts as a kind of attenuator, though still permitting sufficient of the transmitted power to leak through to bias the detector. The received frequency f_r beats with the leaked frequency f_t to produce the Doppler shift f_D, from which the velocity of the oncoming vehicle may be determined.

If the radar is set up at some angle θ to the path of the vehicle, as in Fig. 10.19, and the velocity is calibrated via

$$v = \tfrac{1}{2} c f_D / (f_t \cos \theta) \tag{10.38}$$

with $\theta = 45°$, then for vehicles moving slightly off the anticipated direction by as little as $\pm 5°$, the indicated velocity will be between 92% and 110% of the true velocity. Note

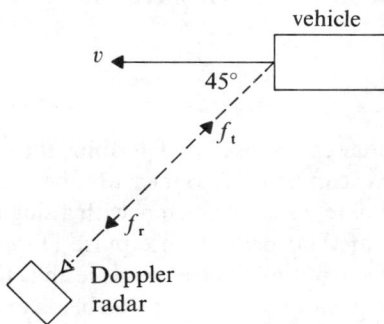

Fig. 10.19 An alternative method of estimating the velocity of moving targets which is critically dependent on the angle of incidence, here 45°.

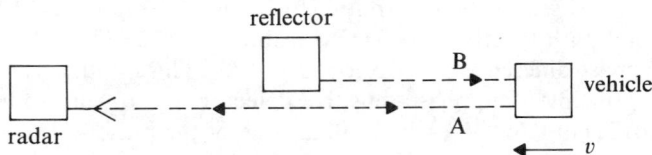

Fig. 10.20 The presence of static or moving reflectors in the vicinity of moving targets can influence the signals returned to the Doppler radar and the signal processing technique needs to be able to eliminate the false returns.

that the path length $p = 2d$ of the radar signal changes at a rate of $\Delta p = c f_D / f_t$ meters per second, or

$$f_D = f_t \Delta p / c \tag{10.39}$$

When the radar is used to indicate the head-on velocity of a vehicle, as in Fig. 10.20, and a large stationary reflector is nearby, it is sometimes possible to receive multiple return echoes. Reflection via path A produces

$$\Delta p = 2v = c f_D / f_t$$

to give us the standard relationship $f_D = f_t (2v/c)$, while reflection via path B produces

$$\Delta p = 2v + 2v = c f_D^* / f_t$$

which gives us a return with $f_D^* = 2 f_D$. Thus the two components would indicate two Doppler frequency shifts and two vehicle velocities, v and $2v$. If the *stationary* reflector is replaced by a moving vehicle, as shown in Fig. 10.20, the situation for path B now changes to

$$f_D^{**} = 2 f_t (2v + 2v)/c > 2 f_D$$

and care must be taken to ensure that these higher Doppler frequencies do not indicate erroneous components of velocity.

Because of the continuous wave nature of this type of radar, the power levels involved with target detection are much reduced and hence the radar range is limited.

10.4.3 Pulse Doppler Radar

We have seen that pulse radar can be used to determine the range d of a target while the carrier-wave Doppler radar can be used to measure the radial velocity v of a moving target. It is possible to estimate both parameters with a single radar in several different ways, one rather obvious method being to use pulse Doppler radar.

Ignoring Doppler effects initially, let us assume that the pulse part of the pulse Doppler radar records the position of the target as being exactly 2000 m (corresponding to the moment the leading edge of the pulse reached the target). The return echo is processed by a range-measuring system and also passes through a very elementary form of phase detector which gives an output voltage proportional to the phase difference (Fig. 10.21). The time delay, with $c = 3 \times 10^8$ m s^{-1} exactly for this example, is $t' = 2d/c = 13.333$ μsec. If the transmitter frequency is exactly $f_t = 10$ GHz or $\lambda_t = 0.030$ m, the total path length of 4000 m corresponds to 133,333.333 wavelengths, or an observable phase difference of $0.333 \times 360° = 120°$. This is much oversimplified but illustrates the principle. The phase-sensitive detector (PSD) will produce an output voltage (Fig. 10.21) of 0.666 V at the leading edge of the pulse.

If the pulse width is 1 μsec and the target has a velocity $v = 150$ m s^{-1} approaching the radar, the target will have traveled a distance of 1.5×10^{-4} m in 1 μsec and the total path length of 3999.9997 m corresponds to 133,333.323 λ_t or an observable phase change of $0.323 \times 360° = 116.4°$. The PSD output will be 0.647 V at the end of the pulse,

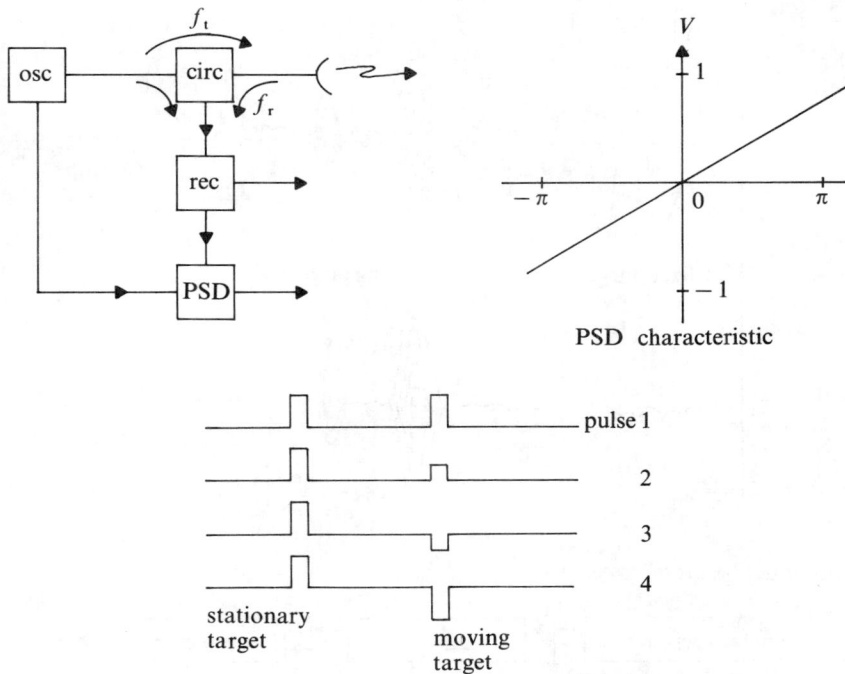

Fig. 10.21 A phase-sensitive detector (PSD) can be used to distinguish between a stationary target and a moving target.

an almost negligible variation. Thus to a first approximation we can assume that the PSD output will be an almost constant pulse of about 0.65 V.

If the pulse repetition frequency $f_R = 800$ pps, the target will move a distance of 0.1875 m in 1.25 msec and the total path length of 3999.625 m corresponds to 133,320.833 λ_t, an observable phase difference of $0.833 \times 360° = 300°$. This will produce a PSD output of -0.33 V, quite different from the previous pulse.

Thus successive pulses will have comparatively small changes in range d but considerably different output magnitudes of the phase-sensitive detector. The situation is illustrated in Fig. 10.21. A radar system which takes advantage of this effect is called a *moving target indication* (MTI) radar. Figure 10.22 illustrates the display pattern produced by several successive pulses from a pulse Doppler radar. Most of the echoes are returned by stationary objects such as hills, trees, buildings, etc., but two moving targets are recognized because the amplitudes of successive pulses are of constantly varying magnitude. By using a suitable kind of signal processing circuit, illustrated in the block diagram in Fig. 10.22, the display pattern is successively subtracted from the previous display pattern. All constant signals then produce zero outputs and only the varying-amplitude pulses produce a finite output. These can then be observed for range measurements of the moving targets, while the usual Doppler arrangement elsewhere in the radar receiver will indicate velocities.

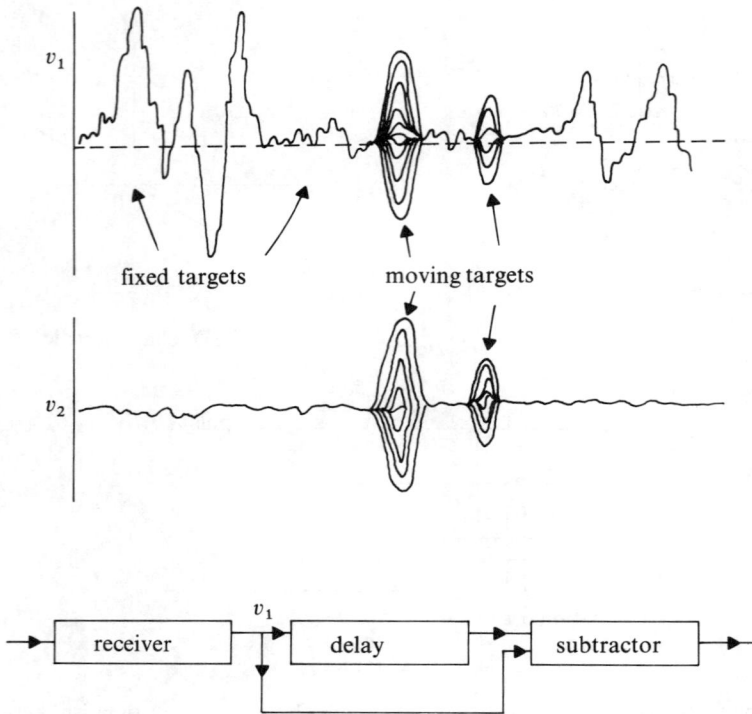

Fig. 10.22 A subtractor can be incorporated in a moving-target indicator (MTI) radar to reduce or eliminate the echoes from stationary targets, leaving only the echoes from moving targets.

10.4.4 FM-CW Radar

As an alternative to the pulse Doppler radar just described, a carrier-wave radar may be modified to measure range as well as radial velocity of a moving target. This is achieved by means of frequency modulation and the system is then called an *FM-CW radar*. There are several versions, but the one which we shall describe is probably the simplest to follow.

If f_c is the reference carrier frequency, the transmitter can be frequency modulated such that the transmitted frequency f_t can vary with time in the manner illustrated in Fig. 10.23. Assume initially that the target is stationary. The return echo has no frequency change due to the Doppler effect but only that due to the transmitter FM. The transmitted and received signals thus have frequency envelopes which are similar in shape but are displaced in time due to the delay $t' = 2d/c$. By comparing these signals the range d can be estimated.

Figure 10.23 shows the envelopes of f_t and f_r and also the envelope of the beat frequency $f_b = f_t - f_r$. By comparing the similar triangles we can see that the ratio f_d/f_b

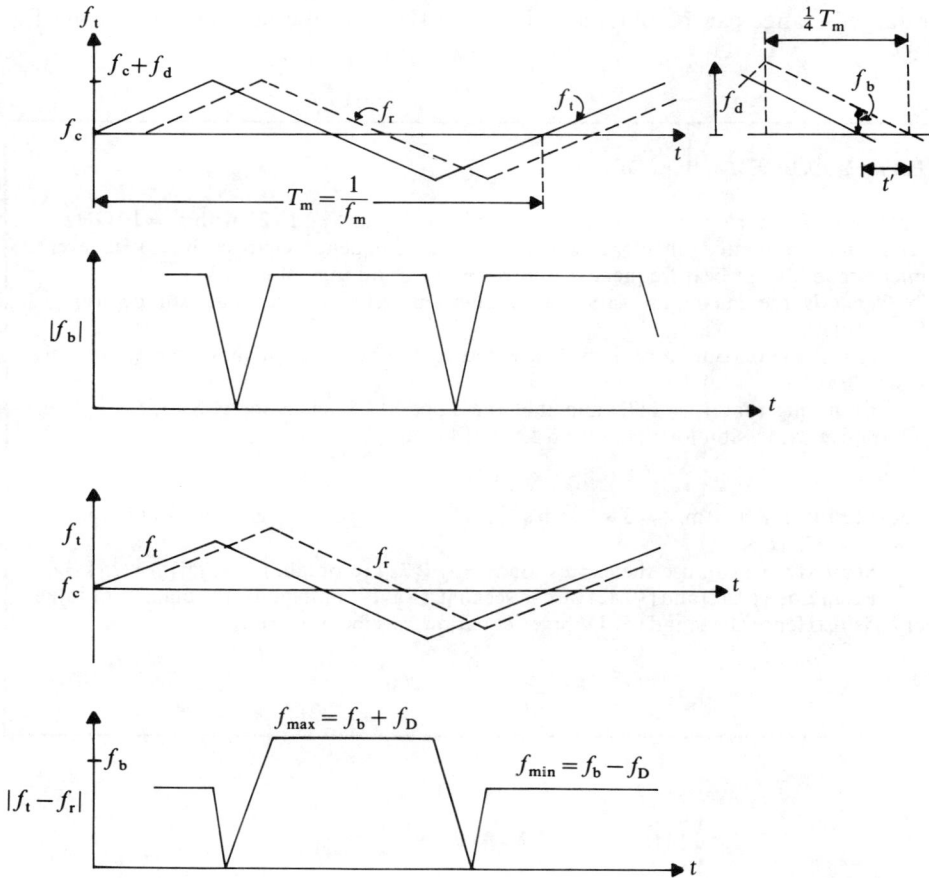

Fig. 10.23 The waveforms associated with an FM-CW radar which can be processed to produce information on both range and radial velocity. The triangular FM used in this illustration is only one of several alternative possibilities.

is equal to the ratio $\frac{1}{4} T_m/t'$, where t' is the delay $2d/c$. Hence

$$d = \frac{c f_b}{8 f_m f_d} \tag{10.40}$$

If the target is moving towards the radar, the received signal frequency f_r is increased by the Doppler frequency shift f_D, and the envelope pattern is changed in the manner shown in Fig. 10.23. The beat frequency now has two particular values, f_{max} and f_{min}, due to the vertical shift of f_r. Thus from $f_{max} = f_b + f_D$ and $f_{min} = f_b - f_D$ we can write

$$f_b = \tfrac{1}{2}(f_{max} + f_{min}) \tag{10.41}$$

and

$$f_D = \tfrac{1}{2}(f_{max} - f_{min}) \tag{10.42}$$

The target's range can be obtained via Eq. (10.40) and the target's velocity via Eq. (10.37).

Example 10.10 FM-CW radar

An FM-CW radar operates in the manner illustrated in Fig. 10.23 with $f_c = 10\,\text{GHz}$.
(a) If the modulation envelope is such that the frequency changes by $1\,\text{kHz}$ every microsecond, what beat frequency f_b corresponds to a range of $d = 2\,\text{km}$?
(b) What is the maximum value of modulation frequency f_m that can be used if $f_d = 10\,\text{kHz}$?
(c) What is the maximum radial velocity that can be detected unambiguously with this system?

By noting that $f_m = 1/T_m$ and that the slope of the envelope is $S = f_d/(\frac{1}{4}T_m)$, an alternative expression for range from Eq. (10.40) is

$$d = cf_b/(2S)$$

Substituting $d = 2000\,\text{m}$, $c = 3 \times 10^8\,\text{m s}^{-1}$ and $S = 10^9\,\text{Hz} - \text{s}^{-1}$ gives us beat frequency $f_b = 13.333\,\text{kHz}$.

Similarly, we can use the expression $S = f_d/(\frac{1}{4}T_m)$ to obtain $f_m = S/(4f_d) = 25\,\text{kHz}$.

From Eqs. (10.41) and (10.42) we can see that, to avoid ambiguity, the maximum value of f_D is f_b. Hence the standard Doppler equation may be written as

$$v = \frac{f_D c}{2f_t} = \frac{f_b c}{2f_c} = 200\,\text{m s}^{-1}$$

PROBLEMS

10.1 A pulse radar system has a duty cycle of 0.1% and a pulse width of $2\,\mu s$. What are the theoretical minimum and maximum limits of the radar range? Is the maximum value realistic if the radar antenna is $100\,\text{m}$ above sea-level and the aircraft target is $1000\,\text{m}$ above sea-level? Plot a graph of maximum line-of-sight range against aircraft height assuming a four-thirds earth radius to account for refraction.

10.2 Determine the reference range for $\text{SNR}(r) = 0\,\text{dB}$ and $\sigma = 1\,\text{m}^2$ for a $3\,\text{GHz}$ pulse radar if the average transmitter power is $1\,\text{kW}$, the antenna gain is $30\,\text{dB}$ and the pulse repetition frequency is $1000\,\text{pps}$. Assume the receiver noise figure $F(r)$ is $6, 8, 10$ and $12\,\text{dB}$ for comparison and that the antenna noise temperature is $290\,\text{K}$.

10.3 A $1300\,\text{MHz}$ pulse radar system has $P(t) = 60\,\text{dBW}$, $\tau = 1\,\mu\text{sec}$, $G(t) = 30\,\text{dB}$, $F(s) = 8\,\text{dB}$, and is used to detect targets with $\sigma = 20\,\text{m}^2$ up to 70 nautical miles. Determine the required $\text{SNR}(r)$ if miscellaneous losses are $5\,\text{dB}$.

10.4 If the signal-to-noise ratio of a pulse radar system is $15\,\text{dB}$ and it is specified that $P_{fa} = 10^{-8}$, determine the probability of detection P_d for the following types of targets: (a) a target which provides a constant-amplitude return echo plus noise; (b) a target whose return echo consists of numerous scattered components plus noise; and (c) a target such as (b) but with one of the components much larger than the others.

10.5 A $\tau = 1\,\mu\text{sec}$ pulse radar system has a mean $P_r = 1\,\text{pW}$ per pulse. What system noise figure $F(s)$ is needed for a mean signal-to-noise ratio of 14 dB?

10.6 A $50\,\Omega$, $G = 30\,\text{dB}$ radar receiver operates with $T_s = 2500\,\text{K}$ at the receiver input. Determine: (a) the rms noise voltage at the receiver output if $\tau = 1\,\mu\text{sec}$; (b) the threshold voltage at the receiver output for $P_{fa} = 10^{-8}$; and (c) the corresponding threshold levels of signal voltage, signal power and signal-to-noise ratio at the receiver input.

10.7 A search radar with $\tau = 2\,\mu\text{sec}$, $f_R = 600\,\text{pps}$ and $G(t) = G(r) = 30\,\text{dB}$ scans a volume covered by solid angle $\psi_s = 0.15$ steradians each second. If it uses non-coherent post-detection integration, what are the values of integration gain $G(i)$ and integration loss $L(i)$?

10.8 A search radar is required to produce $\text{SNR}(i) = 14\,\text{dB}$ when detecting a $\sigma = 10\,\text{m}^2$ target over a range of 40 nautical miles. It scans the search volume at the rate of 0.1 steradians per second and its system noise temperature is $T_s = 3000\,\text{K}$ at the receiver input. Assuming the transmitter has $P_{av} = 1\,\text{kW}$ and miscellaneous losses are 10 dB, determine the effective area of the antenna and its gain at 10 GHz if the illumination efficiency is 55%.

10.9 A large aircraft with $\sigma = 50\,\text{m}^2$ approaches a pulse radar system. The aircraft uses a 100 W, $\Delta f = 60\,\text{MHz}$ radar jammer at the same nominal frequency with $G(j) = 2\,\text{dB}$ in order to interfere with the radar. The ground-based radar uses $P_t = 500\,\text{kW}$, $G(t) = 35\,\text{dB}$, $\tau = 1\,\mu\text{sec}$ and requires $\text{SNR}(r) = 10\,\text{dB}$. If $L(m) - L(j) = 5\,\text{dB}$, estimate within what range $\text{SNR}(r) > 10\,\text{dB}$. If the radar is modified to incorporate post-detection integration with $N = 10$ pulses per train, what is the modified range of the radar with the jammer?

10.10 A carrier-wave Doppler radar at 1 GHz is used to measure the velocity v of an oncoming vehicle. It is required to operate with $0 < v < 150\,\text{mph}$. What is the highest Doppler frequency shift it will need to process? What is the sensitivity of the device in terms of f_D per mph? Why is $f_t = 10\,\text{GHz}$ better for this particular application?

10.11 A short-range, low-speed navigation system uses FM-CW radar with triangular FM envelope. With $f_c = 9\,\text{GHz}$, $f_d = 25\,\text{MHz}$ and $f_m = 120\,\text{Hz}$, the limits of the observed beat frequency are 5390 Hz and 5210 Hz at time t_1; 3390 Hz and 3210 Hz at time t_2; 2845 Hz and 2800 Hz at time t_3; and 800 Hz and 800 Hz at time t_4. What is happening to the target?

REFERENCES

10.1 North, D. O., 'An analysis of the factors which determine signal/noise discrimination in pulsed-carrier systems,' *Proc. IEEE*, 51 (July 1963), 1016–27.

10.2 Blake, L. V., *Radar Range-Performance Analysis*. Toronto: Lexington Books, D. C. Heath and Company, 1980.

10.3 Swerling, P., 'Probability of detection for fluctuating targets,' *Rand Corporation Research Memorandum* RM-1217 (March 1954), reprinted in *Trans. IRE*, IT-6 (April 1960), 269–308.

10.4 Schwartz, M., 'Effects of signal fluctuation on the detection of pulse signals in noise,' *Trans. IRE*, IT-2 (June 1956), 66–71.

Appendix 1

Revision of Electromagnetic Principles

The electrical parameters used in the study of electromagnetic phenomena can be divided into two groups. The first, which contains current I, charge q, resistance R, inductance L, capacitance C, potential difference V, magnetic flux ϕ, electric flux ψ, power P and others, can be described as *lumped* parameters. The second, which contains current density J, charge density ρ, permeability μ, permittivity ε, conductivity σ, electric field strength E, magnetic field strength H, electric flux density D, magnetic flux density B, power flux P_a and others, can be described as *spatial* parameters, the dimensions of which involve the unit of length in the form m^{-n}, where $n = 1, 2$ or 3. The lumped parameters are more common in network theory and the spatial parameters are often preferred in electromagnetic theory.

A1.1 RELATIONSHIPS BETWEEN PARAMETERS

The relationships between the spatial parameters and the lumped parameters are based on definitions or laws. These appear in several mathematical forms, but in this text we shall make use of the integral notation so that transformation into vector algebra is straightforward.

A1.1.1 Volumetric Parameters

If we define ρ as the electric charge per unit volume ($C\,m^{-3}$), the incremental charge dq in incremental volume dv is $\rho\,dv$. When summed over some specific volume v the total charge within that volume is given by the integral equation

$$\int_v \rho\,dv = q \tag{A1.1}$$

There are other volumetric parameters besides ρ, but this is the one of prime interest to us for our present purposes.

A1.1.2 Surface Parameters

By contrast, we are interested in several spatial parameters which involve the dimension m^{-2}. For example, if an incremental amount of electric flux $d\psi$ (C) is perpendicularly incident upon an incremental surface area dS, the spatial parameter $D = \mathrm{d}\psi/\mathrm{d}S$ is called *electric flux density* $(\mathrm{C\,m^{-2}})$. In vector algebra notation, with the dot-product to include angles of incidence other than perpendicular, we can write instead $\mathbf{D}\cdot\mathrm{d}\mathbf{S} = \mathrm{d}\psi$. When summed over a surface area S the total electric flux is given by the integral equation

$$\int_S \mathbf{D}\cdot\mathrm{d}\mathbf{S} = \psi \tag{A1.2a}$$

In a similar manner we can define *magnetic flux density* $\mathbf{B}\,(\mathrm{Wb\,m^{-2}})$, *current density* $\mathbf{J}\,(\mathrm{A\,m^{-2}})$ and *power flux* $\mathbf{P_a}\,(\mathrm{W\,m^{-2}})$ with their corresponding definitions in integral form:

$$\int_S \mathbf{B}\cdot\mathrm{d}\mathbf{S} = \phi \tag{A1.2b}$$

$$\int_S \mathbf{J}\cdot\mathrm{d}\mathbf{S} = I \tag{A1.2c}$$

$$\int_S \mathbf{P_a}\cdot\mathrm{d}\mathbf{S} = P \tag{A1.2d}$$

There are other similarly defined parameters, but these are sufficient for the moment.

A1.1.3 Linear Parameters

Of the various linear parameters, two are of prime interest at this stage. The magnetic field strength $\mathbf{H}\,(\mathrm{A\,m^{-1}})$ and the electric field strength $\mathbf{E}\,(\mathrm{V\,m^{-1}})$ are defined via Ampere's law:

$$\oint_s \mathbf{H}\cdot\mathrm{d}\mathbf{s} = I \tag{A1.3a}$$

and Faraday's law:

$$\oint_s \mathbf{E}\cdot\mathrm{d}\mathbf{s} = -\frac{\mathrm{d}\phi}{\mathrm{d}t} \tag{A1.3b}$$

These indicate that if the dot-product $\mathbf{H}\cdot\mathrm{d}\mathbf{s}$ is summed around a closed loop s, the result is the electric current flowing through the loop; and if the dot-product $\mathbf{E}\cdot\mathrm{d}\mathbf{s}$ is summed around a closed loop s, the result is equal to (minus) the time rate of change of the magnetic flux coupling the loop.

A1.1.4 Stokes's and Green's Theorems

The definitions of ρ, $\mathbf{D}$, $\mathbf{B}$, $\mathbf{J_c}$, $\mathbf{P_a}$, $\mathbf{H}$ and $\mathbf{E}$ are all expressed in integral form using one or other of the diagrams shown in Fig. A1.1. Fortunately, no actual integration is

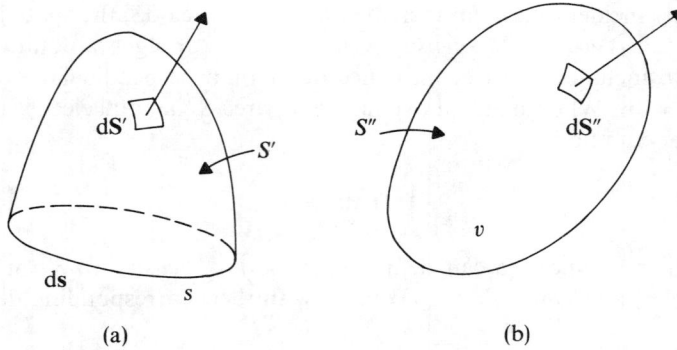

Fig. A1.1 The 'half-eggshell' surface of Stokes's theorem and the 'unbroken egg' volume of Green's theorem.

necessary as linear and surface integrals are related via *Stokes's theorem* and surface and volume integrals are related via *Green's theorem* (or the divergence theorem). The proofs are outside the scope of this text but their applications are straightforward.

Stokes's theorem refers to the '*half-eggshell*' surface S' bounded by closed loop s, as illustrated in Fig. A1.1a. Using vector algebra notation, the theorem indicates that for some vector **F**:

$$\oint_s \mathbf{F} \cdot \mathbf{ds} = \int_{S'} (\operatorname{curl} \mathbf{F}) \cdot \mathbf{dS'} \qquad (A1.4a)$$

where

$$\operatorname{curl} \mathbf{F} \equiv \nabla \times \mathbf{F} \qquad (A1.4b)$$

or

$$\operatorname{curl} \mathbf{F} = \left(\frac{\partial F_z}{\partial y} - \frac{\partial F_y}{\partial z} \right)\mathbf{i} + \left(\frac{\partial F_x}{\partial z} - \frac{\partial F_z}{\partial x} \right)\mathbf{j} + \left(\frac{\partial F_y}{\partial x} - \frac{\partial F_x}{\partial y} \right)\mathbf{k} \qquad (A1.4c)$$

or

$$\operatorname{curl} \mathbf{F} = \frac{1}{r \sin \theta} \left\{ \frac{\partial}{\partial \theta}(F_\phi \sin \theta) - \frac{\partial F_\theta}{\partial \phi} \right\} \mathbf{a}_r$$

$$+ \frac{1}{r} \left\{ \frac{1}{\sin \theta} \frac{\partial F_r}{\partial \phi} - \frac{\partial}{\partial r}(rF_\phi) \right\} \mathbf{a}_\theta + \frac{1}{r} \left\{ \frac{\partial}{\partial r}(rF_\theta) - \frac{\partial F_r}{\partial \theta} \right\} \mathbf{a}_\phi \qquad (A1.4d)$$

Green's theorem refers to the 'unbroken egg' with volume v and surface area S'' illustrated in Fig. A1.1b. Again, using vector algebra notation, this theorem indicates that:

$$\int_{S''} \mathbf{G} \cdot \mathbf{dS''} = \int_v (\operatorname{div} \mathbf{G})\, dv \qquad (A1.5a)$$

where

$$\operatorname{div} \mathbf{G} \equiv \nabla \cdot \mathbf{G} \qquad (A1.5b)$$

Table A1.1 Lumped and spatial parameters

Lumped relationship	*Spatial relationship*	
$\psi = q$	$\operatorname{div} \mathbf{D} = \rho$	(T1a)
$\mathrm{emf} = -n\,\mathrm{d}\phi/\mathrm{d}t$	$\operatorname{curl} \mathbf{E} = -\mathrm{d}\mathbf{B}/\mathrm{d}t$	(T1b)
$\phi = q_\mathrm{m} = 0$	$\operatorname{div} \mathbf{B} = 0$	(T2a)
$\mathrm{mmf} = NI$	$\operatorname{curl} \mathbf{H} = \mathbf{J}_\mathrm{c} + \mathrm{d}\mathbf{D}/\mathrm{d}t$	(T2b)
$\mathrm{mmf} = (L/\mu A).\varphi$	$\mathbf{B} = \mu \mathbf{H}$	(T3)
$V = (L/\varepsilon A).\psi$	$\mathbf{D} = \varepsilon \mathbf{E}$	(T4)
$V = (L/\sigma A).I$	$\mathbf{J}_\mathrm{c} = \sigma \mathbf{E}$	
$I = \mathrm{d}q/\mathrm{d}t$	$\operatorname{div} \mathbf{J}_c = -\mathrm{d}\rho/\mathrm{d}t$	

or

$$\operatorname{div} \mathbf{G} = \frac{\partial G_x}{\partial x} + \frac{\partial G_y}{\partial y} + \frac{\partial G_z}{\partial z} \tag{A1.5c}$$

or

$$\operatorname{div} \mathbf{G} = \frac{1}{r^2}\frac{\partial}{\partial r}(r^2 G_r) + \frac{1}{r\sin\theta}\frac{\partial}{\partial \theta}(G_\theta \sin\theta) - \frac{1}{r\sin\theta}\frac{\partial G_\phi}{\partial \phi} \tag{A1.5d}$$

Using either Stokes's or Green's theorems, as appropriate, the basic laws and definitions can all be expressed in spatial form. Table A1.1 lists some of the comparisons which can be made between well-known lumped equations and the equivalent spatial forms. In particular, note that:

(1) the definitions of ψ and $\mathbf{D}$ are given as Eq. (T1a);
(2) Faraday's law is Eq. (T1b);
(3) the definitions of ϕ and $\mathbf{B}$ are given as Eq. (T2a);
(4) Ampere's law is Eq. (T2b); and
(5) the constitutive equations for a homogeneous isotropic propagation medium are Eqs. (T3) and (T4).

These equations relate the four electromagnetic field vectors $\mathbf{E}, \mathbf{H}, \mathbf{B}$ and $\mathbf{D}$ in terms of four spatial parameters ρ, μ, ε and $\mathbf{J}_\mathrm{c}$. Equations (T1a),(T1b) and (T2a),(T2b) are combined in order to satisfy *Helmholtz's theorem* in vector algebra, which states that both $\operatorname{div} \mathbf{X}$ and $\operatorname{curl} \mathbf{X}$ are necessary to specify vector $\mathbf{X}$ absolutely.

An alternative approach, preferred by some authors, is to use conductivity σ instead of $\mathbf{J}_\mathrm{c}$ via *Ohm's law*, $\mathbf{J}_\mathrm{c} = \sigma \mathbf{E}$. It is also possible to relate $\mathbf{J}_\mathrm{c}$ to ρ via the definition of current, as shown in Table A1.1. The applications of Stokes's and Green's theorems are illustrated in the following two examples.

Example A1.1 Vector algebra forms of charge, flux and current

Derive the expressions relating electric charge, electric flux and electric current in spatial vector algebra form.

Referring to Fig. A1.1b, the electric current leaving volume v is:

$$I_c = -(dq/dt) \qquad (A1.6)$$

Replacing each term by its corresponding spatial parameter gives:

$$\int_{S''} \mathbf{J}_c \cdot d\mathbf{S}'' = -\int_v (d\rho/dt)\, dv$$

We can now apply Green's theorem to change the integrals:

$$\int_v (\operatorname{div} \mathbf{J}_c)\, dv = -\int_v (d\rho/dt)\, dv$$

or

$$\operatorname{div} \mathbf{J}_c = -\frac{d\rho}{dt} \qquad (A1.7)$$

Referring again to Fig. A1.1b, the electric flux ψ emanating from charge q is defined in SI units simply as $\psi = q$. In spatial units:

$$\int_{S''} \mathbf{D} \cdot d\mathbf{S}'' = \int_v \rho\, dv$$

By applying Green's theorem to change the integrals:

$$\int_v (\operatorname{div} \mathbf{D})\, dv = \int_v \rho\, dv$$

or

$$\operatorname{div} \mathbf{D} = \rho \qquad (A1.8)$$

Equation (A1.7) is the spatial definition of current and Eq. (A1.8) is the spatial relationship between electric flux and charge. Some texts use a hypothetical form of magnetic pole density and magnetic current to balance the equations, but the approach here is to assume the net pole density is zero and hence, by similar reasoning:

$$\operatorname{div} \mathbf{B} = 0 \qquad (A1.9)$$

Example A1.2 Spatial form of Ampere's and Faraday's laws

Derive expressions for Ampere's law and Faraday's law in spatial vector algebra form.

Maxwell showed that a more precise form of Ampere's law should include the displacement current as well as the conduction current so that

$$\oint_s \mathbf{H} \cdot d\mathbf{s} = dq/dt + d\psi/dt = \int_{S'} (\mathbf{J}_c + d\mathbf{D}/dt) \cdot d\mathbf{S}'$$

for the arrangement illustrated in Fig. A1.1a. This time we can use Stokes's theorem to convert a linear integral into its surface integral form to give

$$\int_{S'} (\operatorname{curl} \mathbf{H}) \cdot d\mathbf{S}' = \int_{S'} (\mathbf{J}_c + d\mathbf{D}/dt) \cdot d\mathbf{S}'$$

or

$$\operatorname{curl} \mathbf{H} = \mathbf{J}_c + d\mathbf{D}/dt \qquad (A1.10a)$$

This is sometimes given in alternative forms. In time domain, the use of Ohm's law and one of the constitutive equation (for homogeneous isotropic media) gives

$$\text{curl } \mathbf{H} = \sigma \mathbf{E} + \varepsilon \, d\mathbf{E}/dt \tag{A1.10b}$$

In frequency domain, with $E = \hat{E}e^{j\omega t}$,

$$\text{curl } \mathbf{H} = \sigma \mathbf{E} + j\omega\varepsilon\mathbf{E} = (\sigma + j\omega\varepsilon)\mathbf{E} \tag{A1.10c}$$

and this is Ampere's law in its spatial vector algebra form.

Faraday's law has no such modification in our approach and it is written in integral form as

$$\oint_s \mathbf{E} \cdot \mathbf{ds} = - \int_{S'} (d\mathbf{B}/dt) \cdot \mathbf{dS'}$$

for the arrangement illustrated in Fig. A1.1a. Using Stokes's theorem,

$$\int_{S'} (\text{curl } \mathbf{E}) \cdot \mathbf{dS'} = - \int_{S'} (d\mathbf{B}/dt) \cdot \mathbf{dS'}$$

or

$$\text{curl } \mathbf{E} = - d\mathbf{B}/dt = - \mu d\mathbf{H}/dt \tag{A1.11a}$$

The corresponding equation in frequency domain is

$$\text{curl } \mathbf{E} = - j\omega\mathbf{B} = - j\omega\mu\mathbf{H} \tag{A1.11b}$$

This is Faraday's law in spatial vector algebra form.

A1.2 *E*-FIELD AND *H*-FIELD EQUATIONS

The four numbered equations (T1)–(T4) in Table A1.1 may be combined to eliminate two of the field vectors. If we choose to eliminate parameters **B** and **D**, substitution of Eqs. (T3) and (T4) into Eqs. (T1) and (T2) will achieve this purpose:

$$\text{div } \mathbf{E} = \rho/\varepsilon$$
$$\text{curl } \mathbf{E} = - \mu \, d\mathbf{H}/dt = - j\omega\mu\mathbf{H}$$
$$\text{div } \mathbf{H} = 0$$
$$\text{curl } \mathbf{H} = \mathbf{J}_c + \varepsilon \, d\mathbf{E}/dt = \mathbf{J}_c + j\omega\varepsilon\mathbf{E}$$

We can now derive individual equations for **E** and **H** with the aid of the vector algebra identity:

$$\text{curl curl } \mathbf{H} = \text{grad div } \mathbf{H} - \nabla^2\mathbf{H} \quad (\text{with div } \mathbf{H} = 0)$$
$$\text{curl curl } \mathbf{H} = - \nabla^2\mathbf{H}$$
$$\text{curl}(\mathbf{J}_c + j\omega\varepsilon\mathbf{E}) = - \nabla^2\mathbf{H}$$
$$\text{curl } \mathbf{J}_c + j\omega\varepsilon(\text{curl } \mathbf{E}) = \text{curl } \mathbf{J}_c + j\omega\varepsilon(- j\omega\mu)\mathbf{H} = - \nabla^2\mathbf{H}$$
$$\nabla^2\mathbf{H} + \omega^2\mu\varepsilon\mathbf{H} = - \text{curl } \mathbf{J}_c$$
$$\nabla^2\mathbf{H} + k^2\mathbf{H} = - \text{curl } \mathbf{J}_c \tag{A1.12a}$$

and

$$\text{curl curl } \mathbf{E} = \text{grad div } \mathbf{E} - \nabla^2 \mathbf{E} \qquad (\text{with div } \mathbf{E} = \rho/\varepsilon)$$
$$\text{curl}\,(-j\omega\mu\mathbf{H}) = \text{grad}\,(\rho/\varepsilon) - \nabla^2 \mathbf{E}$$
$$-j\omega\mu\,\text{curl }\mathbf{H} = \text{grad}\,(\rho/\varepsilon) - \nabla^2 \mathbf{E}$$
$$(-j\omega\mu)(\mathbf{J}_c + j\omega\varepsilon\mathbf{E}) = \text{grad}\,(\rho/\varepsilon) - \nabla^2 E$$
$$\nabla^2 \mathbf{E} + k^2 \mathbf{E} = \text{grad}\,(\rho/\varepsilon) + j\omega\mu\mathbf{J}_c \qquad (A1.12b)$$

Equations (A1.12a) and (A1.12b) are rather difficult to solve as they stand, but will prove useful later. Note that $k^2 = \omega^2 \mu\varepsilon$.

A1.3 WAVE EQUATIONS

To help solve Eqs. (A1.12a) and (A1.12b), a brief revision of the plane wave solutions of the *wave equation* is useful. The forward and reverse components of a plane wave E-field may be written in terms of an *attenuation factor* (α) and a *phase factor* (β) as

$$E_f(r) = \hat{E}_f(0)e^{j\omega t}e^{-\alpha r}e^{-j\beta r}$$

and

$$E_r(r) = \hat{E}_r(0)e^{j\omega t}e^{-\alpha(-r)}e^{-j\beta(-r)}$$

Using $E_f = \hat{E}_f(0)e^{j\omega t}$, $E_r = \hat{E}_r(0)e^{j\omega t}$ and $\gamma = \alpha + j\beta$, we can combine the equations to produce the resultant

$$E(r) = E_f(r) + E_r(r) = E_f e^{-\gamma r} + E_r e^{+\gamma r}$$

Two successive differentiations of this equation result in

$$\frac{d^2 E(r)}{dr^2} = \gamma^2 E(r)$$

In terms of x, y, z coordinates, with $\nabla^2 \mathbf{E}$ replacing $d^2 E(r)/dr^2$,

$$\nabla^2 \mathbf{E} = \gamma^2 \mathbf{E} \qquad (A1.13a)$$
$$\nabla^2 \mathbf{H} = \gamma^2 \mathbf{H} \qquad (A1.13b)$$

These are the wave equations in terms of α and β. If Eqs. (A1.12a) and (A1.12b) can be reduced to the form of Eqs (A1.13a) and (A1.13b), then the solution is a wave expression of the sort just described.

A1.4 NET-CHARGE-FREE MEDIA

As a special case, though one commonly encountered in practice, we can consider a radio wave propagation medium in which $\rho = 0$. Substitution into Eqs. (A1.12a) and (A1.12b) and replacing curl $\mathbf{J}_c$ first by σ curl $\mathbf{E}$ and then by $-j\omega\mu\sigma\mathbf{H}$, with

$j\omega\mu\mathbf{J}_c = j\omega\mu\sigma\mathbf{E}$, the equations reduce to

$$\mathbf{V}^2\mathbf{H} = (\sigma + j\omega\varepsilon)(j\omega\mu)\mathbf{H} = \gamma^2\mathbf{H} \qquad \text{(A1.14a)}$$

$$\mathbf{V}^2\mathbf{E} = (\sigma + j\omega\varepsilon)(j\omega\mu)\mathbf{E} = \gamma^2\mathbf{E} \qquad \text{(A1.14b)}$$

The solution of Eqs. (A1.14a) and (A1.14b) are then simply the wave equations with

$$\gamma^2 = (\sigma + j\omega\varepsilon)(j\omega\mu) = (\alpha + j\beta)^2$$

We can split this complex expression into real and imaginary components:

$$\alpha^2 - \beta^2 = -\omega^2\mu\varepsilon$$

$$2\alpha\beta = \omega\mu\sigma$$

and these may be solved to obtain

$$\alpha = \omega\sqrt{\tfrac{1}{2}\mu\varepsilon(\sqrt{1 + (\sigma/\omega\varepsilon)^2} - 1)} \qquad \text{(A1.15)}$$

$$\beta = \omega\sqrt{\tfrac{1}{2}\mu\varepsilon(\sqrt{1 + (\sigma/\omega\varepsilon)^2} + 1)} \qquad \text{(A1.16)}$$

$$v = \omega/\beta \qquad \text{(A1.17)}$$

$$n = c/v \qquad \text{(A1.18)}$$

where v is the *velocity of propagation* ($c = 3 \times 10^8 \text{ m s}^{-1}$) and n is the *refractive index* of the medium.

Example A1.3 *Attenuation of net-charge-free medium*

A certain homogeneous, isotropic, net-charge-free medium has values of σ, μ and ε at angular frequency ω such that $\sigma/\omega\varepsilon \ll 1$. What is the attenuation of the radiowaves propagating through such a medium in dB per km?

Starting with the full equation for attenuation rate, Eq. (A1.15),

$$\alpha = \omega\sqrt{\tfrac{1}{2}\mu\varepsilon(\sqrt{1 + (\sigma/\omega\varepsilon)^2} - 1)}$$

and approximating using the binomial expansion,

$$\{1 + (\sigma/\omega\varepsilon)^2\}^{1/2} = 1 + \tfrac{1}{2}(\sigma/\omega\varepsilon)^2$$

the expression for α reduces to

$$\alpha = \omega\sqrt{\tfrac{1}{2}\mu\varepsilon}\{1 + \tfrac{1}{2}(\sigma/\omega\varepsilon)^2 - 1\}^{1/2}$$

Noting that $\sqrt{\mu_0/\varepsilon_0} \approx 120\pi$ ohms,

$$\alpha = 60\pi\sigma\sqrt{\mu_r/\varepsilon_r} \qquad \text{nepers per meter}$$

$$\alpha = (8.686 \times 1000)60\pi\sigma\sqrt{\mu_r/\varepsilon_r} \qquad \text{decibels per km}$$

$$\alpha = 1.64 \times 10^6 \sigma\varepsilon_r^{-1/2} \qquad \text{dB/km} \qquad \text{(A1.19)}$$

as μ_r is commonly unity. Note that σ is measured in s/m (siemens per meter) in this equation.

A1.4.1 Wave Impedance and Power Flow

In order to satisfy the curl $\mathbf{E}$ and curl $\mathbf{H}$ equations given earlier in Eqs. (A1.10) and (A1.11), field vectors $\mathbf{E}$ and $\mathbf{H}$ of plane waves must be mutually perpendicular. The mathematics can often be simplified by choosing the rectangular coordinates such that $\mathbf{E}$ is aligned with E_x and hence $\mathbf{H}$ is aligned with H_y. The propagation is then in the z direction. Faraday's law, Eq. (T1b), may now be written

$$\operatorname{curl}\mathbf{E} = (\partial E_x/\partial z)\mathbf{j} = -(j\omega\mu\mathbf{H})\mathbf{j}$$

In addition, it has been shown earlier in the derivation of the wave equation that

$$\partial E_x/\partial z = -\gamma E_x$$

and hence from the combination of these two equations the ratio of the magnitudes of the mutually-perpendicular field-vectors is

$$\frac{E_x}{H_y} = \frac{|\mathbf{E}|}{|\mathbf{H}|} = \frac{j\omega\mu}{\gamma} = \sqrt{\frac{j\omega\mu}{\sigma + j\omega\varepsilon}} = Z_{\mathrm{w}} \tag{A1.20}$$

This is known as the *wave impedance* of a net-charge-free medium. It is a useful auxiliary parameter though it adds no further information to the basic equations. It simply relates the two field vectors under certain conditions. The two constitutive equations require that the medium is homogeneous and isotropic, and this special case requires that $\rho = 0$. If it is further specified that $\sigma = 0$, $\mu = \mu_0$ and $\varepsilon = \varepsilon_0$, as in free space, the wave impedance is 120π ohms.

An alternative and equally useful relationship between the two field vectors can be derived via another well-known vector identity:

$$\operatorname{div}(\mathbf{E} \times \mathbf{H}) = \mathbf{H}\cdot\operatorname{curl}\mathbf{E} - \mathbf{E}\cdot\operatorname{curl}\mathbf{H} = \mathbf{H}\cdot\left(-\mu\frac{\partial\mathbf{H}}{\partial t}\right) - \mathbf{E}\cdot\left(\sigma\mathbf{E} + \varepsilon\frac{\partial\mathbf{E}}{\partial t}\right)$$

with $\mathbf{E} = E_x$ and $\mathbf{H} = H_y$ as before. This can be rearranged to give

$$-\operatorname{div}(\mathbf{E} \times \mathbf{H}) = \frac{\partial}{\partial t}\left(\frac{\mu H_y^2}{2}\right) + \frac{\partial}{\partial t}\left(\frac{\varepsilon E_x^2}{2}\right) + (\sigma E_x^2) \tag{A1.21}$$

The three terms in brackets represent energy stored magnetically and electrically per unit volume and energy dissipated per unit volume, and hence $-\operatorname{div}(\mathbf{E} \times \mathbf{H})$ represents total input power per unit volume, or $\operatorname{div}(\mathbf{E} \times \mathbf{H})$ represents total output power per unit volume. Thus the output power is either

$$P = \int_v \operatorname{div}(\mathbf{E} \times \mathbf{H})\,dv \qquad \text{or} \qquad P = \int_{S''} \mathbf{P}_{\mathrm{a}}\cdot d\mathbf{S}''$$

If $\mathbf{P}_{\mathrm{a}}$ is the outward power flux. The divergence theorem indicates that

$$\int_v \operatorname{div}(\mathbf{E} \times \mathbf{H})\,dv = \int_{S''}(\mathbf{E} \times \mathbf{H})\cdot d\mathbf{S}'' = \int_{S''} \mathbf{P}_{\mathrm{a}}\cdot d\mathbf{S}''$$

or

$$\mathbf{P}_{\mathrm{a}} = \mathbf{E} \times \mathbf{H} \tag{A1.22}$$

watts per meter-square in the z direction. The mean power flux is the same equation with E and H in rms values, and $\mathbf{E} \times \mathbf{H}$ is known as *Poynting's vector*.

Example A1.4 Wave impedance for low-loss medium

If a certain radio wave propagation medium has $(\sigma/\omega\varepsilon) \ll 1$, so that $(\sigma/\omega\varepsilon)^2$ may be taken as virtually zero, what is a good approximation for its wave impedance?

The full expression for wave impedance is

$$Z_w = \sqrt{\frac{j\omega\mu}{\sigma + j\omega\varepsilon}} = |Z_w|\underline{/\theta}$$

where

$$|Z_w| = \frac{\sqrt{\mu/\varepsilon}}{\sqrt[4]{1 + (\sigma/\omega\varepsilon)^2}} = \sqrt{\frac{\mu}{\varepsilon}}\{1 - \tfrac{1}{4}(\sigma/\omega\varepsilon)^2\} \approx \sqrt{\frac{\mu}{\varepsilon}}$$

and

$$\tan(2\theta) = (\sigma/\omega\varepsilon)$$

or

$$\theta = \tfrac{1}{2}(\sigma/\omega\varepsilon) \text{ radians}$$

The medium, in effect, behaves virtually as a perfect dielectric as θ is going to be very small.

A1.5 MAGNETIC VECTOR POTENTIAL

To solve in general for field vectors $\mathbf{E}$ and $\mathbf{H}$ due to current sources $\mathbf{J}_c$, a method has been devised which makes use of magnetic vector potential $\mathbf{A}$. This auxiliary vector parameter needs to be defined via expressions for curl $\mathbf{A}$ *and* div $\mathbf{A}$, as a vector requires both curl and div for complete specification.

Knowing that Maxwell's equations include div $\mathbf{B} = 0$ and that mathematically div curl $\mathbf{X} \equiv 0$ for any $\mathbf{X}$, we can introduce an auxiliary vector $\mathbf{A}$ such that

$$\text{div curl } \mathbf{A} = 0 = \text{div } \mathbf{B}$$

or

$$\text{curl } \mathbf{A} = \mathbf{B} = \mu\mathbf{H} \tag{A1.23}$$

This new term, with div $\mathbf{A}$ yet to be defined, can be introduced into Faraday's law and Ampere's law.

Firstly, for Faraday's law:

$$\text{curl } \mathbf{E} = -j\omega\mu\mathbf{H} = -j\omega \text{ curl } \mathbf{A}$$

or

$$\text{curl }(\mathbf{E} + j\omega\mathbf{A}) = \mathbf{0}$$

Using the concept of scalar electric potential Φ and the associated electrostatics equation $\mathbf{E}' = -\text{grad } \Phi$, we can use the mathematical identity

$$\text{curl grad } \Phi = 0 = \text{curl grad}(-\Phi)$$

with the previous equation to produce

$$\mathbf{E} = -j\omega\mathbf{A} - \text{grad } \Phi \tag{A1.24}$$

This is Faraday's law in terms of the magnetic vector potential and the scalar electric potential.

Secondly, we can introduce the new term into Ampere's law with the aid of the mathematical identity

$$\text{curl } \mathbf{B} = \text{curl curl } \mathbf{A} = \text{grad div } \mathbf{A} - \nabla^2\mathbf{A}$$

by noting successively that

$$\text{curl } \mathbf{B} = \mu \text{ curl } \mathbf{H} = \mu(\mathbf{J}_c + j\omega\varepsilon\mathbf{E}) = \mu\mathbf{J}_c + j\omega\mu\varepsilon(-j\omega\mathbf{A} - \text{grad } \Phi)$$

and then by equating the two expressions for curl **B** to give

$$\nabla^2\mathbf{A} + \omega^2\mu\varepsilon\mathbf{A} - \text{grad}(\text{div } \mathbf{A} + j\omega\mu\varepsilon\Phi) = -\mu\mathbf{J}_c \tag{A1.25}$$

Here is an opportunity for us to define div **A** and we can choose it such that the third term in the equation is zero. Hence:

$$\text{div } \mathbf{A} = -j\omega\mu\varepsilon\Phi \tag{A1.26}$$

and the combined equation representing both Faraday's and Ampere's laws reduces to

$$\nabla^2\mathbf{A} + \omega^2\mu\varepsilon\mathbf{A} = -\mu\mathbf{J}_c$$

or

$$\nabla^2\mathbf{A} + k^2\mathbf{A} = -\mu\mathbf{J}_c \tag{A1.27}$$

A1.5.1 Solutions for Magnetic Vector Potential

There are several forms of solution for magnetic vector potential **A** from Eq. (A1.27) depending on whether $\mathbf{J}_c$ is a function of coordinates x, y, z, or a function of only two of the coordinates, or simply a function of just one coordinate. For instance, if the current source is simply a linear element carrying current I in the z direction, the magnetic vector potential at some distant point (Fig. A1.2) contains only a component A_z, with $A_x = A_y = 0$. For an element of incremental length dz, the corresponding increment of magnetic vector potential dA is

$$dA_z = \frac{\mu_0}{4\pi}\frac{I dz}{r}\exp(-j\beta r) \tag{A1.28}$$

In other cases the expressions are more complicated but, once having determined **A**, we have at least two ways of working back to the field vectors **E** and **H**. Firstly, from the definition of **A**,

$$\mathbf{H} = \frac{1}{\mu}\text{curl } \mathbf{A}$$

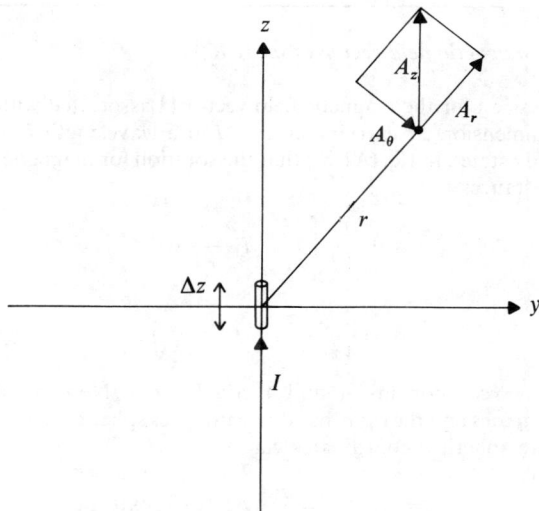

Fig. A1.2 The magnetic vector potential associated with an incremental linear radiator.

and, from Ampere's law, $\text{curl } \mathbf{H} = \mathbf{J}_c + j\omega\varepsilon\mathbf{E}$, so that

$$\mathbf{E} = \frac{1}{j\omega\varepsilon}(\text{curl } \mathbf{H} - \mathbf{J}_c) \qquad (A1.29)$$

Alternatively, using $\mathbf{E} = -j\omega\mathbf{A} - \text{grad } \Phi$, derived earlier, and the second definition of $\mathbf{A}$, viz. $\text{div } \mathbf{A} = -j\omega\varepsilon\mu\Phi$, we obtain

$$\Phi = \frac{\text{div } \mathbf{A}}{-j\omega\mu\varepsilon} \qquad (A1.30)$$

and

$$-\text{grad } \Phi = -j\frac{1}{\omega\mu\varepsilon}\text{ grad div } \mathbf{A}$$

Hence

$$\mathbf{E} = -j\omega\mathbf{A} - j\frac{1}{\omega\mu\varepsilon}\text{ grad div } \mathbf{A} \qquad (A1.31)$$

The use of the magnetic vector potential is of importance in the analysis of antennas used for radio wave propagation. One form of antenna employs current I flowing in a specified direction (e.g. in the z direction) and the mathematics may be simplified by using as reference an *incremental linear radiator* (ILR). This is an elemental length of radiator along which the current amplitude may be taken as constant.

Example A1.5 The magnetic field vector of an ILR

Determine an expression for the magnetic field vector **H** associated with an incremental linear radiator of dimension Δz carrying current I at a wavelength λ (where $\beta = 2\pi/\lambda$).

We have already stated in Eq. (A1.28) that the solution for magnetic vector potential under these circumstances is

$$A_z = \frac{\mu}{4\pi} \int_{-\Delta z/2}^{+\Delta z/2} I \frac{e^{-j\beta r}}{r} \, dz$$

or

$$A_z = \frac{\mu}{4\pi} \frac{I\Delta z}{r} e^{-j\beta r} \tag{A1.32}$$

with the remaining two components, A_x and A_y equal to zero. Note that the left-hand side is in cartesian coordinates and the right-hand side involves spherical coordinate r. Thus we must change A_z into spherical coordinates via

$$A_r = A_z \cos\theta = \frac{\mu}{4\pi} I\Delta z \frac{e^{-j\beta r}}{r} \cos\theta \tag{A1.33a}$$

$$A_\theta = -A_z \sin\theta = \frac{\mu}{4\pi} I\Delta z \frac{e^{-j\beta r}}{r} \sin\theta \tag{A1.33b}$$

$$A_\phi = 0 \tag{A1.33c}$$

The next step is for us to use $\mathbf{H} = (1/\mu)\,\mathrm{curl}\,\mathbf{A}$ in spherical coordinates with $A_\phi = 0$ and with no variation with respect to ϕ. Thus

$$\mathrm{curl}\,\mathbf{A} = \frac{1}{r\sin\theta}\left[\frac{\partial}{\partial\theta}(A_\phi \sin\theta) - \frac{\partial A_\phi}{\partial\phi}\right]\mathbf{a}_r$$

$$+ \frac{1}{r}\left[\frac{1}{\sin\theta}\frac{\partial A}{\partial r} - \frac{\partial}{\partial r}(rA_\phi)\right]\mathbf{a}_\theta$$

$$+ \frac{1}{r}\left[\frac{\partial}{\partial r}(rA_\theta) - \frac{\partial A_r}{\partial\theta}\right]\mathbf{a}_\phi \tag{A1.34}$$

reduces under these conditions to

$$\mathrm{curl}\,\mathbf{A} = \frac{1}{r}\left[\frac{\partial}{\partial r}(rA_\theta) - \frac{\partial A_r}{\partial\theta}\right]\mathbf{a}_\phi \tag{A1.35}$$

and, after appropriate differentiation with respect to r and θ,

$$\mathbf{H} = \frac{\mathrm{curl}\,\mathbf{A}}{\mu} = j\frac{\beta I\Delta z \sin\theta}{4\pi r}\left[1 + \frac{1}{j\beta r}\right]e^{-j\beta r} = H_\phi \tag{A1.36}$$

The other two components, H_r and H_θ, are both zero.

The corresponding solutions for the E-field components require similar mathematical manipulation of the E-field solution given in Eq. (A1.32):

$$\mathbf{E} = -j\omega\mathbf{A} - j(1/\omega\mu\varepsilon)\,\mathrm{grad\ div}\,\mathbf{A}$$

The results are quoted in Table A1.2.

Table A1.2   The solutions of the magnetic vector potential equations for an incremental linear radiator

$$E_r = 30\beta^2 \Delta z[I] 2\cos\theta \left\{ \frac{0}{\beta r} + \frac{1}{(\beta r)^2} - \frac{j}{(\beta r)^3} \right\}$$

$$E_\theta = 30\beta^2 \Delta z[I] \sin\theta \left\{ \frac{j}{\beta r} + \frac{1}{(\beta r)^2} - \frac{j}{(\beta r)^3} \right\}$$

$$H_\phi = \frac{30\beta^2 \Delta z[I]}{z_i} \sin\theta \left\{ \frac{j}{\beta r} + \frac{1}{(\beta r)^2} + \frac{0}{(\beta r)^3} \right\}$$

$$H_r = H_\theta = E_\phi = 0$$

A1.6 BOUNDARY CONDITIONS

If an electromagnetic wave propagates through two media in succession, an interaction takes place at the boundary surface separating the two media. Based on the fundamental equations, we can describe the interactions in terms of a *boundary condition* for each of the field vectors (see Figs. A1.3 and A1.4):

1. The tangential *H*-field is discontinuous across the boundary if there is a surface current $\mathbf{J_s}$.
2. The tangential *E*-field is continuous across the boundary.
3. The normal electric flux density is discontinuous across the boundary if there is surface charge ρ_s.
4. The normal magnetic flux density is continuous across the boundary.

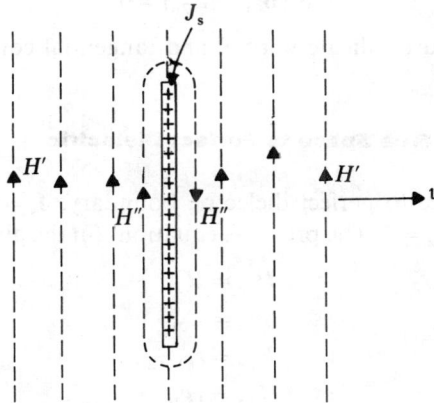

Fig. A1.3 The tangential *H*-field is discontinuous across a boundary if there is a surface current J_s.

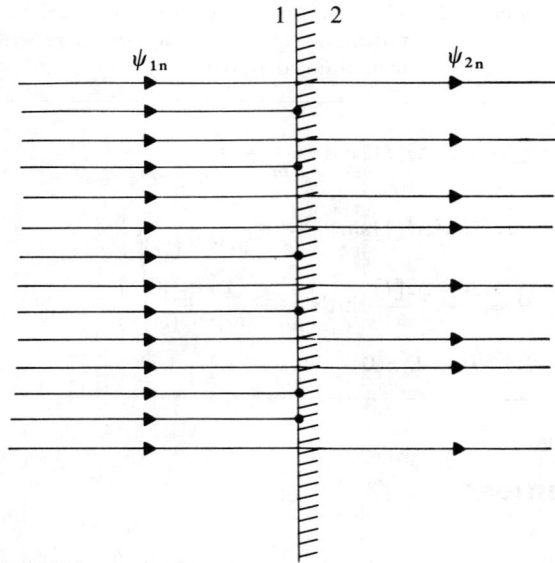

Fig. A1.4 The normal electric flux density is discontinuous across the boundary if there is surface charge ρ_s.

In the above conditions, $\mathbf{J}_s$ is in $A\,m^{-1}$ and ρ_s is in $C\,m^{-2}$. If $\mathbf{n}$ is a unit vector normal to the boundary surface pointing into medium 1:

$$\mathbf{n} \times (\mathbf{H}_{1t} - \mathbf{H}_{2t}) = \mathbf{J}_s \qquad (A1.37a)$$

$$\mathbf{n} \times (\mathbf{E}_{1t} - \mathbf{E}_{2t}) = 0 \qquad (A1.37b)$$

$$\mathbf{n} \cdot (\mathbf{D}_{1n} - \mathbf{D}_{2n}) = \rho_s \qquad (A1.37c)$$

$$\mathbf{n} \cdot (\mathbf{B}_{1n} - \mathbf{B}_{2n}) = 0 \qquad (A1.37d)$$

where the suffixes n and t indicate normal and tangential components.

A1.6.1 Special Case: Free Space to Perfect Dielectric

In the case of free space to perfect dielectric boundary, $\mathbf{J}_s = 0$, $\rho_s = 0$, $\mathbf{D} = \varepsilon\mathbf{E}$, $\mathbf{B} = \mu\mathbf{H}$, $\varepsilon_{r1} = 1$ and $\mu_{r1} = \mu_{r2} = 1$. The previous equations (in the given directions) become:

$$H_{1t} = H_{2t} \qquad (A1.38a)$$

$$E_{1t} = E_{2t} \qquad (A1.38b)$$

$$E_{1n} = \varepsilon_{r2}E_{2n} \qquad (A1.38c)$$

$$H_{1n} = H_{2n} \qquad (A1.38d)$$

These, at first, appear fairly simple but it is not possible for an electromagnetic wave to pass unchanged through the boundary between two different media. For example, if an

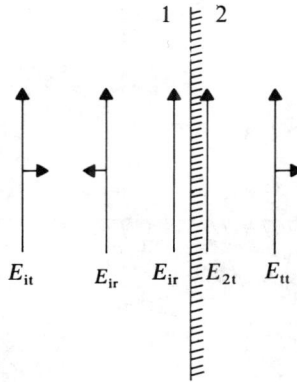

Fig. A1.5 The incident, reflected and transmitted components of a tangential E-field at the boundary between medium 1 and medium 2.

E–H wave is normally incident to a plane boundary, the concept that

$$\frac{E_i}{H_i} = \frac{E_{it}}{H_{it}} = Z_{w1} = \frac{E_{2t}}{H_{2t}} = Z_{w2}$$

is obviously impossible as wave impedance Z_{w1} is not equal to Z_{w2}. To account for this apparent discrepancy, the idea of *incident*, *reflected* and *transmitted* components is introduced so that

$$H_{it} + H_{rt} = H_{tt} \tag{A1.39a}$$

$$E_{it} + E_{rt} = E_{tt} \tag{A1.39b}$$

$$E_{in} + E_{rn} = \varepsilon_{r2} E_{tn} \tag{A1.39c}$$

$$H_{in} + H_{rn} = H_{tn} \tag{A1.39d}$$

Where the first suffix refers to incident, reflected and transmitted components, and the second suffix to normal and tangential components. These, then, are the boundary conditions for a free space to perfect dielectric boundary. Equation (A1.39b) is illustrated in Fig. A1.5.

A1.6.2 Laws of Reflection and Refraction

In the case of the boundary between free space and a perfect dielectric, as illustrated in Fig. A1.6, the incident E-field wavefront AB produces a reflected wavefront ED and a transmitted wavefront CD in time t such that

$$t = \frac{AD}{c} = \frac{BE}{c} = \frac{BC}{v}$$

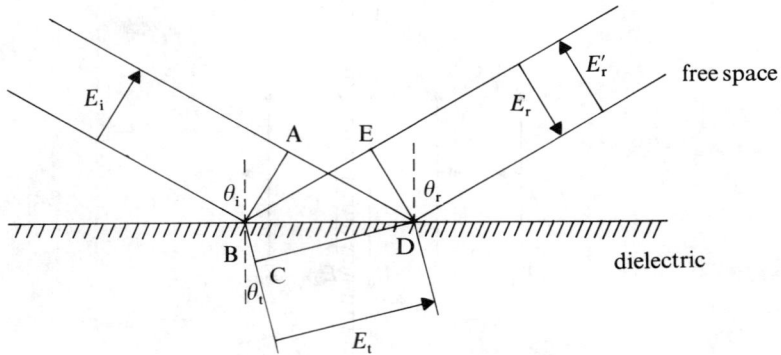

Fig. A1.6 The laws of reflection and refraction at the boundary between free space and a dielectric.

or

$$\frac{BD \sin \theta_1}{c} = \frac{BD \sin \theta_r}{c} = \frac{BD \sin \theta_t}{(c/n)}$$

or

$$\theta_i = \theta_r \qquad \text{law of reflection} \tag{A1.40}$$

and

$$\frac{\sin \theta_i}{\sin \theta_t} = n \qquad \text{law of refraction} \tag{A1.41}$$

Thus the angle of incidence θ_i is equal to the angle of reflection θ_r and the ratio of $\sin \theta_i$ to $\sin \theta_t$ is equal to the refractive index n of the perfect dielectric. These are the well-known laws of reflection and refraction, the latter sometimes being called *Snell's law*.

A1.6.3 Fresnel's Reflection Coefficients

The *reflection coefficients* are defined as the ratio E_r/E_i when either H_i is tangential (for coefficient R_V) or when E_i is tangential (for R_H), assuming a horizontal boundary. Using Fig. A1.6 to illustrate the derivation of reflection coefficient R_V with the H-field horizontal, we note that

$$E_{it} + E_{rt} = E_{tt}$$

and

$$E_{in} + E_{rn} = \varepsilon_{r2} E_{tn}$$

become

$$E_i \cos \theta_i - E_r \cos \theta_r = E_t \cos \theta_t$$

and

$$E_i \sin \theta_i + E_r \sin \theta_r = \varepsilon_{r2} E_t \sin \theta_t$$

Using the laws of reflection and refraction to eliminate θ_r and θ_t, and equating E_t from each equation to eliminate E_t, we obtain

$$R_V = \frac{E_r}{E_i} = \frac{\varepsilon_{r2} \cos \theta_1 - \sqrt{\varepsilon_{r2} - \sin^2 \theta_1}}{\varepsilon_{r2} \cos \theta_1 + \sqrt{\varepsilon_{r2} - \sin^2 \theta_1}} \qquad (A1.42)$$

For reflection coefficient R_H, the boundary conditions in terms of the H-field may be

---

Example A1.6 Brewster angle

Determine the angle of incidence θ_i at which the reflection coefficient $R_V = 0$ for the special case of a boundary between free space and a perfect dielectric with relative permittivity $\varepsilon_{r2} = 4$.

We can obtain the solution from Eq. (A1.42). Obviously $R_V = 0$ when

$$\varepsilon_{r2} \cos \theta_i = \sqrt{\varepsilon_{r2} - \sin^2 \theta_i}$$

or

$$\varepsilon_{r2}^2 \cos^2 \theta_i + \sin^2 \theta_i = \varepsilon_{r2}$$

To eliminate one of the trigonometrical terms we can *force* the expression into the $\cos^2 \theta_i + \sin^2 \theta_i = 1$ identity as follows:

$$\varepsilon_{r2}^2 \cos^2 \theta_i + \varepsilon_{r2}^2 \sin^2 \theta_i + (1 - \varepsilon_{r2}^2) \sin^2 \theta_i = \varepsilon_{r2}$$

This then reduces to

$$\sin \theta_i = \sqrt{(\varepsilon_{r2}^2 - \varepsilon_{r2})/(\varepsilon_{r2}^2 - 1)}$$

and for a dielectric with $\varepsilon_{r2} = 4$ the numerical solution is about $63°$. The angle is known as the *Brewster angle*, θ_B in Fig. A1.7.

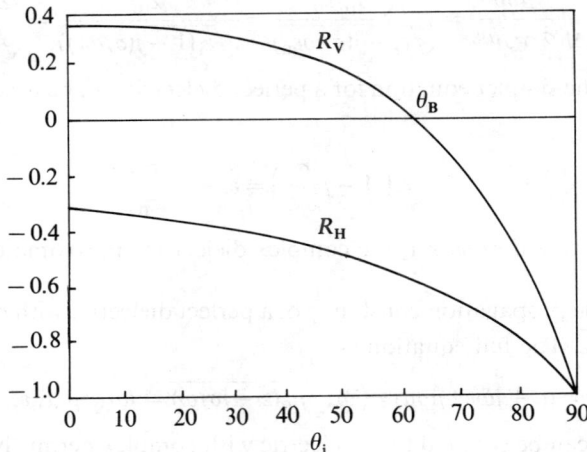

Fig. A1.7 The variation of Fresnel's reflection coefficients with angle of incidence. Note the Brewster angle θ_B at which the reflection coefficient for vertical polarization is zero.

interpreted as

$$H_i \cos \theta_i - H_r \cos \theta_r = H_t \cos \theta_t$$
$$H_i \sin \theta_i + H_r \sin \theta_r = H_t \sin \theta_t$$

We can process these expressions in a similar manner to produce

$$R_H = \frac{E_r}{E_i} = \frac{H_r}{H_i} = \frac{\cos \theta_i - \sqrt{\varepsilon_{r2} - \sin^2 \theta_i}}{\cos \theta_i + \sqrt{\varepsilon_{r2} - \sin^2 \theta_i}} \qquad (A1.43)$$

A1.7 COMPLEX DIELECTRIC

The various equations for attenuation, phase change per meter, velocity of propagation, refractive index, etc., include the term $\sigma/\omega\varepsilon$. If this term is very much greater than unity the medium is primarily conductive. If the term is very much less than unity the medium tends towards a perfect dielectric. If neither of these conditions apply, the full expressions must be used.

However, it has been found that the mathematics can be simplified a great deal if the concept of a *complex dielectric* is used for a medium in which $\sigma/\omega\varepsilon$ is of the order of unity. A complex dielectric is one in which ε_r is replaced by the complex parameter ε_r^*. If the mathematical analysis is first undertaken on the assumption that the medium is a perfect dielectric with relative permittivity ε_r, the solution for the real meduim is the same with ε_r replaced by ε_r^*.

For instance, the wave impedance of a perfect dielectric (with $\sigma = 0$ and $\mu = \mu_0$) is simply $Z_w = Z_0/\sqrt{\varepsilon_r}$, where $Z_0 \approx 120\pi$ ohms. The full equation, by comparison, is

$$Z_w = \sqrt{\frac{j\omega\mu}{\sigma + j\omega\varepsilon}} = \frac{\sqrt{\mu_0/\varepsilon_0}}{\sqrt{\varepsilon_r - j(\sigma/\omega\varepsilon_0)}} = \frac{Z_0}{\sqrt{\varepsilon_r(1 - j(\sigma/\omega\varepsilon))}} = \frac{Z_0}{\sqrt{\varepsilon_r^*}} \qquad (A1.44)$$

We can see that the simpler equation for a perfect dielectric can be used if ε_r is replaced by

$$\varepsilon_r^* = \varepsilon_r \left(1 - j\frac{\sigma}{\omega\varepsilon} \right) = \varepsilon_r - j\frac{\sigma}{\omega\varepsilon_0} \qquad (A1.45)$$

and it is obvious that if $\sigma/\omega\varepsilon \ll 1$, the complex dielectric approximates to the perfect dielectric.

Yet again, the propagation constant γ of a perfect dielectric with $\sigma = 0$ and $\mu = \mu_0$ is simply $j\omega\sqrt{\mu_0\varepsilon}$. The full equation is

$$\gamma = (\sigma + j\omega\varepsilon)(j\omega\mu) = j\omega\sqrt{\mu_0(\varepsilon - j\sigma/\omega)} = j\omega\sqrt{\mu_0\varepsilon_0\varepsilon_r^*} \qquad (A1.46)$$

and the medium can be equated to a dielectric with complex permittivity $\varepsilon^* = \varepsilon_0\varepsilon_r^*$ as before.

Similarly, Ampere's law may be expressed in the form

$$\text{curl } \mathbf{H} = (\sigma + j\omega\varepsilon)\mathbf{E} = j\omega\mathbf{D}^*$$

or

$$\mathbf{D}^* = (\varepsilon - j\sigma/\omega)\mathbf{E} = \varepsilon_0 \varepsilon_r^* \mathbf{E}$$

whence

$$\varepsilon_r^* = \varepsilon_r \left(1 - j\frac{\sigma}{\omega\varepsilon}\right) \tag{A1.47}$$

as before, and various other expressions may similarly be related.

Using this idea, the reflection coefficient at the boundary between free space and a medium in which $\sigma/\omega\varepsilon \simeq 1$ is given by Eqs. (A1.42) and (A1.43) with ε_{r2} replaced by

$$\varepsilon_{r2}^* = \varepsilon_{r2} \left(1 - j\frac{\sigma}{\omega\varepsilon}\right) \tag{A1.48}$$

The resulting equations may then become rather long and complex in form, but the derivation is obviously much easier. Note that both R_V and R_H are now both complex, with magnitude and phase. An example of the complex dielectric is the earth's surface, and charts of R_V and R_H are generally available for typical ground parameters.

Example A1.7 *Reflection coefficient of partially conducting ground*

Determine the reflection coefficient of a vertically polarized radio wave at the boundary between air and a partially conducting horizontal ground plane. The ground parameters are $\varepsilon_r = 15$, $\sigma = 10\,\text{mS/m}$ and $\mu_r = 1$ at frequency $f = 1\,\text{MHz}$. The angle of incidence is $85°$.

To determine the numerical value of R_V we can use Eq. (A1.42) modified by Eq. (A1.48) to obtain

$$R_V = \frac{\varepsilon_r(1 - j\sigma/\omega\varepsilon)\cos\theta_i - \{\varepsilon_r(1 - j\sigma/\omega\varepsilon) - \sin^2\theta_i\}^{1/2}}{\varepsilon_r(1 - j\sigma/\omega\varepsilon)\cos\theta_i + \{\varepsilon_r(1 - j\sigma/\omega\varepsilon) - \sin^2\theta_i\}^{1/2}} \tag{A1.49}$$

It will simplify matters if we first note that

$$\varepsilon_r(1 - j\sigma/\omega\varepsilon) = 15(1 - j12)$$

in this case so that the numerical value of R_V reduces to

$$R_V = \frac{15(1 - j12)(0.0872) - \{15(1 - j12) - 0.9924\}^{1/2}}{15(1 - j12)(0.0872) + \{15(1 - j12) - 0.9924\}^{1/2}}$$

or

$$R_V = \frac{1.308 - j15.696 - \{14.01 - j180\}^{1/2}}{1.308 - j15.696 + \{14.01 - j180\}^{1/2}}$$

Here we note that $14.01 - j180 = 180.5444\underline{/-85.55°}$ and hence its square root is $13.44\underline{/-42.77°}$, or $9.865 - j9.1273$. Substituting this back into the expression for R_V gives

$$R_V = \frac{-8.557 - j6.569}{11.173 - j24.823} = \frac{10.79\underline{/-142.5°}}{27.22\underline{/-65.8°}} = 0.40\underline{/-76.7°}$$

Thus the numerical estimation of Fresnel's reflection coefficient can be quite long-winded and tedious without the aid of a suitable computer. For this reason, several texts contain graphs of the magnitude and angle of the reflection coefficient for various typical values of ground parameters and frequencies.

Appendix 2
Electrical Noise in Radio Systems

When considering a radio communication system, we are concerned primarily with the transmission of a signal from some point A to some other point B. In an ideal system there would be a progressive reduction in the signal strength due to spatial dispersion, or to absorption by the propagation medium, or to attenuation and mismatch by transmission systems, for example, but whatever this reduction it could be overcome by amplifiers with overall gain equal to the overall signal loss. However, this is not the case in reality because in the combined transmission and propagation media between the signal source and the receiver output, additional unwanted fluctuations in electromagnetic energy are superimposed upon the original signal and the overall output is a combination of wanted signal and unwanted *electrical noise*. Whatever type of information is transmitted, the satisfactory level of reception will depend up the ratio of signal power S to noise power N, though the minimum magnitude of this ratio for satisfactory reception will depend on the specific type of signal processing used and the nature of the communication. Good-quality television pictures require a higher signal-to-noise ratio than telephone systems, whereas telegraphic links and radar reception can work perfectly satisfactorily at much lower ratios.

In radio communication systems the various noise contributions can be divided into several categories. Within the radio wave propagation medium itself the most common noise sources are of galactic, atmospheric and man-made origin, whereas within the transmitter and receiver the noise is mainly thermal in origin, with some contribution from shot noise if thermionic devices are used and from other noise sources associated with solid-state devices, etc. Additional forms of unwanted variation, such as co-channel interference and intermodulation noise, are involved with signal processing techniques and we shall not discuss them further.

A2.1 NOISE VOLTAGE

It is not really possible to sketch the full variation of wideband noise voltage with time because all equipment is bandwidth limited and therefore some of the frequencies outside the bandwidth will be omitted. However, for a low-frequency system, as shown in Fig. A2.1, we can indicate roughly the sort of variation which may occur when the bandwidth is restricted to, say, 0 to 5 kHz. Although the waveform is random-looking, there appears to be an underlying maximum pattern of five cycles per millisecond due

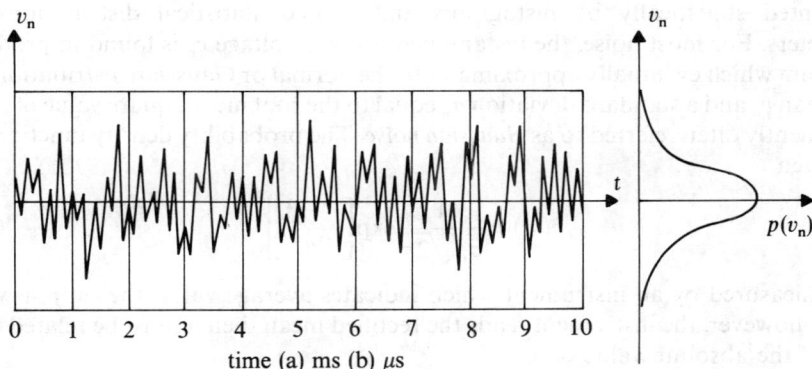

Fig. A2.1 A sketch of the variation of instantaneous noise voltage with time over a period of (a) milliseconds or (b) microseconds, with the Gaussian distribution of its probability density function.

to the 5 kHz bandwidth restriction. A similar sketch resembles the noise waveform if the bandwidth is increased a thousand times to 0 to 5 MHz and the time intervals reduced a thousand times to microseconds. However, relatively speaking, both examples refer to bandwidths of 2 to 1, meaning that the ratio of bandwidth to center frequency is two to one, and this is sufficient to be considered as *wideband noise*.

The noise voltage wave pattern is different for narrow-band or band-limited systems, where the ratio of bandwidth to center frequency is very small. Figure A2.2 gives a rough indication of the variation when the bandwidth is not particularly narrow. For a very narrow bandwidth relative to the center frequency, the higher frequencies appear almost as if modulated by a lower frequency related to the bandwidth.

The two extreme forms of noise waveforms, *wideband* or *white noise* and *narrowband*, *band-limited* or *colored noise*, can be sampled in the standard manner and

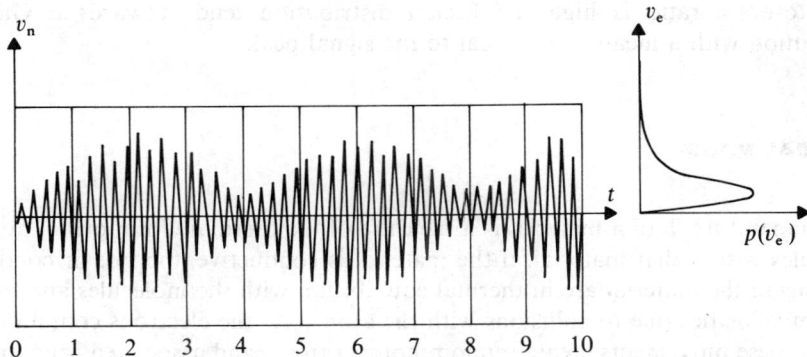

Fig. A2.2 A sketch of the variation of instantaneous noise voltage of narrow-band noise together with the Rayleigh distribution of the probability density function of the noise envelope v_e.

represented statistically by histograms and related statistical distributions and parameters. For most noise, the instantaneous noise voltage v_n is found to produce a histogram which eventually approximates to the normal or Gaussian distribution with zero mean v_n and a standard deviation σ_n equal to the root mean square value of v_n; it is consequently often referred to as *Gaussian noise*. The probability density function may be written

$$p(v_n) = \frac{1}{\sigma_n \sqrt{2\pi}} \exp\left[\frac{-v_n^2}{2\sigma_n^2} \right] \qquad (A2.1)$$

When measured by an instrument which indicates average value, the output will be zero. If, however, the instrument reads the rectified mean then it must be related to the mean of the absolute value of v_n:

$$\overline{|v_n|} = 2 \times \int_0^\infty v_n \frac{1}{\sigma_n \sqrt{2\pi}} \exp\left[\frac{-v_n^2}{2\sigma_n^2} \right] dv_n = \sqrt{\frac{2}{\pi}} \sigma_n \qquad (A2.2)$$

In other words, the ratio of rms to average is $\sqrt{\pi/2}$. An average responding voltmeter measures the average value of a rectified waveform but indicates the rms value of the equivalent sinusoid. When it is used for measuring noise, the meter reading is related directly to the true rms value as follows: the indicated meter reading is 1.11 times the measured average, the average value of the noise is its rms value times $\sqrt{2/\pi}$, and the indicated value on the meter is therefore 1.11 times $\sqrt{2/\pi}$ or 0.886 times the true rms value of the noise voltage. True rms meters are quite common nowadays.

Narrowband or band-limited noise can first be *linearly* detected and then the sampling of the envelope voltage v_e results in a histogram which eventually approximates to the Rayleigh distribution. If the input fluctuations to a narrowband IF amplifier and linear detector consist of a signal of constant amplitude plus white noise, the sampled output voltage of the linear detector produces a Rician distribution with a parameter related to the signal-to-noise ratio. When the signal-to-noise ratio is very small the Rician distribution tends towards the Rayleigh distribution; when the signal-to-noise ratio is high the Rician distribution tends towards a Gaussian distribution with a mean value equal to the signal peak.

A2.2 THERMAL NOISE

The temperature T of a material is a measure of the thermal vibrations of atoms or molecules within that material. If the material is conductive, the free or conduction electrons in the material are in thermal equilibrium with the molecules and move at random velocities due to collisions with the atoms. As the electrons contain electric charge, these movements create random noise currents and associated random noise voltages, the overall effect of which, in a given resistor R at temperature T, may be represented by a noise current source i_n or a noise voltage source e_n and the associated thermal noise power.

Two of the earlier descriptions of thermal noise are given by Johnson and Nyquist in Refs. A2.1 and A2.2. These indicate that as the thermal velocities of the electrons in resistor R are random, the mean current in any direction is zero. However, the mean square values of noise current and associated noise voltage have non-zero values. Nyquist's analysis gives the relationships as

$$\overline{i_n} = 0 \qquad\qquad\qquad\qquad\qquad\qquad (A2.3)$$

$$\overline{i_n^2} = 4kTGB \qquad\qquad\qquad\qquad\qquad (A2.4)$$

where k is Boltzmann's constant, T is the absolute temperature of the source resistor, G is the conductance of the source resistor ($G = 1/R$) and B is the effective noise bandwidth of the device measuring the current.

The corresponding noise voltage is given via

$$\overline{e_n^2} = \overline{i_n^2}R \qquad\qquad\qquad\qquad\qquad (A2.5)$$

or

$$e_n = \sqrt{4kTBR} \qquad\qquad\qquad\qquad (A2.6)$$

and the equivalent Norton's noise current source and Thevenin's noise voltage source are illustrated in Fig. A2.3.

If a resistor R_s, at temperature T_1, representing a noise source is connected to a load resistor R_L at temperature T_2, as shown in Fig. A2.4, noise power is transferred

Fig. A2.3 The Norton's and Thevenin's equivalent circuits of thermal noise sources.

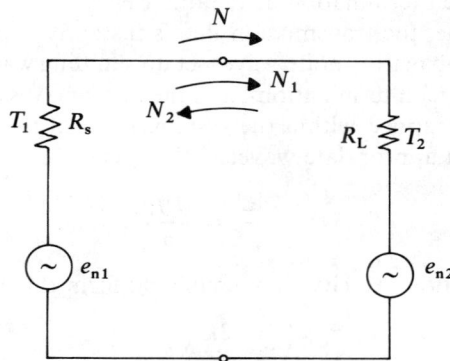

Fig. A2.4 The interchange of thermal noise power between two resistors.

from the source to the load *and* from the load to the source. The noise source provides total noise power

$$N_t = \frac{\overline{e_{n1}^2}}{R_s + R_L}$$

to both resistors in series, or

$$N_1 = \frac{\overline{e_{n1}^2}}{R_s + R_L} \cdot \frac{R_L}{R_s + R_L} = \frac{4kT_1 B R_s R_L}{(R_s + R_L)^2} \tag{A2.7}$$

to the load resistor R_L alone. In reverse, the load supplies the source resistor with

$$N_2 = \frac{4kT_2 B R_L R_s}{(R_s + R_L)^2} \tag{A2.8}$$

and hence the net noise power transferred from source to load is

$$N = N_1 - N_2 = \frac{4kB R_L R_s}{(R_s + R_L)^2}(T_1 - T_2) \tag{A2.9}$$

The maximum noise power which can be transferred from source to load occurs when $R_L = R_s$ and $T_2 = 0$, and this restriction defines the *available noise power* as

$$N = kT_1 B \tag{A2.10}$$

This is a useful reference level in the analysis of noise power in radio communication systems.

A2.2.1 Nyquist's and Planck's equations

To determine the noise equation, Nyquist considered the situation shown in Fig. A2.5 in which a lossless transmission line, length L and characteristic impedance Z_0, is terminated at either end by resistors $R = Z_0$. At temperature T the mean noise power N from each resistor travels along the line as an electromagnetic wave and is completely absorbed in the matched termination at the other end.

If, at some given time, the transmission line is instantly shorted at each end, the forward and reverse propagating noise waves set up standing wave patterns as they are reflected by the short-circuit terminations, and the line behaves as a lossless resonator with the noise energy trapped within the system. Each of the m resonant modes is related to length L and appropriate wavelength λ_m via

$$m = \frac{L}{\lambda_m/2} = \frac{2Lf_m}{v}$$

where v is the propagation velocity. In incremental form this becomes

$$\Delta m = \frac{2L}{v}\Delta f$$

and Δm is then the number of modes within the bandwidth $B = \Delta f$.

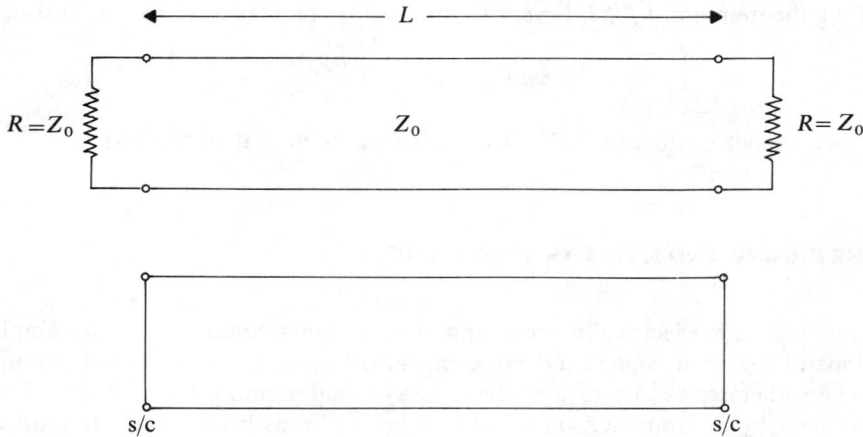

Fig. A2.5 The method used by Nyquist to determine the thermal noise power produced by a resistor R at temperature T. By this technique he determined the well-known equation $N = kTB$.

The mean energy available from *each* resistor is the product of mean noise power N and the time $t = L/v$ for the noise to travel down the length of the line. The *total* energy is twice this amount, viz.

$$W = \frac{2NL}{v} = \frac{N\Delta m}{B}$$

Classical thermodynamic theory indicates that the total energy should be kT joules for each mode and hence

$$kT\Delta m = N\Delta m/B$$

or

$$N = kTB \qquad (A2.11)$$

On the other hand, quantum theory suggests that the energy per mode should be given by Planck's equation,

$$\Delta W = \frac{hf}{\exp(hf/kT) - 1} \qquad (A2.12)$$

which differs from Nyquist's approach but only when $hf/kT \gg 1$ or $f \gg kT/h$. Further modifications are given in Ref. A2.3 which are beyond the scope of this text, but result in

$$\Delta W = \frac{hf}{2} + \frac{hf}{\exp(hf/kT) - 1} \qquad (A2.13)$$

though $hf/2$ is negligible at all radio frequencies. If the exponential term is expanded by

Taylor's theorem and hf/kT is very small, the first two terms of the expansion are

$$\exp(hf/kT) = 1 + \frac{hf}{kT}$$

and then Planck's equation $\Delta W = kT$ is identical with that of Nyquist.

A2.3 NOISE FIGURE AND NOISE TEMPERATURE

All amplifiers are electrically noisy and their outputs contain not only amplified versions of the input signal and noise but also further noise generated within the amplifiers themselves. There are several ways of describing the amount of noise contributed by an amplifier, one of the earlier methods being to attribute to each amplifier a *noise figure* F_r defined by Friis (Ref. A2.4) as the signal-to-available noise ratio at the input terminals, divided by the signal-to-noise ratio at the output terminals when referred to a matched input source at a temperature $T_0 = 290 \, \text{K}$. This is approximately room temperature. Although called noise figure, F_r is really a measure of the degradation of the signal-to-noise ratio due to the amplifier (Fig. A2.6). Numerically we can write

$$F_r = \frac{S_i/N_i}{S_o/N_o} = \frac{S_i}{S_o} \frac{N_o}{N_i} = \frac{1}{G} \frac{Gk(T_0 - T_r)B}{kT_0B} \tag{A2.14}$$

where $G = S_o/S_i$ is the gain of the amplifier, T_r is the effective noise temperature of the amplifier referred to its input terminals, and k is Boltzmann's constant. The *effective input noise temperature* T_r implies that if the amplifier is noiseless, with gain G, then the output noise power due to an input noise source at temperature T_r will equal the noise power from the output of the actual amplifier. This automatically takes into account the effect of noise power flow described in Eq. (A2.9), and the input noise power due to other matched sources can be considered as equal to the available noise power.

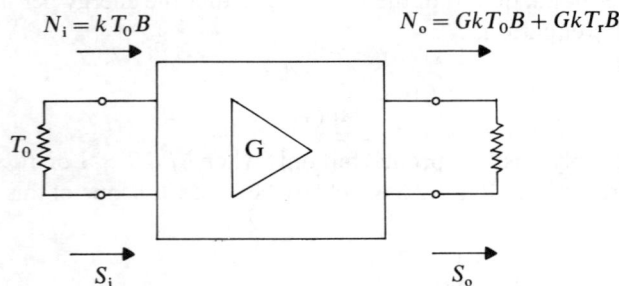

Fig. A2.6 Circuit used to determine the noise figure F_r of an amplifier in terms of its effective input noise temperature T_r.

It is common practice to refer to the noise figure in logarithmic form as

$$F(\mathrm{r}) = 10 \log F_{\mathrm{r}} = 10 \log\left(\frac{T_0 + T_{\mathrm{r}}}{T_0}\right) = 10 \log\left(1 + \frac{T_0}{T_{\mathrm{r}}}\right) \qquad (A2.15)$$

and some authors differentiate between F_{r} and $F(\mathrm{r})$ by referring to them as *noise factor* and *noise figure*, respectively, though this is not universally adopted.

The alternative method of describing the amount of noise generated within the amplifier or receiver is simply to use the effective input noise temperature T_{r} defined previously. By transposing Eq. (A2.15) we can relate this term to the noise figure:

$$T_{\mathrm{r}} = T_0(F_{\mathrm{r}} - 1) \qquad (A2.16)$$

but this is based upon the requirement that the amplifier noise is referred to an input termination at $T_0 = 290$ K. In many instances, particularly if the input termination is an antenna, the effective noise temperature (say T_{a} for the antenna example) need not be 290 K. In this case the effective available input noise power to an equivalent noise-free receiver is

$$N_{\mathrm{i}} = k T_{\mathrm{a}} B + k T_{\mathrm{r}} B = k T_0 B\left(\frac{T_{\mathrm{a}} + T_{\mathrm{r}}}{T_0}\right)$$

or

$$N_{\mathrm{i}} = F_{\mathrm{s}} k T_0 B \qquad (A2.17)$$

This is yet another way of describing the noise contribution of the *system* as a whole, that is, the antenna and amplifier/receiver combined, and F_{s} is called the *system noise figure* with contributions T_{a}/T_0 due to the antenna and T_{r}/T_0 due to the receiver. It is a kind of normalized effective noise temperature. In logarithmic form, combining T_0 with k:

$$N(\mathrm{i}) = F(\mathrm{s}) - 204 + 10 \log B \qquad (A2.18)$$

A2.3.1 Effective Noise Temperature of a Matched Attenuator

An attenuator reduces the magnitudes of both the incident signal and the incident noise but adds its own noise contribution in the process. The effective *input* noise temperature T_{ma} and the corresponding output noise temperature T_{out} for a matched attenuator have been derived by Dicke *et al.* (Ref. A2.5).

Referring to Fig. A2.7, it is seen that if the attenuator (with attenuation $L > 1$) is at $T = 0$ K, the output noise power due to the noise source alone is $P_{\mathrm{out}} = k T_1 B / L$. If, however, both the matched attenuator and the source are at temperature T_1 K, the load sees only the matched output of the attenuator and hence an output noise power $P_{\mathrm{out}} = k T_1 B$, which is obviously larger than the earlier $k T_1 B / L$ as $L > 1$. Dicke concluded that the difference was noise contributed by the matched attenuator and that its effective *output* noise temperature can be found from

$$k T_{\mathrm{out}} B = k T_1 B - k T_1 B / L \qquad (A2.19)$$

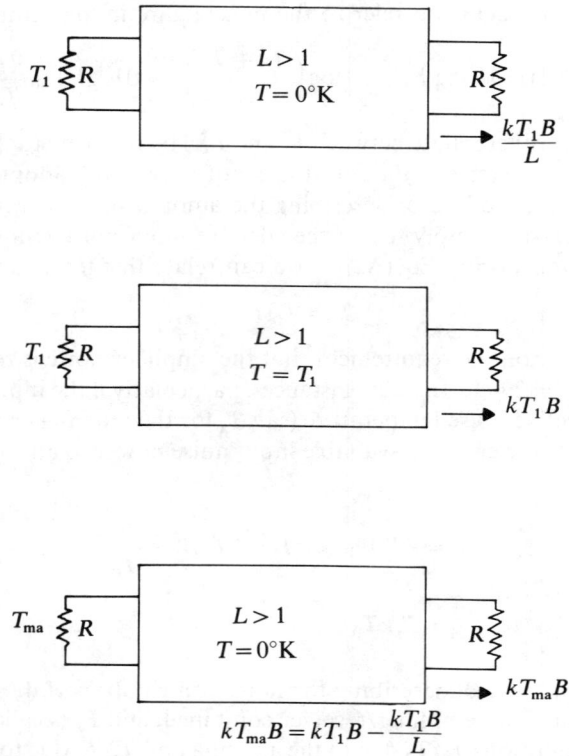

Fig. A2.7 The method used by Dicke to derive the expression $T_{\text{ma}} = 290(L-1)$ for the effective input noise temperature of an attenuator with loss L.

Hence its effective *input* noise temperature T_{ma} can be obtained as

$$T_{\text{ma}} = LT_{\text{out}} = L\left(T_1 - \frac{T_1}{L}\right) = T_1(L-1)$$

Usually the attenuator temperature is close to 290 K and it is customary to define the input noise temperature of a matched attenuator with $T_1 = T_0$:

$$T_{\text{ma}} = T_0(L-1) \tag{A2.20}$$

In radio communication systems an attenuator or its equivalent can arise in several different ways. For example, the feeder connecting the antenna to the first amplifier or receiver has some attenuation and, unfortunately, its contribution to the overall noise in the radio communication system can be considerable, especially in low-noise receiver arrangements. If the antenna noise temperature $T_{\text{a}} = 100$ K, the feeder itself will add $T_{\text{ma}} = 100$ K when

$$T_{\text{ma}} = 290(L-1) = 100$$

or $L = 1.34\,(1.4\,\text{dB})$. If the attenuation is considerably reduced to a fraction of a decibel, the matter is much improved: $L = 0.2\,\text{dB}$ corresponds to $T_{\text{ma}} = 14\,\text{K}$. This is illustrated in Fig. A2.8a.

The atmosphere behaves as an attenuator to signals arriving at the surface of earth from satellites outside the atmosphere (Fig. A2.8b), and also creates its own noise in addition to cosmic, galactic or solar noises originating in outer space. To a first rough approximation, if the atmospheric attenuation at a specified frequency and angle of

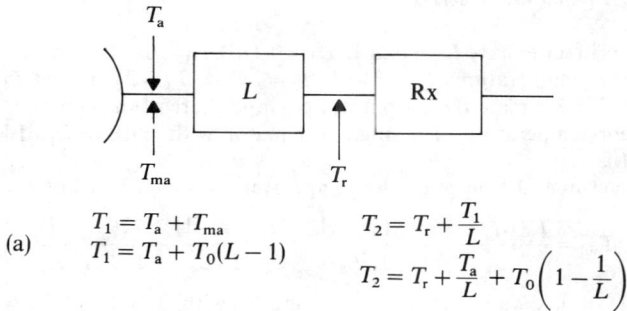

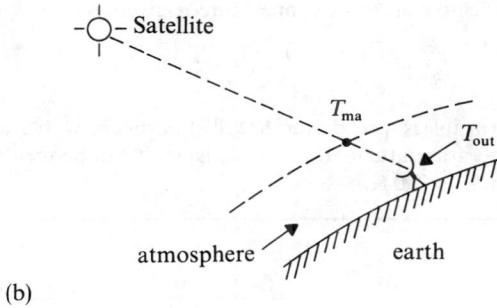

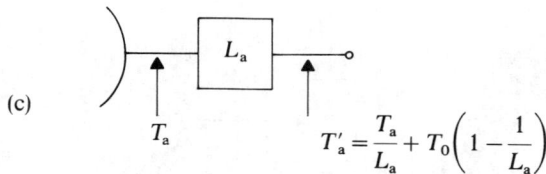

(a)

$$T_1 = T_a + T_{\text{ma}}$$
$$T_1 = T_a + T_0(L-1)$$

$$T_2 = T_r + \frac{T_1}{L}$$
$$T_2 = T_r + \frac{T_a}{L} + T_0\left(1 - \frac{1}{L}\right)$$

(b)

(c)

$$T'_a = \frac{T_a}{L_a} + T_0\left(1 - \frac{1}{L_a}\right)$$

Fig. A2.8 Attenuation produces noise in several different circumstances: (a) the waveguide connecting the antenna to a receiver, (b) the atmosphere surrounding the earth, or even (c) the resistance of the antenna itself.

elevation is (say) 0.5 dB ($L = 1.122$), then the *output* noise temperature is given via Eq. (A2.19) as 31.5 K. This is not strictly correct because both the attenuation rate and atmospheric temperature vary with height and the process should involve an integration along the atmospheric path. The effective temperature resulting from such an integration does not, however, differ too greatly from $T_0 = 290$ K because the bulk of the attenuation and the noise generation occurs in the lower atmosphere.

Example A2.1 Attenuators in series

Four attenuators in series have $L_1 = 0.2$ dB, $L_2 = 0.3$ dB, $L_3 = 0.4$ dB and $L_4 = 0.5$ dB with respective ambient temperatures $T_1 = 140$ K, $T_2 = 190$ K, $T_3 = 240$ K and $T_4 = 290$ K, as shown in Fig. A2.9. Estimate the output noise temperature of the combination and the effective ambient temperature of a single attenuator with $L(\text{dB}) = L_1(\text{dB}) + L_2(\text{dB}) + L_3(\text{dB}) + L_4(\text{dB})$.

We can determine the output noise temperature of such an arrangement via

$$T_{\text{out}} = \frac{T_4(L_4 - 1)}{L_4} + \frac{T_3(L_3 - 1)}{L_3 L_4} + \frac{T_2(L_2 - 1)}{L_2 L_3 L_4} + \frac{T_1(L_1 - 1)}{L_1 L_2 L_3 L_4}$$

Substituting the quoted values of T_4, T_3, T_2 and T_1 with $L_4 = 1.122$, $L_3 = 1.096$, $L_2 = 1.072$ and $L_1 = 1.047$ we obtain

$$T_{\text{out}} = 31.53 + 18.74 + 10.37 + 4.77 = 65.4 \text{ K}$$

If the four attenuators are combined into a single attenuator with $L = 1.047 \times 1.072 \times 1.096 \times 1.122 = 1.380$, the effective ambient temperature is given via

$$T_{\text{out}} = \frac{T_{\text{eff}}(L - 1)}{L} = 65.4 \text{ K}$$

as $T_{\text{eff}} = 238$ K. When this principle is spread over the full atmosphere, with a larger number of increments and full temperature/attenuation relationship, it can be seen that T_{eff} will not be very much less than $T_0 = 290$ K.

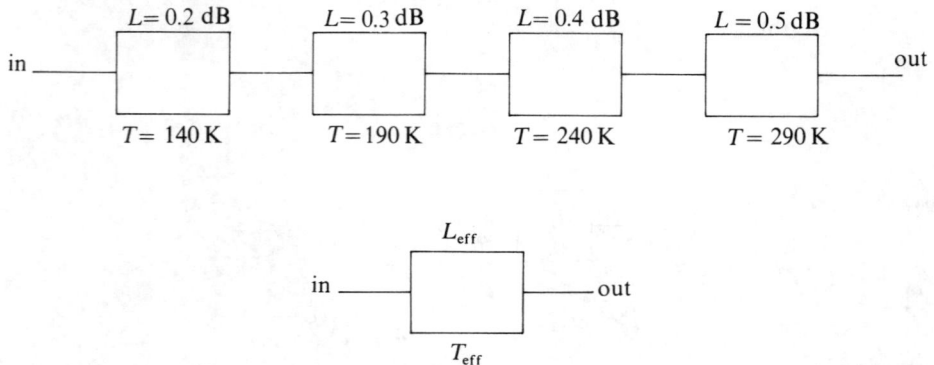

Fig. A2.9 The effective loss and noise temperature of several attenuators in cascade as analyzed in Ex. A2.1.

The antenna's own resistance also produces attenuation and noise. If the external noise temperature of the antenna is T_a when the antenna is considered noiseless, the effective noise temperature including L_a is

$$T_a' = \frac{T_a}{L_a} + T_0\left(1 - \frac{1}{L_a}\right) \tag{A2.21}$$

as illustrated in Fig. A2.8c. If $L_a = 0.1$ dB and $T_a = 70$ K, $T_a' = 75$ K.

A2.3.2 Cascaded Networks

In a radio receiving system, the external and internal noise sources are connected in cascade. If we consider Fig. A2.10 as typical, the antenna temperature T_a combines all the external sources and this is degraded (i.e. made larger) by the cascaded effect of the feeder, preamplifiers and receiver noise contributions. However, the effective noise temperatures of the amplifiers and receivers are usually referred to their *individual* inputs. We can alternatively refer them to some other point in the cascaded system simply by multiplying or dividing by the gain or loss of each cascaded network as appropriate. This is best illustrated by the cascaded circuit shown in Fig. A2.10a in general terms, in Fig. A2.10b in numerical terms, and in Fig. A2.10c in terms of the signal-to-noise ratio.

The numerical illustration clearly demonstrates that the use of high-gain, low-noise preamplifiers is of considerable benefit to the overall signal-to-noise ratio and that in such circumstances the noise figures of later amplifier stages and receivers are not too significant. In addition, very careful consideration of the feeder between antenna and preamplifier is of prime importance.

Example A2.2 Overall noise figure

Two amplifiers with parameters (F_1, G_1) and (F_2, G_2) are connected in cascade to produce an effective single amplifier represented by (F_{12}, G_{12}). Determine F_{12} and G_{12} in terms of F_1, F_2, G_1 and G_2.

Figure A2.11 illustrates the arrangement. We can obtain the effective noise temperature at the input terminals as being either

$$T_{in} = T_0 + (F_1 - 1)T_0 + \frac{(F_2 - 1)T_0}{G_1}$$

or

$$T_{in} = T_0 + (F_{12} - 1)T_0$$

Equating these two alternatives, we obtain

$$F_{12} = F_1 + \frac{(F_2 - 1)}{G_1}$$

while the overall gain is simply $G_{12} = G_1 \times G_2$.

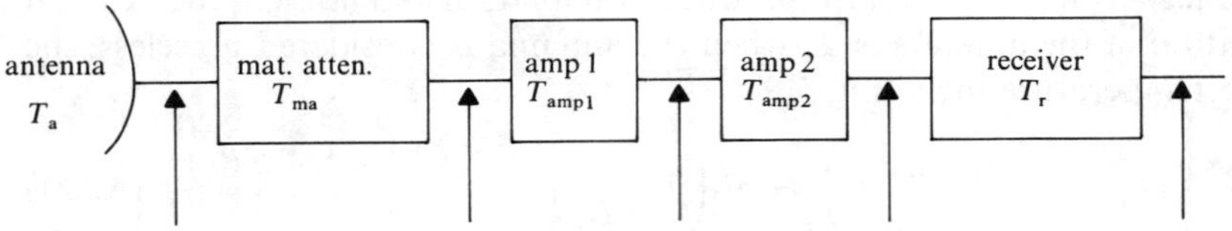

antenna T_a	mat. atten. T_{ma}	amp 1 T_{amp1}	amp 2 T_{amp2}	receiver T_r

	antenna	mat. att.	amp 1	amp 2	recvr
antenna	T_a	$G_{ma}T_a$	$G_1G_{ma}T_a$	$G_2G_1G_{ma}T_a$	$G_rG_2G_1G_{ma}T_a$
mat. att.	T_{ma}	$G_{ma}T_{ma}$	$G_1G_{ma}T_{ma}$	$G_2G_1G_{ma}T_{ma}$	$G_rG_2G_1G_{ma}T_{ma}$
amp 1	T_{amp1}/G	T_{amp1}	G_1T_{amp1}	$G_2G_1T_{amp1}$	$G_rG_2G_1T_{amp1}$
amp 2	$T_{amp2}/G_{ma}G_1$	T_{amp2}/G_1	T_{amp2}	G_2T_{amp2}	$G_rG_2T_{amp2}$
recvr	$T_r/G_{ma}G_1G_2$	T_r/G_1G_2	T_r/G_2	T_r	G_rT_r

(a)

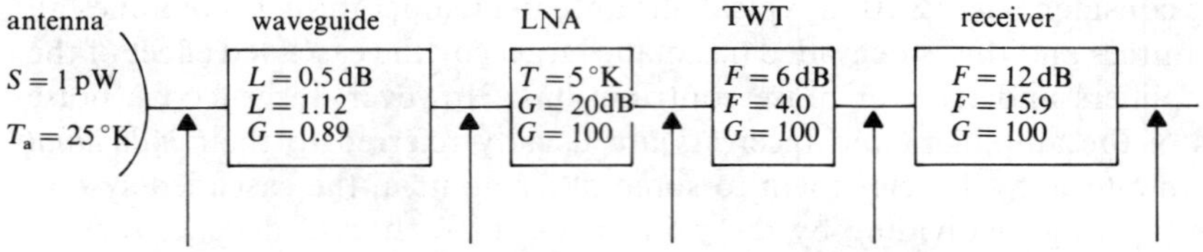

antenna $S = 1\,pW$, $T_a = 25\,°K$ — waveguide $L = 0.5\,dB$, $L = 1.12$, $G = 0.89$ — LNA $T = 5\,°K$, $G = 20\,dB$, $G = 100$ — TWT $F = 6\,dB$, $F = 4.0$, $G = 100$ — receiver $F = 12\,dB$, $F = 15.9$, $G = 100$

antenna	25	22.25	2225	222,500	22,250,000
waveguide	35	31.54	3154	315,400	31,540,000
LNA	5.6	5.00	500	50,000	5,000,000
TWT	9.7	8.65	865	86,500	8,650,000
recvr	0.5	0.43	43	4,300	430,000
T(eff)	75.8	67.9	6787	678,700	678,700,000	K

(b)

T(eff)	75.8	67.9	6787	678,700	678,700,000	K
kT(eff)B	0.0084	0.0075	0.7549	74.9	7490	pW
S	1	0.89	89	8900	890,000	pW
SNR*	20.7	20.7	20.7	20.7	20.7	dB
T(act)	25	53.4	5840	670,500	674,800,000	K
S	1	0.89	89	8900	890,000	pW
$S/kT_{act}B$	363	151	138	120	118	
SNR	25.6	21.8	21.4	20.8	20.7	dB

(c)

Fig. A2.10 A typical receiving system comprising an antenna, coupling, two low-noise amplifiers and the basic receiver. The effective noise temperatures of each item may be referred to any point in the circuit as shown algebraically and numerically. Using the effective temperature at any point (which implies thereafter all ideal amplifiers), the output signal-to-noise ratio is constant at 20.7 dB irrespective of the point of calculation. However, by referring each noise to its individual source, the signal-to-noise ratio progressively degrades from 25.6 dB at the antenna towards the 20.7 dB at the output.

$$T_{in} = T_0 + T_{r1} + \frac{T_{r2}}{G_1} \qquad T_{r2} = (F_2 - 1)T_0$$

$$= T_0 + (F_1 - 1)T_0 + \frac{(F_2 - 1)}{G_1}T_0$$

$$T_{in} = T_0 + (F_{12} - 1)T_0$$

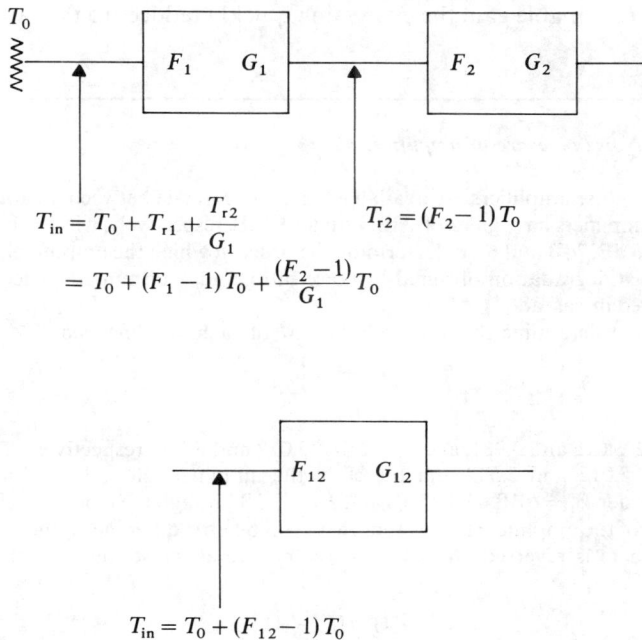

Fig. A2.11 Two networks in cascade can be made equivalent to a single network with equivalent noise figure and gain.

If we have several amplifiers which can be used in cascade, each specified by individual parameters F and G (noise figure and gain), what is the best order in which they should be used to achieve the least degradation in the signal-to-noise ratio? We have shown in Ex. A2.2 that, for two amplifiers in cascade,

$$F_{12} = F_1 + \frac{(F_2 - 1)}{G_1} \tag{A2.22}$$

or, if they were connected in reverse order,

$$F_{21} = F_2 + \frac{(F_1 - 1)}{G_2} \tag{A2.23}$$

As the lesser degradation or the better performance is achieved with the amplifier arrangement with the lower overall noise figure, the rule must be that amplifier 1 is placed before amplifier 2 if $F_{12} < F_{21}$, i.e.

$$\frac{(F_1 - 1)}{(1 - 1/G_1)} < \frac{(F_2 - 1)}{(1 - 1/G_2)} \tag{A2.24}$$

or

$$M_1 < M_2$$

The parameter M is called the *noise measure* of an amplifier, but because most amplifiers have a reasonable gain the expression quickly reduces to the rather obvious decision that $F_1 < F_2$.

Example A2.3   Noise measure of amplifiers

Three small low-noise amplifiers are available for use in cascade between an antenna and a receiver. The amplifiers have gains of 4 dB, 7 dB and 8 dB, respectively, with corresponding noise figures of 5 dB, 7dB and 6 dB. Determine the order in which the amplifiers should be cascaded for least degradation of signal-to-noise ratio, and determine the effective noise figure of all three in cascade.

We must first determine the noise measure M of each amplifier via

$$M = \frac{F-1}{1 - 1/G}$$

where $F = 3.162, 5.012$ and 3.981, and $G = 2.512, 5.012$ and 6.310, respectively. These data give $M = 3.591, 5.012$ and 3.542 and therefore the amplifiers should be connected in cascade in the order $F_1 = 6\,\text{dB}$, $F_2 = 5\,\text{dB}$ and $F_3 = 7\,\text{dB}$, though the closeness of the noise measure of two of the amplifiers means that there will be little difference if the order of the first two amplifiers is reversed. We can obtain the overall noise figure in the manner outlined in Ex. A2.2, viz.

$$F_{123} = F_1 + \frac{(F_2 - 1)}{G_1} + \frac{(F_3 - 1)}{G_1 G_2}$$

$$F_{123} = 3.981 + 0.343 + 0.253 = 4.576\,(6.6\,\text{dB})$$

and the overall gain is 19 dB.

A2.3.3 Noise Field

We have seen that several parameters can be used to give numerical indications of the noise contributed by various sources. Some external noise sources, antennas and low-noise amplifiers are commonly described by their effective noise temperature referred to some point, and receivers are designated by their noise figure $F(\text{r})$; while the system may be represented overall by a system noise figure $F(\text{s})$ or system noise temperature T_s.

An alternative approach is sometimes used at lower radio frequencies to express atmospheric noise. Just as the system has a noise figure F_s defined by

$$F_s = \frac{T_a + T_r}{T_0} = \frac{T_a}{T_0} + \frac{T_r}{T_0}$$

some authors describe the contribution $F_a = T_a/T_0$ as the antenna noise figure. This is related to the noise voltage in a matched system via

$$kT_a B = F_a k T_0 B = \frac{V_n^2}{4 R_{\text{rad}}} \tag{A2.25}$$

where R_{rad} is the radiation resistance of the antenna. For a specified reference antenna, such as a short vertical dipole in which the radiation resistance is $R_{rad} = 40\pi^2 h^2 f^2/c^2$ and $V_n = E_n h/2$, Eq. (A2.25) reduces to

$$E_n^2 = 640\pi^2 k T_0 B F_a f^2/c^2 \qquad (A2.26)$$

In these equations h is the antenna height and E_n is the equivalent *noise field strength*. It is customary to set the effective noise bandwidth to 1 kHz and to refer to frequency f in MHz, and with these specifications the logarithmic form of the equation becomes

$$E(dB\mu) = F(a) - 65.54 + 20 \log f(MHz) \qquad (A2.27)$$

The problem is now reduced to the determination of the parameter $F(a)$ for a given frequency, and worldwide details of $F(a)$ are available in Ref. A2.6 published by the CCIR.

A2.4 MEASUREMENT OF NOISE FIGURE AND NOISE TEMPERATURE

There are several ways in which we can measure the noise figure F_r or noise temperature T_r of an amplifier or receiver in a radio communication system and, with slight adaptions, similar techniques may be used to obtain the antenna noise temperature T_a or even the system noise temperature T_s.

One simple method is illustrated in Fig. A2.12, in which a device known as an *excess noise generator* is inserted between a matched input termination at temperature T_0 and the receiver input. There are several types of these noise sources including, as examples, the temperature-limited diode with *excess noise ratio* (ENR) typically 5 to 6 dB, the argon gas discharge tube mounted in a waveguide with ENR typically 15.5 dB, and solid-state noise sources based on the avalanche effect with ENR typically 15 dB

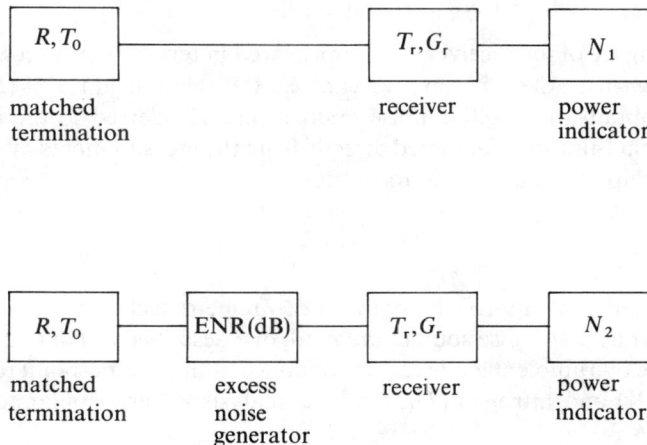

Fig. A2.12 A method of measuring the noise figure of a receiver with the aid of an excess noise generator.

but sometimes up to about 35 dB. In this context the term excess noise ratio is defined logarithmically as

$$\text{ENR(dB)} = 10\log\left(\frac{T_n - T_0}{T_0}\right) = 10\log\left(\frac{T_n}{T_0} - 1\right) \tag{A2.28}$$

where T_n is the output noise temperature of the device, T_0 is the usual reference temperature of 290 K and $T_n - T_0$ is the excess noise temperature.

We need two measurements to obtain the noise parameters F or T, as appropriate. Firstly, with the receiver connected to a matched input termination at temperature T_0, the output noise power level N_1 is measured as being

$$N_1 = kT_0 BG + k(F_r - 1)T_0 BG$$

Secondly, with the appropriate excess noise generator inserted between the termination and the receiver input, the output noise power level N_2 is measured as being

$$N_2 = kT_n BG + k(F_r - 1)T_0 BG$$

With these two measurements we can obtain their ratio $Y = N_2/N_1$ as

$$Y = \frac{F_r T_0 + T_n - T_0}{F_r T_0} \tag{A2.29}$$

and it is easy to transpose this equation into either

$$F_r = \frac{T_n - T_0}{T_0(Y - 1)}$$

or

$$F(r) = 10\log\left(\frac{T_n - T_0}{T_0}\right) - 10\log(Y - 1)$$

i.e.

$$F(r) = \text{ENR(dB)} - 10\log(Y - 1) \tag{A2.30}$$

Thus the noise figure of the receiver can be measured in terms of the excess noise ratio and the *Y-parameter*. Alternatively, we can use the relationship $T_r = (F_r - 1)T_0$ of Eq. (A2.16) to obtain the effective noise temperature T_r referred to the input of the receiver. Note that this can be obtained directly from the measurements by substitution of this relationship into Eq. (A2.29) to produce

$$T_r = \frac{T_n - YT_0}{Y - 1} \tag{A2.31}$$

A slightly different approach to the measurement technique is illustrated in Fig. A2.13a in which two noise sources are used, one described as *hot* and one as *cold*. For example, the two ambient loads can be contained at the boiling-point temperatures of water (373.1 K) and nitrogen (77.3 K). The analysis is very similar to that given previously, with

$$N_c = kT_c BG + k(F_r - 1)T_0 BG$$

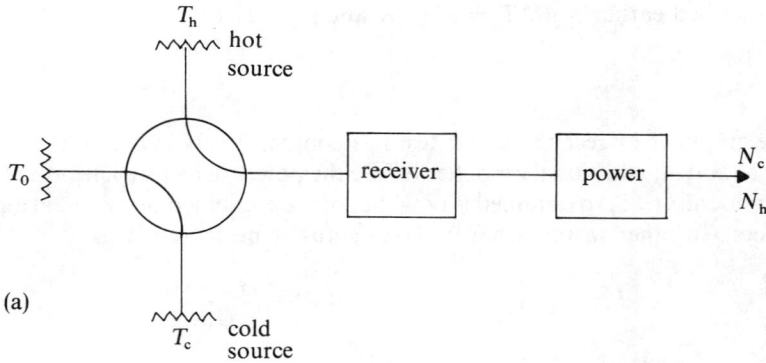

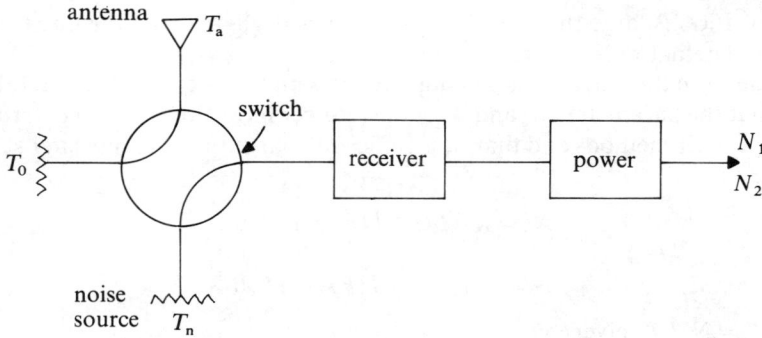

Fig. A2.13 An alternative method of measuring (a) the noise figure of a receiver and (b) the noise temperature of the antenna.

and

$$N_h = kT_hBG + k(F_r - 1)T_0BG$$

With the Y-parameter now defined as $Y = N_h/N_c$ we can easily obtain

$$Y = \frac{T_h + (F_r - 1)T_0}{T_c + (F_r - 1)T_0} \tag{A2.32}$$

$$F_r = 1 + \frac{T_h - YT_c}{T_0(Y - 1)} \tag{A2.33}$$

and

$$T_r = \frac{T_h - YT_c}{Y - 1} \tag{A2.34}$$

Note the resemblance between Eqs. (A2.34) and (A2.31). For the two particular noise sources described earlier, with $T_h = 373.1$ K and $T_c = 77.1$ K, we can write

$$T_r = \frac{373.1 - 77.1Y}{Y - 1} \tag{A2.35}$$

If the amplifier or receiver under test has comparatively low gain G_1, it may be necessary to add an additional amplifier before the power detector. The noise figure F_r' or noise temperature T_r' so obtained is now that of the combined or cascaded amplifiers (or receivers). In other words, what we have actually measured is

$$T_r' = T_{r1} + \frac{T_{r2}}{G_1} = (F_{r1} - 1)T_0 + \frac{(F_{r2} - 1)T_0}{G_1} \tag{A2.36}$$

By transposing this equation we can obtain

$$F_{r1} = \left(1 - \frac{T_r'}{T_0}\right) - \frac{(F_{r2} - 1)}{G_1} \tag{A2.37}$$

Obviously, if G_1 is large the second term becomes negligible and the measured noise figure is the actual noise figure of the first stage.

To measure the antenna noise temperature with the circuit of Fig. A2.13b we can assume that the parameters G_r and T_r of the receiver have already been determined by one of the earlier methods and that G_r is sufficiently large to buffer any later stages. We than note that

$$N_1 = kT_aBG + k(F_r - 1)T_0BG$$

and

$$N_2 = kT_nBG + k(F_r - 1)T_0BG$$

gives us $Y = N_2/N_1$ given by

$$Y = \frac{T_n + T_r}{T_a + T_r} \tag{A2.38}$$

and

$$T_a = \frac{T_n + (1 - Y)T_r}{Y} \tag{A2.39}$$

A2.5 NOISE SOURCES

Cosmic noise is the background noise of space originating from the present temperature of the expanding universe. When the universe was at its smallest its temperature was greater than 10^{10} K but its expansion over millions of years has caused its cooling to the present level of about 2.7 K.

Galactic noise is caused by the hot gases of stars and matter in interstellar space, most intense in the galactic plane and maximum in the direction of the galactic center. The value of noise temperature therefore depends on the direction in which the antenna is pointing and also varies with frequency. One expression (Ref. A2.7) gives

$$T_{\text{gal}} = T_{f = 100\text{MHz}} \left(\frac{100}{f\,\text{MHz}} \right)^{2.5} \tag{A2.40}$$

where T_{100} varies between 500 K and 18,650 K, with a geometric mean of 3050 K. The equation may alternatively be written in logarithmic form as

$$10 \log (T_{\text{gal}}/290) = 10 \log 3050 + 25 \log 100 - 25 \log f - 10 \log 290$$

or

$$10 \log (T_{\text{gal}}/290) = 60 - 25 \log f(\text{MHz}) \tag{A2.41}$$

where the galactic noise temperature is recorded in decibels above 290 K, implying a galactic noise power ratio for a given noise bandwidth. Other authors (for example, Ref. A2.8) use a slightly different empirical expression:

$$10 \log (T_{\text{gal}}/290) = 52 - 23 \log f(\text{MHz}) \tag{A2.42}$$

which corresponds to

$$N_{\text{o}}(\text{gal}) = -152 - 23 \log f(\text{MHz}) \qquad \text{dBW} \tag{A2.43}$$

as the equivalent noise spectral density.

The sun is an obvious source of electrical noise and consequently antennas are orientated to avoid the sun. However, antennas have sidelobes and these can introduce solar noise into the system even when the antenna is aligned in other directions. The effect is not as great as might be expected because the sun subtends an angle of only about 0.5° at the antenna.

Man-made electrical or radio noise is produced by the numerous by-products of modern industrial life, including automobiles (ignition and electrical circuitry), power generation and distribution, industrial machinery, home appliances (vacuum cleaners, electric tools), electric lighting systems, medical equipment, electrical transport systems, and so on. By its origins, it is predominant in industrial and urban areas, reducing in magnitude in the quieter rural areas, though locally higher in the vicinity of farm machinery, etc. The noise pattern covers a wide range of frequencies from about 30 Hz upwards, sometimes extending well into the gigahertz range. This topic is covered extensively in Ref. A2.9.

A2-6 NOISE IN DIGITAL SYSTEMS

A full analysis of the noise in digital systems is outside the scope of this text but a very brief reference to the subject is useful, as several radio communication systems make use of the resulting equations.

In the simplest digital systems the signal is transmitted in pulses, on or off, mark or space, logic '1' or '0', or simply in terms of binary digits. A teleprinter, for example, may use a coded arrangement of binary digits to represent each character on its keyboard. For an elementary coded group of five equal-length elements per character, the word

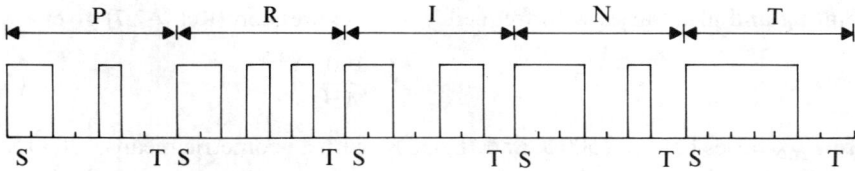

Fig. A2.14 The digital representation of the word PRINT using a teleprinter with five equal-length elements per character plus a 'start' (S) and a 'stop' (T) element at either end. The 'start' and each of the five elements of the character are 20 ms in length; the 'stop' is 30 ms long. The overall length of each character is thus 150 ms.

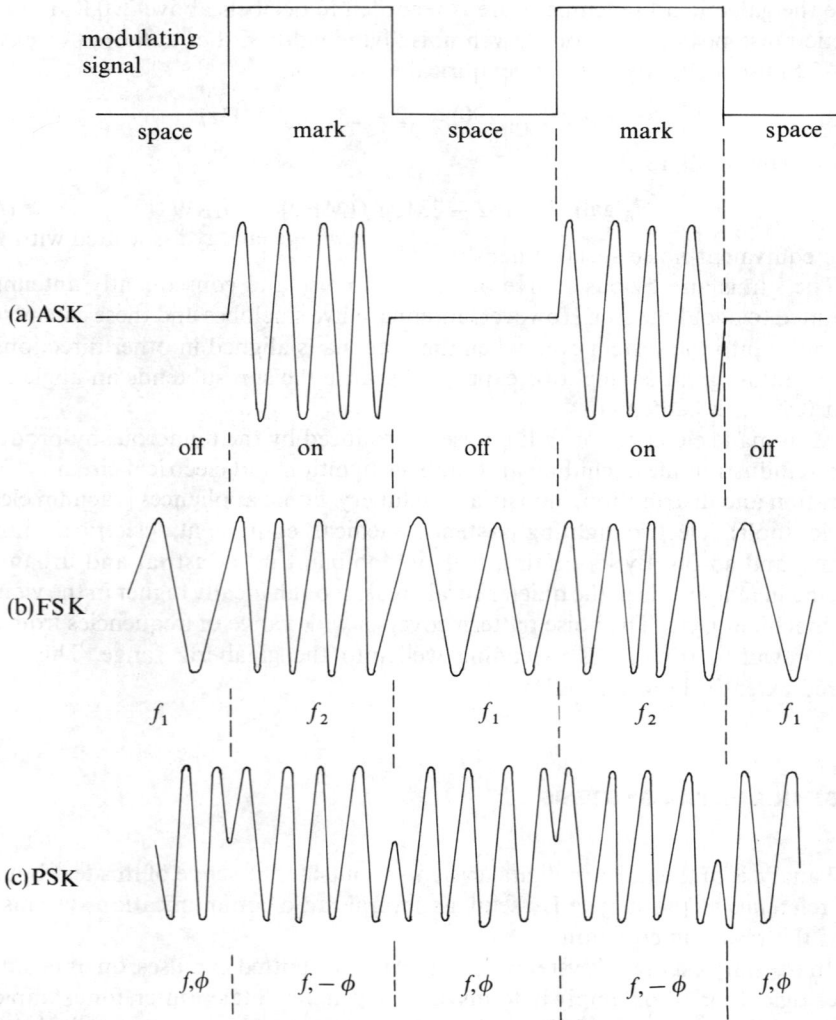

Fig. A2.15 Three simple signal processing techniques in which digital information is represented in terms of the (a) amplitude, (b) frequency or (c) phase of a carrier signal.

PRINT could be represented digitally as shown in Fig. A2.14. If the duration of each element is 20 ms, the transmission rate is 50 elements per second. For telegraphic communication this is usually called 50 bauds and the overall timing involves *start* and *stop* elements as well. However, for general digital communication the transmission rate is usually quoted in binary digits (bits) per second.

Each digit can be represented by its energy content E_b measured in joules, or alternatively by its energy content in units of noise spectral density E_b/N_0, where $N_0 = kT$ and k = Boltzmann's constant. If there are no transmission and/or noise problems between transmitter and receiver, each '1' and '0' is received without error. However, in a system in which the receiver has to decide whether the signal is a '1' or a '0' in the presence of white Gaussian noise, some errors do occur. This is similar to the decision-making processes discussed briefly in the chapter on radar. It can be shown (Ref. A2.10) that the probability of error is given by

$$P_e = \tfrac{1}{2}\text{erfc}\left[\frac{1-\rho}{2}\frac{E_a}{N_0}\right]^{1/2} \tag{A2.44}$$

where ρ is a cross-correlation coefficient between the two alternative elements erfc = complementary error function, and E_a is the average energy associated with the '1' and '0' elements.

Various signal processing techniques may be used in a radio communication link in which the information content is related to either amplitude, frequency or phase, for example.

With a simple on–off process, known as *amplitude-shift-keying* (ASK), the waveform will be similar to that illustrated in Fig. A2.15a. There is no correlation between the signals '1' and '0' ($\rho = 0$) and the average energy E_a is $(E_b + 0)/2$. Thus the bit error rate is

$$\text{BER} = \tfrac{1}{2}\text{erfc}\left[\frac{1}{4}\frac{E_b}{N_0}\right]^{1/2} \tag{A2.45}$$

assuming coherent reception.

With a simple two-frequency process, known as *frequency-shift-keying* (FSK), the waveform will resemble that illustrated in Fig. A2.15b. If the two frequencies are fairly well separated, $\rho = 0$ and $E_a = (E_b + E_b)/2$. The bit error rate with coherent reception is

$$\text{BER} = \tfrac{1}{2}\text{erfc}\left[\frac{1}{2}\frac{E_b}{N_0}\right]^{1/2} \tag{A2.46}$$

In the third process the two signals are represented by a single frequency but with 180° phase shift between the 1's and 0's. This is known as *phase-shift-keying* (PSK) and the waveform will resemble that illustrated in Fig. A2.15c. The two elements have $\rho = -1$ and $E_a = (E_b + E_b)/2$, so that with coherent reception

$$\text{BER} = \tfrac{1}{2}\text{erfc}\left[\frac{E_b}{N_0}\right]^{1/2} \tag{A2.47}$$

Thus if the 'quality' of reception for digital systems is described in terms of bit error rate BER, a direct relationship between BER and E_b/N_0 can be derived in terms of the particular type of signal processing involved in the radio communication link, of which only three of the simplest digital processes have been described here.

Without going into the details of the analysis, we can observe that the carrier-to-noise ratio at the receiver is related to the bit energy E_b joules per bit, the data transmission rate R bits per second, and the noise kT_sB via

$$\text{CNR} = \frac{E_b R}{kT_s B} = \frac{E_b}{N_0}\frac{R}{B} \tag{A2.48}$$

or

$$\text{CN}_0\text{R(dB)} = 10\log\left(\frac{E_b}{N_0}\right) + 10\log R \tag{A2.49}$$

where $10\log(E_b/N_0)$ is commonly expressed in decibels. Thus the term E_b/N_0 is the direct link between our radio wave propagation parameter $\text{CN}_0\text{R(dB)}$ and the quality of reliability of digital reception BER.

REFERENCES

A2.1   Johnson, J. B., 'Thermal agitation of electricity in conductors,' *Phys. Rev.*, 32 (1928), 97–109.

A2.2   Nyquist, H., 'Thermal agitation of electric charge in conductors,' *Phys. Rev.*, 32 (1928), 110–3.

A2.3   Ekstein, H., and N. Rostoker, 'Quantum theory of fluctuations,' *Phys. Rev.*, 100 (November 1955), 1023–9.

A2.4   Friis, H. T., 'Noise figures in radio receivers,' *Proc. IRE*, 32 (July 1944), 419.

A2.5   Dicke, R. H., et al., 'Atmospheric absorption measurements with a microwave radiometer,' *Phys. Rev.*, 70 (September 1946), 340–8.

A2.6   CCIR, 'World distribution and characteristics of atmospheric radio noise,' *Report* 322–1, International Telecommunications Union, Geneva, 1974.

A2.7   Brown, R. H., and C. Hazard, 'A model of radio frequency radiation from the galaxy,' *Phil. Mag.*, 44 (September 1953), 939.

A2.8   Brown, D. W., and H. P. Williams, 'The performance of meteor burst communications at different frequencies,' *AGAARD Conference Proceedings*, No. 244, September 1978, pp. 21–1 to 21–31.

A2.9   Skomal, E. N., *Man-made Radio Noise*. New York: Van Nostrand Reinhold Company, 1978.

A2.10 Lawton, J. G., 'Comparison of binary data transmission systems,' *Proc. 2nd Natl. Conf. Milit. Electronics*, 1958, pp. 54–61.

Appendix 3
Answers to Problems

Note: Not all problems have a numerical solution. More detailed answers are available in a teacher's manual.

1.1 Axial ratio 2 to 1 with left-handed circular polarization.

1.2 $E = 173\,\text{mV/m}$; $H = 0.46\,\text{mA/m}$; $D = 1.53 \times 10^{-12}\,\text{C/m}^2$; $B = 5.8 \times 10^{-10}\,\text{Wb/m}^2$; $\delta W = 1.3 \times 10^{-13}\,\text{J/m}^3$.

1.3 Minimum distance $= 67\,\text{cm}$.

1.4 Voltage at receiver $= 4.8\,\mu\text{V}$.

1.5 Unattenuated field at $1\,\text{km} = 950\,\text{mV/m}$.

1.6 Unattenuated field at $10\,\text{km} = 94\,\text{mV/m}$.

1.7 Antenna gain $= 59\,\text{dB}$.

1.8 Illumination efficiency $= 0.6$.

1.9 Power available at matched receiver $= -111\,\text{dBW}$.

1.10 Voltage at matched receiver $= 170\,\mu\text{V}$.

1.11 Halfpower beamwidths about $90°$, $78°$ and $8°$, respectively.

2.1 Electric field $= 6.7\,\text{mV/m}$.

2.2 Effective height $= 75\,\text{m}$; electric field $= 15.5\,\text{mV/m}$.

2.3 Electric field $= 66\,\text{dB}\mu$ and $83\,\text{dB}\mu$, respectively.

2.4 Electric field $= 1.5\,\text{mV/m}$.

2.5 Electric field is of the order $2\,\text{mV/m}$ (graphical).

2.6 Electric field is of the order $3\,\text{mV/m}$ (graphical).

2.7 Probable range of good reception $= 65\,\text{km}$, $150\,\text{km}$ and $275\,\text{km}$, respectively.

2.8 Approximate boundaries: 0 to $70\,\text{km}$ for ground wave reception, $70\,\text{km}$ to $300\,\text{km}$ for interference zone, and $300\,\text{km}$ to $800\,\text{km}$ for sky wave reception.

2.9 Approximate daytime coverage of $190\,\text{km}$ is reduced to about $30\,\text{km}$ at night.

2.12 Electric field $= 0.9\,\text{mV/m}$.

3.1 Conductivity increases from about $9.5 \times 10^{-9}\,\text{S/m}$ (night) to about $7.6 \times 10^{-7}\,\text{S/m}$ (day); effective relative permittivity drops from almost 1.0 to about 0.9; and attenuation increases from about $0.02\,\text{dB/km}$ to about $1.3\,\text{dB/km}$.

3.2 Conductivity remains low at about $1.1 \times 10^{-10}\,\text{S/m}$ (night) and about $9 \times 10^{-9}\,\text{S/m}$ (day); effective relative permittivity is almost 1.0 night and day; and attenuation is almost negligible at about $0.0002\,\text{dB/km}$ (night) and $0.01\,\text{dB/km}$ (day).

3.3 The complex relative permittivity is about $1.0 - j0.0002$ at night and about $0.9 - j0.02$ during the day.

3.4 Magnetic field $= 42.7\,\text{A/m}$.

3.5 The noon-day zenith angle is about $50°$ (Mar), $26.5°$ (Jun), $50°$ (Sep) and $73.5°$ (Dec).

3.6 $f_0\text{E} = 3.46\,\text{MHz}$, $3.75\,\text{MHz}$, $3.46\,\text{MHz}$ and $2.82\,\text{MHz}$; $f_0\text{F1} = 4.94\,\text{MHz}$, $5.28\,\text{MHz}$, $4.94\,\text{MHz}$ and $4.20\,\text{MHz}$.

3.7 The MUF is about 22 MHz and OWF is about 19 MHz, depending on graphical readings.

3.8 Path length = 2057 km, spatial loss = 124 dB and the angle of launch = 5.3°.

3.9 Angle of incidence = 79°.

3.10 Ionospheric loss = 32 dB.

3.11 Received power level = − 110 dBW.

4.1 Line-of-sight ranges are 61 km, 84 km and 103 km.

4.2 Electric field = 27.8 mV/m.

4.3 Point of reflection $d_1 = 23.6$ km and hence $h_t = 87$ m and $h_r = 42$ m.

4.4 Electric field = 52 mV/m without divergence and 48 mV/m with divergence.

4.6 $E/E_d = 5.9$ dB, 4.5 dB, 0.9 dB and − 18 dB.

4.7 The field strength is likely to vary between 8.3 mV/m and 13.2 mV/m as K varies from 0.6 to 2.0.

4.8 The field strength is likely to vary between 1 mV/m and 3.6 mV/m.

4.9 Receiver voltage = 2 mV/m.

4.11 Minimum transmitter powers of − 7 dBW and + 3 dBW are required for the given reliability levels.

4.12 The median path losses are about 122 dB, 137 dB and 146 dB at 150 MHz, 450 MHz and 900 MHz, respectively.

5.1 Input impedances $Z_1 = Z_2 = 103 + j8$ Ω (graphical); when $V_2 = 0$, $Z_1 = 89 + j62$ Ω (approximately, depending on graphical readings).

5.2 Input currents $I_1 = I_2 = 0.2$ A; electric field = 14 mV/m.

5.4 The input impedances (depending on graphical readings) are about $Z_1 = 89 + j90\,\Omega$, $Z_2 = 97 + j101\,\Omega$, and $Z_3 = 89 + j90\,\Omega$; the gain is almost 4 dB.

5.5 The half-power beamwidths are approximately 126° in each case.

5.7 See Fig. 5.15.

5.10 Relative amplitudes of currents: 1:0.42:1.

6.1 $N = 24$, $\bar{x} = 53$, $\sigma = 14.5$, $X_0 = 53$ and $r = 54.9$.

6.2 See Fig. 6.16.

6.3 $m_1 = \dfrac{b\Gamma\left(\dfrac{n+2}{m}\right)}{\Gamma\left(\dfrac{n+1}{m}\right)}$ and $m_2 = \dfrac{b^2\Gamma\left(\dfrac{n+3}{m}\right)}{\Gamma\left(\dfrac{n+1}{m}\right)}$

6.4 Probability = 99.3%.

6.5 Probability = 93.3%.

6.6 Probability approximately 99.8% depending on graphical reading.

6.7 Hourly median path loss is about 145 dB.

6.8 About 20 kW of transmitter power with one link; reliability about 85% with 100 W on single link.

6.9 Reliability increases to about 84% when power is doubled or to about 96% if dual diversity is used.

7.4 The annual median received power is about − 95 dBW and $P(r)$ exceeds − 117 dBW for 99.99% of time.

7.5 With annual median $N_s = 345$, worst $P(r)$ is of the order − 116 dBW, 9 dB in excess of threshold. With N_s range of the order 70, the summer $P(r)$ is likely to be $0.2 \times 70 = 14$ dB higher than the winter levels or roughly 23 dB above threshold.

7.6   Annual median $P(r) = -117\,dBW$, reliability around 99.9%.
7.7   Range is about 48 km if diffraction is ignored.
7.9   Launch angle = 6.8°, scatter angle = 22.5°.
7.10 $CN_0R(r) = 57\,dB$ and $R = 12\,kbit/s$.

8.1   Height of obstacle = 144 m above reference datum.
8.2 $EIRP(t) = 43\,dBW$.
8.3   Design parameter $N' = 23\,pW$, 87 pW and 3.6 pW, respectively.
8.4   Transmitter power = 515 W, 66 W and 24 W without diversity, or 8 W, 1 W and 0.4 W with diversity for specified loadings.
8.5   (a) $P(r) = -58\,dBW$; (b) probability = 37% and $P(rf) = -60\,dBW$; (c) $P(rf) = -98\,dBW$; (d) $CN_{th}R(dB) = 62\,dB$ and 22 dB; (e) $SN_{th}R(dB) = 86.3\,dB$; (f) $\bar{N}_{rf} = 11.4\,N_o$, $\bar{N}_{rf} = 26.6\,pW$, $N_o = 2.34\,pW$; (g) $N_{rf} = 26,600\,pW$ and $266,000\,pW$, $SNR(dB) = 45.8\,dB$ and 35.8 dB, respectively.
8.6 $SN_{th}R(dB) = 51\,dB$, 61 dB and 66 dB, respectively.
8.7   Eight hops.
8.8 $P(r) = -64\,dBW$, $CN_{th}R(dB) = 56\,dB$, and $BER = 6 \times 10^{-9}$ and 4×10^{-5} under specified fade levels.

9.1   (a) $r = 10,540\,km$, (b) $h = 4164\,km$, (c) $d = 8392\,km$, (d) $v = 6149\,m\,s^{-1}$, (e) $f_D = 5154\,Hz$.
9.2   Elevation = 84.6°.
9.3   Maximum power = 363 W.
9.4 $CN_0R(r) = 105\,dB$.
9.6 $EIRP(t)$ varies between about 32 dBW and 41 dBW, $P(t)$ between 13 dBW and 22 dBW; under reduced reception conditions, P_t is about 8 W.
9.7 $CNIR(S) = 16.3\,dB$.
9.8   The signal-to-noise ratio at each stage is of the order 50 dB, 43 dB, 39 dB, 37 dB and 23 dB.
9.9 $G(T) = 52\,dB$ and $D = 7.4\,m$.
9.10 Yes. $EIRP(t)$ per channel is about $-15\,dBW$.
9.11 Margin is of the order 7 dB.
9.12 Bandwidths are approximately 1445 MHz, 1150 MHz, 57.5 MHz, 575 Hz, 46 Hz and 3.6 Hz, respectively.

10.1 Theoretical $d_{min} = 300\,m$ and $d_{max} = 300\,km$.
10.2 Reference ranges are about 133 km, 119 km, 106 km and 94 km, respectively, for given $F(r)$.
10.3 $SNR(r) = 13.8\,dB$.
10.4 The probability of detection in each case is about 99%, 55% and 68%, respectively, depending on graphical readings.
10.5 Noise figure = 10 dB.
10.6 (a) Noise voltage at receiver output = $41.5\,\mu V$, (b) threshold voltage at receiver output = $252\,\mu V$, (c) $V_r = 8\,\mu V$, $P_r = 1.3\,pW$ and $SNR(r) = 15.6\,dB$ at receiver input.
10.7 Integration gain = 13.4 dB and integration loss = 3.6 dB.
10.8 Effective aperture = $0.04\,m^2$ and gain = 27.4 dB.
10.9 The range is about 8.7 km without integration or about 23 km with integration.
10.10 The highest Doppler shift is about 447 Hz and the sensitivity is about 3 Hz per mph. The sensitivity increases with frequency.
10.11 The target, perhaps a large ship entering dock, has moved from $d = 66\,m$ at $v = 1.5\,m\,s^{-1}$ to $d = 41\,m$, slowing down by $d = 35\,m$ until stopped at $d = 10\,m$ (all from radar antenna).

Appendix 4
Bibliography

Radio Wave Propagation

Bullington, K., 'Radio propagation fundamentals,' *Bell Syst. Tech. J.*, 36 (May 1957), 593–626.

David, P., *Propagation of Waves*. Oxford: Pergamon Press, 1969.

Kirby, R. C., 'International standards in radio communication,' *IEEE Communications Magazine*, 23 (January 1985), 12–17.

Picquenard, A., *Radio Wave Propagation*. New York: John Wiley & Sons, 1974.

Antennas

Burrows, M. L., *ELF Communications Antennas*. Stevenage, Herts.: Peter Peregrinus Ltd. (for IEE), 1978.

Clarke, R. H., and J. Brown, *Diffraction Theory and Antennas*. Chichester, Sussex: Ellis Horwood Ltd., 1980.

Dance, M., 'Advances in antenna design from shf to elf,' *Communications Engineering International*, 6 (September 1984), 10–23.

Kummer, W. H., and E. S. Gillespie, 'Antenna measurements – 1978,' *Proc. IEEE*, 66 (April 1978), 483–507.

Paterson, J. R. T., 'Antennas: the shape of things to come,' *Communications Engineering International*, 2 (December 1980), 10–15.

Rudge, A. W., et al. (Eds.), *The Handbook of Antenna Design*, Vols. 1 and 2. Stevenage, Herts.: Peter Peregrinus Ltd. (for IEE), 1982.

Steinberg, B. D., *Principles of Aperture and Array System Design*. New York: John Wiley & Sons, 1976.

Wharton, W., 'Developments in communication antennas,' *Communications Engineering International*, 3 (February 1981), 8–16.

Woolff, E. A., *Antenna Analysis*. New York: John Wiley & Sons Inc., 1966.

Wood, P. J., *Reflector Antenna Analysis and Design*. Stevenage, Herts.: Peter Peregrinus Ltd. (for IEE), 1980.

Ground Wave Propagation

Barlow, H. M., and J. Brown, *Radio Surface Waves*. International Monographs in Radio. Oxford: Clarendon Press, 1962.

Blackband, W. T., (Ed.) *Propagation of Radio Waves at Frequencies Below 300 kc/s*. Oxford: Pergamon Press, 1964.

Bremmer, H., *Terrestrial Radio Waves*. New York: Elsevier Publishing Co. Inc., 1949.

Budden, K. G., *The Wave Guide Mode Theory of Wave Propagation.* London: Logos Press, 1961.
Burgess, B., and T. B. Jones, 'The propagation of l.f. and v.l.f. radio waves with reference to some system applications,' *The Radio and Electronic Engineer*, 45 (January/February 1975), 47–61.
Farrow, H. E., *Long-Wave and Medium-Wave Propagation.* London: Iliffe & Sons Ltd., 1958.
Galejs, J., *Terrestrial Propagation of Long Electromagnetic Waves.* Oxford: Pergamon Press, 1972.
Millington, G., 'Ground-wave propagation over an inhomogeneous smooth earth,' *Proc. IEE*, 96, Pt. III (1949), 53.
Norton, K. A., 'Low and Medium Frequency Radio Propagation,' *Electromagnetic Wave Propagation.* London: Academic Press, 1960.
Swanson, E. R., 'Omega', *Proc. IEEE*, 71 (October 1983), 1140–55.
Wait, J. R., *Electromagnetic Waves in Stratified Media.* New York: Macmillan, 1962.
Watt, A. D., *VLF Radio Engineering.* Oxford: Pergamon Press, 1967.

Sky Wave Propagation

Bain, W. C., and H. Rishbeth, 'Developments in ionospheric physics since 1957,' *The Radio & Electronic Engineer*, 45 (January/February 1975), 3–10.
Bennington, T. W., *Short-Wave Radio and the Ionosphere.* London: Iliffe and Sons Ltd., 1950.
Budden, K. G., *Radio Waves in the Ionosphere.* Cambridge: Cambridge University Press, 1961.
Davies, K., *Ionospheric Radio Propagation.* New York: Dover, 1966.
Knight, P. 'MF propagation: a wave-hop method of ionospheric field-strength prediction,' *BBC Engineering*, No. 100 (June 1975), pp. 22–4.

Space Wave Propagation

Bean, B. R., 'Atmospheric Bending of Radio Waves,' *Electromagnetic Wave Propagation.* London: Academic Press, 1960.
Bullington, K., 'Radio propagation at frequencies above 30 megacycles,' *Proc. IRE*, 35 (October 1947), 1122–36.
Burrows, W. G., *VHF Radio Wave Propagation in the Troposphere.* Glasgow: International Textbook Co. Ltd., 1968.
Matthews, P. A., *Radio Wave Propagation, VHF and Above.* London: Chapman & Hall, 1965.
Meeks, M. L., *Radar Propagation at Low Altitudes.* Dedham, Mass.: Artech House Inc., 1982.

Microwave Radio-Relay Links

Bray, W. J., 'The standardization of international microwave radio-relay systems,' *Proc. IEE*, 108B (March 1961), 180–200.
Dumas, K. L., and L. G. Sands, *Microwave Systems Planning.* New York: Hayden, 1967.
Györi, A., and F. Tampa, 'Path propagation test for microwave radio routes,' *Communications Engineering International*, 2 (December 1980), 41–7.
Livingston, D. L., *The Physics of Microwave Propagation.* Englewood Cliffs, N.J.: Prentice-Hall Inc., 1970.
Mohamed, S. A., and M. Pilgrim, '29 GHz point-to-point radio systems for local distributions,' *British Telecom. Technol. J.*, 2 (January 1984), 29–40.
Moupfouma, F., 'Model of rainfall rate distribution for radio system design,' *Proc. IEE*, 132H (February 1985), 39–43.
Panter, P. F., *Communication Systems Design.* New York: McGraw-Hill Book Company, 1972.

Scatter Propagation

Booker, H. G., and W. E. Gordon, 'A theory of radio scattering in the troposphere,' *Proc. IRE*, 38 (April 1950), 401–12.

Hill, S. J. 'British Post Office trans-horizon radio links serving off-shore oil/gas production platforms,' *The Radio & Electronic Engineer*, 50 (August 1980), 397–406.

Ince, A. N., 'Communications through em-wave scattering,' *IEEE Communications Magazine*, 20 (May 1982), 27–43.

Ince, A. N., et al., 'A review of scatter communications,' *AGARD Conference Proceedings*, No. 244, October 1977.

Ishimaru, A., *Wave Propagation and Scattering in Random Media*. New York: Academic Press, 1978.

Satellite and Space Communications

Ippolito, L. J., 'Radio propagation for space communications systems,' *Proc. IEEE*, 69 (June 1981), 697–727.

Jansky, D. M., *World Atlas of Satellites*. Dedham, Mass.: Artech House Inc., 1983.

Jansky, D. M., *Communication Satellites in the Geostationary Orbit*. Dedham, Mass.: Artech House Inc., 1983.

Lewis, J. R., 'Factors involved in determining the performance of digital satellite links,' *British Telecommunications Engineering*, 3 (October 1984), 174–9.

Mason, J., 'The role of satellites in the global weather experiment,' *The Radio & Electronic Engineer*, 49 (December 1979), 604–10.

Posner, E. C., and R. Stevens, 'Deep space communications – past, present and future,' *IEEE Communications Magazine*, 22 (May 1984), 8–21.

Pritchard, W. L., 'The history and future of commercial satellite communications,' *IEEE Communications Magazine*, 22 (May 1984), 22–37.

Scales, W. C., 'Air and sea rescue via satellite systems,' *IEEE Spectrum*, March 1984, pp. 48–51.

Spilker, J. J., *Digital Communications by Satellite*. Englewood Cliffs, N.J.: Prentice-Hall Inc., 1977.

Wise, F., 'Fundamentals of satellite broadcasting,' *IBA Technical Review*, 11 (July 1978), 18–26.

Witham, A., and P. Hawker, 'Development of communication and broadcasting satellites,' *IBA Technical Review*, 11 (July 1978), 4–13.

Wright, D., 'International maritime satellite communications,' *Electronics and Power*, 29 (September 1983), 623–6.

Yuen, J. H., *Deep Space Telecommunications Systems Engineering*. New York: Plenum Press, 1983.

Mobile Radio

Jakes, W. C., *Microwave Mobile Communications*. New York: John Wiley & Sons, 1974.

Lee, W. C. Y., *Mobile Communications Engineering*. New York: McGraw-Hill Book Co., 1982.

Radar

Blake, L. V., *Radar Range-Performance Analysis*. Lexington, Mass.: Lexington Books, D.C. Heath and Company, 1980.

DiFranc, J. V., and W. L. Rubin, *Radar Detection*, Dedham, Mass.: Artech House Inc., 1980.
Meeks, M. L., *Radar Propagation at Low Altitudes*. Dedham, Mass.: Artech House Inc., 1982.
Skolnik, M. I., *Introduction to Radar Systems*. New York: McGraw Book Co. Inc., 1962.
Skolnik, M. I., *Radar Handbook*. New York: McGraw-Hill Book Co. Inc., 1970.
Wheeler, G. J., *Radar Fundamentals*. Englewood Cliffs, N.J.: Prentice-Hall Inc., 1967.

Noise

Hansen, R. C., and R. G. Stephenson, 'Communications at megamile ranges,' *Journal Brit. IRE*, 22 (October 1961), 329–44.
Kreutel, R. W., Jr., and A. O. Pacholder, 'The measurement of gain and noise temperature of a satellite communications earth station,' *The Microwave Journal*, 12 (October 1969), 61–6.
Skomal, E. N., *Man-made Radio Noise*. New York: Van Nostrand Reinhold, 1982.

Statistics and Diversity

Brennan, D. G., 'Linear diversity combining techniques,' *Proc. IRE*, 47 (June 1959), 1075–102.
Miller, I., and J. E. Freund, *Probability and Statistics for Engineers*. Englewood Cliffs, N.J.: Prentice-Hall Inc., 1965.

Digital Systems

Clark, A. P., *Principles of Digital Data Transmission*. London and Plymouth: Pentech Press, 1976.

Index

Ampere's law, 331–333, 337–339, 346
Amplitude diagram, 125
Amplitude-shift-keying, 369
Antenna: 6–10, 13–27
  aperture, 6, 23–25
  bat-wing, 171
  beamwidth, 10, 32, 144, 164, 166, 175, 178, 281, 290
  Beverage, 58–59, 95
  biconical, 23–24, 138
  billboard, 218
  cage, 138
  cassegrain, 281, 285
  conical, 23
  corner reflector, 230
  cylindrical, 138
  design factors, 280–283
  discone, 138–139
  flared, 23–25
  folded dipole, 94–95, 137–138, 166–167
  full-wave dipole, 95
  Gregorian, 285
  halfwave dipole, 17–18, 95, 137–139, 142–146, 149–160, 166–171
  helical, 4
  high-frequency, 93–98
  horn, 6, 23
  incremental linear radiator, 6–7, 10–17, 337–341
  interference, 280
  isotropic (isotrope), 6, 9–17
  lens, 6, 25–27
  log-periodic, 96–97, 167–169, 178
  loop, 6, 21–22, 59–61, 97–98
  metal plate lens, 26–27
  monopole, 6, 55–56, 98, 137–138
  paraboloidal reflector, 23–25, 31–32, 218, 243, 261, 280, 315–316
  parasitic, 157–160
  plano-convex dielectric lens, 26
  pyramidal horn, 23, 25, 31
  receiving, 58–60
  reflector, 23, 153–157, 160
  rhombic, 95–97
  sectoral horn, 25

Antenna (cont'd)
  short dipole, 17–18
  short vertical monopole, 11–12, 20–21, 55–58
  shunt-fed dipole, 138
  sidelobes, 162–166, 176, 280–281
  sleeve, 139
  slot, 6, 23–24, 137, 139
  small loop, 21–22
  space wave, 137–139
  superturnstile, 171
  television transmitting, 153–157, 169–171
  transmitting, 55–57
  traveling-wave, 58, 95
  turnstile, 139
  vertical mast, 56
  vertical monopole, 19–21, 32
  whip, 58–59, 97–98
  Yagi-Uda, 139, 166–167
Aperture, maximum effective: 8–9
  halfwave dipole, 18
  incremental linear radiator, 16
  isotrope, 9, 16, 272
  paraboloidal reflector antenna, 25
  short dipole, 18
  short vertical monopole, 21
Array: 142–179
  binomial, 174–175
  broadside, 145, 151–152, 162
  collinear, 149–150
  constant-current, 175–176
  endfire, 152–153, 155, 166
  four-element, 174
  Gaussian, 176–177
  halfwave dipole, 142–171
  of isotropes, 146–149, 161–166
  Kooman, 95
  linear, 160–171
  log-periodic, 167–169, 178
  N-element, 160–178
  parasitic, 157–160
  pine tree, 94
  quadrant, 156
  reflector, 153–156
  seven-element, 161–163

Array (*cont'd*)
    sixteen-element, 170
    ten-element, 165
    three-element, 159–160
    two-element, 146–159
    Yagi-Uda, 139, 166–167
Array beamwidth, 164, 166
Array factor: 149, 163
    normalized, 163, 174
Atmospheric noise factor, 87–89
Attenuation:
    atmospheric, 248, 259, 278–279, 292
    D-layer, 72–74
    fading, 203, 206
    due to fog, snow and rain, 248, 256–257
    of ground wave, 33
Attenuation factor, ground wave, 35–38
Attenuation rates, VLF, 51–52
Automatic gain control, 253
Automatic picture transmission, 286–287
Automatic repeat request, 232–235

Back-off, satellite, 273
Bandwidth: 29–30
    noise, 45, 351–352
BBC World Service, 284–285
Beam-steering, 164
Beamed communication services, 30
Beamwidth of antenna, 10, 32, 144, 164, 166, 175, 178, 290
Binary digits, 367–370
Bit energy, 231, 234, 237, 369–370
Bit error rate, 231, 258–259, 292–293, 369–370
Bit transmission rate, 231, 234–235, 237, 258–259, 292–293
Boundary, land-sea, 41–44, 60
Boundary conditions, 341–347
Brewster angle, 345–346
Broadcasting, 29–30, 32–33, 44, 48–49, 60–62, 87, 289–290

Carrier-to-noise-plus-interference ratio, 275–277
Carrier-to-noise ratio, 273–275, 285, 287, 289
Carrier-to-noise-spectral-density ratio, 231–232, 234, 237, 293–294, 300, 370
Carrier-to-thermal-noise ratio, 246–247, 251–252, 256, 259–262
CCIR, 61, 244, 249–252, 256, 258
CCITT, 244, 258
Chart:
    diversity gain in troposcatter link, 226
    ground wave propagation, 38–41
    prediction for mobile radio, 131
Circuit reliability, 89–93
Circuit reliability factor, 93
Class boundaries, 180, 194, 214
Class intervals, 180, 183, 194
Coastal communication, 30

Coefficient:
    attenuation, 5, 34, 66–67, 71–73, 334–335
    phase-change, 5, 34, 66, 334–335
    propagation, 34, 335
    reflection, 35, 50–51, 101–102, 344–347
Combining techniques, diversity, 202, 208–213
Complex dielectric, 69, 346–347
Composite charts for diversity, 226
Conductivity:
    ground, 34
    ionosphere, 68–71
Coordinate systems, 1
Cornu's spiral, 125–127
Correction factor, ground wave, 40
Critical frequency, 68–69, 76–79, 98
Cross polarization, 283, 285
Cumulative frequency, 194
Curvature of space wave in troposphere, 114–122
Curved-earth reflection, 108–109, 139
Cymomotive force, 36, 40

Deep space communication, 292–293, 295
Design factors, antenna, 280–283
Dielectric, complex, 69, 346–347
Diffraction: 23, 28, 33, 38
    knife-edge, 127–129
    over rounded hill, 129–130
    of space waves, 122–130
Digital links, 30, 231, 234–235, 237
Digital radio relay systems, 256–260, 262
Direct broadcasting from satellite, 289–290
Directivity: 10
    halfwave dipole, 18
    short dipole, 18
    small loop antenna, 22
    short vertical monopole, 21
Director, 160, 166
Distributions (*see* Statistical distributions)
Divergence of reflected waves, 110–111
Diversity: 30, 202–214
    combinations, 208
    composite chart, 226
    equal-gain, 208
    maximal-ratio, 208, 210–213, 215, 226, 232
    scanning, 208
    selection, 208–211, 215
Doppler frequency shift, 130, 268, 319–320, 324–327
Doppler navigation, 296
Dwell time, 312–313

Earth's magnetic field, 69–70, 98
Echoes, 297, 303, 314
Effective:
    height, 9
    isotropically radiated power, 128
    length, 9

Effective (*cont'd*)
  radiated power, 128, 171
  radius of earth, 116–118
Efficiency:
  illumination, 9, 285
  radiation, 10, 56
Electric field strength:
  basic propagation equation, 11, 35, 51
  halfwave dipole, 18, 171
  incremental linear radiator, 6, 15–17, 21, 35,
    143
  inverse distance for mobile radio, 106–108
  isotrope, 13–14
  with knife-edge diffraction, 128, 140
  linear array, 144, 150, 152–156, 162–163,
    171–172
  mobile radio, 130–133
  noise equivalent, 45, 61
  over rounded hill, 129–130, 140
  short vertical monopole, 20, 35–36
  sky wave, 86
  small loop antenna, 22
  vertical monopole, 20
  with VLF propagation, 52–55
Electrical noise (*see* Noise)
Electromagnetic field: 1
  energy equations in free space, 5
Electromagnetic principles, 328–347
Environmental factor, 131–132
Equivalent circuit of antenna, 56–57
European Communication Satellite, 286
European Space Agency, 288
EUTELSAT, 286
Excess-noise generator, 363
Excess-noise ratio, 363–365
Excitation factor, 54
Extraordinary wave, 70, 85, 88

Fade depth, 260
Fading: 87–88, 247–249, 256–257, 261
  fast, 134–137
  slow, 134
Fading medium, 207
False alarms, probability of, 303–307
Far-field, 7
Faraday's law, 331–338
Fast-fading in microwave link, 252–256
FDM/FM radio-relay systems, 241–247, 251,
    261
Field:
  electrostatic, 35
  induction, 35
  radiation, 35
  unattenuated at 1 km, 36, 40, 43, 45
Field-aligned scatter system, 228–229
Figure of merit, 11, 20–21
FM improvement factor, 245–246
Frame time, 312–313

Frequency:
  beat, 325
  of collisions, 68, 70–71
  critical, 68–69, 76–79, 98
  cross-over, 34–35
  electron collision, 70–71
  gyromagnetic, 69
  lower usable, 82
  maximum usable, 79, 81, 83, 89–91, 98
  optimum traffic, 81
  optimum working, 81–83, 98–99
  plasma, 77
Frequency division multiple access, 284
Frequency division multiplexing, 242–243
Frequency-shift-keying, 369
Fresnel-zone clearance, 238–241
Fresnel's ellipsoid, 111–112, 239–241
Fresnel's integrals, 126
Friis' free space equation, 12, 272

Gain (*see* Directivity)
Gamma function, 186–187
Geostationary meteorological satellite, 288
Geostationary operational environmental
    satellite, 288
Geostationary satellite, 268–271
Green's theorem, 329–332
G/T ratio, 247, 273, 275, 285–291, 295
Gyromagnetic frequency, 69

Height, effective, 9–10, 20–21
Height, virtual, 80–81
Helmholtz's theorem, 331
Histogram, 180–181, 183, 214, 223, 350
Huygens' principle, 122–124
Hypothetical reference circuit, 244, 249–252
Hypothetical reference digital circuit, 258

Impedance:
  antenna, 8
  dipole, 144–146
  intrinsic, 6
  mutual, 144–146, 150, 155, 159, 167
  wave, 336–337
INMARSAT, 291
Integration:
  efficiency, 312
  gain, 312
  loss, 312
  post detection, 312
  of pulse trains, 311
INTELSAT, 283–288
Interference, skywave, 48
International Telecommunications Union,
    244
Inversion layer, 248–249
Ionogram, 79
Ionization radiation, 64–65

Ionoscatter radio systems, 228–232
Ionosonde, 79
Ionosphere: 27–30, 62–66
  attenuation, 66–67, 71–73
  conductivity, 68–71
  D-layer, 48–49, 65–66, 71–75
  E-layer, 66, 72–73, 77, 79, 82, 98
  F-layer, 66, 77–79, 82–83, 98
  free electron density, 63–64, 68, 71–72, 77–79, 83
  mechanisms, 64–66
  permittivity, 68–69, 70–71
  reflection coefficient, 50–51
  refractive index, 67–68, 71, 74–76
  sporadic E-layer, 66
  virtual height, 80–81
Ionospheric convergence-gain, 86
Isolation, cross-polar, 283

Jamming, 317
Jupiter, 293

K-factor, 118–119, 134, 139–140, 220
Kooman array, 95

Land loss, 41–44
Launch angle, 81
Laws of reflection and refraction, 343–346
Length, effective, 9
Line-of-sight range, 100–101
Long-term variation, 216, 222–227, 230–231
Loss:
  antenna, 8
  antenna-to-medium coupling, 217, 221
  day-time ionospheric absorption, 85
  excess system, 86
  ionoscatter, 229
  meteor burst, 234–236
  miscellaneous, 218
  night-time ionospheric absorption, 85
  path, 84, 217–223
  polarization coupling, 86
  propagation, 216
  scatter, 217, 220
  spatial, 84
  transmission, 84, 217
Loss effect, 41–44
Lower usable frequency, 82

Magnetic vector potential, 337–341
Map, geological, 41
Mars, 295
Matched attenuator, 355–360
Matched filter, 302
Matched system, 8
Maximum usable frequency, 79, 83, 89–91, 98
Maximum usable frequency factor, 81–82
Maxwell's equations, 6, 333, 337

Mean, 183
Mean square, 183
Mean terrain level, 108
Median, 184
Mercury, 295
Meteor burst communication systems, 229, 234–236
METEOSAT, 288
Microwave link, 31, 105, 118–119
Microwave radio relay systems, 30, 238–262
Mixed path propagation, 41–44, 60
Mobile communication, 30
Mobile radio, 108, 130–133, 140
Modified refractivity, 119–122
Modulation section, 243–244
Moments:
  first, 183
  second, 183
  statistical, 183
Monopulse, 316
Moon link, 32
Multihop fading factor, 250
Multiple propagation paths, 207–208

National Ocean and Atmospheric Association, 286–288
Near-field, 7, 35
Neptune, 293
Net-charge-free media, 334–335
Noise: 348–369
  antenna, 280, 358–359
  atmospheric, 44, 46, 61, 81, 87–89, 277, 357–359
  available power, 352
  band-limited, 302, 349–350
  bandwidth, 45
  bandwidth criterion for radar receiver, 302
  in cascaded networks, 359–362
  colored, 349
  cosmic, 231, 277–279, 357, 366
  current, 351
  in digital systems, 367–370
  in diversity combiners, 212
  electrical, 348
  factor, 355
  field, 362–363
  figure, 45, 354–366
  galactic, 277, 357, 366
  instantaneous, 349
  jammer, 317
  lunar, 277
  man-made, 44, 46, 367
  in matched attenuator, 355–359
  measure, 361–362
  measurement of, 363–366
  natural, 44
  psophometrically weighted, 244–245, 249–252

Noise (*cont'd*)
  receiver, 44
  sky, 278–280
  solar, 277, 357, 367
  sources, 366–367
  temperature, 301, 354–366
  thermal, 45, 244, 301, 350–354
  unweighted, 245, 249–252
  voltage, 301, 348, 351
  white, 349
  wideband, 349
Normal probability paper, 205
Normalized array factor, 163, 174
Null-filling, 170–171
Null-tracking systems, 316
Numerical distance, 37–38, 45, 47–48
Nyquist's equation, 352–354

Ohm's law, 331
Omega navigation system, 7
Optimum traffic frequency, 81
Optimum working frequency, 30, 81–83, 98–99
Orbit of satellite, 263, 286
Orbital period of satellite, 263–265
Ordinary wave, 70, 85, 87

Parameters:
  auxiliary, 34–35, 38, 124
  descriptive, 185
  ground, 34
  linear, 329
  log-periodic antenna, 169
  primary, 4–5, 34, 66–70
  secondary, 4–5, 34, 66–70
  of statistical distributions, 180–185
  Student's-t, 201, 224
  surface, 328–329
  transmitter, 41
  volumetric, 328
Pattern, multiplication, 149, 151
Pattern, radiation (*see* Radiation pattern)
Peak deviation, 246
Peak signal-to-noise ratio, 305–308
Permeability:
  of free space, 4, 66, 69
  relative, 4, 26
Permittivity:
  complex, 69–70, 101–102, 346–347
  effective relative of ionosphere, 68–71
  of free space, 4–5
  relative, 4, 41
Phase sensitive detector, 323
Phase-shift-keying, 369
Plan position indicator, 298
Planck's equation, 352–354
Pluto, 295
Polar diagram (*see* Radiation pattern)
Polarization, 1–4
Polynomials in linear array theory, 171–179

Power profile for satellite link, 274
Poynting's theorem, 6, 336
Poynting's vector, 336–337
Pre-emphasis, 245
Probability: 182
  of detection, 303–314
  of false alarms, 303–310
Probability density functions, 182–201
Probability paper:
  exponential, 209–210
  log-normal, 205
  normal, 205
  Rayleigh, 193–195, 209–210
Propagation:
  ground wave, 27, 29–30, 33–61
  guided electromagnetic wave, 27, 33, 51–55
  ionoscatter, 228–234, 237
  meteor burst, 234–236
  mixed path, 41–44
  scatter, 28–30, 216–237
  sky wave, 27–30, 33, 48–51, 62–99
  space wave, 28, 30–31, 100–141, 272–274
  surface wave, 27
  in urban environment, 130–133
  very low frequency, 51–55
Propagation equations (*see also* Electric field strength):
  deep space, 292–293
  digital radio relay, 259–260
  with fading, 136–137
  Friis' free-space equation, 12, 243–244
  Intelsat, 284–285
  ionoscatter, 228–232
  logarithmic, 12
  marine satellite, 291–292
  microwave link, 243–244, 251–252, 256
  mobile radio, 108, 132
  radar, 299–300, 308–310, 312–313, 316–318
  satellite, 272–275
  sky wave, 84–87
  tropospheric scatter, 216–228
  weather satellite, 287
Protection ratio, 48
Psophometer, 245
Pulse:
  width, 298–299, 302
  repetition frequency, 298–299

Q-factor, 56, 61
q-factor, 34–35

Radar: 296–327
  beacon, 316
  bistatic, 227, 316–318
  collision avoidance, 30
  contact with Venus, 300
  cross-section, 300, 308–310, 318
  Doppler, 318–327
  equation, 227, 299–300
  FM/CW, 324–326

Radar (*cont'd*).
  medium range, 30
  moving target indication, 323–324
  noise jammer, 317–318
  pulse, 296–318
  pulse-Doppler, 322–324
  search, 310–314
  tracking, 315–318
Radiation pattern:
  collinear array, 150–151
  endfire array, 152
  geometric construction, 165
  halfwave dipole, 143
  isotrope, 143
  linear array, 161
  optimum directivity, 165
  reflector array, 155
  television antenna, 170
  two-element array, 148, 173
Radiation resistance: 8
  halfwave dipole, 18
  incremental linear radiator, 15
  short dipole, 17
  short vertical monopole, 21
  small loop antenna, 22
  vertical monopole, 20
Radio detection and ranging, 296
Range, line-of-sight, 100–101
Rayleigh probability paper, 193–195
Receiver, noise-free, 45
Reception levels, 45–46, 87–88, 130
Recovery effect, 41–44
Reflection: 27–28, 34, 248
  flat earth, 101–104
  variable heights, 105
  with variable wavelength, 104–105
Reflection coefficient: 101, 344–347
  ionospheric, 50–51
Reflector: 153–156, 160, 166–167
  corner, 23, 230
  paraboloidal, 23–25
Refraction: 27–28, 34, 74–76
  standard, 133
  sub-standard, 133, 248
  super-standard, 133
Refractive index, 26–27, 249, 335
Refractivity, 236
Refractivity factor, 118–119, 220, 240–241
Repeater, 243
Resistance:
  antenna, 8
  antenna loss, 8
  input, 155
  loop, 19–20
Route planning, 238

Satellite:
  bearing, 270–271
  broadcasting, 289–290
  circumferential velocity, 266

Satellite (*cont'd*)
  commercial, 283–286
  coverage, 271
  Doppler frequency shift, 268
  ECS, 286
  geostationary orbit, 268–269
  GMS, 288
  GOES, 288
  marine, 290–292
  non-synchronous, 264–265
  orbit, 263, 286
  orbital period, 263–265
  power profile, 274
  range, 265–266, 270
  relative velocity, 265–268
  synchronous, 268
  Tiros-N, 286
  weather, 286–288
Satellite communication, 263–292
Satellite link, 4, 31
Scatter angle, 217–221
Scattering:
  cross-section, 227–228
  of radio waves, 216
  volume, 216–217
Sea gain, 41–44
Short-term specifications of HRC, 249–252
Short-term variation, 216, 224–227, 230–231
Signal-to-noise ratio, 87, 90–93, 206, 209, 304–318, 354, 360
Signal-to-thermal-noise ratio, 245–246, 251, 254–256
Single-channel per carrier, 284–287
Skip distance, 49
Sky wave availability factor, 89–91
Sky wave field-strength:
  median, 50–51
  quasi-maximum, 50–51
Smith-Weintraub relationship, 113
Snell's law, 74–76, 80, 114–115, 119–120, 344
Spacecraft communication, 30
Spherical waveguide, 33, 52–55
Standard deviation, 184, 222, 230, 301
Standard test-tone, 246
Statistical distributions: 180–215
  basic equations, 185
  chi-squared, 188, 198, 309
  descriptive parameters, 185
  Erlang, 188
  exponential, 184, 188–191, 206, 225, 309
  gamma, 188, 195–199, 206
  Gaussian, (*see* normal)
  general shapes of, 184
  generalized equation, 213
  generalized exponential, 188
  generalized Rayleigh, 188
  inverse Rayleigh, 254
  log-normal, 200–201, 204, 223, 226, 309
  Nakagami-*m*, 188, 199
  Nakagami-Rice, 200

Statistical distributions (*cont'd*)
  normal, 184, 200–201, 204, 223, 230, 301, 349–350, 369
  normalized equations, 185
  one-sided normal, 188
  Rayleigh, 184, 188, 191–195, 204, 206, 224–226, 230, 252–253, 302, 307, 309, 349–350
  Rician, 304, 309, 315
  Stacey, 188
  Weibull, 188, 309
Stokes's theorem, 329, 332–333
Surface duct, 120–121
Synchronous group working, 48

Tandem fading factor, 255
Target: 227, 296–327
  constant-σ, 307, 310
  fixed, 324
  moving, 321–323
  Rayleigh, 307–310
  Rayleigh plus dominant scatterer, 307–310
Target detection, 296
Telegraphy, 62, 87, 93
Telephony, 62, 87, 93
Television, 30, 104–105, 129–130, 169–171
Threshold signal-to-noise ratio, 305–306
Time between false alarms, 303
Time division multiple access, 4, 284, 286
Tiros-N satellite, 286
Top capacitance, 55–57
Transmitter coverage, 44–48
Troposphere:
  humidity, 112
  permittivity, 113
  pressure, 112
  refractive index, 113
  refractivity, 113

Troposphere (*cont'd*)
  standard models, 114
  temperature, 112
  water vapour pressure, 112
Tropospheric scatter radio relay systems, 30–31, 216–237
Type-A display, 297

Unit circle, 172–173, 177
Uranus, 293

Variance, 184   Van der Pol ~ 37
Velocity:
  group, 70
  phase, 52, 70
  radial, 325
  of propagation, 4, 335
Venus, 295, 300
Virtual height, 80–81

Wave:
  direct, 27–29, 35, 248
  extraordinary, 70
  incident, 101
  ordinary, 70
  reflected, 35, 101, 248
  transmitted, 101
Wave equations, 334
World Administrative Radio Conference, 289

X-rays, 63, 65

Yeh's equation, 220, 236
Y-factor in noise measurement, 364–366

Zenith angle of sun, 78, 85